[redacted]

Springer
Berlin
Heidelberg
New York
Barcelona
Budapest
Hong Kong
London
Milan
Paris
Tokyo

NATO ASI Series

Advanced Science Institutes Series

A series presenting the results of activities sponsored by the NATO Science Committee, which aims at the dissemination of advanced scientific and technological knowledge, with a view to strengthening links between scientific communities.

The Series is published by an international board of publishers in conjunction with the NATO Scientific Affairs Division

A	Life Sciences	Plenum Publishing Corporation
B	Physics	London and New York
C	Mathematical and Physical Sciences	Kluwer Academic Publishers
D	Behavioural and Social Sciences	Dordrecht, Boston and London
E	Applied Sciences	
F	Computer and Systems Sciences	Springer-Verlag
G	Ecological Sciences	Berlin Heidelberg New York
H	Cell Biology	London Paris Tokyo Hong Kong
I	Global Environmental Change	Barcelona Budapest

PARTNERSHIP SUB-SERIES

1. Disarmament Technologies	Kluwer Academic Publishers
2. Environment	Springer-Verlag
3. High Technology	Kluwer Academic Publishers
4. Science and Technology Policy	Kluwer Academic Publishers
5. Computer Networking	Kluwer Academic Publishers

The Partnership Sub-Series incorporates activities undertaken in collaboration with NATO's Cooperation Partners, the countries of the CIS and Central and Eastern Europe, in Priority Areas of concern to those countries.

NATO-PCO DATABASE

The electronic index to the NATO ASI Series provides full bibliographical references (with keywords and/or abstracts) to about 50000 contributions from international scientists published in all sections of the NATO ASI Series. Access to the NATO-PCO DATABASE compiled by the NATO Publication Coordination Office is possible in two ways:

- via online FILE 128 (NATO-PCO DATABASE) hosted by ESRIN, Via Galileo Galilei, I-00044 Frascati, Italy.

- via CD-ROM "NATO Science & Technology Disk" with user-friendly retrieval software in English, French and German (© WTV GmbH and DATAWARE Technologies Inc. 1992).

The CD-ROM can be ordered through any member of the Board of Publishers or through NATO-PCO, Overijse, Belgium.

Series G: Ecological Sciences, Vol. 38

Molecular Ecology of Aquatic Microbes

Edited by

Ian Joint

Plymouth Marine Laboratory
Natural Environment Research Council
Prospect Place, The Hoe
Plymouth, PL1 3DH, UK

Springer

Proceedings of the NATO Advanced Study Institute on Molecular Ecology of Aquatic Microbes, held at Il Ciocco, Lucca, Italy, 28 August - 9 September 1994.

Library of Congress Cataloging-in-Publication Data

Molecular ecology of aquatic microbes / edited by Ian Joint.
p. cm. -- (NATO ASI series. Series G, Ecological sciences ; no. 38)
Includes bibliographical references and index.
ISBN 3-540-60134-1
1. Water--Microbiology. 2. Molecular microbiology. I. Joint, Ian, 1947- . II. Series.
QR105.M65 1995
576'.192--dc20 95-24508
CIP

ISBN 3-540-60134-1 Springer-Verlag Berlin Heidelberg New York

Printed in Germany

Typesetting: Camera ready by authors/editors
Printed on acid-free paper
SPIN 10470312 31/3136 - 5 4 3 2 1 0

Preface

A NATO ASI on "Molecular Ecology of Aquatic Microbes" was held at *Il Ciocco,* Lucca, Italy from 28 August - 9 September 1994. The aims of the ASI were to evaluate the potential for molecular biology to solve some important questions in aquatic microbiology, particularly in relation to biogeochemical cycling and microbial physiology.

Techniques developed by molecular biologists have now been adopted by a wide range of scientific disciplines. In the last 5 years, aquatic microbial ecologists have begun to incorporate these methods into their research and, as a result, are developing a much clearer understanding of phylogenetic diversity, the molecular basis of physiological acclimations and the transduction of environmental signals and organism responses. The aim of this ASI was to assess progress in this new field of research, to compare and describe techniques and experimental approaches, and to foster communication between disciplines. The ASI offered an excellent opportunity to bring together aquatic ecologists with molecular biologists and to encourage efficient technology transfer. The meeting provided a forum for detailed and broad exchange of information on the status and trends of aquatic molecular ecology and to assess how emerging molecular techniques might solve some important problems in ecology which have prove intractable because of lack of appropriate methodologies.

The organising committee was Dr Paul Falkowski (Brookhaven National Laboratory, USA), Professor Noel Carr (University of Warwick, UK) and Dr Luigi Lazzaro (University of Firenzi, Italy) I would like to express my appreciation for the efforts they made to ensure a successful meeting.

I would also like to acknowledge financial support from a number of organisations. NATO provided the majority of the funding which allowed this meeting to take place. The US Department of Energy gave generous assistance through grant number DE-FG02-94ER61896, which enabled us to invite 5 additional lecturers. The UK Natural Environment Research Council also gave a grant from the Special Topic on Molecular and Genetic Advances, which supported 2 lecturers from the UK. The generous support of all these organisations is gratefully acknowledged.

Ian Joint
ASI Director

CONTENTS

The Potential of Molecular Ecology

Ian Joint
NERC Plymouth Marine Laboratory
Prospect Place
The Hoe
Plymouth PL1 3DH
UK

Introduction

The solution of major environmental problems requires a clear understanding of the complex interactions which are part of natural systems. Many important processes, which are poorly understood but which have clear global implications, are mediated by the activity of microbes. Although micro-organisms may have spheres of influence with distance scales of centimetres or less, their combine effect is truly global. However, within any particular environment, microbial biomass and activity can be very variable in both time and space. The challenge to microbial ecologists is to make measurements which are accurate and precise at the scale of a microbial population but which can also be extrapolated to give useful information at local, oceanic basin or global scales.

Marine microbial ecologists have an increasing number of powerful techniques which have allowed great progress to be made in understanding and quantifying microbial activity in the oceans. However, it is important to recognise that many of these techniques also have limitations and that existing methodologies may not be capable of supplying the information which is required to understand ecological interactions. Molecular techniques may offer the solution to some of these problems. In this chapter, I will discuss a number of processes which I have been studying and which I believe are proving intractable with current methodologies. I will present a personal view, derived from my own research experience; however, readers may find parallels in the development of their own research which may convince them, as I am convinced, that molecular biology has a significant role to play in improving our understanding of the functioning of aquatic microbes.

I will begin by explaining why I have come to believe that current methodologies have limitations which are not easy to overcome; I will then give a small number of examples of what I believe is the potential of molecular ecology,

NATO ASI Series, Vol. G 38
Molecular Ecology of Aquatic Microbes
Edited by I. Joint

although others have already done this effectively (Falkowski and LaRoche, 1991). But a wider justification for molecular ecology comes not from this chapter but will be found in the rest of this book.

Phytoplankton biomass and production

Two attributes of phytoplankton, chlorophyll and carbon fixation, are amongst the most commonly measured parameters in oceanography. The simplicity and physical robustness of fluorometers allows routine deployment of these instruments by physical, chemical and biological oceanographers. Chlorophyll fluorescence has become one of the favourite descriptors of the bulk properties of sea water. Similarly, primary production is measured at great frequency. The sensitivity of the ^{14}C technique has resulted in a large body of data which relates to almost every marine province; yet even with frequent use, the ^{14}C method remains one of the most controversial techniques in current use (Williams, 1993).

The ^{14}C method has been important in my own research and, in spite of uncertainties about whether we measure gross or net production (Williams, 1993), I believe valuable information on phytoplankton production has been and will continue to be obtained with this methodology. As an example of what can be achieved, I used the ^{14}C method to estimate seasonal phytoplankton production (Joint & Pomroy, 1993) within the context of a large multidisciplinary study of the southern North Sea (Howarth *et al.*, 1993), a major temperate European shelf sea. Large variations in annual production were found, ranging from 79gC $m^{-2}y^{-1}$ in the eastern North Sea to 261gC $m^{-2}y^{-1}$ for the German Bight. The differences in production were partly a consequence of physical processes, such as the degree of stratification, tidal mixing and depth, but also because of different inputs of nutrients from rivers. The regions of highest annual production were not only those with the shallowest water but also those with the greatest influence of the Rhine and other major rivers. However, the factor which resulted in the greatest difference in phytoplankton production was the presence of a bloom of colony-forming *Phaeocystis*. The reasons for the development of blooms of this phytoplankton are not certain, but a number of workers have suggested that it is direct consequence of increased nutrient supply from rivers (Cadée, 1990; Lancelot, *et al.* 1987; Veldhuis, 1986). Although this is a widely held hypothesis, it has yet to be proven.

Current methods to measure primary production suffer from a number of limitations. Although giving invaluable data about the activities of complete assemblages (or at least those assemblages which survive the *in vitro* assessment), methods such as ^{14}C assimilation or O_2 evolution cannot give any information on the activity of the individual organisms which make up those assemblages. It is unusual for any natural phytoplankton population to be unialgal, although under some bloom conditions one species, such as *Phaeocystis* may reach pre-eminence and greatly outnumber other phytoplankton. Therefore, standard methods to assess primary production estimate the mean, or median, rate of production of all of the phytoplankton which make up an assemblage. For many applications, such as the study of the North Sea (Joint and Pomroy, 1993) this limitation is not important. Yet there may be other occasions when it might be an advantage to have a measure of the growth of individual phytoplankton species.

Measures of community activity, such as the ^{14}C method assume that the average activity of the whole population has some validity in relation to the organisms which make up that assemblage. But is this a realistic assumption? We can readily recognise the heterogeneity of natural populations; different algal species can be identified by standard microscopic observation and we have no difficulty in accepting that organisms are highly variable, with different morphologies, sizes and growth rate. However, neither this heterogeneity nor the ecological consequences of the assumptions behind the ^{14}C method are often considered. Marine phytoplankton exhibit a large range in specific growth rate of over an order of magnitude (Geider and Osbourne, 1992), from 3 d^{-1} for small diatoms to 0.1 d^{-1} for slow growing dinoflagellates. Therefore, it is possible that short term incubations of <1 day do not give a true estimate of growth of an organism which has a generation time of several days. The ^{14}C method may have a distinct bias towards rapidly growing organisms with generation times equal to or less than the incubation period of the experiment. Diel variations in photosynthetic carbon fixation of rapidly growing phytoplankton might be more closely matched to cell growth and division than in phytoplankton cells with doubling times of several days.

Does any of this really matter? Most researchers who use these methods will accept that there are limitations; that the ^{14}C method measures something between net and gross production, and probably close to net production, is probably well accepted and for many purposes, these estimates are more than adequate. However, it could also be argued that the widespread use of the ^{14}C has constrained the development of ecological concepts. The true limitations of current measures of community activity become more apparent as a result of new understandings derived from the study of the organisms which graze the phytoplankton. We are becoming increasingly aware that grazers can be very selective in the phytoplankton cells on which they feed. Copepods demonstrate large differences in growth rate when fed diets comprising different species. In particular, many dinoflagellates are poor food for copepods (Gill and Harris, 1987) and even the physiological state of phytoplankton can affect grazing by zooplankton (Estep *et al.* 1990). Filter feeding bivalves, such as mussels are also sensitive to specific phytoplankton and the presence of dinoflagellates can greatly reduce clearance rate (Bricelj *et al.* 1990). Therefore an understanding of the grazing losses on phytoplankton requires information which is species specific. If the loss terms from a natural assemblage have to be considered at the species level, it may be equally essential to obtain information on biomass and production, also at the species level.

Clearly, standard methods have not been able to differentiate at this level of discrimination. The ^{14}C method has been used with some success to quantify the activity of different sizes of phytoplankton and much of the information of the role of widespread picoplankton assemblages (Murphy and Haugen, 1985) was obtained by the use of size fractionation experiments with natural populations incubated with ^{14}C (Joint, 1986). But the discrimination obtained was very crude, relying on a physical separation - sieving by different pore sizes of filter. Increasingly, instruments such as flow cytometers, have been of value in quantifying the various components of phytoplankton assemblages, with the greatest successes being achieved with the smallest organisms. The discovery of the prochlorophytes is an excellent example of the successful application of this technology (Chisholm *et al.* 1988). Flow cytometers, although useful in determining phytoplankton abundance, give only restricted information on rates; that is, whilst good at quantifying biomass (with dimensions $M^{-1}L^{-3}$) it is difficult to include the dimension (T^{-1}) and hence, to estimate production rates. One very significant recent advance was made by Li (1994) who coupled flow cytometry with ^{14}C labelling to achieve an estimate of primary production of those micro-organisms which could be differentiated and sorted by a flow cytometer. For the first time, advances in instrumentation offer the potential to estimate photosynthetic carbon fixation rates by different phytoplankton species within a natural assemblage. Clearly, the approach has great potential but flow cytometers still have limitations because it is difficult to identify the organism which correspond to the signals derived from the flow cytometer. Clearly, cells which are separated by the flow cytometer can be examined with a microscope to identify the organisms present, but it would be of far greater benefit if the signal generated from fluorescence and forward angle light scatter could be used to identify those organisms. This approach, coupled with the methodology of Li (1994) could then have the potential to measure production of individual and identifiable phytoplankton species.

ALLOMETRIC APPROACH

If the ^{14}C method does not offer the potential to measure the activity of the component species of a natural assemblage, which alternative methods are available? One approach (Joint and Pomroy, 1988) was based on allometry, the relationship between growth rate and cell size. It is generally accepted that smaller microbes grow faster than larger microbes and cell size can explain much of the variance in observed growth rates of different phytoplankton. Fenchel (1974) examined the generation times of a large number of microbes and found a good relationship between maximum growth rate and cell size. It is also well accepted that metabolic activity is related to cell size (Smith and Kalff, 1982). Specific relationships for microalgal growth rate, based on laboratory culture, were published by Banse (1976) and formed the basis of an allometric investigation of a natural assemblage (Joint and Pomroy, 1988) in the Celtic Sea, a temperate European shelf sea.

The method was applied to a patch of water which was followed for 4 days; the weather was calm, the surface mixed layer was well defined and dispersion was minimal. It was assumed that any changes in phytoplankton species composition and biomass were largely the result of biological processes rather than brought about by mixing or advection. During the experiment, the change in the abundance of each of the phytoplankton species present in the surface mixed layer was measured at daily intervals. The approach was applied to two groups of phytoplankton which could be easily characterised; cells >5µm in diameter (Joint and Pomroy, 1988), identified by standard microscopic procedures (Hasle, 1978) and picoplankton (<2µm), which was dominated by cyanobacteria, and could be readily identified from phycoerythrin fluorescence (Joint, 1991). The cell dimensions of each phytoplankton species were determined using computerised image analysis and these dimensions were converted to volume by assuming idealised shapes and applying standard formulae to determine volume; published values for carbon/volume relationships allowed the estimation of cell carbon content. These biomass data were used to determine the specific growth of each of the phytoplankton species by applying the allometric equation

$$\mu = a\mathbf{W}^{b}$$

where µ is specific growth rate, W is cell biomass, *a* and *b* are constants, with *a* having the dimension T^{-1}. Values for *a* and *b* were taken from the literature (*cf* Joint and Pomroy, 1988) and were different for groups of algae, such as diatoms, which have well characterised rapid growth rates (Geider and Osbourne, 1992). The procedure, therefore, resulted in an estimation of the biomass of each of the species which comprised the phytoplankton assemblage as well as an estimate of the specific growth rate of each of those phytoplankton species. With a further assumption that phytoplankton at 10m depth was growing close to its maximum potential growth rate, it was possible to compare this estimate of assemblage production with an estimate by the standard ^{14}C method. Variance in the allometric estimate explained 90% of the variance in the ^{14}C estimate for both phytoplankton >5µm (Joint and Pomroy, 1988) and picophytoplankton (Joint, 1991). In contrast, chlorophyll concentration explained only 59% of the variance.

Fig. 1a shows a comparison of the ^{14}C and allometric estimates for the population at 10m depth within a surface mixed layer of 15m. Good agreement is apparent between the two methods; the production estimate for the whole assemblage by ^{14}C uptake, which is a measure of community activity, was very similar to that estimated by the allometric model, which sums the estimated growth of all the individuals within the assemblage. That is, an estimate based on the sum of potential growth rates for all of the phytoplankton species within a natural assemblage gave a very similar estimate of production to the ^{14}C method, which was applied to the same assemblage incubated *in vitro*. If it is assumed that the estimate for the whole assemblage is correct, then it follows that the estimated growth rates for all the individuals species which comprise that assemblage are also correct. Therefore, the approach allows an estimate of production, at the species level.

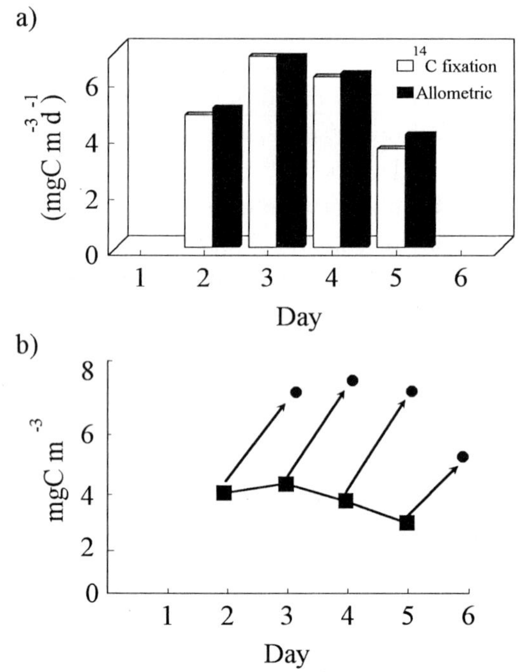

Fig. 1(a) Comparison of phytoplankton production at 10m on 4 consecutive days in the Celtic Sea in July determined by ^{14}C fixation and the allometric model.
(b) The biomass (■) of an unidentified photosynthetic flagellate measured at 10m on consecutive days at a drifting station in the Celtic Sea in July; the biomass of the flagellate (●) which would be expected the next day based on the allometric model if there had been no losses.

But the allometric approach could also be used to measure losses to the assemblage. Since the water mass was followed over a 4 day period, it was possible to measure changes in the biomass of each of the phytoplankton species which made up that assemblage. Application of the allometry model allowed an estimate of the potential growth of each of those species; so any difference between the observed and expected biomass at daily intervals, was a measure of the losses of that population due to sedimentation and grazing. Fig. 1b shows an example for an unidentified flagellate from the Celtic Sea; although this flagellate could be readily distinguished from other phytoplankton species in the assemblage, and hence was easy to count, it could not be identified unequivocally to genus or species (Joint and Pomroy, 1988). Biomass was determined from cell counts and image analysis; the specific growth rate, estimated by allometry, was then used to estimate how much increase in each species there should have been each day and compared this was compared with the observed biomass. It is assumed that any difference between the observed biomass and that which would be expected by applying the specific growth rate to the biomass present on the previous day, would be due to a loss of cells, either by grazing or by sedimentation.

Therefore, this single non-destructive procedure, which relies only on the estimation of biomass at the species level, has the potential not only to estimate the growth of each of those species but also to measure the loss of cells from the population. However, the disadvantage of the approach is the length of time taken to make one estimation. Even the application of computerised image analysis does not remove the need for tedious microscopic examination and skilled identification of each of the species present. Nevertheless, this approach is unusual because to allows an investigation of primary production at the species level within a natural assemblage.

Bacterioplankton biomass and production

If current methods cannot easily measure the growth of individual phytoplankton species, the situation with bacterioplankton is much worse because, until very recently, it was not possible even to think in terms of species of marine bacteria. Early studies of marine bacteria were largely ineffective because of the failure of standard microbiological methods. The majority of bacterioplankton do not grow in microbiological culture medium; as a result early attempts to quantify the numbers of marine bacteria using classical microbiological approaches resulted in estimates of a few hundred bacteria ml^{-1} sea water. ZoBell (1946) gave monthly average plate counts, over a 10 year period from 1932 for Californian coastal water, which varied between 350 and 620 colony forming units ml^{-1} seawater. It was only with the introduction of epifluorescence microscopy (Hobbie *et al.*, 1977) that the true number of bacteria in surface mixed layers was revealed to be close to 10^6 bacteria ml^{-1}. Interestingly, ZoBell (1946) remarked on the unusual constancy in the numbers of colonies throughout the 10 year period. This constancy is still remarkable because epifluorescence microscope methods reveal little variation in the numbers of bacteria from a wide range of provinces and seasons. The difference is, that we now believe that the numbers of bacteria in seawater are 100's of thousands ml^{-1} and not just hundreds ml^{-1}.

If such large numbers of bacteria were present in sea water, it became more important to understand what these bacteria were doing, how many of them were active and what was feeding on them (Joint and Morris, 1982). The seminal paper of Azam *et al.* (1983) did much to stimulate research on the role of bacteria in the oceans by providing a simple method to determine bacterial production from the incorporation of ^3H thymidine (Fuhrman and Azam, 1982). Yet, still the study of bacterioplankton was severely limited by lack of appropriate methods. Since bacterioplankton have few morphological features with which to distinguish one cell from another and since the standard methods of microbial taxonomy have largely developed around metabolism of highly defined culture media, it is not surprising that methods were not available to study the diversity of bacteria in natural assemblages and that many fundamental problems remain to be solved.

As an illustration of the frustrations which have resulted because we cannot distinguish one bacterium from another, I show some unpublished data (Fig. 2) which were obtained from the coastal zone of the south-west coast of England about 5 years ago. I was interested to discover any seasonal differences in bacterioplankton activity which might be consistent with species succession. At the time that the study was done, the parameters which could be measured in my laboratory were biomass, using epifluorescence microscopy, and bacterial production, from the incorporation of ^3H thymidine; both approaches had been applied successfully in a study of autotrophic and heterotrophic picoplankton activity in a temperate shelf sea (Joint and Pomroy, 1983, 1987). Since metabolic activity is dependent on temperature, I argued that the temperature range of activity might be one way to distinguish one species assemblage from another; the incorporation of ^3H thymidine was, therefore, determined in a temperature gradient apparatus at regular intervals throughout the development of seasonal stratification in the spring.

Fig. 2 shows that, although there had been an increase of water temperature of 2°C between sampling dates in March and May, there was only a small difference in the response to temperature of the two assemblages. It was also surprising that the bacterial assemblage showed activity over such a wide range of temperatures. Typically the maximum temperature of this coastal water is 18 to 20°C and the minimum is ca 6°C. Therefore, the temperature optimum of the assemblage in March was slightly greater than the maximum temperature which these bacteria would encountered in late summer. However, little could be concluded from this study, since there was nothing, other that temperature response, to distinguish the bacteria which made up these assemblages. Techniques were not available at that time to make progress in characterising the community.

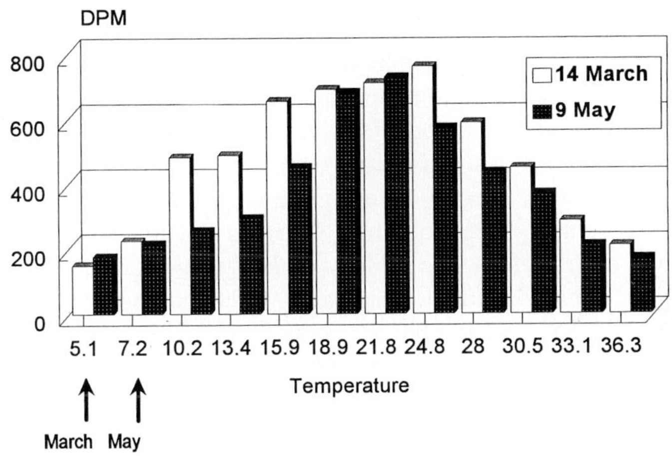

Fig. 2 Rate of incorporation of ^{3}H thymidine of samples taken on 14 March and 9 May at different temperatures; the incubation period was 1 hour and the method was that used by Joint and Pomroy (1987).

The effect of temperature on bacterial activity has not actually been the subject of much investigation. Fuhrman and Azam (1983) made one of the few investigations to date of the effect of temperature when they compared the temperature response of bacteria from below the surface mixed layer with those in the euphotic zone. These authors found that uptake of ^{3}H thymidine by bacteria from deeper water, where the ambient temperature was *ca* 10°C, showed a similar temperature response to bacteria from surface waters where the temperature was >15°C. Interestingly, the temperature response of these bacteria from the Pacific Ocean was very similar to that presented in Fig. 2 for bacteria from the north-east Atlantic. Fuhrman and Azam found greater differences in the temperature response when leucine uptake was measured; bacteria from the surface did not show any levelling off in response at the highest temperatures measured (*ca* 25°C) but bacteria from subsurface water showed maximum leucine uptake at 15-20°C. They suggested that bacteria attached to particles in the deeper water had temperature responses similar to bacteria from the surface layer and that there had been no time for significant adaptation to the lower temperature as the particles sank through the water column.

Ducklow (1993) also considered the effect of temperature on bacterial activity in the Indian Ocean. He needed to extrapolate from his measurements at surface sea water temperatures to estimate the activity of bacterioplankton in deep water where the temperature was <10°C. Ducklow suggested that, because of uncertainties in the Q_{10} for growth, which might vary from 2 to *ca* 4, it was not possible to make a reliable extrapolation. He suggested that the lack of a good understanding of the effect of temperature on the growth of bacteria in natural assemblages is inhibiting a real understanding of bacterial activity, particularly in the deep ocean, where pressure and temperature effects combine.

These examples, like many other studies of bacterioplankton, demonstrate that understanding of processes was greatly restricted because it was not possible to distinguish one bacterial species, or even one functional group, from another. Given the very small size of many bacterioplankton, morphological characteristics are clearly of no value. Many of the bacteria present in the oceans cannot readily be cultured and the approach which has worked so well for medical microbiology has little to offer the study of marine bacteria. The need for novel approaches was undeniable and, in the last 5 years, it has become clear that molecular biology has cleared the log jam. It is now possible to demonstrate that the oceans are dominated by bacteria which have not been cultured, but which can be characterised by rRNA gene sequencing (Giovannoni *et al*, 1990: Britschgi and Giovannoni, 1991: Giovannoni *et al.*, 1995) and even to demonstrate the widespread occurrence of Archaea within the oceans (Fuhrman *et al.*, 1992).

Can molecular approaches aid the determination of phytoplankton activity?

If the methods which have been developed over the last half century had not enabled investigation of the activity of individual phytoplankton species, then the introduction of molecular approaches promised fundamental changes. For the first time it was possible to think about studying not only individual species but also to consider the measurement of specific gene expression. For example, if it was thought that a process such as nitrogen fixation might be occurring, it is now possible to look for the expression of nitrogenase genes within natural assemblages (Zehr, 1995). Although the basic problem remains that expression of a specific gene product does not give information of the rate of activity of that enzyme, nevertheless the approach is by far the most process-specific that has yet been possible in biological oceanography. In an increasing number of cases, it is now possible to show unequivocally that an organism is present in a water sample and that it has the potential to carry out a specific biochemical function.

Molecular techniques are clearly having a large impact on many aspects of phytoplankton biology, including improved understanding of phytoplankton evolution. Certain phytoplankton, such as *Phaeocystis*, lack strong morphological features which could be used to distinguish different species. As a result, there has been confusion about how many species of *Phaeocystis* existed (Sournia, 1988) or whether it was a single species, with word-wide distribution. Analysis of rRNA genes from different cultures of the organism have shown that there are a number of different species, with characteristic geographical distributions (Medlin *et al.* 1994). Other prymnesiophytes, such as the coccolithophores, have also benefited from detailed molecular taxonomy, which have resolved some uncertainties of this important group of phytoplankton (Medlin *et al.* 1995).

In the study of prokaryotic phytoplankton, molecular techniques have also found rapid acceptance. The taxonomic position of the marine *Synechococcus* was established early by RFLP analysis (Douglas and Carr, 1988). When the existence and abundance of small, divinyl chlorophyll-containing prokaryotes was demonstrated by Chisholm *et al.* (1988) there were suggestions that all organisms containing this pigment should be grouped into the "Prochlorophyta". The evolutionary origin of the prochlorophytes has been established from the sequence of a subunit of RNA polymerase by Palenik and Haselkorn (1992) who demonstrated close clustering of *Prochlorococcus* with oceanic *Synechococcus*. Other divinyl chlorophyll containing prokaryotes, such as *Prochlorothrix* show greater divergence and clearly the establishment of the Prochlorophyta was not justified. However, these analyses do raise the interesting question of the origin of divinyl chlorophyll and if it arose at several times within evolutionary time.

Other benefits are accruing from the application of molecular techniques to the study of specific metabolic processes in phytoplankton. The process of photosynthesis is increasingly well understood at a molecular level (*cf* Falkowski, 1995) and this has allowed the development of novel instrumentation, such as the fast repetitive rate fluorometer which offer novel ways to measure photosynthesis in natural assemblages. The usefulness of the method was dramatically demonstrated during the IRONEX experiment in the equatorial Pacific (Martin *et al.*, 1994). This experiment tested the hypothesis that phytoplankton growth in certain provinces is limited by the availability of iron. When iron was added to a patch of water, the first measurement showing an effect of the iron enrichment was the FRR fluorometer, which indicated an early stimulation of photosynthesis (Kolber *et al.*, 1994).

The problem of nutrient limitation of phytoplankton assemblages is an other area where rapid progress is being made (Mann, 1995). When phytoplankton cells become nutrient limited, they frequently synthesis enzyme systems to maximise nutrient assimilation. LaRoche *et al.* (1993) studied the effect of limitation of growth in cultures of *Phaeodactylum tricornutum* by phosphorus, nitrogen and iron and measured the induction of specific proteins which were expressed under limiting conditions. Phosphorus limitation resulted in the synthesis of 55kDa protein and iron in the expression of 23kDa protein. Mann (1995) describes similar results for *Synechococcus* which expressed a 32kDa protein when grown under phosphate limitation. The expression of specific proteins in response to nutrient stress offers scope to develop molecular probes which could be applied to natural assemblages. For example, *in situ* hybridisation, coupled with automated systems such as flow cytometers, would allow direct study of natural assemblages; it should be possible to determine the proportion of a population possessing a specific protein and which could be assumed to be nutrient limited. However much work and strict controlled experiments will be required to achieve that goal.

Molecular approaches to the determination of bacterioplankton activity

Molecular biology has probably had an even greater impact on the study of bacterioplankton than on phytoplankton. The analysis of rRNA has provided the first practicable way to characterise natural bacterial assemblages (Giovannoni *et al.* 1995). The inability to culture the majority of bacteria in the sea, which has perhaps been the greatest barrier to progress, is no longer a significant problem.

Unexpected discoveries have been made. It was surprising to find that Archaea were a significant proportion of the bacterial population (Fuhrman *et al.*, 1992). The Archaea had previously been associated with extreme environments, such as hypersaline environments and hydrothermal vents; it was, therefore, surprising that they comprised a significant proportion of the bacterial population in oxygenic surface waters. Although analysis of rRNA has shown that the Archaea are abundance, their metabolic status has yet to be determined and we know nothing of their role in biogeochemical cycling.

Immunological approaches have also shown promise within natural assemblages of specific bacteria. For example, fluorescent antibodies were used by Ward and Cockcroft (1993) to quantify denitrifying bacteria in Monterey Bay. Numbers were about 0.1% of the total heterotrophic bacteria. However, Ward and Cockcroft commented that neither oxygen nor nitrate were depleted within the water column where denitrifying bacteria were found and denitrification conditions were not present. Therefore, although immunology was invaluable in demonstrating the presence of specific bacteria, the approach could not be used to demonstrate a particular metabolic function, only the presence of bacteria which are capable of that function if conditions are favourable. Subsequent research by Ward (1995) has demonstrated when specific genes are switched on. If the enzyme is being synthesised, then it becomes more likely, although be no means certain, that the bacteria are actively denitrifying.

Can molecular techniques determination rates?

One of the debates about molecular ecology is whether these methods can ever give information of the rate of activity of micro-organisms. Clearly molecular approaches have begun to make an important impacts in the study of natural communities. It is possible to distinguish which bacterial groups are present; it is possible to improve taxonomic discrimination between bacteria and phytoplankton; and it is possible to demonstrate how cells regulate their function in response to changes in the environment and stresses, such as nutrient limitation. However, can the demonstration of a gene or protein do any more than indicate the potential of an organism to carry out a physiological process?

It is possible to show expression of specific genes; but can metabolic activity be inferred? Enzyme activity is subject to allosteric modifications and phosphorylation and so the presence of an enzyme may be poorly correlated with *in situ* activity. However, I believe that progress is possible. The allometric model demonstrates that it is possible to add the dimension of time to an estimate of biomass and hence to estimate a rate. With a good fundamental understanding of how gene expression is coupled to physiological responses to fluctuating environments, and with careful controls in each experiment, it is possible to think that the molecular approach will be useful in determining the rate of activity of microbes involved in biogeochemical processes. This is the great potential of molecular ecology.

References

Azam F, Fenchel T, Field JG, Gray JS, Meyer-Reil LA, Thingstad F (1983) The ecological role of water-column microbes in the sea. *Mar Ecol Prog Ser* **10:**257-263

Banse K (1976) Rates of growth, respiration and photosynthesis of unicellular algae as related to size - a review. *J Phycol* **12:** 135-140

Bricelj VM, Lee JH, Cembella AD, Anderson DM (1990) Uptake kinetics of paralytic shellfish toxins from the dinoflagellate *Alexandrium fundyense* in the mussel *Mytilus edulis*. *Mar Ecol Prog Ser* **63:** 177-188

Britschgi TB and Giovannoni SJ (1991) Phylogenetic analysis of a natural marine bacterioplankton population by rRNA gene cloning and sequencing. *Appl Environ Microbiol* **57:** 1707-1713

Cadée GC (1990) Increased blooms. *Nature* **346:** 418.

Chisholm SW, Olsen RJ, Zehler ER, Goericke R, Waterbury JB, Welschmeyer NA (1988) A novel free-living prochlorophyte abundant in oceanic euphotic zone. *Nature* **334:** 340-343

Douglas SE and Carr NG (1988) Examination of genetic relatedness of marine *Synechococcus* spp. by using restriction fragment length polymorphisms. *Appl Environ Microbiol* **54:** 3071-3078

Ducklow HW (1993) Bacterioplankton distributions and production in the northwestern Indian Ocean and Gulf of Oman, September 1986. *Deep-Sea Res* II **40:** 753-771

Estep KW, Nejstgaard JC, Skjoldal HR, Rey F (1990) Predation by copepods upon natural populations of *Phaeocystis pouchetii* as a function of the physiological state of the prey. *Mar Ecol Prog Ser* **67:** 235-249

Falkowski PG (1995) Towards understanding the molecular ecology of phytoplankton photosynthesis. *This Volume* 17-38

Falkowski PG and LaRoche J (1991) Molecular biology in studies of ocean processes. Int Rev Cytol 128: 261-303

Fenchel T (1974) Intrinsic rate of natural increase: the relationship with body size. *Oecologia* **14**: 317-326

Fuhrman JA and Azam F (1982) Thymidine incorporation as a measure of heterotrophic bacterioplankton production in surface waters: evaluation and field results. *Mar Biol* **62:** 1-12

Fuhrman JA and Azam F (1983) Adaptations of bacteria to marine subsurface waters studied by temperature response. *Mar Ecol Prog Ser* **13:** 95-98

Fuhrman JA, McCallum K, Davis AA (1992) Novel major archaebacterial group from marine plankton. *Nature* **356:** 148-149

Geider RJ and Osborne BA (1992) *Algal photosynthesis; the measurement of algal gas exchange* Chapman and Hall, London pp256

Gill CW and Harris RJ (1987) Behavioural responses of the copepods *Calanus helgolandicus* and *Temora longicornis* to dinoflagellate diets. *J mar biol Assoc UK* **67:** 785-801

Giovannoni SJ, Britschgi TB, Moyer CL, Field KG (1990) Genetic diversity in Sargasso Sea bacterioplankton. *Nature* **345:** 60-63

Giovannoni SJ, Mullins TD, Field KG (1995) Microbial diversity in oceanic systems: rRNA approaches to the study of unculturable microbes *This Volume* 217-248

Hasle, GR (1978) The inverted microscope method. In: *Phytoplankton manual* (ed. A Sournia) UNESCO, Paris pp88-96

Hobbie JE, Daley R, Jasper S (1977) Use of Nuclepore filters for counting bacteria by epifluorescence microscopy *Appl Environ Microbiol* **33:** 1225-1228

Howarth MJ, Dyer KR, Joint IR, Hydes DJ, Purdie DA, Edmunds H, Jones JE, Lowry RK, Moffat TJ, Pomroy AJ and Proctor R (1993) Seasonal cycles and their spatial variability. *Phil Trans R Soc Lond A* **343:** 383-403.

Joint IR (1986) Physiological ecology of picoplankton in various oceanographic provinces. *In* Platt T and Li WKW (Eds) Photosynthetic picoplankton. *Can Bull Fish Aquat Sci* **214:** 287-309

Joint I (1991) The allometric determination of pelagic production rates. *J Plankt Res* **13:** Supplement 69-81.

Joint IR and Morris RJ (1982) The role of bacteria in the turnover of organic matter in the sea. *Oceanogr Mar Biol Ann Rev* **20:** 65-118

Joint IR and Pomroy AJ (1983) Production of picoplankton and small nanoplankton in the Celtic Sea. *Mar Biol* **77:** 19-27

Joint IR and Pomroy AJ (1987) Activity of heterotrophic bacteria in the euphotic zone of the Celtic Sea. *Mar Ecol Prog Ser* **41:** 155-165

Joint IR and Pomroy AJ (1988) Allometric estimation of the productivity of phytoplankton assemblages. *Mar Ecol Prog Ser* **47:** 161-168

Joint I and Pomroy A (1993) Phytoplankton biomass and production in the southern North Sea. *Mar Ecol Prog Ser* **99**: 169-182

Kolber ZS, Barber RT, Coale H, Fitzwater SE, Greene RM, Johnson KS, Lindley S, Falkowski PG (1994) Iron limitation of phytoplankton photosynthesis in the Equatorial Pacific Ocean. *Nature* **371:** 145-149

Lancelot C, Billen G, Sournia A, Weisse T, Colijn F, Veldhuis MJW, Davies AG, Wassman P (1987) *Phaeocystis* blooms and nutrient enrichment in the continental coastal zones of the North Sea. *Ambio* **16:** 38-46

LaRoche J, Geider RJ, Graziano LM, Murray H, Lewis K (1993) Induction of specific proteins in eucaryotic algae grown under iron-, phosphorus-, or nitrogen-deficient conditions. *J Phycol* **29:** 767-777

Li WKW (1994) Primary production of prochlorophytes, cyanobacteria, and eucaryotic ultraphytoplankton: Measurements from flow cytometric sorting. *Limnol Oceanogr* **39:** 169-175

Mann NH (1995) How do cells express nutrient limitation at the molecular level? *This Volume* 171-190

Martin JH and 43 other authors (1994) Testing the iron hypothesis in ecosystems of the equatorial Pacific Ocean. *Nature* **371:** 123-143

Medlin LK, Lange M, Baumann MEM (1994) Genetic differentiation among three colony-forming species of *Phaeocystis*: further evidence for the phylogeny of the Prymnesiophyta. *Phycologia* **33:** 199-212

Medlin LK, Lange M, Barker GLA (1995) Can molecular techniques change our ideas about the species concept? *This Volume* 133-152

Murphy LS and Haugen EM (1985) The distribution and abundance of phototrophic ultraplankton in the North Atlantic. *Limnol Oceanogr* **30:** 47-58

Palenik B and Haselkorn R (1992) Multiple evolutionary origins of prochlorophytes, the chlorophyll b-containing prokaryotes. *Nature* **355:** 265-267

Sournia A (1988) *Phaeocystis* (Prymnesiophyceae): how many species? *Nova Hedwigia* **47:** 211-217

Smith REH and Kalff J (1982) Size-dependent phosphorus uptake kinetics and cell quota in phytoplankton. *J Phycol* **18:** 275-284

Veldhuis MW, Venkamp LAH, Ietswaart T (1986) Availability of phosphorus for blooms of *Phaeocystis pouchetii* (Haptophyceae) in the North Sea: impact of the river Rhine. *Neth J Sea Res* **21:** 219-229

Ward BB and Cockcroft AR (1993) Immunofluorescence detection of the denitrifying strain *Pseudomonas stutzeri* (ATCC 14405) in seawater and intertidal sediment environments. *Microb Ecol* **25:** 233-246

Ward BB (1995) Functional and taxonomic probes for bacteria in the nitrogen cycle *This Volume* 73-86

Williams PJLeB (1993) Chemical and tracer methods of measuring plankton production. *ICES mar Sci Symp* **197:** 20-36

Zehr JP (1995) Nitrogen fixation in the sea: why only *Trichodesmium*? *This Volume* 335-364

ZoBell CE (1946) *Marine microbiology: a monograph on hydrobacteriology* Chronica Botanica Co, Waltham, Mass, pp 240

Towards Understanding the Molecular Ecology of Phytoplankton Photosynthesis

Paul G. Falkowski
Oceanographic and Atmospheric Sciences Division
Brookhaven National Laboratory, Upton,
NY 11973,
USA

Introduction

Many of the basic questions related to the magnitude, control, and interspecific variation of phytoplankton productivity confronting aquatic ecologists during the last decade of the twentieth century are virtually identical to those posed in the first three decades (Mills 1989). For example, determining which factor limits community production, let alone the growth rates of the component species, is a time-honoured endeavour (Johnstone 1911; Harvey 1926; Riley 1965; Ryther 1969; Eppley 1980; Barber 1992; Joint 1995). The contemporary discussions about the importance of nutrient fluxes, trace metals, light, temperature and grazers in regulating the abundance and distribution of primary production, phytoplankton biomass and individual species are incarnations of historical debates dating to the inception of aquatic science (Cooper 1935; Harvey 1957; Banse 1992; Martin 1992). That these issues remain, for the most part, unresolved, bespeaks profoundly of our inability to quantitatively relate ecophysiological responses to population genetics and *vice versa*. The general problem may be rephrased as a search for a way of predicting the ecological manifestation of the evolutionary concept of "fitness". Even given knowledge of all the genomes of all the organisms in the world, this Gordian knot cannot be cut by reductionist experimentation (Gillespie 1991), yet that approach is so ingrained in our training as scientists that virtually no other is explored. Aquatic ecologists often resort to the safe harbour of laboratory or mesocosm experiments on model organisms or systems, from whence new-found "paradigms" are projected to the responses of organisms or assemblages in nature. While clearly not wholly satisfactory, this philosophy underlies the nascent field of molecular ecology (Falkowski and LaRoche 1991), and the approach can be insightful if it is delimited in scope to specific responses (Kilham and Kilham, 1980).

NATO ASI Series, Vol. G 38
Molecular Ecology of Aquatic Microbes
Edited by I. Joint

Because the photosynthetic apparatus in oxygenic organisms is highly conserved from both evolutionary and functional perspectives (Barber and Andersson 1994), and photosynthesis plays such an important role in determining both community production and specific growth rates of most aquatic photoautotrophs (Falkowski and Woodhead 1992), molecular and biophysical studies of photosynthetic responses have been among the few biogeochemically oriented areas of phytoplankton research to be successfully translated to natural aquatic ecosystems (Falkowski *et al.* 1992). These avenues have proven tractable because experimental molecular biology has provided a framework to test specific hypotheses of photosynthetic responses using biophysical tools. The tools allow measurements of the rate of light absorption, electron transport and photochemical energy conversion efficiency in natural phytoplankton communities. With knowledge of the molecular architecture of the photosynthetic apparatus, inferred from laboratory studies with model organisms, such measurements can be used to interpret both the diversity photosynthetic responses and patterns of change of these responses to environmental cues (Falkowski 1992). Here I will examine, from an undoubtedly biased perspective, some of the issues and progress in understanding the molecular basis of phytoplankton photosynthesis and the projection of that understanding to the responses of phytoplankton to their environment.

Background

Defining photosynthetic responses

A major role of phytoplankton, both from ecological and biogeochemical viewpoints, is the fixation and chemical reduction of inorganic carbon by oxygenic photosynthesis, and thus a great deal of attention has been paid to the physiological ecology of photosynthesis in these organisms. While the genetic diversity of phytoplankton is extremely high (Wood and Leatham, 1992; Medlin 1995), and the evolutionary pressure for the maintenance of such diversity remains enigmatic (Huchinson, 1961: Smayda, 1980: Brand 1981), the basic machinery of the photosynthetic apparatus is highly conserved. Thus, although there are approximately 18 classes of prokaryotic and eukaryotic algae, the molecular mechanisms responsible for the oxidation of water and reduction of carbon are basically identical in all classes (Falkowski and Raven 1995). Superimposed on the mechanisms are variations in rate constants for light harvesting, electron transport and enzyme activity, resulting primarily from slow genetic drift and selection in key proteins or enzymes responsible for secondary gene products (*eg* lipids). Early attempts to codify the genetic diversity in terms of the functional photosynthetic responses for specific algal classes (Ryther and

Menzel 1959; Steemann Nielsen 1962; Prézelin and Alberte 1978), have mostly been superseded by an appreciation that the physiological plasticity and genetic variations within an individual may totally mask any interphylogenetic differences (Gallagher and Alberte 1985).

Most of the information about photosynthetic responses in phytoplankton has been deduced from mathematical analyses of photosynthesis-irradiance curves. For pragmatic reasons, carbon fixation is usually normalized to chlorophyll *a*, and so-called P *vs* I curves are usually based on the irradiance-dependent rate of incorporation of radiolabelled inorganic carbon into acid stable (mostly particulate) organic material and the simultaneous measurement of phytoplankton biomass as chlorophyll *a* (Geider and Osborne 1992; Falkowski 1994). Numerous reviews, critiques, commentaries and diatribes have been written about the methods, and it is not my purpose or intention to discuss these issues here. The salient features of P-I curve models is that they can be used to predict: (a) the initial slope, α^B, or "light utilization efficiency"; (b) the maximum photosynthetic rate, $P^B{}_{max}$, or "photosynthetic capacity", and (c) the intercept of these two parameters, I_k (= $P^B{}_{max}$), or "light saturation parameter". Sometimes a fourth parameter, related to the supraoptimal irradiance-dependent depression of photosynthesis, β, is included. The most commonly used models employ either a hyperbolic tangent or exponential function to fit the data, although there is no experimentally-validated basis for believing that either of the functions are biologically meaningful (Chalker 1980; Fasham and Platt 1983; Dubinsky *et al.* 1986; Henley 1993).

An underlying belief in the analysis of P *vs* I curves is that a systematic investigation of the behaviour of its parameters can be related to environmental variations. Thus, for example, acclimation to low or high irradiance might be expected to lead to a change in α^B or I_k, while nutrient stress might lead to a reduction in $P^B{}_{max}$. It is reasoned that if the responses of each of the parameters can be predicted or described, these might be related to specific environmental regimes; in biological oceanography these regimes are sometimes called "provinces" (Platt and Sathyendranath 1988; Morel 1991). In fact, many hundreds of papers have been published describing phytoplankton responses to one or more environmental variable in terms of α^B, I_k and $P^B{}_{max}$. Essentially, these mathematical constructs became part of the *lingua franca* of contemporary phytoplankton ecology, and while they have been extremely useful in rationalizing photosynthetic responses, attempts to extend the rationalization to biophysical or molecular responses or their causes, has not been forthcoming (Falkowski 1980; Falkowski 1992).

By the early 1980's, the biophysical concept of a "photosynthetic unit" (which emerged in the early 1930's) had been resurrected and explored in laboratory studies of phytoplankton photosynthesis (Falkowski and Owens 1980; Falkowski *et al.* 1981; Perry *et al.* 1981). The photosynthetic unit concept could be related to the mathematical parameters used to describe the P *vs* I curve by a combination of dimensional analysis and syllogistic argument (Herron and Mauzerall 1971; Myers and Graham 1971). Thus, the initial slope could be defined:

$$\alpha^B = \sigma_{PSII}\, \mathbf{n} \qquad (1)$$

where σ_{PSII} is the effective absorption cross section of photosystem II, and **n** is the concentration of PSII reaction centres (*ie* the "number" of photosynthetic units). Similarly, P_{max} could simply be related to the number of photosynthetic units:

$$P^B{}_{max} = n / \tau \qquad (2)$$

where τ is the minimum time required for whole chain electron transport from water to the terminal electron acceptor (usually CO_2). It follows that:

$$I_k = 1/(\tau\, \sigma_{PSII}). \qquad (3)$$

This system of interrelated equations relates the abstract mathematical constructs, α^B, I_k, and $P^B{}_{max}$ to three biophysical parameters, n, τ, and σ_{PSII}. These parameters cannot be derived solely from P *vs* I analyses, hence most of our research on the molecular ecology of phytoplankton photosynthesis over the past decade has been directed towards elucidating the changes in these parameters and understanding the molecular bases of their responses to environmental changes.

MOLECULAR ARCHITECTURE AND RESPONSES OF REACTION CENTRES

While the molecular structure of the reaction centre complexes in oxygenic photosynthetic organisms is not yet completely elucidated on an atomic level, their general features are inferred from analogous structures in photosynthetic bacteria and molecular biological and biophysical studies on isolated membrane complexes (Barber 1992; Goldbeck 1992). Oxygen evolution occurs in PSII reaction centres, which contain an electron donor chlorophyll molecule (P_{680}), a phaeophytin acceptor molecule (Phaeo), and a secondary acceptor, a bound quinone, (Q_A). The photochemical oxidation of P_{680} leads to the formation of the strongest known biological oxidant, $P_{680}{}^+$, which has a midpoint potential of about +1.2volts. The three electron carriers, together with another mobile carrier, plastoquinone, are localized in a heterodimer containing a 32 and 33 kDa protein, D1 and D2 respectively; the two proteins are linked by a non-heme iron which does not participate in electron transfer reactions (Fig. 1). A comparison of

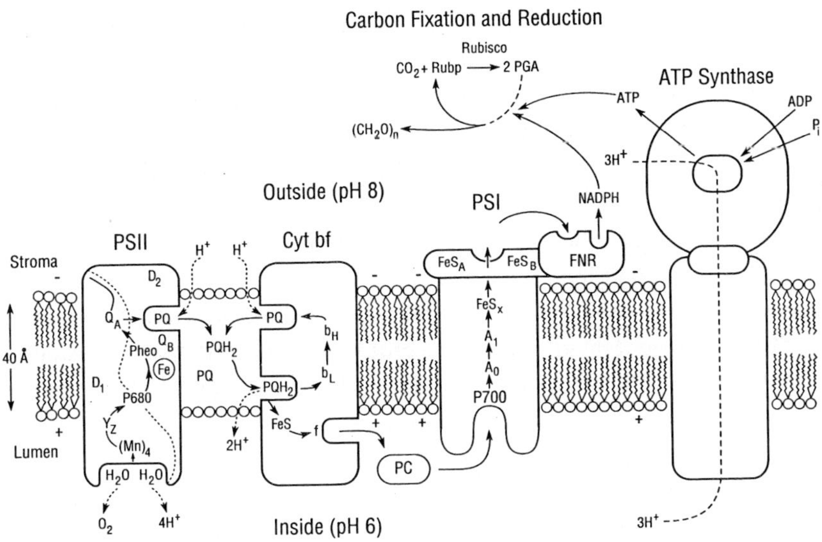

Fig 1 A schematic representation of the photosynthetic apparatus. The electron transport systems are associated with the thylakoid membrane. Water is oxidized by the photochemical activity of photosystem II, which receives excitation from membrane associated light harvesting chlorophyll-protein complexes (not shown). PSII core proteins include D1 and D1 (as well as cytochrome b559), that are bound by a non-heme iron. The photochemical oxidation of P_{680} leads to the transient reduction of phaeophytin, that subsequently donates its electron to a quinone, QA. P_{680}^+ is rereduced by an electron donated from a tyrosine, Y_Z, that is an integral part of D1. Y_Z, in turn, is reduced by an electron donated from a Mn atom that is ultimately derived from the oxidation of water. On the acceptor side of PSII, the electrons from Q_A are used to reduce a second quinone, Q_B, that is identical to plastoquinone, PQ. Upon receiving 2 electrons and 2 protons, PQH_2 dissociates from its binding pocket and migrates to the cytochrome b_6/f complex, where it is oxidized by cytochrome f. The electrons from cytochrome f are passed through a mobile carrier (either plastocyanin or a c-type cytochrome, depending upon species) to PSI. In PSI a second photochemical process, energized by light absorbed by its light harvesting pigment-protein complex (not shown), is used to successively reduced a series of iron-containing carriers, leading to the reduction of ferredoxin. Ferredoxin reduces NADP via the activity of FNR. The proton gradient established by the oxidation of water in PSII is partially dissipated by the ATP synthase, that reversibly couples the vectorial flux of the ions to the scalar production of ATP. Both ATP and NADPH diffuse from the membrane surface into the stromal fluid, where they are used to fix and reduce CO_2 to carbohydrate and to regenerate the intermediates in the Calvin cycle.

deduced amino acid sequences of D1 and D2 reveals that both of these hydrophobic proteins are extremely highly conserved throughout the photosynthetic world; in eukaryotic cells they are encoded in the chloroplast. The evolutionary template for the PSII reaction centre proteins was purple bacteria; *ie* anaerobic photosynthetic bacteria that are presently more of an evolutionary curiosity than biogeochemical linchpin. Superimposed on this template is the oxygen evolving complex, which appeared about 2.7 billion years ago. The oxygen evolving complex contains 4 Mn atoms, of which 2 undergo successive 2 electron reductions, resulting in the photochemically catalyzed oxidation of H_2O.

Electrons generated in PSII are sequentially passed, *via* a series of electron carriers, to PSI. The sequence homology of the PSI core proteins suggests that they originated from green sulphur bacteria. The specific carriers differ in some details between algal divisions, however their redox potentials remain relatively constant. In PSI, electrons are photochemically extracted from one chlorophyll molecule, P_{700}, and used momentarily to reduce a second, before being sent off to a vitamin K derivative and subsequently iron, and iron-sulphur complexes. The photochemical reactions in PSI are extremely rapid, hence the opportunity for radiative dissipation is relatively low; *ie* the room temperature fluorescence signal from PSI is negligible. As in PSII, the prosthetic groups in PSI reaction centres are contained in a heterodimer containing two proteins of 82 and 84 kDa, encoded in the chloroplast genome. The two proteins are about 45% identical in amino acid sequence and both are highly hydrophobic. Like PSII reaction centre proteins, the amino acid sequences for PSI are highly conserved, and approximately 90% of the residues are identical (or substituted with a conservative replacement) between cyanobacteria and higher plants.

PSII and PSI reaction centres are energetically coupled to antenna pigment-protein complexes that harvest and transfer excitation energy to the respective photosystem. There are myriad pigment protein complexes, differing markedly between phytoplankton classes (and consequently the pigment composition is often used as a taxonomic tool). In green algae the antenna complexes contain chlorophylls *a* and *b* as well as accessory carotenoids and are called light harvesting chlorophylls (LHC's). In some chromophyte algae the complexes contain, in addition to chlorophyll *c*, the accessory pigment fucoxanthin and hence are called fucoxanthin chlorophyll complexes (FCP's). In dinoflagellates, where peridinin is used as an accessory pigment, the antenna complexes are called the peridinin-chlorophyll protein (PCP). The molar ratio of antenna pigments (usually measured as chlorophyll *a*) to reaction centres is often taken as an index of the "size" of the photosynthetic unit. This biochemical ratio has limited biophysical meaning (Falkowski *et al.* 1981; Mauzerall and Greenbaum 1989). By "size" it is meant the effective absorption cross section, σ;

ie the ability of the antenna to absorb incident radiation at a specific wavelength. The effective cross section can be determined by following the intensity dependent change in photochemical yield for either PSII or PSI. For the former, the photochemical yield is usually (and conveniently) measured as the intensity dependent change in fluorescence, or (less conveniently) as oxygen evolution (Ley and Mauzerall 1982; Falkowski *et al.* 1986; Kolber *et al.* 1988). For PSI, the yield is usually measured as the rate of oxidation of the reaction centre observed by following the kinetics of the change in absorption at 700 or 820nm or from ESR spectra (Weaver and Weaver 1969; Zipfel and Owens 1991). The latter measurement is based on the fact that when a radical, such as P_{700}^{+} is formed, it provides a electron spin signal that can be measured as a function of actinic light intensity. As both absorption based changes and ESR spectra require relatively large amounts of photosynthetic material for the analyses, measurements of PSI cross sections are not routinely feasible in aquatic systems, consequently we know little about the variation and acclimation of the antenna of PSI in phytoplankton. Fluorescence-based measurements of PSII cross sections are feasible, however, and have proven informative about photosynthetic responses of natural phytoplankton communities (Kolber and Falkowski 1993).

FLUORESCENCE-BASED DERIVATIONS OF KEY PHOTOSYNTHETIC PARAMETERS

From an operational perspective, the photochemical reactions in PSII are extremely useful, as they can be quantitatively assessed by measuring changes in variable chlorophyll fluorescence (Falkowski *et al.* 1986; Genty *et al.* 1989; Krause and Weis 1991; Kolber *et al.* 1993). When Q_A is oxidized, fluorescence is at a minimal level, designated F_o; conversely, when Q_A is reduced, fluorescence is rises to a maximal level, designated, F_m. It can be shown that the ratio of variable fluorescence, F_v (= F_m- F_o) to F_m is a measure of the quantum efficiency of photochemistry in PSII. Precise measurements of these fluorescence yields can be made in real-time, non-destructively in natural phytoplankton communities; such measurements have served to help interpret how physical and chemical processes influence phytoplankton photosynthetic responses at a molecular level.

Using either a classical pump-and-probe fluorescence method (Falkowski *et al.* 1992; Kolber and Falkowski, 1993) or a newly developed fast repetition rate (FRR) fluorescence method, it is possible to follow the rate of photon absorption by PSII from the intensity dependent rise in variable fluorescence (Kolber *et al.* 1993). The profile of the flash intensity saturation curve of variable fluorescence is often described by a cumulative one-hit Poisson function (the complement of an exponential), such as:

$$\mathbf{F_v/F_{vmax} = 1 - e^{-\sigma E}} \qquad \mathbf{(4)}$$

where $\mathbf{F_v}$ is the variable fluorescence yield at flash energy $\mathbf{E}$, $\mathbf{F_{vmax}}$ is the maximum variable fluorescence yield induced by a saturating flash, and σ_{PSII} is the effective absorption cross section of PSII. The effective absorption cross section for PSII (σ_{PSII}) can be numerically derived from the fluorescence saturation profile (Mauzerall 1978; Ley *et al.* 1982; Falkowski *et al.* 1986).

Simultaneous measurements of the photochemical efficiency of PSII and the fraction of functional reaction centres permit very rapid analysis (<100ms) of photosynthetic parameters without a need for incubation. Instrumentation is such that continuous mapping of all these parameters can be made as a ship steams (Kolber *et al.*, 1994), or vertical profiles can be obtained from CTD-based machines.

In the late 1980s, laboratory studies with a variety of eukaryotic phytoplankton suggested that the maximal photochemical energy conversion efficiency, inferred from the change in maximum quantum yield of variable fluorescence, was remarkably constant in nutrient-replete cells grown over a wide range of irradiance levels, but was a hyperbolic function of growth rate if growth was limited by nitrogen (Kolber *et al.* 1988). Moreover, the absolute value of the maximum change in variable fluorescence (*ie*, the quantum efficiency of PSII photochemistry) was remarkably constant; F_v/F_m values averaged about 0.65-0.70, independent of species (Falkowski, 1992; Kolber *et al.* 1988). There is some suggestion that this efficiency may be lower for phycobilisome containing cyanobacteria, although a systematic study of this class has not been conducted (Krause *et al.* 1991). Measurements of the effective absorption cross section of PSII, obtained from the flash intensity saturation curve, revealed that some organisms acclimate to changes in irradiance through changing σ_{PSII}, while others alter n (Kolber *et al.* 1988). Note, that from equation 1, the effect of either of these "strategies" is the same with respect to α^B.

The two strategies are not related in any obvious way to algal class. For example, some centric diatoms, such as *Skeletonema costatum*, photoacclimate by changing σ_{PSII}, while other centric diatoms, such as *Thalassiosira weissflogii*, keep σ_{PSII} remarkably constant, but alter n. Moreover, as cells become nutrient limited, σ_{PSII} increases, even as the cells become chlorotic. This latter phenomenon seems to occur in all species examined, and at first sight, might appear counterintuitive.

Close examination of the flash intensity saturation profiles reveals that at low flash intensities variable fluorescence often lags slightly behind a true Poissonian and then rises more steeply. This deviation from a true Poissonian can be interpreted in the context of energy transfer between PSII reaction centres (Ley and Mauzerall 1986). As a reaction centre becomes closed, excitation energy, that normally would be reradiated as fluorescence, can instead be transferred, *via* a shared antenna system, to an open PSII reaction centre, where it is absorbed. The probability of energy transfer, P, can be calculated from the rate of rise of the flash intensity profile (Fig. 2).

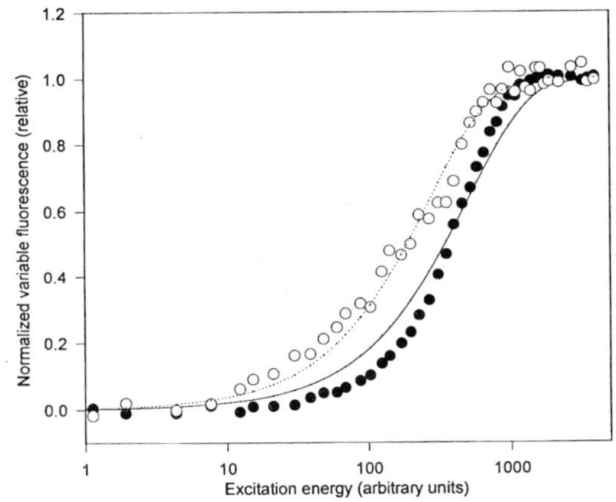

Fig. 2 An example of the flash saturation profiles of variable fluorescence from irondeficient cells (O) and iron replete cells (●). In the iron deficient cells, the profile saturates at lower excitation energy, indicative of a larger functional cross section for PSII, and closely follows a cumulative one-hit Poisson function (dotted line). The iron replete cells have a smaller cross section, but the saturation profile rises more steeply than the true Poissonian (solid line is a best fit of the experimental data to a cumulative one-hit Poisson function). The increase in steepness of the curve is indicative of the transfer of excitation energy between PSII reaction centers in iron replete cells.

The transfer of excitation energy between PSII reaction centres is not infinite. Imagine that a cell had a single reaction centre and 10^6 chlorophyll molecules in the antenna. The excitation absorbed by any chlorophyll molecule could wander in the antenna, visiting each of the 10^6 molecules, before being absorbed by the reaction centre. A probability that an excitation is localized to a single chlorophyll molecule is thus $1/10^6$. The natural lifetime of the excitation in the antenna is about 10^{-9} sec. Each "visit" occurs on average at 10^{-12} sec. Thus, under such conditions, the probability that a photon absorbed by the antenna would actually get to the reaction centre before its energy is lost by reradiation or heat is vanishingly small, on the order of 10^{-7}.

Rather, the antenna are thought to provide excitation energy to a few reaction centres, to form a "domain". When all reaction centres are functional, they compete with each other for excitation energy from the antenna; the effective absorption cross section per reaction centre is small. As reaction centres become non-functional within a domain, the remaining functional reaction centres are serviced by an apparently larger antenna; the effective absorption cross section per reaction centre increases. This biophysical model of a domain explains the effect of nutrient limitation on the simultaneous reduction in photochemical energy conversion efficiency and increase in the effective absorption cross section of PSII.

With this very brief background description of some of the molecular structures and biophysical signals of the photosynthetic apparatus, let us now examine how the information may be used to explore the ecology of phytoplankton photosynthesis.

Applications of Biophysical Tools to Phytoplankton Photosynthesis

MOLECULAR BASES OF NUTRIENT LIMITATION

1. Nitrogen

Two nutrients we have studied in some detail at a molecular level are nitrogen and iron. While the effects of the limitation of either of these elements can probably be found in all metabolic pathways in phytoplankton, I will focus here only on how the limitation affects photosynthesis and the photosynthetic apparatus.

Nitrogen limitation is effectively a from of translational control (Herzig and Falkowski 1989). Nitrogen-limited cells have low pool sizes of free amino acids, and the incorporation of nitrogen leads to marked increases in the pools of all amino acids relatively rapidly (Zehr *et al.* 1988; Flynn *et al.* 1989). Pulse labelling studies of proteins under nitrogen limitation are technically difficult to carry out because the rate of protein synthesis is often very low. Consequently, the specific activity of the nascent proteins requires either very high radiative pulses and/or long exposure times. Thus, most of the information we have on the biochemical effects of nitrogen limitation on proteins is related to the pool sizes inferred from SDS polyacrylamide gel electrophoresis. Densitometric scans of stained gels of proteins from cells grown in chemostats with nitrogen as the limiting element reveals few obvious differences in the differential accumulation of specific proteins. While overall protein abundance per cell can be markedly lower than in nutrient-replete cells, the only large change in a specific protein that is usually obvious is a disproportionate decrease in the Calvin cycle protein, RUBISCO (Falkowski *et al.* 1989). Except in cyanobacteria, the decrease in light-harvesting complexes appears to be proportional to the overall reduction in cell N.

Western blots of proteins associated with light-harvesting antenna functions and the reaction centre of PSII or PSI usually reveal that nitrogen and iron limitation leads to a reduction in the reaction centre complexes per unit antenna. This biochemical/immunological analysis is supported by biophysical measurements of reaction centre density using oxygen flash yields and P_{700} absorption differencc assays for PSII and PSI, respectively (Herzig *et al.* 1989).

Nitrogen limitation of protein synthesis does not appear to be equally distributed between nuclear-encoded proteins and plastid-encoded proteins. There appears to be a subtle, but general tendency towards a more marked reduction in the levels of plastid-encoded proteins. We hypothesize that this difference is a consequence of the compartmentalization of amino acid biosynthetic machinery. While the plastid contains the ammonium assimilatory enzymes, glutamine synthetase (GS) and glutamate synthase (GOGAT), the transamination reactions and carbon skeletons for the synthesis of many of the essential amino acids are localized in the cytoplasm. Thus, tRNAs in the cytoplasm have a higher probability of being charged with their designated amino acids than the tRNAs in the chloroplast. The latter must rely on the diffusive influx of amino acids from the cytoplasm. While the differential synthesis rate of proteins may then be uncoupled, the loss of plastid-encoded proteins inevitably leads to an overall reduction in electron transfer potential in both photosystems (Falkowski *et al.* 1989).

2. Iron

Iron limitation is of interest because the element potentially limits phytoplankton photosynthesis in three important areas of world oceans, namely, the eastern equatorial Pacific, the subArctic Pacific, and the Southern Ocean (Martin 1992). Iron is required for heme synthesis, and for the formation of iron-sulphur complexes. These two prosthetic groups are important in electron transfer reactions in the photosynthetic apparatus. Autoradiographs of SDS-denatured proteins that have been separated by one-dimensional electrophoresis reveals that ^{59}Fe is rapidly incorporated into cytochrome b_{559} and the cytochrome b_6/f complex (Greene *et al.* 1992). Western blots of iron-limited cells reveal that, like nitrogen limitation, reaction centre proteins are markedly reduced relative to the light-harvesting chlorophyll protein complexes (LaRoche *et al.* 1993). One exception is the α-subunit of cyt b_{559}, which is overexpressed by an order of magnitude, compared with iron-replete cells (Vassiliev *et al.* in press). This 9 kDa protein is chloroplast encoded. Biophysical measurements clearly show that the protein does not contain a functional heme; *ie*, it lacks the prosthetic group.

Heme biosynthesis is regulated by a metabolic pathway in the cytoplasm, and controlled by nuclear-encoded genes. It appears that in iron-limited eukaryotic phytoplankton, the synthesis of some proteins that normally would bind a heme group can become uncoupled from heme synthesis. This is a form of post-translational limitation; it specifically reduces the number of functional PSII reaction centres. While PSI activity is also reduced under iron-limited conditions, the loss of photochemical activity appears to be comparatively lower than that for PSII.

The reduction in functional PSII reaction centres under both nitrogen and iron limitation reduces the competition for excitation energy within the shared antenna of the domains. Simply put, these two types of limitation result in fewer reaction centres in each domain. Consequently, as long as excitation is shared by the antenna within the domain, the effective absorption cross section of the remaining antenna increases (Vassiliev *et al.* in press). The increase in σPSII tends to lead to lower I_k values. Interestingly, however, because the reduction in the number, n, of reaction centres is compensated by an increase in σ_{PSII}, their product, α, can remain remarkably constant over a wide range of limitation. Thus, α is a poor indicator of nutrient limitation.

PHOTOACCLIMATION

When phytoplankton are nutrient replete they physiologically acclimate to growth irradiance by altering the abundance of light-harvesting and electron transport components (Falkowski and LaRoche 1991). Cells grown at low irradiance contain higher levels of chlorophyll proteins than cells grown at low irradiance. Other proteins, especially those involved in photosynthetic electron transport, are also similarly affected by irradiance levels. These changes require the cell to perceive irradiance level and to transduce that signal *via* a biochemical process to affect gene expression. In eukaryotic phytoplankton, there do not appear to be irradiance-dependent changes in RUBISCO; the relative abundance of the enzyme per cell is remarkably constant (Sukenik *et al.* 1987; Fisher *et al.* 1989). Thus, the irradiance-dependent signal is not broadcast to all genes in the photosynthetic apparatus.

There are numerous reviews on the ecology, physiology, biochemistry and some aspects of the molecular biology of photoacclimation (Richardson *et al.* 1983; Falkowski and LaRoche 1991). Until recently, however, the molecular signalling system for the irradiance-dependent responses has not been understood. I will try to summarize what we know at the present time.

As mentioned earlier, the light-harvesting chlorophyll proteins serving as the antenna for PSII are encoded in the nucleus by the *cab* genes. Nuclear run-on transcription assays and actinomycin-D studies together establish that the irradiance-dependent increase in chlorophyll/cell is primarily a consequence of changes in gene transcription rather than post-transcriptional processing (Escoubas *et al.* 1995). Moreover, post-translational stabilization of the nascent proteins through pigment binding does not appear to be a major control route

(LaRoche *et al.* 1990; Mortain-Bertrand *et al.* 1990). Thus, irradiance levels can affect gene transcription in the nucleus in a fully developed eukaryotic cell.

An increase in *cab* mRNA can be elicited in high light by sublethal levels of 3-(3,4- dichlorophenyl)-1,1-dimethylurea (DCMU), which restricts electron transfer from Q_A to the plastoquinone pool, thereby keeping the pool oxidized. Alternatively, if the oxidation of plastoquinone pool is partially blocked with 2,5-dibromo-3-methyl-6-isopropyl-p-benzoquinone (DBMIB), there is a reduction in *cab* transcripts. Thus, when the PQ pool is oxidized, cells increase chlorophyll levels as if they were in low light, while, conversely, when the pool is reduced, chlorophyll levels decline as if the cell were acclimating to high light. Uncouplers of ATP synthesis, such as methylamine or carbonyl cyanide-*m*-chlorophenylhydrazone (CCCP), and inhibitors of oxygen evolution on the donor side of PSII, have no effect on the relative abundance of chlorophyll or *cab* message levels. These results strongly suggest that the primary signal transduction pathway for regulating *cab* gene transcription in response to irradiance levels is mediated by the redox state of the plastoquinone pool; *ie* the changes in irradiance are signalled by the redox state of the plastoquinone pool (Escoubas *et al.* 1995).

The redox state of the plastoquinone pool is a well-documented modifier of a protein kinase in the chloroplast (Allen *et al.* 1981; Bennett 1991). Inhibitors of protein phosphatases, such as okadaic acid or tautomycin, block the derepression of *cab* genes at low light. That is, when a high light-adapted cell, with low chlorophyll content, is placed in low light, in the presence of a phosphatase inhibitor, the cells do not acclimate to the low irradiance nearly as effectively as the control cells without the inhibition. These inhibitors only affect protein phosphatases in the cytoplasm. We postulate that the redox regulation of PQ is transduced by a protein kinase (Escoubas *et al.* 1995). At high irradiance levels, when the PQ pool is reduced, a signal protein is phosphorylated. The phosphorylated signal peptide exits the chloroplast and either directly targets the promotor region of the *cab* genes or, more likely, mediates a second phosphorylation. The phosphorylated signal peptide, *cab* repressor factor (CRF), reduces the transcription of *cab* genes at high light intensities, but at low irradiance levels, can be dephosphorylated, leading to an increased expression level. A similar phenomenon may occur for other genes, including chloroplast-encoded genes.

Thus, the signalling system involved in photoacclimation is effectively a feedback between photosynthetic electron transport and the genes encoding for light harvesting (Fig. 3). The oxidation of plastoquinone is the slowest step in the electron transport chain connecting PSII to PSI (although it is not the slowest step in the overall electron transport from water to carbon dioxide), and the

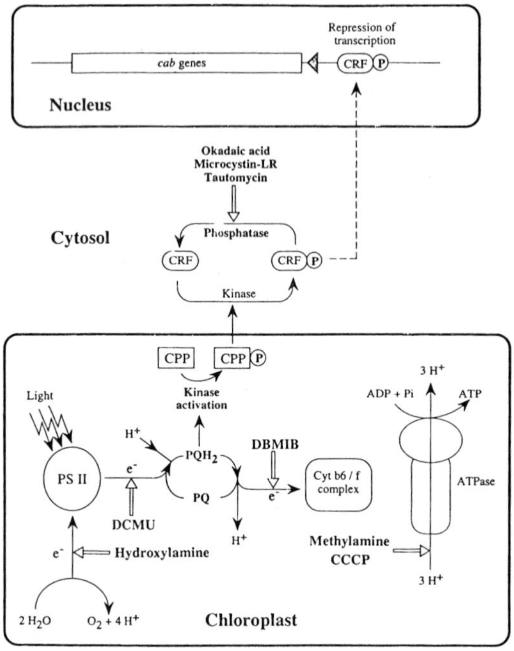

Fig 3 Schematic representation of the signal transduction pathway for photoacclimation. Excitation energy absorbed by PSII leads to the reduction of plastoquinone (see Fig. 1). The oxidation of PQH_2 is the slowest step in the electron transport chain between PSII and PSI. If the rate of excitation of PSII exceeds the rate of oxidation of PQH2, the plastoquinone pool will become increasingly reduced. Under such circumstances, a kinase is activated, leading to the phosphorylation of a chloroplast phosphoprotein (CPP), that can exit the plastid. The phosphoprotein leads to a second (and possibly more) phosphorylation in the cytoplasm and/or nucleus of a *cab* repressor factor (CRF). CRF binds upstream of the *cab* coding region, and downregulates transcription of the gene, leading to acclimation of a "high-light" type of cell with reduced chlorophyll-protein complexes. This process can be chemically mimicked by the partial inhibition of the oxidation of PQH_2 with DBMIB, or can be reversed by the preventing the reduction of the plastoquinone pool with sublethal concentrations of DCMU. The Uncouplers of ATP synthesis, methylamine or CCCP, and the inhibition of water splitting by hydroxylamine, have no effect on *cab* transcription. The derepression of *cab* transcription can be blocked by serine-threonine phosphatase inhibitors such as okadaic acid, microcystin or tautomycin. (This figure is reproduced from Escoubas *et al.* 1995, with permission).

stoichiometry between plastoquinone and PSII reaction centres varies from between 5 and 30 on a molar basis. The reduction level of the PQ pool is balanced by the rate of photon absorption by PSII, which is given by $\sigma_{PSII} \cdot I$, and the rate of electron output, which is given by τ. Inspection of equation 3 reveals that the irradiance level corresponding to this balance is I_k. Cells growing at irradiance levels lower than I_k have a largely oxidized pool, while those growing at levels greater than I_k have a largely reduced pool.

TEMPERATURE ACCLIMATION

As the output of electrons from the photosynthetic electron transport chain is determined by enzyme activity in the Calvin Cycle, photosynthesis is a temperature-dependent process (Eppley 1972; Li 1980). Over a period of a few days, many species of phytoplankton and macrophytes acclimate to changes in temperature. Cells grown at low temperature have lower levels of chlorophyll proteins, analogous to the acclimation to high light. In fact, the similarity between temperature acclimation and photoacclimation (Davison 1991), may be a consequence of the same redox mechanism that regulates gene transcription. At low temperatures, τ is reduced, consequently I_k will be lower. The reduced I_k is a proxy for the relative redox poise of the PQ pool. I suggest that at irradiance levels above I_k there is a feedback from the PQ pool to reduce the expression of chlorophyll and photosynthetic electron transport components. Thus, while a cell may appear to be adapted to low light at high temperatures, if the temperature is reduced but irradiance is kept constant, the cell will adjust to a high light type of condition if the effect is to bring the photosynthetic response to a level above I_k. This thermal acclimation hypothesis may explain the extremely low I_k values found in polar diatoms.

P_{MAX}

The maximum photosynthesis rate, obtained at light saturation, is one of the least understood variables in phytoplankton ecology (Eppley 1972; Li 1980; Falkowski 1981; Geider *et al.* 1992). While it might appear from equation 2 that an increase in the number of reaction centres per antenna will lead to an increase in P^B_{max}, or P_{max}/cell, respectively, more often the exact opposite occurs. Phenomenologically, the contrary response is (must be) due to a decrease in τ (Myers *et al.* 1971). Thus, cells that are grown, for example, at low light, and have high reaction centre levels, have lower τ values, and consequently P^B_{max} is either lower or the same as cells grown at high irradiance. In a careful study of the stoichiometry of individual components of the photosynthetic apparatus during photoacclimation, it emerged that the rate constant, $1/\tau$, was a linear function of the ratio of the reaction centre components to RUBISCO (Sukenik *et al.*

1987). As the number of electron transport chain components increased relative to RUBISCO (the low light condition), $1/\tau$ was low (τ was high) and *vice versa*. This apparent paradox makes inference of the "strategy" of photoacclimation (increased size or numbers of photosynthetic units) from P *vs* I curves virtually impossible; cells with more reaction centres almost always have lower P_{max} values per unit chlorophyll. Moreover, when cells become nutrient limited, chlorophyll per cell decreases while P^B_{max} often does not decrease as markedly; thus, like α, P^B_{max} is a poor index of nutrient limitation (Herzig *et al.* 1989).

PHOTOINHIBITION

Continuous exposure of photosynthetic organisms to high light levels can lead to a reduction in photosynthetic capacity. This phenomenon is commonly taken in the phytoplankton literature as "photoinhibition", although the term has alternative definitions amongst higher plant physiologists (Neale 1987; Long *et al.* 1994). Photoinhibition appears to be a common phenomenon in the surface ocean, especially in the tropics and subtropics (Falkowski *et al.* 1994). A simple diagnostic of photoinhibition is a hysteresis in the photosynthesis-irradiance response. In phytoplankton, part of this hysteresis is a consequence of increased respiratory activity at high light (Weger *et al.*, 1989); however, part may also be ascribed to a reduction in photosynthetic electron transport capacity. The latter phenomenon has been extensively studied on a molecular level, and it is clear that most of the damage resulting from supraoptimal irradiance exposure is localized to PSII reaction centres. The exact nature of the "downregulation" in the reaction centre remains elusive; there are advocates for the degradation of D1 and advocates for damage on the donor side (Prasil *et al.* 1992; Osmond 1994). Biophysical evidence from natural phytoplankton clearly indicates that a reduction in photochemical efficiency in the reaction centre is largely reversible within a few hours, and is more rapid at low irradiance levels than in darkness (Falkowski *et al.* 1994; Vassiliev *et al.* 1994).

The tendency for photoinhibition is related to rate of excitation delivery to PSII reaction centres (determined by the product of σ_{PSII} and incident irradiance) and the rate of repair of damage. The former is related to the degree of nutrient and light limitation, and the latter is related to nutrient supply and temperature. Both rates are under genetic control and thus species dependent as well. Consequently, no credible mathematical model predicting the degree of photoinhibition in natural phytoplankton communities (that may undergo considerable variations in irradiance throughout the day) and based on an understanding of the processes, has thus far been developed. One difficulty is that while numerous laboratory experiments on molecular bases of photoinhibition have been published (Baker and Bowyer 1994), for the most part the experimental conditions are so far removed from the realm of ecological relevance as to make impossible the quantitative application of the molecular information to natural aquatic ecosystems.

Conclusions and Prognosis

I began the discussion of the molecular ecology of phytoplankton photosynthesis by describing the philosophical inadequacies of a reductionist approach to understanding ecological processes and then proceeded to take that approach in an effort to obtain some understanding of photosynthetic responses to ecologically relevant stimuli. While not wholly satisfying, I believe that the combination of biophysical and molecular biological tools has provided some insight into mechanisms of photosynthetic responses. However, the ecological interpretation of these responses has been made possible by intensive laboratory studies on model organisms.(Kolber *et al.* 1988; Greene *et al.* 1991; Geider *et al.* 1993). I maintain that relating a biophysical parameter to a tangible molecular entity permits experimental testing of mechanistic hypotheses; without such a relationship we are at the whim of nature to provide clues to its rules. The focus in aquatic ecology in general, and biological oceanography in particular, on quantitative measurements of photosynthetic rates, in my opinion, has distracted us from focusing on more relevant and intellectually challenging issues of understanding phenomena (Falkowski *et al.* 1992; Falkowski 1994). For example, the obsession with deriving ever more accurate (and usually higher) measurements of primary production has not permitted an answer to the relatively simple question of whether phytoplankton photosynthesis in the eastern Equatorial Pacific is limited by the availability of iron (Chisholm and Morel 1991). Molecular and biophysical approaches applied to that question have yielded clear answers (Greene *et al.* 1994; Kolber *et al.* 1994). Similarly, the coupling between geophysical forcing of ocean circulation, as it affects the distribution of nutrients, and photosynthetic energy conversion efficiency has been clearly described from the phenomenological measurements of biophysical processes and a molecular interpretation thereof (Falkowski 1994).

A challenge, as I see it, is to explore in detail the molecular architecture of the photosynthetic apparatus of individual species in an effort to understand how and why genetic drift has led to increased or decreased efficiency in light utilization and carbon fixation, and to examine if and how that response affects the growth and success of individual species of phytoplankton in nature. This process requires detailed investigations of a wide range of species, leading to an understanding of the rates of mutation and how the mutations affect protein function (Gillespie, 1992). On a community level, this issue develops into the question of whether the assemblage of phytoplankton affects the overall light utilization efficiency of a water column, or do inter and intraspecific variations in photosynthetic responses converge towards a constant quantum efficiency (Falkowski *et al.* 1994)? If so, what has been the selection pressure for variations in photosynthetic responses?

Acknowledgements
This research has been supported by the US. Dept. of Energy under contract DE-AC02- 76CH00016. I thank John Berges, Zbigniew Kolber and Jean-Michel Escoubas for comments and discussions.

References

Allen JF, J. Bennett J, Steinback KE, Arntzen CJ (1981) Chloroplast protein phosphorylation couples plastoquinone redox state to distribution of excitation energy bewteen photosystems. *Nature* **291:** 25-30

Baker NK and Bowyer JR (eds) 1994. *Photoinhibition of Photosynthesis: From Molecular Mechanisms to the Field.* Bios Scientific Publishers, Oxford. pp471

Banse K. (1992) Grazing, Temporal Changes of Phytoplankton Concentrations, and the Microbial Loop in the Open Sea. In: *Primary Productivity and Biogeochemical Cycles in the Sea* Falkowski PG and Woodhead AD (eds) Plenum Press, New York and London 409-440.

Barber J (ed) (1992) *The Photosystems: Structure, Function and Molecular Biology.* Topics in Photosynthesis. Elsevier, New York. pp557

Barber J and Andersson B (1994) Revealing the blueprint of photosynthesis. *Nature* **370:** 31-34

Barber RT (1992) Geological and climatic time scales of nutrient availability. In: *Primary Productivity and Biogeochemical Cycles in the Sea.*, Falkowski PG and Woodhead A (eds) Plenum Press, New York pp89-106

Bennett J (1991) Protein phosphorylation in green plant chloroplasts. *Ann Rev Plant Physiol* **42:** 281-311

Brand LE (1981) Genetic variability in reproduction rates in marine phytoplankton populations. *Evolution* **35:** 1117-1127

Chalker BE (1980) Modeling light saturation curves for photosynthesis: an exponential function. *J theor Biol* **84:** 205-215

Chisholm SW and Morel FMM (eds) (1991) What Control Phytoplankton Production in the Nutrient-Rich Areas of the Open Sea? *Limnol Oceanogr* **36:** 1507-1970

Cooper LHN (1935) Iron in the sea and in marine plankton. *Proc R Soc Lond* **118B:** 419-438

Davison IR (1991) Environmental effects on algal photosynthesis: temperature. *J Phycol* **27:** 2-8

Dubinsky Z, Falkowski PG, Wyman K (1986) Light harvesting and utilization in phytoplankton. *Plant Cell Physiol* **27:** 1335-1349

Eppley RW (1972) Temperature and phytoplankton growth in the sea. *Fish Bull* **70:** 1063-1085

Eppley RW (1980) Estimating phytoplankton growth rates in the central oligotrophic oceans. In: *Primary productivity in the sea.* Falkowski PG (ed) Plenum Press, New York pp231-242

Escoubas J-M, Lomas M, LaRoche J. Falkowski PG (1995) Light intensity regulation of cab gene transcription is signaled by the redox state of the plastoquinone pool. *Science* (in press):

Falkowski PG (1980) Light-shade adaptation in marine phytoplankton. In: *Primary productivity in the sea.* (Falkowski PG ed) Plenum Press, New York pp99-119

Falkowski PG (1981) Light-shade adaptation and assimilation numbers. *J Plankt Res* **3:** 203-216

Falkowski PG (1992) Molecular ecology of phytoplankton photosynthesis. In: *Primary productivity and biogeochemical cycles in the sea.* Falkowski PG and Woodhead A (eds) Plenum Press, 47-67

Falkowski PG (1994) The role of phytoplankton photosynthesis in global biogeochemical cycles. *Photosyn. Res* **39:** 235-258

Falkowski PG, Greene R, Geider R (1992) Physiological limitations on phytoplankton productivity in the ocean. *Oceanography* **5(2):** 84-91

Falkowski PG, Greene R, Kolber Z (1994) Light utilization and photoinhibition of photosynthesis in marine phytoplankton. In: *Photoinhibition of Photosynthesis: From molecular mechanisms to the field* Baker N and Bowyer J (eds) Bios Scientific, Cambridge 407-432

Falkowski PG and LaRoche J (1991) Acclimation to spectral irradiance in algae. *J Phycol* **27:** 8-14

Falkowski PG and LaRoche J (1991) Adaptation to spectral irradiance in unicellular algae. *J Phycol* **27:** 8-14

Falkowski PG and LaRoche J (1991) Molecular biology in studies of ocean processes. *Int Rev Cytol* **128:** 261-303

Falkowski PG and Owens TG (1980) Light shade adaptation: two strategies in marine phytoplankton. *Plant Physiol* **66:** 632-635

Falkowski PG, Owens TG, Ley AC, Mauzerall DC (1981) Effects of growth irradiance levels on the ratio of reaction centers in two species of marine phytoplankton. *Plant Physiol* **68:** 969-973

Falkowski PG and Raven JA (1995) *Aquatic Photosynthesis.* Blackwell Scientific Publishers, Oxford, (in press)

Falkowski PG, Sukenik A, Herzig R (1989) Nitrogen limitation in *Isochrysis galbana* (Haptophyceae). II. Relative abundance of chloroplast proteins. *J Phycol* **25:** 471-478

Falkowski PG and Woodhead AD (1992) *Primary Productivity and Biogeochemical Cycles in the Sea.* Plenum Press, New York, pp 550

Falkowski PG, Wyman K, Ley AC, Mauzerall DC (1986) Relationship of steady state photosynthesis to fluorescence in eucaryotic algae. *Biochim Biophys Acta* **849:** 183-192

Fasham MJR and Platt T (1983) Photosynthetic response to light: a physiological model. *Proc. R. Soc. Lond. Ser B* **219:** 355-370

Fisher T, Shurtz-Swirski R, Gepstein S, Dubinsky Z (1989) Changes in the levels of ribulose-1, 5-bisphosphate carboxylase/oxygenase (Rubisco) in *Tetraedon minimum* (Chlorophyta) during light and shade adaptation. *Plant Cell Physiol* **30:** 221-228

Flynn KJ, Dickson MJ, Al-Almoudi OA (1989) The ratio of glutamate:glutamine in microalgae: a biomarker for N-status suitable for use at natural cell densities. *J Plankton Res* **11:** 165-170

Gallagher JC and Alberte RS (1985) Photosynthetic and cellular photoadaptive characteristics of three ecotypes of the marine diatom, *Skeletonema costatum* (Grev.) Cleve. *J Exp Mar Biol Ecol* **94:** 233-250

Geider RJ, Greene RM, Kolber Z, MacIntyre HL, Falkowski PG (1993) Fluorescence assessment of the maximum quantum efficiency of photosynthesis in the western North Atlantic. *Deep-Sea Res* **40:** 1205-1224

Geider RJ and Osborne BA (1992) *Algal Photosynthesis: The Measurement of Algal Gas Exchange*. Chapman and Hall, New York, pp256

Genty B, Briantais J-M, Baker NR (1989) The relationship between the quantum yield of photosynthetic electron transport and quenching of chlorophyll fluorescence. *Biochim Biophys Acta* **990:** 87-92

Gillespie JH (1991) *The Causes of Molecular Evolution.* Oxford University Press, New York, pp 336

Goldbeck JH (1992) Structure and function of photosystem I. Ann. Rev. Plant Phyisol. *Plant Mol Biol* **43:** 293-324

Greene RM, Geider RJ, Falkowski PG (1991) Effect of iron limitation on photosynthesis in a marine diatom. *Limnol Oceanogr* **36:** 1772-17782

Greene RM, Geider RJ, Kolber Z, Falkowski PG (1992) Iron-induced changes in light harvesting and photochemical energy conversion processes in eucaryotic marine algae. *Plant Physiol* **100:** 565-575

Greene RM, Kolber ZS, Swift DG, Tindale NW, Falkowski PG (1994) Physiological limitation of phytoplankton photosynthesis in the eastern equatorial Pacific determined from the variability in the quantum yield of fluorescence. *Limnol Oceanogr* **39:** 1061-1074

Harvey HW (1926) Nitrates in the sea. *J Mar Biol Ass UK* **14:** 71-88

Harvey HW (1957) *The Chemistry and Fertility of Sea Water.* Cambridge University Press, Cambridge, pp 234

Henley WJ (1993) Measurement and interpretation of photosynthetic light-response curves in algae in the context of photoinhibition and diel changes. *J Phycol* **29:** 729-739

Herron HA and Mauzerall D (1971) The development of photosynthesis in a greening mutant of *Chlorella* and an analysis of the light saturation curve. *Plant Physiol* **50:** 141-148

Herzig R and Falkowski PG (1989) Nitrogen limitation of *Isochrysis galbana.* I. Photosysnthetic energy conversion and growth efficiencies. *J Phycol* **25:** 462-471

Huchinson GE (1961) The paradox of the plankton *Am Nat* **95:** 137-145

Johnstone J (1911) *Life in the Sea.* Cambridge University Press, London, pp 150

Joint I (1995) The potential of molecular ecology. *This volume* 1-16

Kilham P and Kilham SS (1980) The evolutionary ecology of phytoplankton. In *The physiological ecology of phytoplankton.* Morris I (ed) Univ California Press, Los Angeles 571-597

Kolber Z and Falkowski PG (1993) Use of active fluorescence to estimate phytoplankton photosynthesis in situ. *Limnol Oceanogr* **38:** 1646-1665

Kolber Z, Zehr J, Falkowski PG (1988) Effects of growth irradiance and nitrogen limitation on photosynthetic energy conversion in Photosystem II. *Plant Physiol* **88:** 72-79

Kolber ZS, Barber RT, Coale H, Fitzwater SE, Greene RM, Johnson KS, Lindley S, Falkowski PG (1994) Iron limitation of phytoplankton photosynthesis in the Equatorial Pacific Ocean. *Nature* **371:** 145-149

Krause GH and Weis E (1991) Chlorophyll fluorescence and photosynthesis: The basics. *Ann Rev Plant Physiol Plant Mole Biol* **42:** 313-349

LaRoche J, Geider RJ, Graziano LM, Murray H, Lewis K (1993) Induction of specific proteins in eucaryotic algae grown under iron-, phosphorus-, or nitrogen-deficient conditions. *J Phycol* **29:** 767-777

LaRoche J, Mortain-Bertrand A, Falkowski PG (1990) Light-intensity changes in ab mRNA and LHC II apoprotein from the unicellular marine chlorophyte *Dunaliella tertiolecta. Plant Physiol* **97:** 147-153

Ley A and Mauzerall D (1986) The extent of energy transfer among photosystem II reaction centres in *Chlorella. Biochim Biophys Acta* **850:** 234-248

Ley AC and Mauzerall D (1982) Absolute absorption cross sections for photosystem II and the minimum quantum requirement for photosynthesis in *Chlorella vulgaris. Biochim Biophys Acta* **680:** 95-106

Li WKW (1980) Temperature adaptation in phytoplankton: cellular and photosynthetic characteristics. In: *Primary Productivity in the Sea* Falkowski PG (ed), Plenum Press, New York 259-279

Long SP, Humpries S, Falkowski PG (1994) Photoinhibition of photosynthesis in nature. *Ann. Rev. Plant Physiol. Plant Mol Biol* **45:** 655-662

Martin JH (1992) Iron as a limiting factor in oceanic productivity. In: *Primary Productivity and Biogeochemical Cycles in the Sea* Falkowski P and Woodhead A (eds) Plenum Press, New York 137

Mauzerall D (1978) Multiple excitations and the yield of chlorophyll a fluorescence in photosynthetic systems. *Photochem Photobiol* **28:** 991-998

Mauzerall D and Greenbaum NL (1989) The absolute size of a photosynthetic unit. *Biochim Biophys Acta* **974:** 119-140

Medlin LK, Lange M, Barker GLA, Hayes PK (1995) Can molecular techniques change our ideas about the species concept? *This volume* 133-152

Mills EL (1989) *Biological Oceanography. An Early History, 1870-1960.* Cornell University Press, Ithaca, pp 378

Morel A (1991) Light and marine photosynthesis: A spectral model with geochemical and climatological implications. *Prog Oceanogr* **26:** 263-306

Mortain-Bertrand A, Bennett J, Falkowski PG (1990) Photoregulation of LHCII in *Dunaliella tertiolecta:* Evidence that apoprotein abundance but not stability requires chlorophyll synthesis. *Plant Physiol* **94:** 304-311

Myers J and Graham J-R (1971) The photosynthetic unit in *Chlorella* measured by repetitive short flashes. *Plant Physiol* **48:** 282-286

Neale PJ (1987) Algal photoinhibition and photosynthesis in the aquatic environment. In: *Photoinhibition* Kyle DJ, Osmond CB, Arntzen CJ (eds) Elsevier, New York 39-65

Osmond CB (1994) What is photoinhibition? Some insights from comparisons of shade and sun plants. In: *Photoinhibition of Photosynthesis: From Molecular*

Mechanisms to the Field Baker NR and Bowyer JR (ed) Bios Scientific, Oxford 1-24

Perry MJ, Talbot MC, Alberte RS (1981) Photoadaptation in marine phytoplankton: response of the photosynthetic unit. *Mar Biol* **62:** 91-101

Platt T and Sathyendranath S (1988) Oceanic primary production: estimation by remote sensing at local and regional scales. *Science* **241:** 1613-1620

Prasil O, Adir N, Ohad I (1992) Dynamics of photosystem II: mechanisms of photoinhibition and recovery processes. In: *The Photosystems: Structure, Function and Molecular Biology*, Barber JR (ed) Elsevier, New York 295-348

Prézelin BB and Alberte RS (1978) Photosynthetic characteristics and organisation of chlorophyll in marine dinoflagellates. *Proc Nat Acad Sci USA* **75:** 1801-1804

Richardson K, Beardall J, Raven JA (1983) Adaptation of unicellular algae to irradiance: an analysis of strategies. *New Phytol* **93**: 157-191

Riley GA (1965) A mathematical model of regional variations in plankton. *Limnol Oceanogr* **10:** R202-215

Ryther JH (1969) Photosynthesis and fish production in the sea. *Science* **166:** 72-76

Ryther JH and Menzel DW (1959) Light adaptation by marine phytoplankton. *Limnol Oceanogr* **4:** 492-497

Smayda TJ (1989) Phytoplankton species succession. In: *The physiological ecology of phytoplankton* Morris I (ed) Univ California Press, Los angeles 493-570

Steemann Nielsen E (1962) The adaptation to different light intensities in *Chlorella vulgaris* and the time dependence on transfer to a new light intensity. *Physiol Plant* **15:** 505-517

Sukenik A, Wyman KD, Bennet J, Falkowski PG (1987) Light-saturated photosynthesis-limitation by electron transport a carbon fixation? *Biochim Biophys Acta* **891:** 205-215

Vassiliev IR, Kolber ZS, Mauzerall D, Shukla VK, Wyman KD, Falkowski PG (1995) Effects of iron limitation on Photosystem II composition and energy trapping in *Dunaliella tertiolecta. Biochim. Biophys. Acta* in press.

Vassiliev IR, Prasil O, Wyman KD, Kolber Z, Hanson AK Jr, Prentice JE, Falkowski PG (1994) Inhibition of PSII photochemistry by PAR and UV radiation in natural phytoplankton communities. *Photosyn Res* **42:** 51-64

Weaver EC and Weaver HE (1969) Paramagnetic unit in spinach subchloroplast particles; estimation of size. *Science* **165:** 906-908

Wood AM and Leatham TL (1992) The species concept in phytoplankton ecology. *J Phycol* **28:** 723-729

Weger HG, Herzig R, Falkowski PG, Turpin DH (1989) respiratory losses in a marine diatom: measurements by short-term mass-spectrometry. *Limnol Oceanogr* **34:** 1153-1161

Zehr JP, Falkowski PG, Fowler J, Capone DG (1988) Coupling between ammonium uptake and incorporation in a marine diatom: Experiments with the short-lived radioisotope ^{13}N. *Limnol Oceanogr* **33:** 518-527

Zipfel W and Owens TG (1991) Calculation of the absolute Photosystem I absorption cross-sections from P_{700} photooxidation kinetics. *Photosyn Res* **29:** 23-32

Bacteria in Oceanic Carbon Cycling as a Molecular Problem

Farooq Azam, David C. Smith,
Richard A. Long, Grieg F. Steward
Marine Biology Research Division
Scripps Institution of Oceanography
University of California, San Diego
La Jolla, California 92093
USA

Introduction

This paper is written to facilitate dialogue between molecular biologists and marine microbial ecologists on research strategies in understanding the role of marine microbes in the oceanic carbon cycle. We think molecular biology can revolutionize marine microbial ecology and biogeochemistry, but at this early stage the challenge is to articulate fundamental unsolved problems in a way that shows how molecular approaches might be applied. In this paper we discuss one of the central problems in oceanic carbon cycling, namely bacteria-organic matter coupling. We will explain why bacteria-organic matter coupling is an important problem, stress the need for studying the problem in a suitable ecosystem context, formulate hypotheses for future research, and suggest how molecular approaches might be of value.

Statement of the problem

THE IMPORTANCE OF UNDERSTANDING CARBON FLUX PATHWAYS.

Biogeochemists seek to understand how biological forces create the spatial-temporal patterns of distribution of carbon, nitrogen, phosphorus and iron that are observed in the ocean. A major question concerns the fate of carbon incorporated into organic matter by phytoplankton. This seemingly esoteric goal has fundamental as well as practical importance. For instance, there is interest in predicting whether the ocean, with potential to sequester materials in its sediments for aeons, will respond to increasing atmospheric CO_2 by acting as net CO_2 sink. Biodiversity and fisheries management also requires predicting spatial-temporal patterns of carbon fluxes into various biomass and species pools. In

NATO ASI Series, Vol. G 38
Molecular Ecology of Aquatic Microbes
Edited by I. Joint

order to make such prediction we need to know, and model, how the ocean's biological and physical-chemical forces move carbon atoms in space and in time.

THE IMPORTANCE OF BACTERIA-ORGANIC MATTER COUPLING.

Organic matter synthesized by phytoplankton can go into three main pathways. It can be eaten by protozoa and metazoa, be utilized by bacteria, or it can aggregate and sediment. Historically, it was thought that metazoa dominate organic matter utilization, that the release of their metabolic wastes (NH_4, urea, PO_4, CO_2) is the main mechanism to regenerate nutrients, and that their faecal pellets are the dominant vehicle of downward flux of organic matter. Bacteria were thought to use little primary production (<1%; ZoBell, 1973) and ecosystem models did not assign a significant role to them in carbon fluxes (Steele, 1974). Since bacteria directly use only dissolved organic matter (DOM), low bacterial carbon demand (BCD) also meant that carbon flux through DOM was negligible and that most primary production flowed within the particulate phase (phytoplankton → herbivores → carnivores).

This view has changed dramatically and we now know that metabolism of bacteria and protozoa strongly influences carbon fluxes. Measuring bacterial metabolism *in situ* is difficult and previous studies had grossly underestimated it. Improved methods have shown that, even in the euphotic zone, bacterial biomass is often comparable to phytoplankton biomass and BCD is often equivalent to about one-half of the primary production (Cole et al., 1988). Since bacteria directly use only DOM, high BCD means that major fluxes of organic matter occur through the DOM pool in the pathway: *phytoplankton → DOM → bacteria → protozoa* (Microbial loop; Azam *et al.*, 1983). These findings have fundamentally changed how we think of the "ocean's metabolism" (*ie* greater role of bacterial metabolism), its regulation and its consequences for the ocean's chemistry. They have also changed our ideas of food web structure and carbon flux pathways (greater emphasis on the pathway: *phytoplankton → DOM → bacteria*). Since bacteria mediate substantial carbon fluxes, understanding bacterial utilization of organic matter is a significant problem in oceanography.

Organic matter flux into bacteria: A molecular problem?

Bacterial utilization of organic matter has now been estimated in many environments. It turns out that the fraction of *co-local* primary production used by bacteria is not always ~50%, as generalized above, but varies widely, from 0% (bacteria not able to use primary production; Pomeroy *et al.*, 1991) to >100% (bacteria use some primary production from a different time or place; Azam *et al.*, 1994). This large variability can strongly influence the partitioning of primary production among competing pathways (bacteria, animals and aggregation), and hence, can change the overall patterns of carbon flux through the ecosystem. Therefore, if we are to go beyond descriptive measurements of fluxes to *predicting variations* in flux patterns, we must discover the mechanisms which mediate and regulate organic matter flux into bacteria and the strategies by which bacteria compete for organic matter in the ecosystem. This is a "molecular problem", *ie* to find out which bacteria species are present, what phenotypes they express that are important for bacteria-organic matter coupling, and how these phenotypes are regulated. The ecosystem functions of bacteria (organic matter flux, nutrient cycling and food web structure) can then be seen as consequences of bacterial strategies for organic matter utilization which are consistent with optimizing their survival in the ecosystem. The strategies addressed in this paper are "digestion" (by ectoenzymes), substrate transport, and behaviour (sensing and motility). The exciting challenge is to do this molecular biology, or molecular ecology, in the ecosystem context.

BACTERIA IN SEAWATER.

In order to study the strategies of bacteria for utilizing organic matter, we need to know what bacterial genotypes are present, but this had not been possible until recently. Microscopy shows that there are generally ~ 10^6 ml^{-1} bacteria in the photic layer, but typically only 10^3 ml^{-1} form colonies. This has been a problem in isolate-based phylogenetic studies because we did not know whether the isolates represented true species diversity (*ie* few individuals of every species grew up) or whether the isolates were a minute fraction of the species present. Phylogeny based on 16S rRNA gene sequences of clones from pelagic bacterial DNA libraries has made it possible to do phylogenetic analysis without culturing (see Giovannoni, this volume, for a treatment of phylogeny). PCR amplification of natural assemblage DNA followed by sequence isolation by DGGE (denaturing gradient gel electrophoresis) is making it much easier to do phylogenetic analysis (Murray and Hollibaugh, 1994). It also makes it possible to obtain diversity "finger-prints" and to determine spatial-temporal variations in diversity. Since the utilizability and utilization rate of organic matter might depend on the dominant species (discussed later), phylogeny and diversity analyses might help explain and eventually predict bacteria-organic matter coupling on the basis of specific strategies of dominant genotypes . We stress that phylogenetic analysis is necessary, but clearly not sufficient; environmental expression of the relevant phenotypes is what underlies bacteria-organic matter coupling and this must also be studied.

ORGANIC MATTER IN THE ENVIRONMENT OF BACTERIA.

Before further discussing bacteria-organic matter coupling we need to review current ideas on organic matter in photic zone seawater. It is a complex and largely uncharacterized mixture of monomers, polymers and particles. DOM concentration in C units is ~100 µg-at l^{-1}, most of which (generally 90-95%;) is hundreds of years old (Williams and Druffel, 1987) including humic-like material not utilizable by bacteria. Most utilizable DOM is polymeric (*eg* polysaccharides, proteins, nucleic acids) requiring hydrolysis before uptake by bacteria. The monomers of these polymers are apparently the main conduit for organic matter flux into bacteria. Their pools turn over rapidly (generally 10-100h) and their concentrations are kept low (generally 0.1 -10nmol l^{-1} for individual amino acids and 1-100nmol l^{-1} for individual sugars) by the high affinity uptake system of bacteria (Azam and Cho, 1987). The large polymer pool (an order of magnitude larger than the direct substrates pool) may "buffer" fluctuation in monomer pools.

Organic matter has been operationally defined based on filtration: that which passes an 0.45µm filter is DOM and that which is retained is POM. Historically, however, "particles" were typically thought of as readily observed phytoplankton cells and large detritus (*eg*, >10µm) which are present at 10^2-10^3 ml^{-1}. Recently, our view of POM has become more sophisticated as new techniques have revealed other particles at many orders of magnitude higher abundances. Koike *et al.* (1990) found 10^7 particles, 0.3-1.3µm in size; Wells and Goldberg (1991; 1993) found 0.05-0.12µm colloidal particles at 10^7-$10^{10}ml^{-1}$ and 0.2- 2µm colloidal aggregates at 10^4-10^5 ml^{-1}; Alldredge *et al.* (1993) found 10^2-10^4 ml^{-1} polysaccharide particles, including strings, sheets and bundles, 3 - >100µm in size. Thus, the organic matter pool is a size-continuum, a thin gel of colloids in which larger particles are embedded and through which polymers and small molecules diffuse. Bacteria, phytoplankton, protozoa and viruses are embedded in the gel, interacting with it and among themselves, and may be organized as communities at microscale. On average the inter-bacteria distances are <50µm and bacteria to algae distances are <100µm. Colloids and small particles would be few to 10µm apart and phytoplankton-size detritus would be ~100µm apart. So, bacteria need not travel far to find a particle.

The concept of organic continuum is important in thinking about molecular strategies of bacteria to sense and interact with components of the organic matter field. But it does not imply that organic matter is distributed *homogeneously* in the bacterium's environment. On the contrary, particles such as phytoplankton and larger detritus would actually create significant *heterogeneity* of organic matter distribution or "hot spots". Since the "hot spots" are microscale features within the organic continuum the bacterium will always be interacting with the organic matter field (with monomers, polymers and colloids), but its metabolism and growth might change depending on its location in the field (*eg* whether or not it is in a hot spot). Particles harbour unique bacterial communities (DeLong *et al.*, 1993) and are sites of activities not expected in the bulk phase (*eg* sulphate reduction, an anaerobic process, occurs in aerobic waters; Shanks and Reeder, 1993). Thus, the "hot spots" may be the sites of quantitatively and qualitatively different bacterial activities and thus influence bacteria-organic matter coupling. Interactions of bacteria with protozoa and viruses in the microzone would further influence bacteria-organic matter coupling.

STRATEGIES FOR DOM UTILIZATION.

As the above description of DOM-bacteria coupling shows, most organic matter flux into bacteria occurs in the pathway: *polymeric substrates* → *direct substrates* → *bacteria*. The first step is mediated by bacterial hydrolytic ectoenzymes (Hollibaugh and Azam, 1983; Chróst, 1990) and the second step, which is the final step in *all* organic matter flux into bacteria, is mediated by bacterial permeases. There has been much work on bacterial ectoenzymes (Chróst, 1990) and more limited work on monomer transport mechanisms (discussed later). The good news is that a few classes of polymers and monomers may be responsible for most organic matter flux, so we need not study the hydrolysis and uptake of the hundreds or thousands of organic compounds present in the DOM. There is hope that a few general adaptive strategies of polymer hydrolysis and transmembrane transport will be found to work best for organic matter acquisition in the ocean with different bacteria using variations on the same basic molecular themes. Those involved in comparative studies of the molecular bases of bacterial adaptations might be interested in the "digestion" and uptake strategies of bacteria in the ocean.

Some biochemical strategies have been discovered and are ripe for molecular analysis. Since bacterial biovolume is 10^{-7} of seawater volume it would be energetically prohibitive for free bacteria to hydrolyze polymers by releasing enzymes. Many studies have found that the enzymes (protease, glucosidase, nuclease, nucleotidase, chitobiase and lipase) are bound to bacterial cell surface or occur in the periplasmic space (Hollibaugh and Azam, 1983; Paul *et al.* 1988; Martinez and Azam, 1993). Thus, bacteria must touch the polymers (and particles; discussed later) to cause hydrolysis and this creates a monomer-rich microzone on bacterial surface. The next challenge is to achieve some hydrolysis-uptake coupling, so bacteria can take advantage of the monomer patch. There is kinetic evidence in the cases of utilization of protein (Hollibaugh and Azam, 1983) and nucleotides (Ammerman and Azam, 1985) that significant hydrolysis-uptake coupling is achieved. These two strategies for polymer utilization, cell-bound enzymes and hydrolysis-uptake coupling, have profound implications for bacteria-organic matter coupling. At present we know nothing about the mechanism of hydrolysis-uptake coupling, whether it involves, *eg*, periplasmic binding proteins, multimeric proteins with hydrolysis and binding domains, or some other strategy. This can be studied by molecular genetics using cultures.

Whether bacteria experience high DOM patches has been considered. Wright and Burnison (1979) suggested that evolution had pushed the transport K_m of pelagic bacteria to match the nmol l^{-1} substrate concentrations in seawater. Studies have since shown that marine bacteria have multiple transport systems for D-glucose with K_m from nmol l^{-1} to mmol l^{-1} (Azam and Hodson, 1981; Nissen *et al.*, 1984). Nissen *et al.* (1984) found a marine isolate (LNB 151; Carlucci and Shimp, 1974) had multiphasic transport system with K_m from 10^{-9} to 10^{-3} mol l^{-1}. The low K_m systems had low V_{max} (low flow systems) and the high K_m systems had high V_{max} (high flow systems). Simultaneous occurrence of low flow and high flow systems would enable the bacterium to take advantage of high substrate concentrations in microzones, yet efficiently take up substrates at the low bulk-phase concentrations. Such an adaptation suggests that pelagic bacteria do experience large variations in substrate concentrations in seawater, thus providing indirect evidence for the existence of "hot spots".

Logan and Fleury (1993) suggest, on theoretical grounds, that non-saturable Fickian diffusion can explain multiple or multiphasic kinetics in bacteria but, at low substrate concentrations in seawater, dependence on simple diffusion would be a disadvantage for competition with bacteria using facilitated and active transport systems (Hodson and Moran, in press). Molecular genetic analysis can definitively test whether multiple/multiphasic transport systems exist and, in view of their importance in DOM flux, their molecular study should be a priority. We should also find out whether this transport strategy is common among pelagic bacteria. This is also a good example of an ecological problem which is best resolved by using molecular biology. Much remains to be learned about transport strategies of marine bacteria, and, indeed, it is possible that novel transport systems uniquely designed for the ocean conditions remain to be discovered.

STRATEGIES FOR POM UTILIZATION.

Primary production and its demise mainly generates particles, *ie*, phytoplankton cells and detritus (for the purpose of our discussion, phytoplankton are included in the POM pool). This POM can sediment or be directly ingested by animals, but it can also be enzymatically solubilized and then consumed by bacteria in a pathway analagous to that described for DOM (polymeric substrate $\rightarrow$ direct substrate $\rightarrow$ bacteria). Therefore, even though bacteria can directly use only DOM, the use of ectohydrolases to solubilize particles allows bacteria to compete with animals for POM. The conversion of POM to DOM by bacteria not only renders it inaccessible to animals, but also converts it from a sinking to a non-sinking pool, thus conserving carbon in the upper ocean where it can be recycled. The rate of solubilization of particles by bacteria is thus an important variable in determining the fate of particulate carbon in the food web and we need to understand the mechanisms which mediate and regulate the POM → DOM transition.

Despite much research we do not know the dominant mechanisms of DOM production from POM. Most studies have examined phytoplankton exudation (Bjørnson, 1988), zooplankton "sloppy feeding" (Lampert, 1978) and DOM release from animal egesta (Jumars *et al.*, 1989). However, these releases can be small relative to BCD (Smith *et al.*, submitted) and, while animals are important in processing POM (*eg* disruption or partial digestion), they probably do not *directly* produce enough DOM to satisfy the BCD. It has been proposed that bacteria must actively compete with particle-eating animals for POM by rapidly solubilizing it to DOM (Azam, 1984; Pomeroy and Wiebe, 1988; Smith *et al.*, submitted). What strategies are used by bacteria to solubilize POM and to utilize the resulting DOM, and do they solubilize the particles fast enough to compete with animals? It is clear that, by using their ectohydrolases, bacteria *can* solubilize particles. In order to compete with particle eating animals, it would seem adaptive for bacteria to find and stay with sustained sources of high quality nutrition or "hot spots" (*eg* phytoplankton and rich detritus). Below, we consider two particle types as sustained sources of organic matter, phytoplankton and detritus of the size of phytoplankton or larger.

BACTERIA-PHYTOPLANKTON INTERACTION.

The question has been raised whether bacteria, endowed with ectohydrolases, attack and solubilize healthy algae (Pomeroy and Wiebe, 1988; Azam and Smith, 1991). It is generally thought that healthy algae use chemicals to inhibit bacterial approach and attack (Waksman *et al.*, 1937; Pratt *et al.*, 1944; Steeman-Nielsen, 1955; Sieburth, 1960; Trick *et al.* 1984). Although these studies invoke the production of antibiotics by phytoplankton, this does not appear to be a general strategy. How then do phytoplankton protect themselves? We have previously suggested (Azam and Smith, 1991) that they shield the cell wall by secreting mucus which bacteria use in a symbiotic relationship. In a mesocosm study of diatom-bacteria interaction, Smith *et al.* (submitted) found that the diatoms (mainly *Chaetoceros* spp) became heavily colonized yet remained healthy and grew rapidly ($>1d^{-1}$) as did attached bacteria (generally $5\text{-}10d^{-1}$). Robust diatoms growth occurred even though attached bacteria expressed extremely high protease, glucosidase and chitobiase activities. Bacteria appeared embedded in diatom mucus and this layer may have protected the diatom as well as provided the energy to support rapid bacterial growth (Smith *et al.*, submitted). Thus, bacteria-diatom association created hot-spots with intense ectohydrolase activities, and rapid bacterial growth (growth rates of pelagic assemblages are only ~ 0.1-1d^{-1}) and perhaps also high rates of remineralization.

However, most phytoplankton in the sea are not seen heavily colonized and many are not colonized at all. We think bacteria-phytoplankton symbioses are common, given the close proximity and trophic interdependence of bacteria and phytoplankton, but that most associations are delicate and are disrupted by sampling. Since attached bacteria can be eaten along with the phytoplankton, firm attachment may be a specialization (*eg* attached bacteria may release progeny into seawater; Smith *et al.*, submitted) while delicate symbioses may be the norm in the fluid environment of seawater. We note, parenthetically, that current models of marine symbioses involving bacteria are physically robust systems (*eg* symbiosis between flash-light fish and luminous bacteria; "hydrothermal vents symbioses"; Cavanaugh *et al.*, 1981, Distel *et al.*, 1988, Haygood and Distel 1993). We may be missing interesting but physically delicate symbioses in seawater which could provide new insights into the molecular strategies in microbial symbioses. Molecular biology should be invaluable in discovering and studying such symbioses.

How bacteria-phytoplankton associations are established is unclear despite much interest, probably because most studies used oversimplified ecosystem scenarios. Most studies considered algal release of diffusible molecules as the basis of the association (Bell and Mitchell, 1972; Azam and Ammerman; 1984) but several models (Mitchell *et al.*, 1985; Jackson *et al.*, 1987; Jackson, 1989) predict such exudation could not create bacterial clusters except under special conditions. The model of Bowen *et al.* (1993) suggests that random motility and relative motion could bring bacteria and algae close, and chemokinesis could then prolong bacterial residence time in the exudate-rich microzone. This model predicts bacteria could accumulate to 10^8 -10^9 ml^{-1} in the vicinity of the alga. We think bacterial clustering cannot be understood if we consider only algae and bacteria and ignore other particles in the organic matter field of the bacterium. For instance, exudation of diffusible molecules might not be a good strategy to establish symbiosis because bacteria touching the multitude of small particles with their ectoenzymes would create new DOM loci which might generate so much "noise" as to obscure the exudation signal. The Bowen *et al.* (1993) mechanism, since it does not require bacteria to follow exudate gradient to find the alga, would get around the noise problem, but a mechanism for maintaining the symbiosis is still required.

We think that *cell-bound* and *polymeric* "exudates" (Azam and Smith, 1991) would be a better strategy to establish symbiosis. First, being cell-bound, the exudates will become utilizable only when bacteria physically contact algal surface with their ectoenzymes. Some hydrolysate may be lost and its diffusion gradient will increase the residence time of chemokinetic bacteria near the alga, as needed by Bowen *et al.* (1993) model. For creating bacterial clusters, this strategy of bacteria-mediated slow-release of diffusible molecules is superior to direct exudation (Smith *et al.* submitted). Slow mucus hydrolysis balanced by its production will also protect the alga. Second, the mucus might act like "marine snow", scavenging particles and DOM (*eg* colloids, nucleic acids and proteins) from seawater. A polysaccharide layer could thus become N, P rich and complex medium supporting rapid bacterial growth and remineralization. This would expose the alga to high local levels of regenerated nutrients and support higher primary production than would otherwise be possible. Bacterial clusters would attract protozoa, creating a concentrated microbial loop with intense carbon fixation and nutrient regeneration. Mucus may further serve as an interface between bacteria and the alga where the alga creates conditions to select specific bacterial species or bacteria expressing specific phenotypes. Simply by secreting surface-bound mucus, then, the alga can create symbiosis with bacteria with profound implications for organic matter cycling, primary production, nutrient cycling, microbial species and community structure and spatial distribution of microbes and their metabolism. F. Rassoulzadegan (per. comm.) has remarked of this symbiosis strategy, that if it does not exist then it should be invented.

Evidence for such delicate symbioses is circumstantial and indirect and new molecular and optical methods are needed to test the hypothesis. We do not know enough about bacterial behaviour in seawater to model bacteria-algae interactions. For instance, while most pelagic isolates are flagellated and motile we do not know under what natural conditions this phenotype is expressed. Cho and Azam (unpubl.) using a capillary assay found that 3-35% of bacteria in water off Scripps pier were motile and Mitchell (1994) using dark-field microscopy found extremely high swimming velocities, (144μm sec^{-1} with bursts up to 407μm s^{-1}). However, electron microscopy of natural assemblages rarely shows flagellated bacteria (G. Steward, unpubl. obs.). Perhaps flagellation is only transiently expressed in seawater, in response to microenvironmental conditions. We need to develop a way to observe bacterial behaviour in the sea. Molecular biology will be valuable in discovering molecular bases of bacterial sensing and behaviour. We should find out how a sensing system works in such a "noisy" and variable environment; whether the energy cost of motility might be so high for the slow-growing bacteria (doubling time 1-10d) that motility is expressed only in response to specific conditions; whether, and in what time frame, a bacterium switches between being motile and non-motile and what cues in the sea induce motility; and, whether marine bacteria use novel mechanisms of sensing and chemokinesis. Further, to understand bacteria-phytoplankton symbioses we need to establish whether phytoplankton elicit chemokinesis in bacteria.

BACTERIA-DETRITUS INTERACTION.

Here, we discuss whether bacteria-detritus interaction, when considered in the context of the overall organic matter field as well as other organisms, can create bacterial "hot spots". As mentioned, the assemblage-average growth rate of bacteria in surface waters is generally 0.1 -1d^{-1}. Jobörn *et al.* (unpubl.) tested the growth response of pelagic bacteria to inputs of fresh detritus by adding freeze-thaw-killed cells of the diatom *Nitzschia angularis* to 1µm-filtered seawater. Heavy colonization and rapid growth occurred with a growth rate of 15d^{-1} within 20h. Colonies developed on detritus which contained morphologically similar bacteria in a gelatinous-looking matrix. These results suggest pelagic bacteria can opportunistically utilize fresh algal detritus to achieve bursts of growth. Pelagic isolates typically grow fast in rich media (an isolate, S8, from California coastal waters has a doubling time of 13 minutes in culture at 17°C [J. Martinez, unpubl.]). If pelagic bacteria always grow slowly in the sea , then it is surprising that they carry genetic potential to grow at 15d^{-1} on detritus-enriched seawater and even faster in culture. We speculate that the potential for rapid growth is an adaptation to an environment where bacteria episodically experience "hot spots" supportive of rapid growth.

Alldredge and Gotschalk (1990) found that marine snow is heavily colonized, at 10^8-10^9 ml^{-1} of the aggregate. However, unlike bacteria on live or dead diatoms (above) these grow slowly, at 0.1-1d^{-1}. Smith *et al.* (1992) found that, while bacteria on marine snow grew slowly, they expressed high activities of protease, phosphatase, chitobiase, a- and ß- glucosidase which rapidly solubilized the POM. Poor hydrolysis-uptake coupling (tested in the case of protein solubilization) caused > 97% hydrolysate to be released. Why these populations do not grow fast is unclear, but they may be like stationary cultures which slow down or stop growing due to deteriorating growth conditions despite high levels of residual nutrient (Smith *et al.*, 1992; Azam *et al.*, 1993). Despite slow growth, marine snow bacteria cause rapid POM $\rightarrow$ DOM transition and are "hot spots" of bacterial activity (*ie* solubilization). Surprisingly, marine snow collected in the upper ~ 20 m is always heavily colonized despite its short residence time (<1day) due to sinking. Azam *et al.* (1993) suggest that bacteria grow fast on diver-invisible "source particles" (*eg* living and dead algae; above) developing dense populations and aggregation of "source particles" to diver-visible snow "instantaneously" produces heavily colonized snow.

These examples of bacteria-detritus interactions highlight two points. First, bacteria in the euphotic zone can greatly vary their *in situ* growth rates (>100-

fold). This has implications for bacterial ecology and the significance of hot-spots for rapid growth. Second, a major gap in our understanding of bacterial ecology and organic matter cycling is that we know almost nothing about the regulation of bacterial growth in the sea. This problem must be addressed if we are to understand the variability in bacteria-organic matter coupling. Molecular biology can help in understanding the regulation of bacterial growth in response to environmental fluctuation, perhaps by a combination of laboratory model systems and adapting molecular techniques to the complex field conditions.

Variability in bacterial utilization of organic matter: Significance of ectoenzyme expression and species succession.

Bacterial abundance in surface waters varies narrowly, generally 0.5-2 $x10^6$ ml^{-1}, but might similar bacterial concentrations express quite different absolute and relative levels of different ectohydrolase. This could happen if different species were dominant or if the same species expressed enzymes at a significantly different per-cell level under different conditions. Billen (1991) found that the biomass-specific exoprotease activity in pelagic bacteria was constant, which is surprising since it suggests a lack of metabolic responsiveness in using an important nutrient namely protein. However, Smith *et al.* (submitted) found, in the diatom bloom experiment mentioned earlier, that the per-cell protease activity of attached bacteria was 3 orders of magnitude greater than free bacteria and varied by an order of magnitude during the course of the bloom. This shows that bacterial ectoenzyme activities, and hence bacteria-organic matter coupling, can vary temporally as well as at the microscale (*ie* whether bacteria were attached or free).

Martinez *et al.* (1994) tested whether changes in species composition would change the "enzyme profile" of the assemblage. They tested 44 pelagic isolates for activities of protease, alkaline phosphate, lipase, chitobiase, a-and ß-glucosidase. The per-cell activities varied over ~2.5 (protease) to ~4.5 (chitobiase) orders of magnitude. So, species shifts could strongly affect hydrolysis of particles or polymers. Martinez *et al.* (1994) further asked 1) whether different species "specialize" in using different polymer pools by expressing more of a particular enzyme ("enzyme specialists") and 2) whether among the isolates there were any "super bugs" expressing high levels of all enzymes. They found that while "enzyme specialists" were common there were no superbugs among the tested isolates. Taken together these finding show that the enzyme activity which a bacterial assemblage brings to bear on the particulate and polymeric phases in seawater can be dramatically different depending on the dominant species and the level of enzyme expression. This could be a significant variable in bacteria-organic matter coupling.

The question then arises whether major shifts in species dominance do occur in seawater. Only few studies have been done using 16S rRNA probes, but they show significant shifts. Rehnstam *et al.* (1993) found sequence-specific bloom of bacteria in waters off the Scripps Pier and Rehnstam, Azam and Hagström (unpubl.) found spatial variation of sequence-specific bacteria in a 15 x 20 km grid in California coastal waters. Giovannoni (1995) found depth-stratification of specific genotypes in the Sargasso Sea. DeLong *et al.* (1993) found marine snow harboured bacteria closely related to *Cytophaga* which were not detected in the surrounding water. Thus, there is evidence that variation in dominant species do occur spatially and temporally at mesoscale and, in the case of the observations on marine snow, also at microscale. Martinez *et al.* (1994) tested for variability in assemblage enzyme profiles by measuring the several enzyme activities in seawater from Scripps pier on four days during a single week. They found that enzyme ratios varied strongly from day to day.

Since species shifts can change the assemblage's enzyme profile similar bacterial abundances could exert quite different biochemical forces on the particulate and polymeric phases in the sea. Also, bacteria with enzyme repertoires which make them particle specialists may accumulate at particle loci, *ie* phytoplankton and detritus, and play roles in bacteria-organic matter coupling.

Implications for ecosystem functions of bacteria

We have shown that understanding bacterial molecular strategies for organic matter uptake can help in understanding the control of carbon flux into bacteria. It could also help in understanding the mechanisms of variability in other ecosystem functions of bacteria which are coupled to organic matter utilization. However, it requires that we not study each ecosystem function of bacteria piece-meal, but rather develop a unified view based on bacterial adaptive strategies at the microscale. This point is illustrated below.

First, the idea of "hot spots" dispenses with the "link or sink" dichotomy (Pomeroy, 1974), *ie* whether C assimilated by bacteria is used directly by metazoa (link) or recycled in the mineralization-efficient microbial loop. Both paths may operate simultaneously with varying relative magnitudes (link *and* sink). Protozoa feed on free and attached bacteria and metazoa feed on particles. The relative accessibility of bacteria to metazoa and protozoa will depend on strategies of bacteria-particle interactions (*eg* firm or delicate attachment). Bacteria- particle interactions can thus help understand factors structuring the food web.

Second, whether to model bacteria as net-producers or net-consumers of regenerated nutrients is a serious problem. This role reversal can be understood in terms of the variations in bacterial metabolism in rich microzone (net-mineralization) or poor microzone (net-consumption). Such response to composition of growth environment is known (Wheeler and Kirchman, 1986). Oligotrophic waters may have a lower abundance of "hot spots", so bacteria there would be dominantly in the competitive mode. The ecosystem function of bacteria would be to concentrate nutrients, while protozoa feeding on bacteria would be the mineralizers. In eutrophic waters, because of abundant "hot spots' a greater fraction of remineralization would be due directly to bacteria. Once again, we see that bacteria-organic matter interactions at microscale can help understand and model the mechanism of an important ecosystem function namely nutrient regeneration.

Third, the DOM and POM pools have very different biogeochemical behaviours and bacteria mediate DOM $\rightarrow$ POM as well as POM $\rightarrow$ DOM transitions. Bacterial production causes DOM $\rightarrow$ POM flux and this ecosystem function can be understood in terms of regulation of bacterial growth at the microscale, as discussed. Bacterial ectohydrolases cause POM $\rightarrow$ DOM transition and this can be understood in terms enzyme regulation in response to environmental conditions. We thus see that large-scale phase transitions of organic matter in the ocean can be understood in terms of the strategies of bacteria for organic matter utilization and growth.

Finally, the traditional models of food web organization do not work well when applied to pelagic microbes, due to the role variabilities and reversals of the type just discussed (also see: Turner and Roff, 1993). There is a need for new approaches. Instead of piece-meal studies of specific microbial activities, we should study intact microbial consortia in their natural setting to understand the molecular strategies which dominantly shape the communities and their metabolic exchanges with the environment. This will require new methods and molecular biology should play an important role. We do not advocate dispensing with large space-scale studies. Although production and decomposition are mediated by unicellular organisms at microscale we do need also to understand them at the ocean basin scale. We should study production-decomposition coupling at several space scales with different techniques and levels of biological resolution. Such a unified view should impart predictive power to models of the ocean's ecology and biogeochemistry.

References

Alldredge AL and Gotschalk CC (1990) The relative contribution of marine snow of different origins to biological processes in coastal waters *Cont Shelf Res* **10:** 41-58

Alldredge AL, Passow U, Logan BE (1993) The abundance and significance of a class of large transparent organic particles in the ocean *Deep-Sea Res* **40:** 1131-1140

Ammerman JW and Azam F (1985) Bacterial 5'-nucleotidase in aquatic ecosystems: a novel mechanism of phosphorus regeneration. *Science* **227:** 1338-1340

Azam F (1984) The radioecological role of marine bacterioplankton. In: *The role of microorganisms in the behaviour of radionuclides* Bonnyns-Van Gelder E and Kirchmann R (eds) Institut Royal des Sciences Naturelles de Belgique, Brussels, 2-7

Azam F and Ammerman JW (1984) Cycling of organic matter by bacterioplankton in pelagic marine ecosystems: microenvironmental considerations. In: *Flows of Energy and Material* Fasham MJR (ed) Plenum Publishing, New York, 45-360

Azam F and Cho BC (1987) Bacterial utilization of organic matter in the sea. In: *Ecology of microbial communities* Cambridge University Press, Cambridge, 261-281

Azam F, Fenchel T, Field JG, Gray JS, Meyer-Reil LA, Thingstad F (1983) The ecological role of water-column microbes in the sea. *Mar Ecol Prog Ser* **10:** 257-263

Azam F and Hodson RE (1981) Multiphasic kinetics for D-glucose uptake by assemblages of natural marine bacteria. *Mar Ecol Prog Ser* **6:** 213-222

Azam F, Martinez J, Smith DC (1993) Bacteria-organic matter coupling on marine aggregates. In: *Trends in microbial ecology* Guerrero R, Pedrós-Alió C (eds) Spanish Society for Microbiology, Barcelona, 410-414

Azam F and Smith DC (1991) Bacterial influence on the variability in the ocean's biogeochemical state: A mechanistic view. In: *Particle analysis in oceanography* (Demers S ed) Springer-Verlag, 213-236

Azam F, Steward GF, Smith DC, Ducklow HW (1994) Significance of bacteria in carbon fluxes in the Arabian Sea. *Proc Indian Acad Sci (Earth Planet Sci)* **103:** 341-351

Bell WH and Mitchell R (1972) Chemotactic and growth responses of marine bacteria to algal extracellular products. *Biol Bull* **143:**265-277

Billen G (1991) Protein degradation in aquatic environments. In: *Microbial enzymes in aquatic environments* Chróst RJ (ed) Springer-Verlag, New York, 123-143

Bjørnsen PK (1988) Phytoplankton exudation of organic matter: Why do healthy cells do it? *Limnol Oceanogr* **33:** 151-154

Bowen JD, Stolzenback KD, Chisholm SW (1993) Simulating bacterial clustering around phytoplankton cells in a turbulent ocean. *Limnol Oceanogr* **38:** 36-51

Carlucci AF and Shimp SL (1974) Isolation and growth of a marine bacterium in low concentrations of substrate. In: *Effect of the ocean environment on microbial activities.* Colwell RR and Morita (RY, eds) University Park Press, Baltimore, 363-367

Cavanaugh CM, Gardiner SL, Jones MJ, Jannasch HW, Waterbury JB (1981) Prokaryotic cells in the hydrothermal vent tube worm *Riftia pachyptila* Jones: possible chemoautotrophic symbionts. *Science* **213:** 340-342

Chróst RJ (1990) Microbial ectoenzymes in aquatic environments. In: *Aquatic microbial ecology: biochemical and molecular approaches* (Overbeck J and Chróst RJ eds) Brock/Springer, New York, 47-78

Cole JJ, Findlay S, Pace ML (1988) Bacterioplankton production in fresh and saltwater ecosystems: a cross-system overview. *Mar Ecol Prog Ser* **43:** 1-10

DeLong EF, Franks DG, Alldredge AL (1993) Phylogenetic diversity of aggregate-attached vs. free-living marine bacterial assemblages. *Limnol Oceanogr* **38:** 924-934

Distel DL, Lane DJ, Olsen GJ, Giovannoni SJ, Pace B, Pace NR, Stahl DA, Felbeck H (1988) Sulfur-oxidizing bacterial endosymbionts: Analysis of phylogeny and specificity by 16S rRNA sequences. *J Bacteriol* **170:** 2506-2510

Giovannoni S, Mullins TD, Field KG (1995) Microbial diversity in oceanic systems: rRNA approaches to the study of unculturable microbes *This volume* 217-248

Haygood MG and Distel DL (1993) Bioluminescent symbionts of flashlight fishes and deep-sea anglerfishes form unique lineages related to the genus *Vibrio*. *Nature* **363:** 154-156

Hodson RE and Moran M (in press) degradation kinetics of simple and complex dissolved organic matter in aquatic ecosystems. In: *The role of non-living organic matter in the earth's carbon cycle*. Dahlem Workshop. Berlin.

Hollibaugh JT and Azam F (1983) Microbial degradation of dissolved proteins in seawater. *Limnol Oceanogr* **28:** 1104-1116

Jackson GA (1987) Simulating chemosensory responses of marine microorganisms. Limnol Oceanogr **32:** 1253-1266

Jackson GA (1989) Simulation of bacterial attraction and adhesion to falling particles in an aquatic environment. *Limnol Oceanogr* **34:** 514-530

Jumars PA, Penry DL, Baross JA, Perry MJ, Frost BW (1989) Closing the microbial loop: dissolved carbon pathway to heterotrophic bacteria from incomplete ingestion, digestion and absorption in animals. *Deep-Sea Res* **36:** 483-495

Koike I, Shigemitsu H, Kazuki T, Kogure K (1990) Role of sub-micrometre particles in the ocean. *Nature* **345:** 242-244

Lampert W (1978) Release of dissolved organic carbon by grazing zooplankton. *Limnol Oceanogr* **23:** 831-834

Logan BE and Fleury RC (1993) Multiphasic kinetics can be an artifact of the assumption of saturable kinetics for microorganisms. *Mar Ecol Prog Ser* **102:** 115-124

Martinez J and Azam F (1993) Periplasmic aminopeptidase and alkaline phosphatase activities in a marine bacterium: Implications for substrate processing in the sea. *Mar Ecol Prog Ser* **92:** 89-97

Martinez J, Smith DC, Steward GF, Azam F (1994) Ectohydrolytic activity profiles of marine bacteria. *EOS Trans Amer Geophys Union* **75:** (3):86

Mitchell JG, Okubo A, Fuhrman JA (1985) Microzones surrounding phytoplankton form the basis for a stratified marine microbial ecosystem. *Nature* **316:** 58-59

Mitchell JG, Pearson L, Dillon S (1994) Do high speed bacterial motility bursts indicate a new chemotactic mechanism that enhance the likelihood of phycospheres? *ASLO & PSA 1994 Joint Meeting Book of Abstracts*, p. a51

Murray AE and Hollibaugh JT (1994) A new molecular approach for describing diversity in coastal bacterioplankton communities. *EOS, Trans Amer Geophys Union* **75:** 210

Nissen H, Nissen P, Azam F (1984) Multiphasic uptake of D-glucose by an oligotrophic marine bacterium. *Mar Ecol Prog Ser* **16:** 155-160

Paul JH, DeFlaun MF, Jeffrey WH (1988) Mechanisms of DNA utilization by estuarine microbial populations. *Appl Environ Microbiol* **54:** 1682-1688

Paul JH, DeFlaun MF, Jeffrey WH (1988) Mechanisms of DNA utilization by estuarine microbial populations. *Appl Environ Microbiol* **54:** 1682-1688

Pomeroy LR (1974) The ocean's food web changing paradigm *Bioscience* **24:**499-504

Pomeroy LR, Wiebe WJ (1988) Energetics of microbial food webs. *Hydrobiologia* **159:** 7-18

Pomeroy LR, Wiebe WJ, Deibel D, Thompson RJ, Rowe GT, Pakulski JD (1991) Bacterial responses to temperature and substrate concentration during the Newfoundland spring bloom. *Mar Ecol Prog Ser* **75:** 143-159

Pratt R, Daniels RC, Eiler JJ, Gunnison JB, Kumler WD, Oneto JF, Strait LA (1944) Chlorellin, an antibacterial substance from *Chlorella. Science* **99:** 351-352

Rehnstam A-S, Bäckman S, Smith DC, Azam F, Hagström Å (1993) Blooms of sequence-specific culturable bacteria. *FEMS Microb Ecol* **102:** 161-166

Shanks AL and Reeder ML (1993) Reducing microzones and sulfide production in marine snow. *Mar Ecol Prog Ser* **96:** 43-47

Sieburth JM (1960) Acrylic acid, an antibiotic principle in *Phaeocystis* blooms in Antarctic waters. *Science* **132:** 676-677

Smith DC, Simon M, Alldredge AL, Azam F (1992) Intense hydrolytic enzyme activity on marine aggregates and implications for rapid particle dissolution. *Nature* **359:** 139-142

Smith DC, Steward GF, Long RA, Azam F (1995) Bacterial mediation of carbon fluxes during a phytoplankton bloom in a mesocosm. *Deep-Sea Res* (submitted)

Steele JH (1974) *The structure of marine ecosystems* Harvard Univ Press 128pp

Steemann-Nielsen E (1955) The production of antibiotics by plankton algae and its effects upon bacterial activities in the sea. Paper in Marine Biology and Oceanography *Deep-Sea Res* **3 (suppl)**:281-286

Trick CG, Andersen RJ, Harrison PJ (1984) Environmental factors influencing the production of an antibacterial metabolite from a marine dinoflagellate *Prorocentrum minimum. Can J Fish aquat Sci* **41:** 423-432

Turner JT and Roff JC (1993) Trophic levels and trophospecies in marine plankton: lessons from the microbial food web. *Mar Microb Food Webs* **7:** 225-248

Waksman SA, Stokes JL, Butler MR (1937) Relation of bacteria to diatoms in sea water. J Mar Biol Assoc U K **22:** 359-373

Wells ML and Goldberg ED (1991) Occurence of small colloids in sea water. *Nature* **353:** 342-344

Wells ML and Goldberg ED (1993) Colloid aggregation in seawater. *Mar Chem* **41:** 353-358

Wheeler PA and Kirchman DL (1986) Utilization of inorganic and organic nitrogen by bacteria in marine systems. *Limnol Oceanogr* **31:** 998-1009

Williams PM and Druffel ERM (1987) Radiocarbon in dissolved organic matter in the North Pacific Ocean. *Nature* **330:** 246-248

Wright RT and Burnison BK (1979) Heterotrophic activity measured with radiolabeled organic substrates. In: *Native aquatic bacteria: enumeration, activity and ecology* Costerton JW, Colwell RR (ed) Am. Soc. for Testing Materials, Philadelphia, 140-155

ZoBell CE (1973) In: *Estuarine Microbial Ecology*, (Stevenson LH, Colwell RR eds) Columbia, South Carolina, 178-179

The Role and Regulation of Microbes in Sediment Nitrogen Cycle

Henry Blackburn
Department of Microbial Ecology
Institute of Biological Sciences
University of Aarhus
DK 8000 Aarhus C
Denmark

Abstract

Microbial activity in sediment is regulated principally by the quantity, quality (C:N), distribution of degradable particulate organic matter and the diffusibility of the products of degradation. These four factors not only determine the total amount of organic nitrogen mineralised, but also determine the relative amounts of the products (NH_4^+, NO_3^-, N_2 and dissolved organic-N) leaving the sediment. Other factors *eg* the concentration of O_2, NO_3^- and NH_4^+ in the overlying water, also determine the ratios of these products and the total denitrification rate. These effects were examined in 24 simulations, with and without sulphide diffusion. Increased NH_4^+ and O_2 in the water increased nitrification and coupled denitrification; an increased availability of sulphide was inhibitory. Sulphide availability (anoxia) and NO_3^- in the water stimulated non-coupled denitrification. These contrary stimulations and inhibitions produced complex patterns of total denitrification in the various simulations.

Introduction

The nitrogen cycle in sediments is essentially the same as the nitrogen cycle in other environments, with some differences imposed by the stratified nature of sediments. The most important steps are the hydrolysis and degradation of particulate organic matter (POM) resulting in the liberation of NH_4^+. In sediments which have a high rate of organic matter decomposition and a resulting high rate of O_2 consumption, O_2 does not penetrate far below the sediment surface and there is no oxidation of NH_4^+ until it diffuses into this oxic zone, close to the sediment-water interface. The rate of NH_4^+ production depends on both the rate of POM degradation and the C/N ratio of the degraded substrate:

NATO ASI Series, Vol. G 38
Molecular Ecology of Aquatic Microbes
Edited by I. Joint

when the C/N ratio is high little NH_4^+ may be released immediately, as it is incorporated into microbial cells.

The oxidation of NH_4^+ to NO_3^- is complex, with the production of various intermediates, depending on whether the process of nitrification is autotrophic or heterotrophic. It is probable that autotrophic nitrification is the dominant process in sediments, with NO_2^- as the principal intermediate. When NO_3^- or NO_2^-, produced in the sediment by nitrification or coming from the overlying water, is transported into the anoxic sediment layers, denitrification can occur. The types of bacteria responsible for these main processes (mineralization, nitrification and denitrification) are presumably physiologically and biochemically similar to those associated with similar processes in other ecosystems, but this is not certain, as comprehensive isolation studies have not been made. Since most bacteria active in the sediment nitrogen cycle have not been isolated in pure culture, little is known of their molecular biology. The best known aspect of the nitrogen cycle concerns measurement of the rates of mineralization, nitrification and denitrification. These studies are typically more concerned with the ecological significance of these rates than with the factors which control them. One way to examine these control factors is to make simulation models of sediments, in which all major processes are allowed to interact, so that interdependencies and inhibitions may be identified. It is possible to determine which parameter changes exert the greatest control on the nitrogen cycle.

Some of the early models of nitrogen cycling indicated the importance of coupled nitrification and denitrification, but these models lacked the integration of carbon, nitrogen, sulphur and oxygen cycles (Billen, 1982, Billen and Lancelot, 1988, Vanderborght and Billen, 1975). I attempted to do this, making a model in STELLA (High Performance Systems, New Hampshire, USA) which had all these processes interacting in 4 sediment layers (Blackburn, 1990). This was followed by other models in a more sophisticated modelling environment (Blackburn and Blackburn, 1993) which was used initially to simulate the O_2, NO_3^- and NH_4^+ profiles measured with high resolution in an Arctic sediment (Blackburn and Blackburn, 1993c). This modelling environment, Cellmatic, was also used to investigate some of the factors with the most influence on coupled nitrification-denitrification reactions (Blackburn and Blackburn, 1992). These studies emphasised the obvious importance of the quantity and quality (C:N ratio) of degradable POM on sediment biogeochemical processes in general and on the nitrogen cycle in particular. The role of the distribution of POM within the sediment was also explored, indicating that dissolved organic carbon and nitrogen (DOC and DON) could be lost to the overlying water, when the POM was degraded close to the sediment-water interface. This aspect was further investigated in a comprehensive series of simulations in which different quantities of POM were distributed in three different patterns (Blackburn and Blackburn, 1993b). These three patterns were 1) mostly at the sediment surface, 2) distributed in a linear gradient and 3) completely mixed, as might have resulted from active bioturbation. It was predicted that the mixed distribution would result in high rates of sulphate reduction, of nitrification and of denitrification. Subsequent simulations confirmed these predictions and suggested that this combination of effects would only be observed when sulphide (or other soluble reduced products of organic carbon oxidation) was not allowed to diffuse (Blackburn and Blackburn, 1993a). The explanation is as follows: mixing of POM to the lower sediment layers resulted in high rates of carbon oxidation by sulphate and NH_4^+ was released but not oxidised until it diffused to the oxic zone. Here, it did not have to compete for O_2, as there was relatively little carbon in this zone; the result was high rates of nitrification. Some of the NO_3^- diffused downwards and was denitrified. The flux downward was quite high as the depth of O_2 penetration, and thus the zone of nitrification, was well below the sediment water interface, giving a relatively shallow diffusional gradient of NO_3^- to the surface. When sulphide was allowed to diffuse freely (no Fe), the sulphide competed with the NH_4^+ for O_2, the depth of O_2 penetration was small, and nitrification and denitrification were also relatively small.

For a model to be accepted as a reflection of reality, it must be validated, or shown to be able to reproduce experimental data. This had been a problem with sediment models as, with the exception of some deep sea sediments (Goloway and Bender, 1982, Jahnke *et al.*, 1982) data was lacking on the distributions of O_2, NH_4^+ and NO_3^- which could be used for validation. Recently some high precision data on the distribution of these molecules and of the rates of nitrification, denitrification and O_2 consumption became available (Jensen *et al.*, 1994), and have been used to make a model of these processes in a freshwater sediment. The model generated profiles of O_2, NO_3^-, NH_4^+, nitrification and denitrification which were essentially identical with the experimental data, thus confirming that this particular situation can be modelled. However, this does not mean that the model can be used to reproduce all types of sediments, and the concept of validation may be of little relevance in this context.

The freshwater sediment model re-emphasised the importance of selecting appropriate rate constants for nitrification and denitrification. This had been investigated earlier (Blackburn and Blackburn, 1992, Blackburn and Blackburn, 1993c) and its significance observed. The freshwater model introduced a new factor which was of great significance in determining the rate of denitrification of NO_3^- which had been produced within the sediment (Ds). This new factor was the degree to which denitrification was inhibited by O_2. If denitrification was inhibited by low concentrations of O_2 ($<5\mu M$), then the zones of nitrification and denitrification were spatially separated and denitrification could only occur well below the sediment-water interface. If denitrification was relatively insensitive to O_2, then the zones could partially overlap. There was also a short diffusional path for NO_3^- to reach the zone of denitrification, resulting in a high proportion of the NO_3^- being denitrified.

The sediment nitrogen cycle is, therefore, mostly dependent on the quantity of degradable POM reaching the sediment and on its quality, or nitrogen content. The proportions of the degradation products (DON, NH_4^+, NO_3^- and N_2) effluxing to the overlying water depend principally on the distribution of the POM within the sediment and on the diffusibility of sulphide and other reduced products. These four factors can conveniently be remembered as quantity, quality, distribution and diffusion or as QQDD (or QD2). Additional factors eg rate constants and stimulation or inhibition by O_2 or other molecules must be taken into account.

There are other factors which are also important in determining the rates of denitrification and denitrification, and thus of the ratio of N-products leaving the sediment. Among these are the concentration in the overlying water of O_2, NO_3^- and NH_4^+. The interaction of these three factors has been investigated in our present model, under conditions where sulphide diffused or did not diffuse, and where other parameters (POM, C:N, rate constants *etc*) were kept constant.

Methods

Cellmatic, the System Dynamics modelling environment, allows for the definition of a number of model variables or agents (Blackburn and Blackburn, 1993). These agents are present in cells, which are arranged in a vertical array, representing different layers in the sediment. The number of cells is variable as is the individual cell height. The models are reaction-diffusion in character. The equations defining the reactions and diffusions are present in every cell, as are all the agents. Diffusion of agents between cells is defined by Fick's first law, while rates of reaction between agents is controlled by Michaelis-Menten kinetics.

The selection of the appropriate cell height is very important to the correct functioning of a model. If the cell height is too large rates of diffusion may be too low, because of the weak gradient over the long diffusional path. This is illustrated in Table 1, where the maximum rate of O_2 diffusion from the water to the sediment is calculated, assuming that the concentration of O_2 in the water is 200µM.

Table 1 Maximum flux of oxygen from water at decreasing cell heights

Cell height (mm)	Oxygen flux* (mmol m^{-2} d^{-1})
4.0	8.2
2.0	16.4
1.0	32.8
0.4	82.0
0.2	164.0
0.1	328.0

*The concentration of O_2 is 200 μM in the overlying water and 0 μM in the sediment below the topmost cell. The flux was calculated with a diffusion coefficient of 1.64 cm^2 d^{-1}

When the height of the topmost sediment cell was 1 mm, the maximum diffusional flux was only 32.8 mmol m^{-2} d^{-1}, a value characteristic of a moderately active marine sediment. Many inshore sediments have oxygen consumption rates greater than 30 mmol m^{-2} d^{-1}, suggesting that cell heights of <1.0 mm should be used to simulate the activities of such systems. Values of 0.2 mm should be sufficient for most situations. The disadvantage of selecting small cell heights is that many cells have to be used (50 0.2 mm cells for 1 cm sediment), and very small time steps are also necessary. The result is a model which runs extremely slowly. A compromise can be made by selecting small cells at the sediment-water interface, where gradients are steepest, and larger cell heights lower in the sediment, where gradients tend to be shallow.

In the present model, there were 23 cells of 0.2 mm at the top and 6 cells of 1.0 mm below these. The uppermost 3 cells were in the overlying water and the sediment column was thus composed of 26 cells with a total height of 1.0 cm. Porosity was 1.0 in the water and 0.95 in the topmost sediment cell, decreasing by 0.05 units per cell to 0.70, the value for all lower cells. The sulphate concentration was initialised at 20 mM in the sediment pore water; all other agents were allowed to reach their own sediment concentrations. The maximum time step, dt, for the cell height of 0.2 mm was 0.00005 d.

The main components of the model and their interconnections are illustrated in Fig.1. There were two types of ammonium (NH4w and NH4s) designated w or s, depending on whether they came from water or sediment. There were similarly two nitrate pools. This diagram may be used in association with the equations (Table 2), which define reactions and diffusions. Every cell contained a complete model; all equations were evaluated and updated at each time step. Each simulation was run for a minimum of 1 day or until steady state was reached.

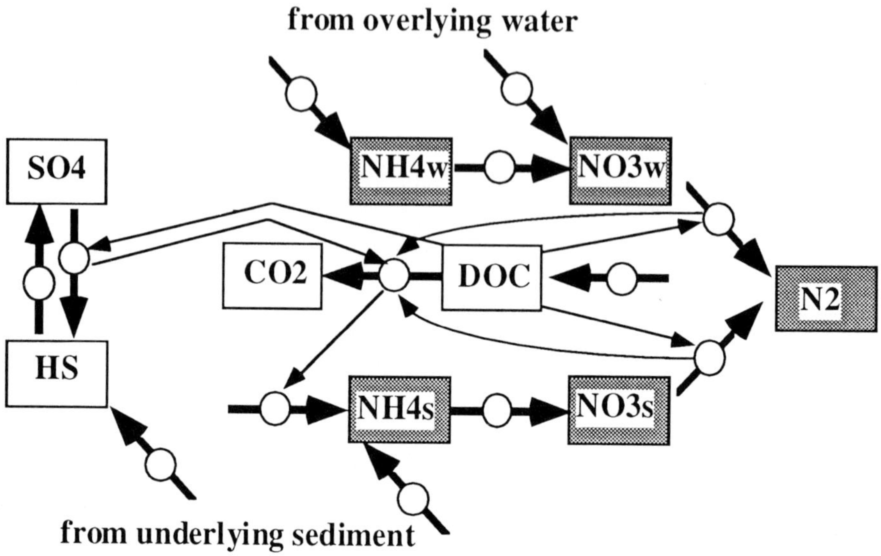

Fig. 1. A System Dynamics diagram of the main components of the model. the boxes represent masses of the agents. The heavy arrows indicate the flow between boxes in the S-, C- and N-cycles. Oxygen is not shown. The circles on these arrows are rate controls. The thin arrows show where there are dependencies. Sulphide (HS) and ammonium (NH4s) diffused from below, as well as being produced. Ammonium (NH4w) and nitrate (NO3w), like O_2, could diffuse from the water.

The agents are listed at the top of Table 2; they were dissolved organic carbon (DOC), carbon dioxide (CO2), oxygen (O2), water (H2O), sulphate (SO4), sulphide (HS), ammonium (NH4s and NH4w), nitrate (NO3s and NO3w), and dinitrogen (N2). In addition, porosity (Por) and cell height (_h) were agents. Model variables are listed under the heading "Formula". The first 4 were the variables which were changed in the present simulations. The agents represent stocks or quantities and were expressed in nmol per cell. For this reason, 0.6 nmol NO_3^- in the topmost water cell whose height was 0.02 cm and area was 1 cm^2, gave a concentration of 30µM. Changes in these parameters gave a matrix of simulations, which will be discussed later. The first order rate constants for denitrification, nitrification, sulphide oxidation, sulphate reduction and carbon oxidation by O_2 are listed in that order. The carbon/nitrogen ratio (CN) of the particulate organic matter undergoing decomposition was 10. The model was

Table 2. Model construction

AGENTS		=	DOC, CO2, O2, H2O, SO4, HS, NH4s, NH4w, NO3s, NO3w, N2, Por, Δh
FORMULA	cNO3w	=	0, 0.2 or 0.6; {0, 10 or 30 μM in overlying water}
	cNH4w	=	0 or 0.2; {0 or 10 μM in overlying water}
	cO2	=	1 or 4; {50 or 200 μM in overlying water}
	cHS	=	0.05 or 0.3; {0.05 or 0.3 of HS available}
	k_denit	=	30; {day $^{-1}$}
	k_nit	=	30; {day $^{-1}$}
	k_hs	=	200; {day $^{-1}$}
	k_sr	=	5; {day $^{-1}$}
	k_cox	=	30; {day $^{-1}$}
	CN	=	10; {mole ratio}
	doc_flux	=	750; {half of total 1500 nmol cm^{-2} d^{-1} }
	tort	=	3; {tortuosity factor is 2 for sand, 3 for mud}
	stim_O2	=	if O2/Δh/Por < 10 then O2/Δh/Por/10 else 1;
	stim_NO3s	=	if NO3s/Δh/Por < 10 then NO3s/Δh/Por/10 else 1;
	stim_NO3w	=	if NO3w/Δh/Por < 10 then NO3w/Δh/Por/10 else 1;
	stim_SO4	=	if SO4/Δh/Por < 10 then SO4/Δh/Por/10 else 1;
	inhib_O2	=	if O2/Δh/Por < 10 then -O2/Δh/10 + 1 else 0; {complete inhibition level set at 10 μM O2}
	dh	=	if Δh < nbr Δh then Δh else nbr Δh;
	mPor	=	(Por + nbr Por)/2;
REACTIONS	NH4s,NO3s	=	(y>2)*NH4s*k_nit*stim_O2;
	NH4w,NO3w	=	(y>2)*NH4w*k_nit*stim_O2;
	NO3s,N2	=	(y>2)*DOC*k_denit*inhib_O2*stim_NO3s;
	NO3w,N2	=	(y>2)*DOC*k_denit*inhib_O2*stim_NO3w;
	SO4,HS	=	(y>2)*DOC*k_sr*inhib_O2*stim_SO4*cHS;
	HS,SO4	=	(y>2)*HS*k_hs*stim_O2;
	O2,H2O	=	(y>2)*DOC*k_cox*stim_O2;
	DOC,CO2	=	NO3s,N2*1.25 + NO3w,N2*1.25 + SO4,HS*2/cHS + O2,H2O; {sum of processes in DOC oxidation}
OUT	out.CO2	=	-(y=28)*doc_flux + (y=0)*CO2/dt;
	out.NH4s	=	-(y=28)*doc_flux/CN - (y>2)*doc_flux/CN*Δh + (y=0)*NH4s/dt;
	out.HS	=	-(y=28)*doc_flux/2*cHS + (y=0)*HS/dt; {entry of NH4, CO2 and HS at bottom cell: removal at top}
	out.O2	=	NH4s,NO3s*2 + NH4w,NO3w*2 + HS,SO4*2 + O2,H2O -(y=0)*(cO2-O2)/dt;
	out.NO3s	=	(y=0)*NO3s/dt;
	out.NO3w	=	-(y=0)*(cNO3w-NO3w)/dt;
	out.NH4w	=	-(y=0)*(cNH4w-NH4w)/dt;
	out.DOC	=	(y=0)*DOC/dt - (y>2)*doc_flux*Δh;
	out.N2	=	(y=0)*N2/dt;
	out.SO4	=	-(y=0)*(20000*0.02-SO4)/dt;
DIFFUSION	toNbr.X	=	(X/Δh/Por-nbr(X/Δh/Por))/dh*D*mPor/ (1+tort*(1-mPor))/2;

controlled by the availability of DOC. In total, 1500 nmol DOC was added per cm^2 (15 mmol m^{-2} d^{-1}). Half of this quantity was distributed to the cells in the upper 1 cm of sediment; the other half had been completely degraded by sulphate reduction below 1 cm depth, resulting in the production of NH_4^+, CO_2 and HS^-, which were added to the bottom sediment cell. This procedure avoided having a pool of particulate organic carbon, but more importantly it reduced the number of cells needed to represent processes in the lower, largely anoxic sediment layers. Tortuosity (tort) was used in the diffusion equations (Iversen and Jørgensen, 1993). There were 4 stimulation (stim_) functions, used to mimic Michaelis-Menten kinetics in reactions, where O_2, NO_3^- and SO_4^{2-} concentrations affected reaction rates. The quantity of agent was first converted to a pore water concentration. Above a certain concentration, eg 10μM, zero order kinetics occurred, while at lower concentrations the stimulation was proportional to the concentration. A similar inhibition (inhib_) function was used to inhibit reactions sensitive to O_2. The term "dh" was used in the diffusion equations to ensure that, when exchange occurred between two cells (neighbour = nbr), the smaller of the two cell heights was used. Previously, the smaller of two porosities had been selected in a similar manner (Blackburn *et al.*, 1994), but here the mean of the two neighbouring porosities (mPor) was found to give a smoother simulation.

There are 8 equations defining reactions in Table 2: two nitrification reactions (NH4s,NO3s and NH4w,NO3w), two denitrification reactions (NO3s,N2 and NO3w,N2), sulphate reduction (SO4,HS), sulphide oxidation (HS,SO4), DOC oxidation by O_2 (O2,H2O) and the sum of DOC oxidations to CO2 (DOC,CO2). The equations are quite simple to understand, with the exception of those containing cHS. Multiplication by cHS reduced the amount of free HS which was produced, while division by cHS corrected the amount of DOC oxidised by sulphate reduction. The top cell in the water was 0 in the y-array, the top sediment cell was 3 and the lowest sediment cell was 28. These specifications were used to stipulate that a process should occur only in specified cells *eg* in sediment (y>2).

The "out." prefix indicated the agents entered or exited specified cells, from or to unspecified sources or sinks. The agents CO2, NH4s and HS entered the bottom sediment cell, proportional to the amount of organic matter degraded below this level, and were removed from the top water cell. O2 was removed in the oxidation of DOC, NH4 and HS, but its concentration was maintained at a constant level in the top water cell. NO3s, which was produced by nitrification of NH4s, was removed from cell 0. The concentrations of agents NO3w and NH4w were kept constant in cell 0. Agent DOC was removed from cell 0, but was added to each sediment cell, in proportion to the cell height, at a total rate of 750 nmol cm^{-2} d^{-1}. Dinitrogen was removed from, and SO_4^{2-} added to, cell 0.

Diffusion was controlled by the generalised equation, given at the bottom of Table 2, where X was any agent. The pore water concentration of the agent in each cell was calculated (division of the mass by cell height and porosity), and the gradient determined (division of the difference by the lesser of the two cell heights). The gradient was multiplied by the mean porosity, divided by a term to correct for tortuosity and multiplied by the diffusion coefficient, D. The diffusion coefficient for O_2 was 1.84 cm^2 d^{-1}, 0.5 for DOC and 1.64 for all other molecules.

A number of simulations (24) were run with different parameters, each starting with the control (cont) values for the agents. The control simulation was made without NO_3^- or NH_4^+ in the overlying water, with O_2 maintained at 50µM and with only 5% of the sulphide being free to diffuse. The control simulation started with no agent pools in the sediment, with the exception of SO_4^{2-}. This simulation was run for 8 days, to steady state. The simulations at 5% sulphide availability were labelled according to the scheme in Table 3. Three concentrations of NO_3^- (0, 10 and 30µM), two of O_2 (50 and 200µM), and two of NH_4^+ (0 and 10µM) were used. The 12 simulations with 30% sulphide availability were labelled similarly, with a "S1" prefix.

Table 3. Labels used to identify simulations

	-	NH4w (10µM)	O2 (200µM)	NH4w+O2 (10 and 200µM)
NO3w (0µM)	cont.	A	O	O.A
NO3w (10µM)	N1	A.N1	O.N1	O.A.N1
NO3w (30µM)	N2	A.N2	O.N2	O.A.N2

A summary of the rates of nitrification, denitrification and the budgets from the 24 simulations are summarized in Fig.2. Nitrification at 5% sulphide availability (A) indicated that increasing NO_3^- concentration resulted in slight increases in nitrification for the simulations with NH_4^+, O_2, or NH_4^+ plus O_2, possibly due to a sparing effect on O_2. This effect was less obvious at 200µM O_2. The presence of 10µM NH_4^+ in the overlying water resulted in an increase in Nw, an effect increases at 200µM O_2. The higher oxygen concentration resulted in ~2-fold higher rates for Ds. The highest combined rate (Ns + Nw) was close to 1.5 mmol m^{-2} d^{-1}, the rate of NH_4^+ production. At 30% availability of sulphide (Fig. 2A), NO_3^- had no effect, the general pattern of nitrification was very similar to that for Fig. 6A, but the rates were much smaller, for both Ns and Nw.

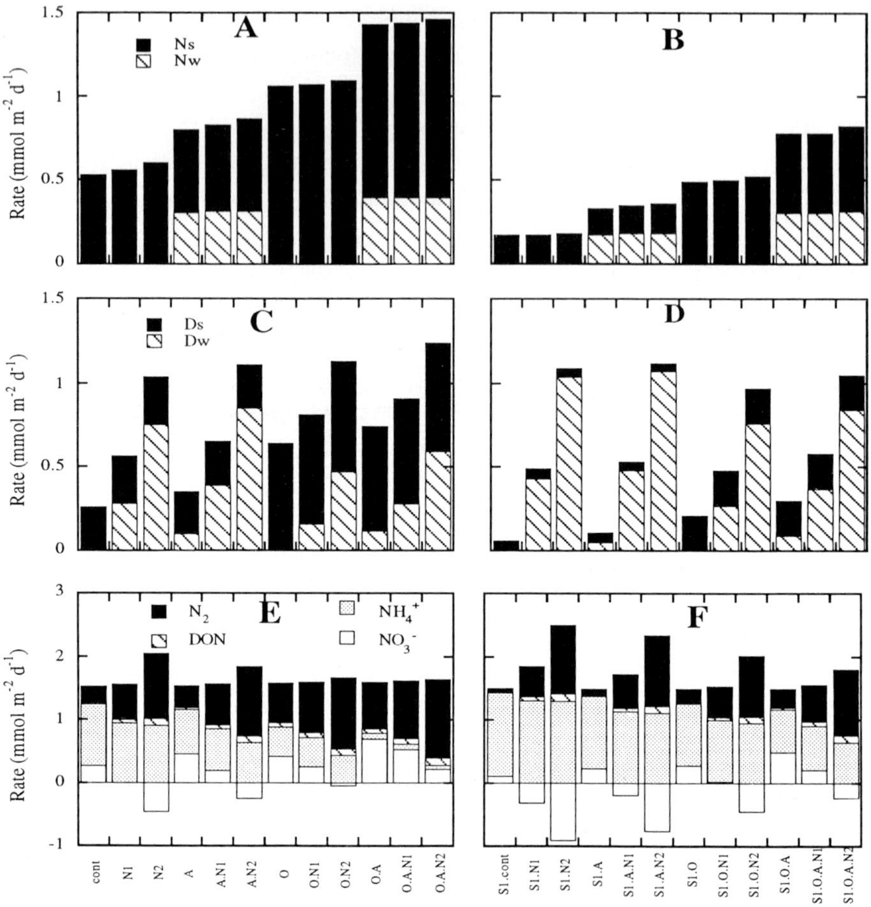

Fig. 2. Rates of nitrification, denitrification and net nitrogen budgets with 5% sulphide availability (A, C and E) and with 30% sulphide availability (B, D, and F), respectively. Ns and Nw denote nitrification of $NH4^+$ from sediment or water; Ds and Dw denote denitrification of NO_3^- from sediment or water. The budgets are composed of net fluxes of N_2, NH_4^+, DON and NO_3^-; negative values indicate influx from water to sediment.

Denitrification rates of NO_3^- produced in the sediment (Ds) were essentially the same for the first 6 treatments, with 5% sulphide availability (Fig.3C). These rates were doubled with 200μM O_2. Nitrate in the overlying water greatly increased the Dw rate, external NH_4^+ at 10μM having a lesser effect. The Dw rates were repressed by the higher $O2$ concentration, but the effect of external NH_4^+ was slightly increased. The same trends occurred when 30% sulphide was available, but the Ds rates were smaller and the Dw rates were higher (Fig.3D).

The NH_4^+ production rate in the sediment was 1.5 mmol m^{-2} d^{-1}, which was close to the budget for most of the simulations in which 5% sulphide was available (Fig.3E). This indicated that NO_3^- at 10µM in the overlying water (the N1 simulations) did not affect the total budget, although the ratio of effluxing molecules changed significantly. In all simulations, the efflux of NO_3^- became less positive, with increasing concentration of external NO_3^-. This trend was diminished by increasing concentrations of external NH_4^+ and O_2, due to the increased production of sediment NO_3^-.The presence of high O_2 and NH_4^+ completely counteracted the effect of high NO_3^- (simulation O.A.N2), resulting in a budget of 1.5 mmol m^{-2} d^{-1}. High availability of sulphide (Fig.6F) produced the same trends, but with higher fluxes of NH_4^+, consistent with a more anoxic character.

The increase in the sediment nitrogen budget, by the influx of NO_3^- from the overlying water was thus progressively diminished by the presence of NH_4^+ and O_2 in the overlying water. This resulted in the progressive decrease of the NO_3^- effect in the simulation series N2, A.N2, O.N2 to O.A.N2 and also in the second series (Fig. 2F) of S1.N2, S1.A.N2, S1.O.N2 to S1.O.A.N2. This can be explained by stimulation of nitrification (and in turn denitrification) by increased O_2 and/or NH_4^+ concentrations, which also resulted in decreased net influx of NO_3^-.

The plots of total denitrification with sediment depth, for low sulphide availability (Fig. 3A) and high sulphide availability (Fig. 3B), showed increasing rates with increasing NO_3^- concentration (cont., N1 and N2), with an additional small increment due to the presence of NH_4^+ (10µM) in the water. The presence of O_2 in the water resulted in a displacement of the zone of denitrification downward, but otherwise the two sets of plots (with and without O_2) were similar. Fig. 3B had the same patterns as Fig. 3A, the difference being that both sets of plots were displaced upward, due to the removal of O_2 by sulphide.

The zones of nitrification lay above those of denitrification (Figs. 4A and 4B). They were again grouped in sets of 3 (*eg* cont., N1 and N2 overlapping in Fig. 6A). The presence of NH_4^+ increased nitrification; the presence of increased O_2 pushed the zones deeper into the sediment. Higher availability of sulphide forced the zones of nitrification closer to the sediment surface (Fig. 4B).

The O_2 profiles were in two groups (Figs. 2C and 2D), depending on whether the external concentration was 50 or 200µM. Ammonium in the water had little effect. Higher availability of sulphide resulted in decreased O_2 penetration and steeper gradients.

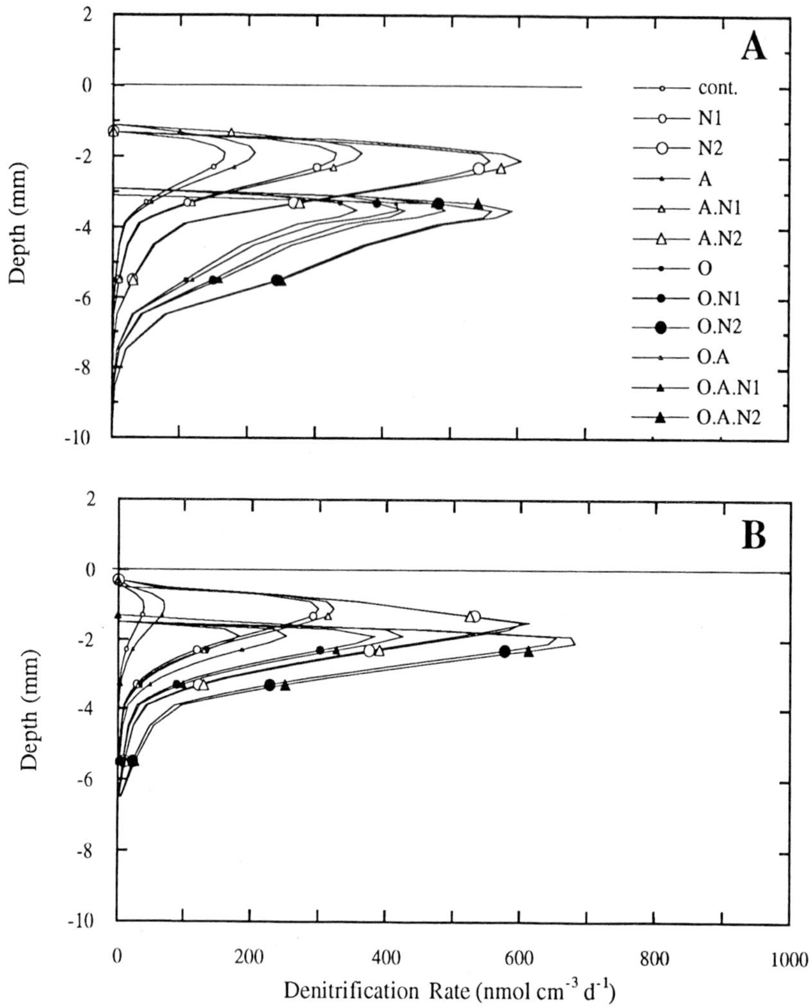

Fig. 3. Profiles of denitrification rates at 5% sulphide availability (A) and at 30% sulphide availability (B).

The summed NO_3^- concentrations in pore waters were in three groups, depending on the concentration of NO_3^- in the water. When the NO_3^- concentration in the water was low (0 or 10μM) sediment-produced NO_3^- resulted in quite sharp nitrate peaks, generating fluxes to the overlying water and the zone of denitrification in the lower sediment. Higher availability of sulphide resulted in smaller NO_3^- concentrations in the sediment and lower penetrations.

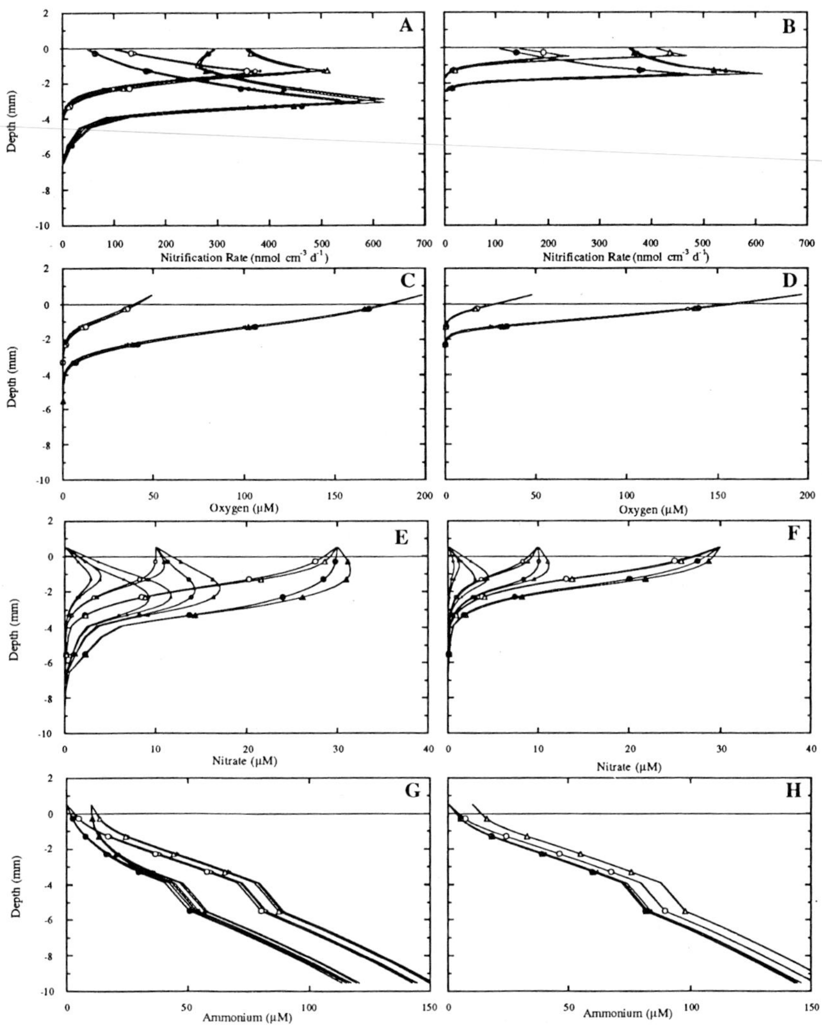

Fig. 4. Rates of nitrification, concentrations of O_2, NO_3^- and NH_4^+ with 5% sulphide availability (A, C, E and G) and with 30% sulphide availability (B, D, F and H). The symbols are as in Fig.3.

The NH_4^+ profiles were in two main groups (Fig. 4G), the simulations with 200µM O_2 had lower pore water NH_4^+ concentrations, because of nitrification. Ammonium in the water (10µM) gave slightly increased concentrations, particularly at the sediment surface. The same general pattern was found in the simulations with increased sulphide availability (Fig. 4H), but the concentrations were higher and the curves closer together, due to the decreased oxygen penetration and decreased nitrification.

Interest is often expressed in the extent to which net NH_4^+ production in sediments can be denitrified. The rates of total denitrification, as predicted in the 24 simulations, were plotted as percentages of the rate of organic nitrogen mineralization (Fig. 5). At the low level of sulphide availability, there was an increase in denitrification at 0, 10 and 30μM NO_3^- concentrations, as NH_4^+, O_2 and NH_4^+ plus O_2 were increased. The relative amount of the increase was least at the highest external NO_3^- concentration. At the high level of sulphide availability (designated "+"), the same response was observed at 0μM NO_3^-, but the values were very much lower. This pattern was not seen at 10 and 30μM NO_3^-, where there was a much more complex response to external NH_4^+ and O_2. The range was from 32 - 39% of N-mineralization at 10μM NO_3^-, and 65 - 75% at 30μM NO_3^-. The highest proportion of the N-mineralization rate to be denitrified was 83% at the low sulphide availability, 10μM NH_4^+ and 30μM NO_3^-. Less than 17% was denitrified in the low sulphide, with no external NH_4^+ and 50μM O_2 and only 4% was denitrified in the parallel high sulphide simulation.

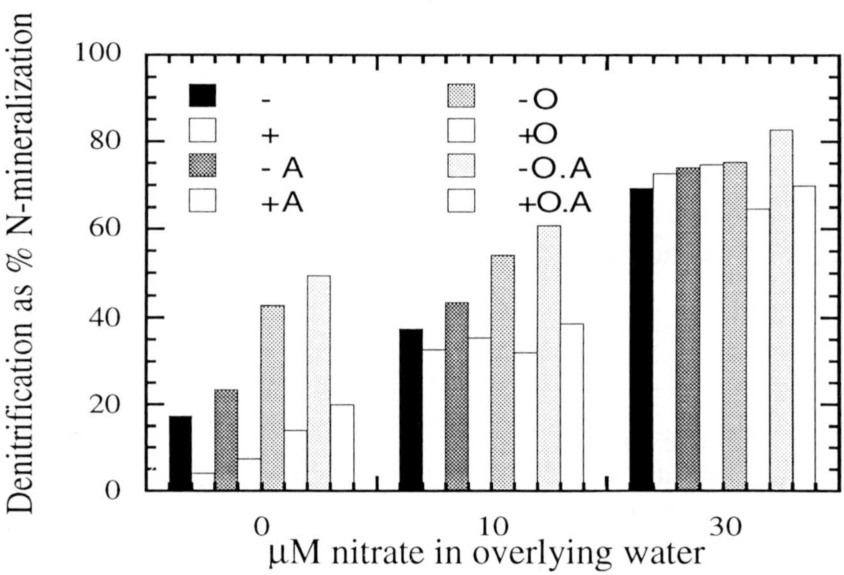

Fig. 5. Denitrification expressed as a percentage of N-mineralization in sediment (1.5 mmol m^{-2} d^{-1}) in 24 simulations. The "+" and "-" indicates those with low and high sulphide availability, respectively. "A" indicates 10 μM NH_4^+, "O" indicates 200μM O_2, in the overlying water.

Conclusions

The parameters in the controls were: 15 mmol DOC (1.5 mmol DON) available m^{-2} d^{-1}, half of the DOC oxidised by sulphate reduction below 1 cm depth, 50μM O_2 but no NH_4^+ nor NO_3^- in the overlying water, and 5% availability of the sulphide produced by sulphate reduction. This would be typical of a moderately active sediment, covered by oxygen-depleted water low in nutrients. The distribution of organic matter (sites of DOC production) would indicate some bioturbation or physical mixing of the sediment.

Nitrate concentrations of 10μM in the water would not be unusual in coastal water, before the onset of the spring bloom, but 30μM NO_3^- would be more typical of deeper water masses. Ammonium concentrations of 10μM would be unusually high for any natural water body. High concentrations of both NO_3^- and NH_4^+ can build up rapidly in sediment core incubations, which are often used to study fluxes across the sediment-water interface (Seitzinger, 1988), and to measure rates of nitrification (Sloth *et al.*, 1992) and of denitrification (Nielsen, 1992). It was thus of some interest to evaluate the effect that relatively high concentrations of NH_4^+ and NO_3^- might have on the sediment N-cycle, in particular on nitrification and denitrification, since the net rate of NH_4^+ production in the sediment was constant. It was also important to determine the effect of increased O_2 concentration in the water and of increased sulphide availability in the sediment on these two processes. Sulphide would react with Fe in the sediment, producing black amorphous ferrous sulphide, but if there was a limited amount of free Fe, sulphide would be free to diffuse. Two situations were examined: sulphide availability at 5 and 30%.

The simulations confirmed that increases in the water NO_3^- concentration did not affect coupled nitrification-denitrification reactions in the sediment. Increased NH_4^+ concentration in the water increased coupled nitrification-denitrification in the sediment, as did high O_2 concentrations. The reverse was true for increases in the availability of sulphide. This is very reasonable as increased availability of NH_4^+ and O_2 would increase nitrification (and thus Ds), while increased availability of sulphide would decrease the availability of O_2, decreasing nitrification and Ds.

However, high concentrations of NO_3^- in the water led to diffusion of this NO_3^- into the sediment, stimulating denitrification (Dw). This process was highest in reducing sediments (when sulphide availability was high), it was inhibited by O_2 and insensitive to NH_4^+.

Total rates of denitrification were thus dependent on contrary processes; Ds was dependent on nitrification (oxic), whereas Dw was dependent on an anoxic sediment and a high concentration of NO_3^- in the water. It was not surprising that these total rates were quite irregular (Fig. 2C and 2D). The highest total denitrification rate (83% of N-mineralization) was achieved with low sulphide availability, high O_2, high NH_4^+ (high Ds) and with high NO_3^- (high Dw). It was not possible to achieve 100% denitrification, as has been cited for some freshwater and marine systems (Seitzinger, 1988).

Acknowledgements
The Danish Ministry of the Environment HAV project 3-18 and the Commission of the European Communities' STEP programme under contract number STEP-CT90-0080/DSCN) are thanked for financial assistance.

References

Billen G (1982) An idealized model of nitrogen recycling in marine sediments. *Am. Jour. Sci.* 282: 512-541

Billen G and Lancelot C (1988) *Modelling benthic nitrogen cycling in temperate coastal ecosystems*. John Wiley and Sons, Chichester, UK

Blackburn ND and Blackburn TH (1993) A reaction diffusion model of C-N-S-O species in a stratified sediment. *FEMS Microbiol. Ecol.* **102:** 207-215

Blackburn TH (1990) Denitrification model for marine sediment. In: *Denitrification in Soil and Sediment.* Blackburn TH and Sørensen J (ed.) John Wiley and Sons, Chichester, 323-337

Blackburn TH and Blackburn ND (1992) Model of nitrification and denitrification in marine sediments. *FEMS Microbiol. Lett.* **100:** 517-522

Blackburn TH and Blackburn ND (1993a) Coupling of cycles and global significance of sediment diagenesis. *Mar. Geol.* **113:** 101-110

Blackburn TH and Blackburn ND (1993b) Rates of microbial processes in sediments. *Phil. Trans. R. Soc. Lond. A* **344:** 49-58

Blackburn TH and Blackburn ND (1993c) A reaction diffusion model of C-N-S-O species in some Arctic sediments. *FEMS Microbiol. Ecol.* 102: 197-205

Blackburn TH, Blackburn ND, Jensen K, Risgaard-Petersen N (1994) Simulation model of the coupling between nitrification and denitrification in a freshwater sediment. *Appl. Environ. Microbiol.* (in press):

Goloway F and Bender M (1982) Diagenetic models of interstitial nitrate profiles in deep sea suboxic sediments. *Limnol. Oceanogr.* **27:** 624-638

Iversen N and Jørgensen BB (1993) Diffusion coefficients of sulphate and methane in marine sediments: Influence of porosity. *Geochim. Cosmochim. Acta* **57:** in press

Jahnke RA, Emenson SR, Murray JW (1982) A model of oxygen reduction, denitrification, and organic matter mineralization in marine sediments. *Limnol. Oceanogr.* **27:** 610-623

Jensen K, Sloth NP, Risgaard-Petersen N, Rysgaard S, Revsbech NP (1994) Estimation of nitrification and denitrification from nitrate and oxygen microprofiles in model sediment systems. *Appl. Environ. Microbiol.* **60:** 2094-2100

Nielsen LP (1992) Denitrification in sediment determined from nitrogen isotope pairing. *FEMS Microbiol. Ecol.* **86:** 357-362

Seitzinger SP (1988) Denitrification in freshwater and coastal marine ecosystems: ecological and geochemical significance. *Limnol. Oceanogr.* **33:** 702-724

Sloth NP, Nielsen LP, Blackburn TH (1992) Measurement of nitrification in sediment cores using acetylene inhibition. *Limnol. Oceanogr.* **37:** 1108-1112

Vanderborght J-P and Billen G (1975) Vertical distribution of nitrate concentration in interstitial water of marine sediments with nitrification and denitrification. *Limnol. Oceanogr.* **20:** 953-961

Functional and Taxonomic Probes for Bacteria in the Nitrogen Cycle

Bess B. Ward
Marine Sciences Program
University of California
Santa Cruz, CA 95064
USA

Abstract

The contrasting physiology and phylogeny of nitrifying and denitrifying bacteria require different methods for detection and quantification in the environment. rRNA probes are appropriate for nitrifiers. Denitrifiers are more diverse; the conserved enzymology of denitrification provides a functional approach for probes. Although quantification is still rudimentary, comparison among immunofluorescence enumeration, PCR amplification and quantitative hybridization allow an assessment of genotype vs phenotype abundances. Nitrifiers are a very small fraction of the total bacterial population, but a few serotypes and rRNA groups are ubiquitous in natural waters. The denitrification genotype is apparently abundant, but individual denitrifying strains are present in very low numbers.

Introduction

Nitrification and denitrification are essential parts of the nitrogen cycle in aquatic and terrestrial environments because they determine the relative and absolute abundance of various nitrogen compounds. When coupled across an oxic/anoxic interface, such as in soil microsites or microbial mats or anoxic marine basins, the net result is loss of fixed nitrogen from the system. Thus, although nitrification may be seen as simply rearranging the concentrations of reduced and oxidized species of nitrogen (NH_4, NO_2, NO_3), its role in producing the substrates for denitrification, which leads directly to removal of fixed nitrogen, often necessitates consideration of the two processes together. As shown clearly in aquatic sediment systems (Nishio *et al.*, 1983; Risgaard *et al.*, 1993) the rate of denitrification often depends on the rate of nitrification, and neither can be adequately characterized or quantified in isolation.

NATO ASI Series, Vol. G 38
Molecular Ecology of Aquatic Microbes
Edited by I. Joint

Due to the dependence of both processes on oxygen concentration, the entire sequence of nitrification- denitrification can be considered to be inducible. Thus, predicting the rate of the net process, or whether it will occur in a particular environment, depends not only on the presence of the organisms, but on the environmental conditions present on scales appropriate for interaction with the microorganisms. Direct measurement of the rates of processes using ^{15}N tracers in sediment (see above) and pelagic environments (*eg* Lipschultz *et al.*, 1990; Ward 1987) is a powerful biogeochemical approach to understanding the processes. On microbiological scales, regulation of the processes can be approached by identifying the organisms responsible and evaluating the effect of environmental variables of the biochemical pathways involved. This ecological approach has potential for investigations on the physical scale of individual microorganisms and detection of effects due to differential species composition and activity, which may modify the regulation and interpretation of overall processes. A combination of the two approaches is necessary and is becoming increasingly possible.

In this paper, the development and initial application of probes for quantification of denitrifying bacteria are summarized and illustrated with examples from aquatic environments. While the focus is on denitrification, nitrification is also considered in less detail. The contrasting physiologies and phylogenies of nitrifying and denitrifying bacteria make them an excellent comparison for the usefulness of different kinds of probes. Based on serological (Ward and Carlucci, 1985; Josserand and Cleyet-Marel, 1979) and rRNA sequence data (Woese *et al.*, 1984; Head *et al.*, 1993), ammonium oxidizers and nitrite oxidizers each constitute highly related, tightly linked groups of genetically and physiologically similar bacteria. Autotrophic nitrification as a way of life constrains a cell to energy production solely via the oxidation of single nitrogen compounds (either ammonium or nitrite) and to biomass construction solely from the fixation of CO_2. It's a "hard way to make a living" and dictates essentially the entire biochemistry of the cell. Although DNA hybridization studies supported the definition of several genera and the existence of distinct species within the morphologically defined genera (Koops and Harms, 1985), subsequent rRNA sequencing discerned only two genera within the β subclass *Proteobacteria* group of ammonium oxidizers and a third genus in the γ subclass (Head *et al.*, 1993)

Denitrification, on the other hand, is simply a set of loosely linked respiratory reactions which enable the cell to respire in the absence of oxygen. For most denitrifiers, a vast array of organic compounds are acceptable as carbon and heterotrophic energy sources and the use of oxygen as an electron acceptor is the preferred respiratory mode. The ability to denitrify is simply an optional accessory and its inclusion in the genetic make up of the cell is not diagnostic of any other particular combination of genetic traits or phylogenetic associations. Denitrifying bacteria are found throughout the Bacterial kingdom and the ability also occurs in the Archaea (Zumft, 1992).

These characteristics have implications for the kinds of probes that will be useful in identifying, quantifying and studying nitrifiers and denitrifiers. It is likely that rRNA based PCR primers and phylogenetic probes will be powerful approaches in the study of nitrifying bacteria, but much less so for denitrifiers. The functionally defined group nitrifying bacteria roughly coincides with a small number of phylogenetically defined groups. Probes based on rRNA phylogeny are predicted to tell the same story as those based on the key enzymes of autotrophic nitrogen oxidation, ammonium mono-oxygenase and nitrite oxidase.

The functionally defined group denitrifying bacteria overlaps almost the entire range of bacterial diversity. In order to develop probes which identify the denitrification genotype, the enzymes of the denitrification pathway itself must be targeted. Biochemical data (Bryan, 1983) implies that there is significant conservation among denitrifying bacteria of the main enzymes in the pathway, such that probes for a small suite of functional genes might suffice to identify a large array of denitrifying bacteria which are not related by rRNA phylogeny.

Methods

STUDY SITES

Samples were collected from 2 stations in the Southern California Bight, both in approximately 900m water depth. The Santa Monica Basin was sampled in June 1988 for total bacteria counts and in April 1994 for immunofluorescence enumeration. The Santa Barbara Channel was sampled in July 1992 for quantitative hybridization. The Monterey Bay samples for immunofluorescence were collected in March 1993 from a station near the centre of the bay in 900m depth. The Plußsee, a small kettle lake in northern Germany (Overbeck and Chrost, 1993), was sampled in November 1993.

The probes used here have been previously described and the methodological details of their production and application will not be repeated here. The probes and sources are listed in Table 1.

PCR

DNA extracts were amplified using three sets of primers. Universal eubacterial (EUB) rRNA primers (Liesack *et al.* 1991) were used as controls to verify the existence of amplifiable DNA in all the genomic DNA extracts tested for amplification with the NitAB and NitDB primers (see below). The EUB primers (5' positions 9-27 and 3' positions 1525-1542 in the *E. coli* sequence) amplified a region of approximately 1530 bp. Direct amplification of every sample was achieved with the eubacterial primers.

Table 1. Kinds and sources of probes used in this investigation

Probe	Target Organism or Molecule	Reference
PCR primers	beta subclass ammonium oxidizing nitrifiers	Voytek and Ward, submitted
	N. europaea	Ward *et al.*, in preparation
	eubacterial universal	Liesack *et al.*, 1991
immunofluorescence assays	*N. marina* serotype	Ward and Carlucci, 1985
	P. stutzeri (ATCC 14405)	Ward and Cockcroft, 1993
721 bp DNA probe	nitrite reductase cd_1 heme type (NiR gene)	Ward *et al.*, 1993

The second primer set, NitAB, selectively amplifies a 1080 bp fragment from the rDNA of the β-subclass group of ammonium oxidizers (*Nitrosomonas europaea, Nitrosomonas eutropha, Nitrosomonas marina, Nitrosospira briensis, Nitrosovibrio tenuis, Nitrosococcus mobilis* and *Nitrosolobus multiformis*). Characterization of these primers in terms of specificity, sensitivity and optimum amplification conditions has been described elsewhere (Voytek and Ward, submitted). A third primer set, NitDB, amplifies an 800 bp fragment internal to the NitAB fragment and the D primer makes this amplification specific for *Nitrosomonas europaea* (Ward *et al.*, in preparation).

The standard amplification reaction for all three sets of primers was performed using a Perkin-Elmer Cetus DNA Thermocycler with the following cycle: 10 min at 95°C, pause at 82°C, then 35 cycles of 2 min at 60°C, 2.5 min at 72°C, 1 min at 94°C, followed by a final annealing step and a 5 min elongation at 72°C. One µl of sample DNA extract was routinely amplified (see below).

SAMPLE PRESERVATION AND DNA EXTRACTION

For immunofluorescence, samples of seawater or lake water were preserved with formalin (2% v/v) and stored in the dark at 12°C or less. Replicate 20ml subsamples were stained by the filter method (Ward, 1982) and enumerated by epifluorescence microscopy. For DNA samples from the Southern California Bight, 1 litre samples were concentrated immediately after collection by direct filtration onto 0.2µm pore size polysulfone filters (Gelman). The filters were placed in cryovials with 0.8ml 0.5M EDTA and frozen until extraction by routine methods based on Ausubel *et al.* (1987). Extract (up to 50µl) representing 500ml of seawater was blotted for probing using a Hoefer slot blot apparatus. For the Plußsee DNA extracts, 41 samples were concentrated by tangential flow (Amicon) and finally pelleted in a refrigerated microcentrifuge. The pellet was resuspended in 0.8ml 0.5M EDTA and frozen until extraction.

Quantitative hybridization: The 721 bp NiR DNA probe (part of the structural gene for nitrite reductase as described by Jungst *et al.*, 1991) was prepared by PCR as described previously (Ward *et al.*, 1993) and labelled by random priming using the Genius method (Boehringer Mannheim). Slot blots were prepared from known concentrations of standard DNA (DNA extracted from *P. stutzeri* ATCC 14405 and previously quantified by spectrophotometry) and DNA extracts representing known volumes of environmental samples. The blots were developed by the colorimetric method (Boehringer Mannheim) and stored in the dark for up to 2 years before quantitative analysis. The blots were digitized by scanning (HPScanJet II) and subsequently analyzed using Image (NIH) or ScanAnalysis (Biosoft, Ferguson, MO, USA) software.

Results

DETECTION OF NITRIFYING BACTERIA

Using an antiserum produced against *Nitrosomonas marina* (Ward, 1982), ammonium-oxidizing cells were enumerated in Plußsee. In culture, the reaction of *N. europaea* with this antiserum is as strong as the homologous reaction with *N. marina*. Due to limitations on the amount of antiserum available, single counts were made of each sample. Previous determinations of the error of such counts (Ward and Carlucci, 1985; Ward 1987) resulted in a coefficient of variation of about 30%. The depth profile (Fig. 1) shows a maximum (4898 cells ml^{-1}) in the oxic layer at about 5m, and many fewer cells (536 cells ml^{-1}) at the oxic/anoxic interface. Samples from below the interface were not enumerated.

DNA samples collected at the same time were amplified using the the EUB primers and the two sets of primers designed for nitrifying bacteria. Using the standard assay (amplifying 1µl of DNA extract, representing about 13ml of lake water) nitrifier DNA was detected at all depths (Table 2), including in the anoxic layer. At two depths (5m, 15m) nitrifier DNA could not be detected by direct amplification, but was detected when the fragment resulting from amplification with the EUB primers was amplified with the NitAB primers. Direct amplification with the NitDB primers was successful at only two depths, indicating that *N. europaea* is either not present or is a minor part (undetectable) of the total ammonium oxidizing population at the other depths.

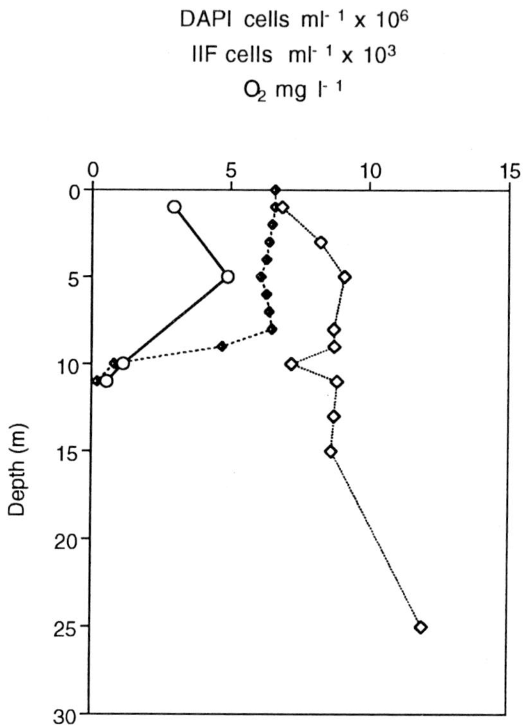

Fig. 1. Distribution of total bacterial abundance (DAPI ◊), *N. marina* serotype cells (IIFO), oxygen (mg l^{-1} ◆) in the Plußsee, November 1993.

QUANTIFICATION OF NITRITE REDUCTASE GENES IN SEAWATER

Quantitative hybridization with colorimetric detection of NiR DNA was evaluated by hybridization of a dilution series of genomic DNA purified from *P. stutzeri* (Fig. 2). The limit of detection was estimated visually to be between 0.3 and 3.0 pg of NiR DNA (between slots 4 and 5). Integration of the blot shown in Fig. 2 yielded the scan in Figure 2B, and the linear regression shown in Fig. 3. Slot number 4 containing 3pg NiR DNA is not above background, placing the actual detection limit above 3pg. The relationship between area and DNA concentration above 3pg is a reproducible standard curve, however, which allows the calculation of NiR concentrations in unknown samples.

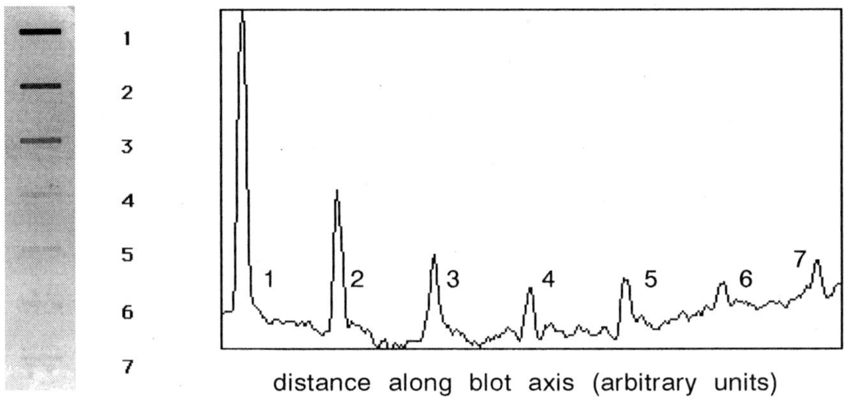

Fig. 2. Quantitative hybridization of NiR fragment in *P. stutzeri* DNA. A) Slot blot of dilution series. Slots, pg NiR fragment DNA: 1, 123 pg; 2, 61.6 pg; 3, 30.8 pg; 4, 3.1 pg; 5, 0.3 pg; 6, no DNA; 7, 2 µg salmon sperm DNA; 8, no DNA. B) Scan of dilution series slot blot obtained using IMAGE software.

Table 2. Direct amplification of DNA extracts from Plußsee with bacterial universal (EUB), beta-subclass ammonium oxidizer-specific (NitAB) and *N. europaea*-specific (NitDB) primers, except * denotes NitAB amplification was successful only after amplification with EUB primers.

	Amplification		
Depth (m)	EUB	NitAB	NitDB
1	+	+	+
5	+	+*	-
10	+	+	+
12	+	+	-
15	+	+*	-
20	+	+	-
25	+	+	-

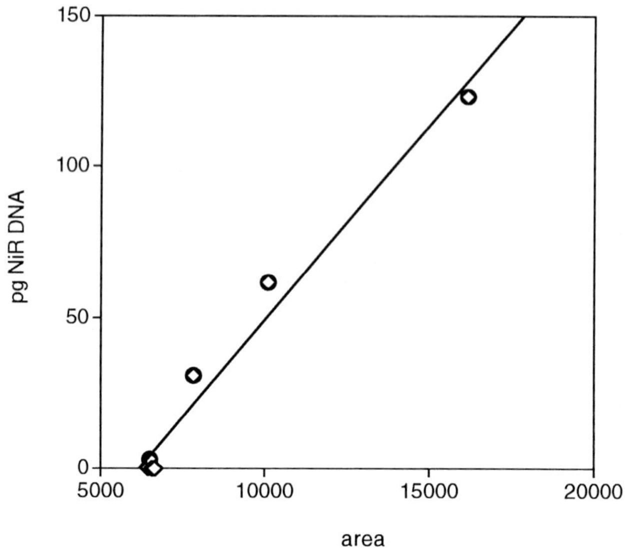

Fig. 3 Linear regression of dilution series DNA concentration and arbitrary area units from quantification of slots using ScanAnalysis software.
pg NiR DNA = (0.013 x area) - 79.54 ($r^2 = 0.976$)

The Nir gene fragment was quantified in seawater collected in the Santa Barbara Channel (Southern California Bight) in July 1992. The target was detected in DNA extracted from filtered seawater using colorimetric detection in slot blots, in which a sample of previously quantified DNA from *P. stutzeri* was also included (Fig.4). After subtraction of the background (average of negative control slot 2, and no DNA control slot 10, not shown) from the integrated areas of all slots, the amount of NiR DNA in each slot was calculated from the ratio of NiR DNA to area for the standard (slot 1). Assuming one copy of the NiR gene per genome, the number of genomes (= cells) present in the extract was calculated and normalized to volume filtered. The abundance of NiR fragments varied from 3 x 10^4 ml^{-1} at 250m to 2.1 x 10^5 ml^{-1} at the surface (Fig. 5).

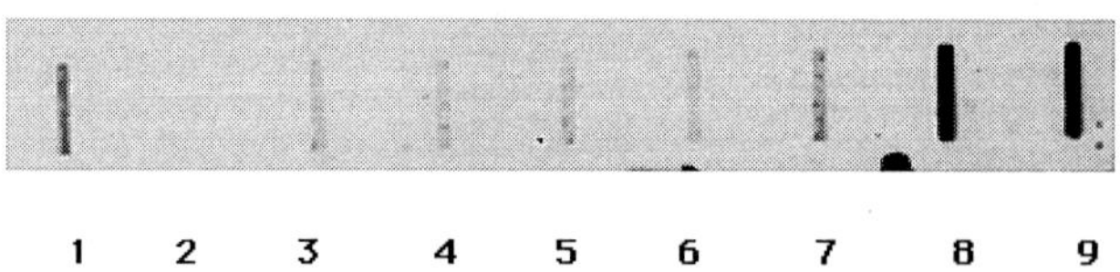

Fig. 4 Slot blot of total DNA extracts from the Santa Barbara Channel. Slots: 1, 264ng *P. stutzeri* genomic DNA; 2, 2μg salmon sperm DNA; 3, 250m; 4, 200m; 5, 160m; 6, 120m; 7, 80m; 8, 40m; 9, 0m

Total bacterial abundance was determined at a similar site in the Santa Monica Basin (Fig. 5) during June of 1988. These data were collected at about the same time of year, although not simultaneously, with the NiR samples. Previous data on the bacterial abundance of this region is plentiful (Azam 1986), and while seasonal and even diel variations have been detected, the relative magnitude and distribution shown here can be taken as representative.

P. stutzeri (ATCC 14405) was enumerated in preserved samples from the Southern California Bight and Monterey Bay (Fig. 5) using previously described methods (Ward and Cockcroft, 1993). The maximum abundance (over 500 cells ml^{-1}) was detected at 100m in Monterey Bay and the abundance averaged just over 100 cells ml^{-1} between 100m and the bottom at 900m. *P. stutzeri* abundance in the Santa Monica Basin in April 1994 was generally less than 100 cells ml^{-1} (Fig. 5).

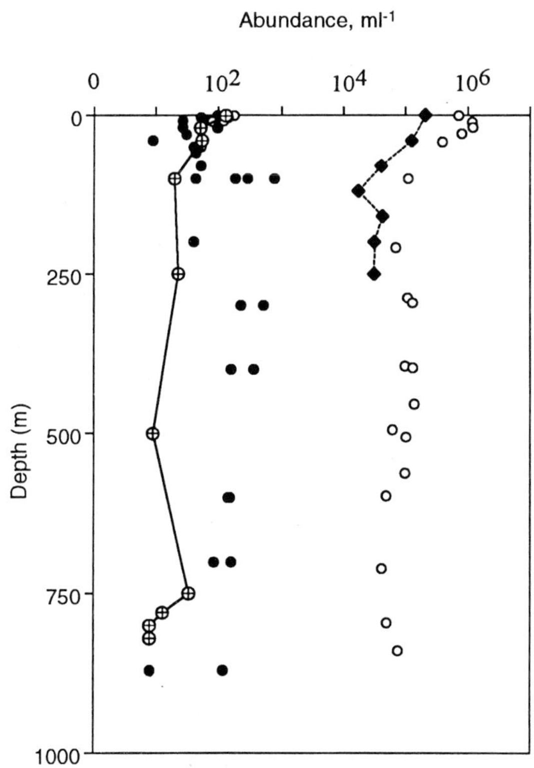

Fig. 5 Depth distribution of NiR fragment (◆), total bacterial abundance (DAPI, O), and *P. stutzeri* (IIF) cells in Monterey Bay (●) and Southern California Bight (⊕).

Discussion

The abundance of ammonium-oxidizing nitrifiers as estimated by enumeration with the *N. marina* antiserum in Plußsee represented a maximum of 0.04% of the total bacterial population. The absolute abundance of *N. marina* serotype cells was much higher than generally reported for oceanic environments (Ward and Carlucci, 1985; Ward, 1987), where abundances tend to be on the order of 200 cells ml^{-1}. The proportion that this serotype contributes to the total population is, however, similar to the range estimated for the contribution of the *N. marina* serotype in seawater (Ward and Carlucci, 1985), because the total cell abundance is also higher. This antiserum has not been used previously in freshwater environments and, as is the case in seawater, it necessarily underestimates the total number of ammonium oxidizers due to the presence of non-reactive serotypes.

Using selective amplification by PCR, ammonium oxidizers were detected at all depths in the Plußsee. All DNA extracts were suitable for direct amplification with the eubacterial universal primers, indicating the lack of inhibitory substances in the purified DNA extracts. Only two of the extracts required two stage amplification in order to detect the NitAB sequence, as was reported previously (Voytek and Ward, submitted). Using serial dilutions of the DNA extracts, the sensitivity of the assay was investigated. For EUB primers, it was possible to amplify a 10^4 dilution of the DNA extract from the surface sample (representing 1.3μl). The total cell number in this sample was 6.85×10^6 cells ml^{-1}, so 1.3 μl represents amplification of about 9133 cells. At this depth, a 10-fold dilution of the DNA extract could be amplified with the NitDB primers (1.3ml). If the NitDB primers amplified with the same efficiency as the EUB primers, this implies the presence of 6850 ammonium oxidizing cells ml^{-1}. Nearly 3000 cells were detected by immuno-fluorescence. It is unlikely that the two primer sets amplify with the same efficiency, given the vast variability in the target DNAs and optimum annealing conditions: one might expect that the Nit primers should be more efficient since they target a very small group of nearly identical sequences. However, the interference from non-target DNA would be more serious for the Nit primers, probably reducing their efficiency (Sykes *et al.*, 1992). Thus, it seems likely that amplification efficiency varies with depth as the composition of the bacterial assemblage varies. While more sophisticated methods for quantification of PCR may be possible (*eg* Wang and Mark, 1990, Gilliland *et al.*, 1990), this exercise illustrates the practical limits and complications arising from the nature of environmental samples.

In the case of the use of probes for the functional gene, nitrite reductase, to identify the phenotypically defined, but phylogenetically diverse group, denitrifying bacteria, similar kinds of calculations can be made to evaluate the usefulness of the approach. Although the samples for enumeration and quantification were not collected simultaneously, the abundances can be taken as representative values and used for qualitative comparisons.

Immunofluorescence enumeration of a single strain of denitrifying bacteria, *P. stutzeri* ATCC 14405, detected an average of 100 cells ml^{-1} in Monterey Bay, a more productive region than the Southern California Bight, where the abundance in surface waters was about 50 cells ml^{-1}. *P. stutzeri* constituted 0.004% to 0.07% of total cells in the Southern California Bight, similar to the range previously estimated for Monterey Bay (0.02 to 0.08%, Ward and Cockcroft, 1993).

Using 40m as representative of the surface layer in the Southern California Bight, *P. stutzeri* cells contributed 0.01% of the total cell number. The number of NiR fragments estimated for this depth is equal to 33% of the total cell number. Thus, if NiR is present at one copy per genome, a third of the total cells possessed the genetic capability for denitrification. This value varied from 16 to 41%, using data from equivalent depths for the two measurements. Given the uncertainty in various parts of this comparison, the exact values are less important than the scale. It implies that a large fraction of the total population possesses the denitrification genotype and that this genotypic group is probably composed of a large number of species, each present at low numbers. Another form of nitrite reductase, a copper enzyme, is also common in denitrifying isolates. It remains to be determined whether it is also abundant in the environment.

Conclusions

This comparison of several different kinds of probes designed to detect or quantify bacteria involved in biogeochemical transformations in the environmental nitrogen cycle provides both information specific to the mircoorganisms investigated here and more general ideas on what kind of approaches might be most useful in future applications. Immunofluorescence assays tend to be specific for the species or even the strain for which they are produced. This may be less the case for nitrifying bacteria, for it appears that the *N. marina* serotype includes strains from many environments and many strains classified as various species of *Nitrosomonas*. The advantage of immunofluorescence is the ability to visualize and enumerate individual cells, but the disadvantage is the serotype, rather than genotype, specificity. Thus, ecological questions concerning population structure and diversity may be addressed using immunofluorescence to enumerate ecologically or phenotypically defined parts of the population.

The versatility of the rRNA molecule as a target for PCR amplification means that PCR as a detection method for closely related strains can be optimized for specificity in ecologically relevant terms. It appears that there are only a few phylogenetically defined groups of ammonium-oxidizing nitrifiers, such that a small number of primers might suffice to detect them all. The NitAB primers used here detect all the known ammonium-oxidizing strains in the β subclass as well as a great many isolates (Voytek and Ward, submitted). The other main group, *Nitrosococcus oceanus* types in the γ subclass, has fewer known members. Another set of primers for that group, currently under development, will then make it possible to detect all ammonium oxidizers in the environment, unless completely different unknown types remain to be discovered. The success of detecting target DNA in various environments implies that known nitrifying genotypes are ubiquitous in nature. Primers with greater selectivity (*eg* NitDB) can provide information on species composition and population structure within the overall ammonium oxidizer population.

In the case of denitrifying bacteria, immunofluorescence and molecular methods based on ribosomal RNA should both be useful in detection and enumeration of individual strains. The DNA probe for the key denitrification enzyme, nitrite reductase, however, should be more useful in detecting cells which possess the denitrification genotype, regardless of their strain affiliation or whether or not they are performing denitrification at the time. The contrast between abundance of *P. stutzeri* estimated with the strain specific immunofluorescent assay and the abundance of NiR genotype in the environment demonstrates the difference between taxonomic and functional probes for the phylogenetically diverse group, denitrifying bacteria. The individual strain is a very small part of the total bacterial population, while the functional probe implies that a much larger fraction of the population is genetically capable of denitrification. While the absolute values for both of these estimates contain unresolved sources of error and approximation, their relative scales suggest high diversity within natural bacterial populations.

Acknowledgments

M. T. Geissler, D. G. Bard and C. Burghardt provided assistance in immunofluorescence enumeration. The work in the Plußsee was done in collaboration with M. A. Voytek and K.-P. Witzel and was supported by a visiting scientist grant from the Max-Planck-Gesellschaft for work performed at the Max-Planck Institut für Limnologie, Plön, Germany. The rest of the research was supported by NSF grants to BBW.

References

Azam F (1986) Nutrient cycling and food web dynamics in the Southern California Bight: The microbial food web. In: *Plankton Dynamics of the Southern California Bight* Eppley RW (ed) Springer-Verlag, Berlin New York 274-288

Ausubel FM, Brent R, Kingston RE, Moore DD, Smith JA, Sideman JG, Struhl K (eds) (1987) *Current protocols in molecular biology*. Wiley, New York

Bryan BA (1981) Physiology and biochemistry of denitrification. In: *Denitrification, Nitrification and Atmospheric Nitrous Oxide.* Delwhiche CC ed) Wiley, New York 67-84

Gilliland G, Perrin S, Bunn HF (1990) Competitive PCR for Quantitation of mRNA. In: *PCR Protocols: A Guide to Methods and Applications.* Innis MA, Gelfand DH, Sninsky JJ, and White TH (eds) Academic Press, San Diego 60-69

Head, IM, Hiorns WD, Martin T, McCarthy AJ, Saunders JR (1993) The phylogeny of autotrophic ammonia-oxidizing bacteria as determined by analysis of 16S ribosomal RNA gene sequences. *J Gen Microbiol* **139:** 1147-1153

Koops H-P and Harms H (1985) Deoxyribonucleic acid homologies among 96 strains of ammonia-oxidizing bacteria. *Arch microbiol* **141:** 214-218

Josserand A and Cleyet-Marel. JC (1979) Isolation from soils of *Nitrobacter* and evidence for novel serotypes using immunofluorescence. *Microb Ecol* **5:** 197-205.

Jungst A, Wakabayashi S, Matsubara H, Zumft WG (1991) The *nir*STBM region coding for cytochrome cd_1 dependent nitrite respiration of *Pseudomonas stutzeri* consists of a cluster of mono-, di- and tetraheme proteins. *FEBS Letters* **279:** 205-209

Liesack W, Weyland H, Stackebrandt E (1991) Potential risks of gene amplification by PCR as determined by 16S rDNA analysis of a mixed-culture of strict barophilic bacteria. *Microb Ecol 21:* 191-198.

Lipschultz F, Wofsy SC, Ward BB, Codispoti LA, Friederich GE, Elkins JW (1990) Bacterial transformations of inorganic nitrogen in the oxygen deficient waters of the eastern tropical south Pacific Ocean. *Deep-Sea Res* **37:** 1513-1541

Overbeck J and Chrost RJ (1993) (eds) *Microbial Ecology of Lake Plußsee.* Springer-Verlag New York Berlin

Nishio T, Koike I, Hattori A (1983) Estimates of denitrification and nitrification in coastal and estuarine sediments. *Appl Environ Microbiol* **45:** 444-450

Risgaard S, Rysgaard-Petersen H, Nielsen LP, Revsbech NP (1993) Nitrification and denitrification in lake and estuarine sediments measured by the ^{15}N dilution technique and isotope pairing. *Appl Environ Microbiol* **59:** 2093-2098

Sykes PJ, Neoh SH, Brisco MJ, Highes E, Condon J, Morley AA (1992) Quantitation of targets for PCR by use of limiting dilution. *Biotech* **13:** 444-449

Voytek MA and Ward BB Detection of ammonium-oxidizing bacteria of the beta-subclass Proteobacteria in aquatic samples using the polymerase chain reaction (submitted to *Appl Environ Microbiol*)

Wang AM and Mark DF (1990) Quantitative PCR. In: *PCR Protocols: A Guide to Methods and Applications*. Innis MA, Gelfand DH, Sninsky JJ, White TH (eds). Academic Press, San Diego 70-75

Ward BB and Carlucci AF (1985) Marine ammonia- and nitrite-oxidizing bacteria: Serological diversity determined by immunofluorescence in culture and in the environment. *Appl Environ Microbiol* **50:** 194-201

Ward BB (1987) Nitrogen transformations in the Southern California Bight. *Deep-Sea Res* **34:** 785-805

Ward BB and Kilpatrick KA (1993) Methane oxidation associated with mid-depth methane maxima in the Southern California Bight. *Cont Shelf Res* **13:** 1111-1122

Ward BB and Cockcroft AR (1993) Immunofluorescence detection of denitrifying bacteria in seawater and intertidal sediment environments. *Microbial Ecol* **25:** 233-246

Ward BB, Cockcroft AR, Kilpatrick KA (1993) Antibody and DNA probes for detection of nitrite reductase in seawater. *J Gen Microbiol* **139:** 2285-2293

Woese CJ, Wesiburg WG, Paster BJ, Hahn CM, Tanner RS, Krieg NR, Koops H-P, Harms H, Stackebrandt E (1984) The phylogeny of the purple baceria: the beta subdivision. *System Appl Microbiol* **5:** 327-336

Woese CJ (1987) Bacterial Evolution. *Microbiol Rev* **51:** 221-271

Zumft WG (1992) The denitrifying prokaryotes. In: *The Prokaryotes: A Handbook on the Biology of Bacteria: Ecophysiology, Isolation, Identification, Applications.* Balows A, Truper HG, Dworkin M, Harder W, Schleifer K-H (eds.), 2nd ed. Springer-Verlag New York 554-582

The Role of Mixotrophy in Pelagic Environments

Bo Riemann[1], Harry Havskum[1], Frede Thingstad[2], and Catherine Bernard[3]

[1] The International Agency for ^{14}Carbon Determination, VKI
Agern Allé 11,
DK-2970 Hørsholm,
Denmark

[2] Dept. of Microbiology,
University of Bergen,
Jahnebakken 5,
N-5020 Bergen,
Norway.

[3] Marine Biological Laboratory,
University of Copenhagen
Strandpromenaden 5,
DK-3000 Helsingør,
Denmark.

Abstract

This paper reviews the occurrence and ecological importance of mixotrophic flagellates and ciliates in pelagic environments, particularly in marine ecosystems. Mixotrophy is here defined as the combination of photoautotrophic and heterotrophic nutrition in a single individual, often used in the restricted sense of combining photosynthesis and phagotrophy. Mixotrophic protists represent an alternative strategy that allows a shortcut between the traditional food web and the microbial loop. A large number of reports have been published on the ecological importance of mixotrophic flagellates in freshwater, yet only a few studies have been carried out in seawater. In contrast, most of the knowledge of mixotrophic ciliates comes from marine environments. Results from field studies have demonstrated that both mixotrophic flagellates and ciliates are commonly found in many marine environments, and mixotrophic flagellates can dominate the biomass of photoautotrophs and be responsible for the entire grazing of bacteria or protists. Results from laboratory experiments on factors controlling the degree of photoautotrophy/phagotrophy in flagellates are presented. Finally, we present a hypothesis for a growth strategy of bacterivorous mixotrophic flagellates.

Introduction

Aquatic ecologists often describe the flow of materials and energy in a simplified

NATO ASI Series, Vol. G 38
Molecular Ecology of Aquatic Microbes
Edited by I. Joint

compartmentalization of the water column into a classical food chain and the microbial loop. In the classical food chain, organic material is assumed to be transported from phytoplankton directly to zooplankton and fish larvae. In the microbial loop (Azam *et al.* 1983, Azam and Fuhrman 1984), part of the primary production is released as dissolved organic material and subsequently taken up by bacteria (Hobbie and Williams 1984, Riemann and Søndergaard 1986). The bacteria are grazed by flagellates, ciliates, and crustaceans before part of the energy is returned to the fish level in the classical food chain.

This scenario is based on a number of assumptions, of which several have been questioned (Jumars *et al.* 1989), and some have been proved to be wrong. For example, one fallacy is that the microbial consortia can be strictly identified as photoautotrophs or heterotrophs. Mixotrophy is commonly found among flagellates (Fenchel 1991, Sanders 1991a), ciliates (Laval-Peuto 1992, Dolan 1992), and other protists (Laval-Peuto 1992). A mixotrophic organism would in many cases be able to survive periods with insufficient light, a scarcity of inorganic nutrients, or a depletion of food particles, conditions that are commonly found in many aquatic environments.

Although mixotrophy has been recognized for a long time (Gourret 1883, Pascher 1917), a general stimulation of research on the topic first occurred in the mid-1980s. Mixotrophs were enumerated, and preliminary details of their ecological importance in a number of environments were presented (*eg* Bird and Kalff 1986, Estep *et al.* 1986, Ishida and Kimura 1986).

Studies on algal phagotrophy in natural environments are hampered by tedious and often insufficient methods. The content of food vacuoles in larger protists can be identified, but this technique is not adequate when small flagellates are considered (Bockstahler and Coats, 1993a). Alternatively, fluorescently labelled particles of artificial or natural origin have been used as tracers to measure phagotrophy in algae (McManus and Fuhrman 1986a, Berninger *et al.* 1992, Hall *et al.* 1993).

The present consensus is based on a large number of autecological studies and some synecological studies on mixotrophy in freshwater lakes and only a few reports concerning marine environments. A major outcome of this review is that mixotrophs are indeed commonly found in a variety of freshwater and marine environments. Little is known, however, about the factors controlling photosynthetic and phagotrophic nutrition modes in natural microbial assemblages. The lack of suitable models further limits the possibility of predicting the conditions that favour growth of mixotrophs rather than strictly photoautotrophs or strictly phagotrophs.

The review is organized around descriptions of mixotrophy in flagellates, factors controlling phagotrophy and photoautotrophy, mixotrophic flagellates in natural environments, some theoretical aspects of the strategy of mixotrophs, mixotrophy in ciliates and finally, a few ecological implications of mixotrophy.

Mixotrophy in Flagellates

PHAGOTROPHY IN PIGMENTED FLAGELLATES

Normally, pigmented flagellates are not suspected to have phagotrophic abilities, whereas non-pigmented flagellates are often studied to find out how they take up food particles. Many observations on phagotrophy by pigmented flagellates are thus made by chance. Nevertheless, phagotrophic pigmented flagellates (mixotrophs) are known from many algal classes *eg* Cryptophyceae, Dinophyceae, Chrysophyceae, Prymnesiophyceae and Pedinellophyceae (Table 1).

The prey of mixotrophic flagellates that has been observed most frequently are bacteria, which are often present in cultures of pigmented flagellates, many species of which have been revealed to be bacterivorous *eg Chroomonas pochmanni* (Kugrens and Lee 1990). In natural environments, most tracer experiments to determine particle uptake rates by flagellates have been performed with bacteria-sized particles. The use of fluorescently labelled latex beads or bacteria makes it possible to discriminate between uptake by pigmented and non-pigmented flagellates (Berninger *et al.* 1992, Hall *et al.* 1993). These experiments revealed that the most important bacterivorous, pigmented flagellates in freshwater belong to the Chrysophyceae, *eg* species belonging to the genera *Catenochrysis*, *Chromulina*, *Chrysococcus*, *Chrysosphaerella*, *Chrysostephanosphaera*, *Dinobryon*, *Ochromonas*, *Uroglena* (Salonen and Jokinen 1988, Bird and Kalff 1987, 1989, Sanders *et al.* 1989, Bennett *et al.* 1990, Berninger et al 1992). Data on bacterivorous, pigmented flagellates from seawater are less numerous. As grazers of bacteria-sized particles, bloom-forming species belonging to the Prymnesiophyceae (*eg Chrysochromulina* spp., *Prymnesium* spp.) seem to be the most important (Green 1991, Nygaard and Tobiesen 1993, Havskum and Riemann, unpublished). Marine Chrysophyceae, *eg* species belonging to the genera *Ochromonas* and *Chrysamoeba*, can also be abundant in the sea and have been shown to be bacterivorous (Estep *et al.* 1986).

Possible herbivory or carnivory by pigmented flagellates is very rarely revealed. Culturing normally does not involve feeding with other eukaryotic organisms, and in field studies, uptake experiments with eukaryotic tracers (Rublee and Gallegos 1989) are not commonly performed. Only few investigators have tried to analyse the food vacuole content of pigmented flagellates in natural environments (*eg* Bockstahler and Coats 1993a,b). Nevertheless, many herbivorous, pigmented flagellates are known, *eg Chrysochromulina brevifilum*, which ingests small green flagellates (Jones *et al.* 1993), *Cryptomonas borealis*,

Table 1. Prey organisms of mixotrophic flagellates

Prey	Algal class + example	Reference
Picoplankton-sized particles <2µm	Cryptophyceae: *Chroomonas pochmanni*	Kugrens and Lee 1990
	Dinophyceae: *Ceratium hirundinella*	Dodge and Crawford 1970
	Chrysophyceae: *Uroglena americana*	Ishida and Kimura 1986
	Prymnesiophyceae: *Prymnesium parvum*	Nygaard and Tobiesen 1993
	Pedinellophyceae: *Pedinella hexacostata*	Swale 1969
Plankton organisms 2-5µm	Dinophyceae: *Ceratium hirundinella*	Dodge and Crawford 1970
	Chrysophyceae: *Poterioochromonas malhamensis*	Sanders *et al.* 1990
	Prymnesiophyceae: *Chrysochromulina brevifilium*	Jones *et al.* 1993
Plankton organisms >5µm	Cryptophyceae: *Cryptomonas borealis*	Wawrik 1970
	Dinophyceae: *Amphidinium cryophilum*	Wilcox and Wedemeyer 1991
	Chrysophyceae: *Poterioochromonas malhamensis*	Pringsheim 1952
	Prymnesiophycea: *Chrysochromulina ericina*	Parke *et al.* 1956
	Pedinellophyceae: *Cyrtophora pedicellata*	Pascher 1917
Ciliates	Dinophyceae: *Gymnodinium sanguineum*	Bockstahler and Coats 1993a,b

which feeds on smaller cryptophytes (Wawrik 1970), or the cf. *Ochromonas* sp., which has been observed to significantly reduce a bloom of small centric diatoms (Olrik and Nauwerck 1993). Examples of carnivorous, pigmented flagellates are the dinoflagellates *Gyrodinium pavillardii*, *Gymnodinium sanguineum*, *Gyrodinium uncatenatum*, and *Ceratium furca*, which have been observed to prey preferentially on oligotrichous ciliates (Biecheler 1952, Bockstahler and Coats 1993a).

The uptake of prey organisms by pigmented flagellates is in most cases by phagocytosis *sensu stricto*, where the whole prey organism is engulfed into a nascent food vacuole inside the predator's cell body (Table 2). Mixotrophic flagellates whose particulate food uptake mechanism is by phagocytosis *sensu stricto*, *eg Chroomonas pochmanni*, which feeds on bacteria (Kugrens and Lee 1990), generally feed on organisms smaller than themselves. In many mixotrophic flagellates, the flagella seem to be involved in particle capture by creating a water current towards the cell body, *eg Epipyxis*, *Ceratium*, *Ochromonas*, *Poterioochromonas*, (Pringsheim 1952, Cachon *et al.* 1991, Moestrup and Andersen 1991, Wetherbee and Andersen 1992). Other cell extensions, like the tentacles in *Cyrtophora* or the haptonema in *Chrysochromulina*, may also be involved in food capture (Pascher 1917, Kawachi *et al.* 1991). When the food particle comes into contact with the cell surface, some kind of selection or rejection seems to be possible. If selected, it is either directly engulfed, or a pseudopodial structure surrounds it and draws it subsequently into the cell body. Particles too large or too heavy cannot normally be taken up by this kind of feeding. Only in rare cases does prey size equal or slightly exceed the predator's size, as in *Gyrodinium pavillardii*, which feeds on *Strombidium* spp. (Biecheler 1952).

Other uptake mechanisms, such as pallium feeding and myzocytosis (Table 2), which have been described in detail for many colourless dinoflagellates in recent years, are much more difficult to observe (Spero 1982, Schnepf *et al.* 1985, Jacobson and Anderson 1986, Hansen 1991, Wilcox and Wedemeyer 1991).

In myzocytosis, the prey organism's periplast is punctured by the predator's feeding tube. Some or most of the prey's protoplasm is subsequently taken up through this feeding tube into a nascent food vacuole. In the mixotrophic dinoflagellate *Amphidinium cryophilum*, the feeding tube is formed *de novo* at the onset of each feeding event. It is not associated with any microtubules; it pierces and fuses with the prey cell membrane, and after the feeding event, it is left *in situ* (Wilcox and Wedemeyer 1991). The feeding tube of many colourless dinoflagellates, on the other hand, is a column of cytoplasm that contains one or several rows of microtubules. This organelle, often termed a peduncle, is permanent and can be extruded and retracted (Spero 1982, Schnepf *et al.* 1985). After the feeding event, the prey's periplast, which is often difficult to digest, is left behind. The prey in the predator's food vacuole is thus difficult to recognize. This type of feeding was first observed by Biecheler (1952) in the pigmented dinoflagellate *Gyrodinium vorax*, which feeds on other dinoflagellates by taking up only some of the prey's cell content.

In pallium feeding, the predator attaches itself to the prey by a capture filament and subsequently encloses it by a cytoplasmic extension, a pallium. The prey is partly digested in the pallium, outside the predator's cell body. Since the pallium is often lost during routine fixations, this type of feeding has only rarely been observed (Jacobson and Anderson 1986). By studying the pigmented

Table 2. Phagotrophy in mixotrophic flagellates

Phagotrophic modes of nutrition	Definitions	Examples	References
Phagotrophy *sensu stricto*	Uptake of an entire food particle (incl. the cell membrane) into a food vacuole inside the cell body.	*Chroomonas pochmanni*	Kugrens and Lee 1990
		Chrysamoeba sp.	Estep *et al.* 1986
		Chrysochromulina brevifilum	Jones *et al.* 1993
		cf. *Ochromonas* sp.	Olrik and Nauwerck 1993
		Gyrodinium pavillardii	Biecheler 1952
Myzocytosis	Partial uptake of the prey's protoplasm through a feeding tube into a food vacuole inside the cell body. The periplast is left behind.	*Gyrodinium vorax*	Biecheler 1936, 1952.
		Amphidinium cryophilum	Wilcox and Wedemeyer 1991
Pallium feeding*	Uptake and partial digestion of particulate food in a food vacuole outside the cell body.	*Ceratium hirundinella* ?	Hofeneder 1930

* Described with certainty only for colourless dinoflagellates, *eg Protoperidinium* spp. (Jacobson and Anderson, 1986).

dinoflagellate *Ceratium hirundinella*, Hofeneder (1930) observed diatoms and small flagellates covered by cytoplasmic extensions resembling such a pallium, but pallium feeding has only been with certainty described for colourless dinoflagellates, *eg Protoperidinium* spp. (Jacobson and Anderson 1986). For flagellates with uptake mechanisms like myzocytosis or pallium feeding, where either only parts of the prey are taken up, or the prey is digested outside the predator's cell body, the prey size can considerably exceed the predator's own size.

While it is recognized that some mixotrophs can switch between various particulate food sources, the range in size and quality of these particles is known only for a few mixotrophs. *Chrysochromulina brevifilum* ingests both live algae and inert particles. *Gymnodinium sanguineum* does not only ingest ciliates, but has occasionally also been found with small diatoms in its food vacuoles (Bockstahler and Coats 1993a). *Ceratium hirundinella* feeds on bacteria, cyanobacteria, and diatoms (Dodge and Crawford 1970). *Poterioochromonas malhamensis* ingests bacteria, yeasts, inert particles, and algal cells almost as big as itself, and in cultures without any other organisms or solid particles, *Poterioochromonas* is even cannibalistic (Pringsheim 1952). *Chrysochromulina ericina* ingests a wide variety of prey, but there must be some kind of selection, since some chrysophytes, diatoms, and green algae are ingested, and others are not (Parke *et al.* 1956).

Mixotrophic flagellates have a common function as primary producers. The size and quality of their particulate prey, however, can vary between species. Among the mixotrophic flagellates, there are detrivores, bacterivores, herbivores, carnivores, and omnivores, which perform different roles within the microbial food web.

Plastid-retaining flagellates

Certain colourless flagellates take up photosynthetic prey without immediate digestion of the plastids. Since the plastids continue to function, the plastid-retaining flagellates are functional mixotrophs, as they derive nutrition from both photosynthesis and ingested particulate food (Larsen 1988, Schnepf *et al.* 1989). For example, the colourless dinoflagellate *Amphidinium poecilochroum* sequesters plastids of cryptophytes and cannot be grown in cultures without its prey organisms. The occurrence of starch grains within the predator's cytoplasm demonstrates that these plastids remain photosynthetically active, and that they supply the predator with energy. In some food vacuoles, the plastids appear normal, but undoubtedly they are eventually digested, since plastids in different stages of reduction are also found. In this way, they are used twice: first indirectly, as a supplier of photosynthetic products, and later directly, as food. In another functional mixotrophic dinoflagellate, *Gymnodinium aeruginosum*, the plastids are not digested before some weeks after uptake, and during cell division, they are distributed into the daughter cells.

As is the case in plastid-retaining ciliates (section 6), one could imagine that the alien functional plastids enhance the survival of the plastid-retaining flagellates during times of low particulate food abundance.

Factors controlling phagotrophy and photoautotrophy

Mixotrophic flagellates include a gradient of algae that ranges from almost entirely phagotrophic to almost completely photo-autotrophic (Sanders et al. 1990, Sanders 1991b). Several factors, often species-specific, govern the switching between photoautotrophic and phagotrophic feeding modes. Important ones, briefly presented below, include particle density, light, inorganic nutrients, pH, and dissolved organic matter.

EFFECTS OF PARTICLE DENSITY

Particle density was most important in determining the balance between phagotrophy and photoautotrophy in the ochromonad *Poterioochromonas malhamensis*, (Sanders *et al.* 1990, Caron *et al.* 1990). When heat-killed bacteria were added to axenic cultures of *P. malhamensis* growing photoautotrophically, the cellular chlorophyll content decreased from a range of 160-350fg cell^{-1} to 50fg cell^{-1}. Phagotrophically growing cells showed an increased chlorophyll content when bacteria were grazed below 10^6 ml^{-1}, indicating an increase in photosynthetic nutrition (Fig. 1). The switching time between mainly phagotrophic to photoautotrophic nutrition was about 24 hours, indicating that the ecological importance of this phenomenon is restricted to periods longer than diel cycles (Sanders *et al.* 1990). In fact, this period is probably even longer, since the cellular chlorophyll content reached only 20-60% of the cellular chlorophyll content of cells growing photoautotrophically. The change from photoautotrophy to phagotrophy in *P. malhamensis* occurred almost simultaneously as soon as sufficient bacteria were added, suggesting that the cells are always capable of bacterivory, or that phagotrophy is highly inducible (Sanders *et al.* 1990). When bacteria concentrations were above 10^6 ml^{-1}, the contribution of photoautotrophy to the carbon budget did not exceed 7%.

Mixotrophic ochromonads have been encountered in great numbers in natural environments. The cf. *Ochromonas* sp. observed by Olrik and Nauwerck (1993) fed effectively on a spring bloom of small unicellular, centric diatoms in Hjarbæk Fjord, Denmark. Mineral nutrient concentrations were high and light was probably not a limiting factor for photosynthesis. Like the ochromonads studied in the laboratory, this cf. *Ochromonas* sp. may have an opportunistic feeding response to high particulate food abundance irrespective of factors supporting photosynthesis.

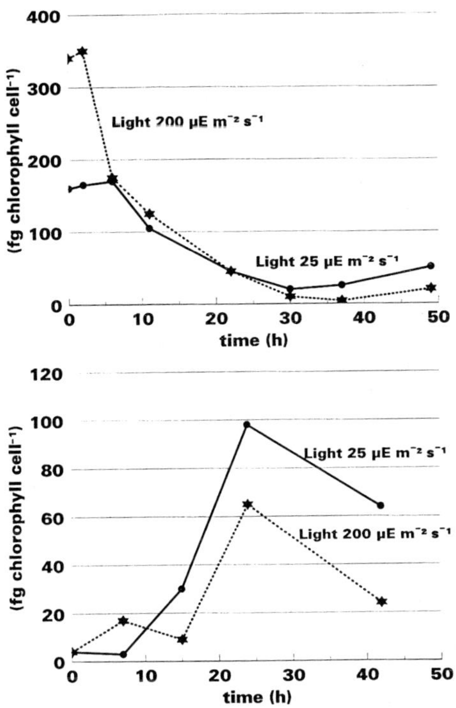

Fig. 1. Changes in chlorophyll per cell of *P. malhamensis* during a time-course experiment designed to examine the shift from photoautotrophy to phagotrophy (top) and from phagotrophy to photoautotrophy (bottom). Redrawn from Sanders *et al.* (1990).

EFFECTS OF LIGHT

At present, there is no straightforward relationship between phagotrophic behaviour and light. Wilcox and Wedemayer (1991) mentioned (no data presented) that phagotrophic feeding by the mixotrophic dinoflagellate *Amphidinium cryophilum* was reduced in light. A similar pattern was found in the mixotrophic chrysophyte *Dinobryon cylindricum* (Caron *et al.* 1993). Jones *et al.* (1993) found also an inverse relationship between light intensity (0-225μmol $m^{-2}s^{-1}$) and ingestion of a small green flagellate by the prymnesiophyte *Chrysochomulina brevifilum*. Even considering that many other species of the genus *Chrysochromulina* also have phagotrophic feeding modes (Green, 1991), others show no signs of phagotrophy even under low light conditions (Hällfors and Niemi 1974, Wehr *et al.* 1985). Consequently, it is not possible to generalize, not even within the same genus.

Mixotrophic flagellates, especially members of the Chrysophyceae, have been reported as important grazers in natural environments with reduced light intensities. Significant grazing impacts on bacteria-sized particles by mixotrophs were observed in frontal zones between nutrient-poor and nutrient-rich water masses, where light intensities were reduced (Bird and Kalff 1986, 1987, 1989, Hall *et al.* 1993), and at the bottom end of stratified water masses, where photosynthesis is limited by light rather than by mineral nutrients (Sanders *et al.* 1989). In winter, when light limits photosynthesis, and when nutrients are well mixed throughout the water column, mixotrophic flagellates have been observed as important grazers in upper water masses as well (Sanders *et al.* 1989, Bennett *et al* 1990, Berninger *et al.* 1992). In light-limited environments, the acquisition of carbon through phagotrophy may explain the success of these mixotrophs.

In contrast to the above mentioned mixotrophic flagellates that can have a significant grazing impact on bacteria or protists, a great number of other mixotrophic flagellates have only occasionally been observed with ingested food particles. These species are probably almost completely photoautotrophic and may never play an important role as grazers. The cryptophytes investigated by Tranvik *et al.* (1989) grazed less than 2% per day on a natural bacteria population. Although *Cryptomonas* sp. in nature had increased uptake rates at reduced light intensities, no increase could be induced in complete darkness in the laboratory as investigated for *Cryptomonas ovata* and *Cryptomonas erosa*. Tranvik *et al.* calculated that these cryptophytes at most incorporated 2% of their cell carbon daily through bacterivory. Even though these uptake rates of particulate food are low, they may still enhance their survival in environments with low light intensities and give them a competitive advantage over strictly photoautotrophic organisms.

No relationship was found between light intensity (up to 1100μmol $m^{-2}s^{-1}$) and particle uptake rates for *Poterioochromonas malhamensis* (Sanders *et al.* (1990). Growth of *P. malhamensis* in the dark was comparable to growth in light as long as sufficient bacteria were present. Porter (1988) demonstrated for the same species that photosynthesis was maximal and particle ingestion rates lowest above 400μmol $m^{-2}s^{-1}$. Below that level, photosynthesis was reduced and particle ingestion increased. The discrepancies between these results may be due to differences in acclimation to the light intensities and in the particle densities used.

Andersson *et al.* (1989) demonstrated that the cell number of a marine *Ochromonas* sp. remained constant over weeks without bacteria in light; the photosynthetic capacity just allowed the cells to survive. On the other hand, the flagellate grew well when fed bacteria, and photosynthesis in combination with phagotrophy did not enhance its growth rate.

EFFECTS OF INORGANIC NUTRIENTS

It has been hypothesized that mixotrophic flagellates in nutrient-depleted environments might incorporate a large part of their nutrients from particulate food, and that high concen-trations of inorganic nutrients might depress the rate of phagotrophy (Caron *et al.* 1990). Although this seems to be the case for mixotrophic flagellates that are primarily photoauto-trophs, neither of these hypotheses were verified using *Poterioochromonas malhamensis* (Caron *et al.* 1990). Andersson *et al.* (1989) reported that ingestion of bacteria by a marine *Ochromonas* species did not compensate for inefficient uptake of nitrogen and phosphorus. They argued that if the bacteria should compensate for inefficient nutrient uptake, one would suspect an increase in the growth rate of *Ochromonas* in the light compared with growth in the dark with the same amount of bacteria food.

In contrast, a clear relationship between nutrient limitation and increased phagotrophy was found in other mixotrophic species. Jones *et al.* (1993) reported a clear relationship between increased phagotrophy and phosphate starvation (Fig. 2). Initial ingestion rates of a small green flagellate were 10 times higher when the prymnesiophyte *Chrysochromulina brevifilum* was starved of phosphate than when phosphate was in excess. Similar results were obtained by Nygaard and Tobiesen (1993) who measured uptake rates of radiolabeled bacteria by prymnesiophytes, chrysophytes, pedinellophytes and dinoflagellates in batch cultures with excess phosphate and in chemostats with reduced phosphate levels. Seven of the eight species tested had higher uptake rates in phosphate-reduced conditions.

There are only a few reports of mixotrophic flagellates as important grazers from natural environments with low inorganic nutrient concentrations. Uptake of bacteria by prymnesiophytes observed in summer in the upper photic zone of systems limited by nutrients (Nygaard and Tobiesen 1993, Havskum and Riemann, unpublished) could potentially supply the need for mineral nutrients by photosynthesis. Other prey organisms as well could potentially supply limiting nutrients, *eg* in N-limited surface mixed layer of Chesapeake Bay, USA, pigmented dinoflagellates ingested small oligotrichous ciliates. Bockstahler and Coats (1993b) hypothesized that one of these dinoflagellates, *Gymnodinium sanguineum*, could meet approximately 15% of its nitrogen requirement through the ingestion of ciliates. Cells of two other mixotrophic dinoflagellates, *Ceratium furca* and *Gyrodinium uncatenatum* were less frequently found with food vacuoles, indicating a lower potential for meeting their nitrogen requirements through phagotrophy (Bockstahler and Coats 1993a).

EFFECTS OF pH

For *P. malhamensis*, pH has opposite effects on photosynthesis compared with the effects on phagotrophy. At pH 7photoautotrophic growth of *P. malhamensis* was measurable in continuous light without bacteria, while slow but measurable growth has been observed at pH 4.0. Substantially higher growth rates were found at all pH values when sufficient bacteria were added, and low pH decreased the uptake of bacteria irrespective of light conditions (Sanders *et al.* 1990).

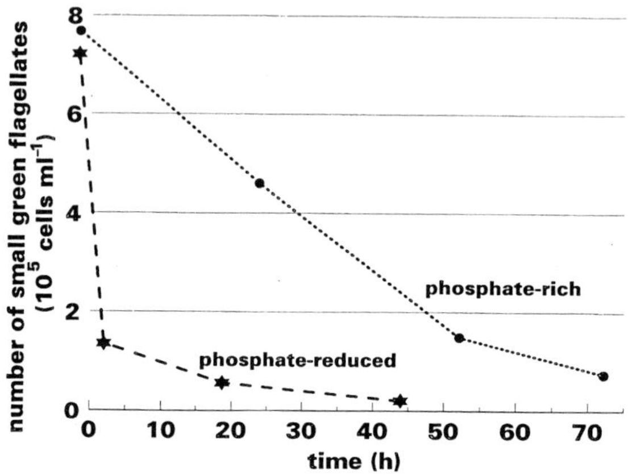

Fig. 2. Decrease in the population of a small green flagellate when fed to *Chrysochromulina brevifilum* grown under phosphate-rich or phosphate-reduced conditions. Redrawn from Jones *et al.* (1993).

EFFECTS OF DISSOLVED ORGANIC CARBON

It has been hypothesized that dissolved organic carbon would affect phagotrophy. Yet there seems to be no clear-cut consensus. Porter (1988) demonstrated that addition of glucose (from 0 to 0.1%, corresponding to 0 to 400 mg C $liter^{-1}$ carbon equivalents) gradually reduced ingestion rates of 0.57 μm^3 beads by *P. malhamensis.* On the other hand, Sanders *et al.* (1990) found no suppression of the bacterivory rates in the same species with the addition of dissolved organic carbon. The reasons for these differences are not clear.

Mixotrophic Flagellates in Natural Environments

Many competitive advantages have been attributed to mixotrophic flagellates, as they do not rely on only one trophic mode of nutrition. In environments with low concentrations of mineral nutrients, mixotrophic flagellates could potentially meet their requirements for photosynthesis and growth through phagotrophy. There are only a few reports of mixotrophic flagellates as important grazers in nutrient-limited environments. These observations are from upper water masses in summer, when light did not limit photosynthesis. Uptake of particulate food like bacteria (which are especially rich in N and P) by the prymnesiophytes observed in Scandinavian waters, could potentially supply the need for mineral nutrients in photosynthesis (Nygaard and Tobiesen 1993; Havskum and Riemann, unpublished). Other prey organisms as well, like the small oligotrichous ciliates ingested by pigmented dinoflagellates in Chesapeake Bay, USA, could potentially supply mineral nutrients for photosynthesis. Bockstahler and Coats (1993b) hypothesized that *Gymnodinium sanguineum* could meet approximately 15% of its nitrogen requirement through the ingestion of ciliates.

In light-limited environments, the acquisition of carbon through phagotrophy may explain the success of some mixotrophs. Mixotrophic flagellates as important grazers in environments with reduced light intensity have been reported several times. These observations were made in water masses where photosynthesis is typically not limited by mineral nutrients. Mixotrophic flagellates were observed as important grazers in frontal zones between nutrient-poor and nutrient-rich water masses (Bird and Kalff 1986, 1987, 1989, Hall *et al.* 1993) or at the bottom end of stratified water masses (Sanders *et al.* 1989). Phagotrophy by pigmented flagellates could thus be an explanation for the often observed chlorophyll layers below the photic zone. In winter, when light limits photosynthesis, and when nutrients are well mixed throughout the water column, mixotrophic flagellates have been observed as important grazers in upper water masses as well (Sanders *et al.* 1989, Bennett *et al.* 1990, Berninger *et al.* 1992).

Furthermore, mixotrophic flagellates have been encountered in great numbers in environments where neither light nor mineral nutrients are limiting photosynthesis. These mixotrophs are probably closer to the phagotrophic extreme. Like colourless flagellates, they may have an opportunistic feeding response to high particulate food abundance, but in contrast to their colourless competitors, they may survive during times of low particulate food concentration through photosynthesis. The cf. *Ochromonas* sp. observed by Olrik and Nauwerck (1993) fed effectively on a bloom of small unicellular, centric diatoms in Hjarbæk Fjord, Denmark. The mineral nutrient concentration was high, and light in spring was probably not a limiting factor. The observation by Wawrik (1970) of *Cryptomonas borealis* feeding on a bloom of small Cryptophyceae was also probably an opportunistic feeding response, since the observation was made on a sunny summer day in a highly eutrophied lake in Austria. A positive correlation of mixotrophic flagellates with the abundance of their prey was also found by Bockstahler and Coats (1993a) in an investigation in Chesapeake Bay, USA.

In contrast to some mixotrophic flagellates that have been found to have a significant grazing impact on bacteria or protists, a great number of other mixotrophic flagellates have only occasionally been observed with ingested food particles. In the latter flagellates *eg Cryptomonas ovata* and *Cryptomonas erosa* as investigated by Tranvik *et al.* (1989), phagocytosis may be difficult to induce. These mixotrophs are probably closer to the photoautotrophic extreme and may never play an important role as grazers, but in environments with low mineral nutrient concentrations or low light intensities, even these mixotrophic flagellates could have a competitive advantage over strictly photoautotrophic organisms, since their phagotrophic abilities may enhance their survival.

It becomes increasingly clear that mixotrophic flagellates can be abundant in different aquatic habitats, and from the field studies mentioned, no single mixotrophic strategy emerges for all.

Theoretical Aspects of the Strategy of Bacterivorous Pigmented Flagellates. In strongly nutrient-limited systems, most of the mineral nutrients seem to be taken up by bacteria rather than by phytoplankton (Currie and Kalff 1984, Suttle *et al.* 1990, Thingstad *et al.* 1993), an observation leading to the question of where the phytoplankton might obtain their mineral nutrients (Currie and Kalff 1984). In such an environment, bacterivory by pigmented flagellates (mixotrophy) would seem to be a wise algal strategy for acquiring mineral nutrients. Looking at the mixotroph-bacteria interaction in isolation (Fig. 3a), such a strategy emerges as a way of "eating the competitor"; a strategy which would appear to allow the mixotroph to "kill two birds with one stone". While eating bacteria provides the mixotroph with the mineral nutrients already scavenged by the efficient bacterial uptake system, the mixotroph at the same time removes its superior competitors. Formulation of the system shown in Fig. 3 in terms of Lotka-Volterra-type equations gives an interesting set of equations combining the properties of the classical equations for predation and competition (for an account of the classical theories see, *eg* Wilson and Bossert 1971).

Advancing one step towards the complexity of the real world, one may ask for the advantages of being a mixotroph in a world where the mixotroph also has to cope with the competition from protists that have specialized as either photoauto/osmo-trophs or as hetero/phago-trophs. One could make the assumption that these specialized protists can become more efficient in their respective specialized niches than the generalist mixotroph trying to master both. In this perspective, the strategy of mixotrophy looks more complicated. Relevant questions seem to be: Suppose there is a certain "price" to be paid for mixotrophy in terms of lower affinities for both mineral nutrients and for bacterial prey than those achievable as a specialist? How high could this price be, before the mixotrophic strategy becomes unrewarding? Or, as a corollary in the language of everyday experience, how can you survive when there is someone better than you in all individual disciplines?

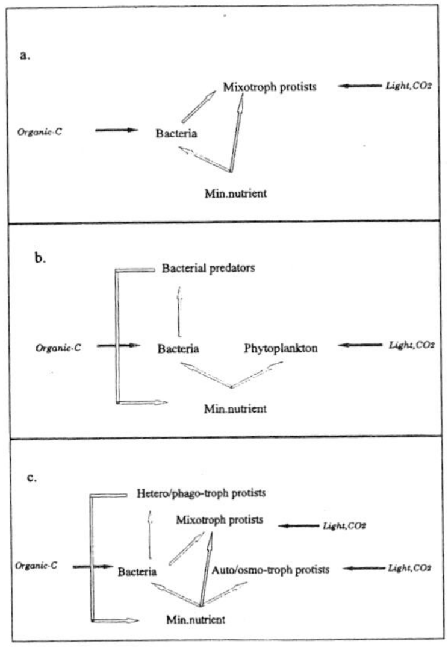

Fig. 3. Simplified microbial food web structures discussed: a) Mixotrophic protists preying on their bacterial competitor for mineral nutrients. b) "Traditional" separation of the two trophic functions of bacterial predation and photosynthesis/osmotrophic nutrient uptake into two populations of specialized protists, and c) combination of a) and b) into a food web with three protist populations.

Even without mixotrophs, the system of bacteria, colourless bacterial predators, and strictly photoautotrophic phytoplankton (Fig. 3b) has some counter-intuitive properties that become apparent as system complexity is increased in steps (Thingstad and Pengerud 1985, Pengerud *et al.* 1987): Compared to the community with 1) "bacteria alone", the community with 2) "bacteria + bacterivorous colourless protists" where the protists feed on bacteria may degrade more organic carbon material because of remineralization of the limiting nutrient otherwise immobilized in bacterial biomass (Fig. 4). In the community 3) "bacteria + strictly photoautotrophic algae", the algae will be outcompeted owing to the superior bacterial affinity for mineral nutrients. Combining communities 2) and 3) into 4): "bacteria + colourless bacterial predators + strictly photoautotrophic algae does, however, give an exactly opposite effect to what could be expected from naively "adding" the two separate cases: Glucose degradation is reduced relative to the "bacteria alone" situation and strictly photoautotrophic algae may dominate, leaving little of the available nutrients to be incorporated into the biomass of bacteria and colourless bacterial predators. Qualitatively, this may be explained as the combined effect of the bacterial growth rate being limited by mineral nutrients kept low by competition from strictly photoautotrophic algae, and bacterial biomass being kept low by predation. The situation illustrates one of the challenges facing a mixotroph trying to establish itself in such a community (Fig. 3c): If it is too good a bacterial predator, it will remove so much of the bacterial population that the next best osmotroph, *ie* the specialized photoautotrophs, may become dominant. Using a simple Lotka-Volterra formulation, one may actually show (given a fixed environment) that there is an optimum mixotrophic choice of affinity for bacteria. A lower affinity will leave the advantage to the specialized phagotrophs, a higher affinity will lead to dominance by specialized photoautotrophs. Within the framework of simple microbial food-web models such as the one in Fig. 3c, a proper mixotrophic strategy can be shown to minimize the three other populations, channelling the majority of the available nutrients into the mixotroph biomass.

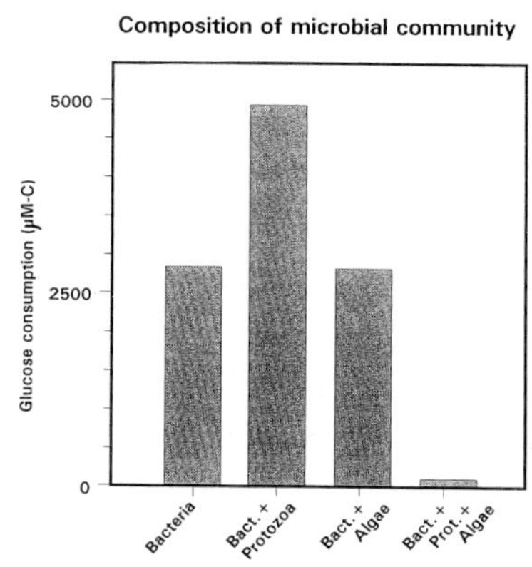

Fig. 4. Degradation of glucose in chemostats with different microbial communities (data from Pengerud *et al.* 1987). Bacterial degradation of glucose was •stimulated by adding bacterivorous colourless protists, •unaffected by adding strictly photoautotrophic algae alone, because these were outcompeted for phosphate, and •almost completely inhibited by adding both bacterivorous colourless protists and strictly photoautotrophic algae to the bacteria because of the simultaneous growth rate (competition) and biomass (predation) control of the bacteria.

Mixotrophy in ciliates

Ciliates are phagotrophic protists (primarily heterotrophic), which in different groups have developed a wide range of associations, permanent or not, with algae or plastids (Lindholm and Mürk 1989, Laval-Peuto 1992, Dolan 1992) or with bacteria (Finlay and Fenchel 1989, Fenchel and Bernard 1993). These ciliates can thus be mixotrophic, mixotrophy meaning "mixed nutrition". As mentioned above, mixotrophy in ciliates is generally understood as a combination of heterotrophy and photoautotrophy (Laval-Peuto 1992), by association with photoautotrophic organisms. The association can be permanent, for example, as in the freshwater ciliates *Paramecium bursaria* or *Platyophrya chlorelligera* that harbour intact algae (*Chlorella*) (Droop 1963, Weis 1982, Reisser 1986, Finlay *et al.* 1987, Kawakami 1991, Dolan 1992), or as in *Mesodinium rubrum* with a modified cryptophyte (Taylor 1982, Lindholm 1985, Lindholm *et al.* 1988). *M. rubrum*, however, is the only known non-phagotrophic ciliate. The association can also be temporary, as is the case with the ciliates, which retain functional plastids that have to be renewed regularly (Blackbourn *et al.* 1973, Laval-Peuto and Febvre 1986, McManus and Fuhrmann 1986b, Jonsson 1987, Stoecker *et al.* 1987, 1988). In his recent review, Laval-Peuto (1992) followed the general definition of plastid-retaining protists, plastid retention being not a particularity of the ciliates, since it is encountered as well in other groups of protists like flagellates (see above), foraminifers (Lopez 1979) and heliozoa (Patterson and Dürrschmidt 1987. The rest of this section concerns these plastid-retaining ciliates, which are both phagotrophic and photoautotrophic.

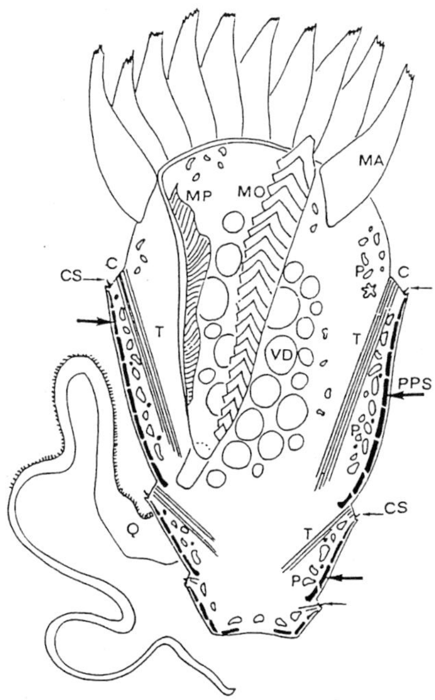

Fig. 5. Longitudinal section of *Tontonia appendiculariformis* (plastid-retaining strombidiid) showing the general organization of the ciliate and the localisation of the plastids retained at the periphery of the cell. MA: adoral membranelles, MO: oral membranelles, MP: paroral kinety, C: helicoidal belt, Q: tail (from Laval-Peuto *et al.* 1994, with permission).

Plastid-retaining ciliates have been known since the description by Blackbourn *et al.* (1973) of a *Prorodon* and a *Strombidium* containing intact algal plastids. The plastid-retaining ciliates have since triggered numerous studies (Dolan 1992, Laval-Peuto 1991, 1992, Sanders 1991b). The majority of these studies concerned the plastid-retaining ciliates' ultrastructure (Laval-Peuto *et al.* 1986, Laval-Peuto 1992) and metabolism (photosynthesis, respiration - McManus and Fuhrman 1986b, Jonsson 1987, Putt 1990a, Stoecker and Michaels 1991), while less work tackled their ecology (Stoecker *et al.* 1987, Bernard and Rassoulzadegan 1994 and references therein). Except for *Prorodon* (Holophryidae) (Blackbourn *et al.* 1973), all the plastid-retaining ciliates belong to the family Strombidiidae (Oligotrichida) (Fig. 5), which dominates the marine planktonic populations of ciliates (Smetacek 1981, Stoecker *et al.* 1987, 1989, Verity and Vernet 1992, Nielsen and Kiørboe 1994) and is also present in freshwater environments (Finlay *et al.* 1988, Rogerson *et al.* 1989).

The plastid-retaining ciliates can be easily recognized with fluorescence microscopy by the natural fluorescence of the surviving plastids (Haas 1982), which should not be mistaken for prey ingested (Laval-Peuto and Rassoulzadegan 1988, Bernard and Rassoulzadegan 1994). Nevertheless, this way of characterization has some limits, because it has been shown that fluorescence is not always equivalent to plastid retention, and that plastid retention is not always equivalent to photosynthetic activity (e.g., *Strombidium vestitum* - Jonsson 1987). The plastid-retaining ciliates seem to be spread over the photic zone of the oceans, from the Antarctic (Stoecker *et al.* 1992) to the Arctic Oceans (Auf dem Venne 1990, Putt 1990b), while they appear mostly in temperate and subtropical regions (Laval-Peuto and Rassoulzadegan 1988, Stoecker 1991, and references therein). They may form dense populations and may represent a significant percentage of the populations of ciliates present in the water column: 58-65% and 14-28% of the ciliate numerical abundance at the surface and at 50 m, in drift ice in polar areas, respectively (Putt 1990b), and 51% on an annual average of the volume of the planktonic ciliates in the Ligurian Sea (Fig. 6) - NW Mediterranean Sea (Bernard and Rassoulzadegan 1994).

Their seasonal distribution has been reported in only two ecosystems: the waters of Great Harbour, USA (Stoecker *et al.* 1987) and the Ligurian Sea (Bernard and Rassoulzadegan 1994). A clear seasonal pattern was shown, plastid-retaining ciliates being less abundant in winter and autumn than in spring and summer. During the latter, they can represent 100% of the volume of the planktonic ciliates in Mediterranean Sea (Bernard and Rassoulzadegan 1994), and a maximum of 81% in Great Harbour (Stoecker *et al.* 1987). This ecological success might be explained by the advantages that the plastid-retaining ciliates get from association with the plastids.

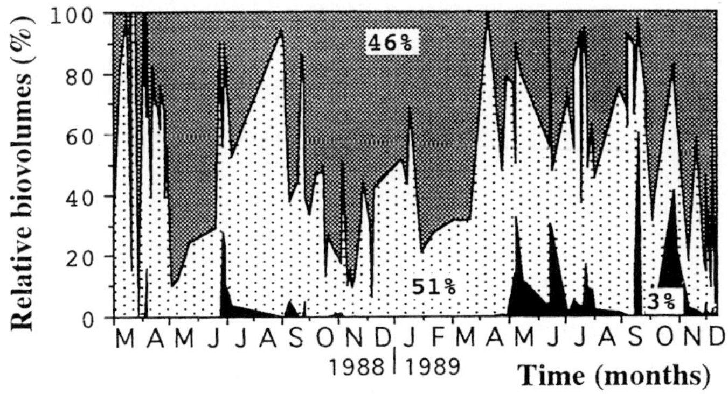

Fig. 6. Seasonal distribution of the relative biovolumes of oligotrichs (strict heterotrophs [medium shading] and mixotrophs [dotted shading]) and of *Mesodinium rubrum* [dark shading] in surface waters of NW Mediterranean Sea, from March 1988 to December 1989. From Bernard and Rassoulzadegan (1994 with permission).

In addition to using the phagotrophic pathway, plastid-retaining ciliates are capable of photoautotrophy. They can isolate the plastids from the ingested prey and transport them to the periphery of the cell, where they keep up their photosynthetic activity for a while (Laval-Peuto 1991). To date, the mechanisms of retention and isolation of the plastids are not known. Most attention has been given to the ultrastructure of the plastids by transmission electron microscopy (Blackbourn *et al.* 1973, Laval-Peuto and Febvre 1986, Laval-Peuto *et al.* 1986, Stoecker and Silver 1987, Laval-Peuto and Salvano 1991), the photosynthetic pigments they contained (McManus and Fuhrman 1986b), their photosynthetic activity (Jonsson 1987, Stoecker *et al.* 1987, 1988, 1989, Putt 1990a, b) and the photosynthates produced (Laval-Peuto 1991, Laval-Peuto and Salvano 1991). The number of plastids varies from cell to cell, but can be a few hundred (Laval-Peuto and Febvre 1986). The plastids show diversity in size and shape. Plastids predominantly originate from dinophytes, prymnesiophytes and in a less extent from diatoms and chrysophytes (Blackbourn *et al.* 1973, Laval-Peuto and Febvre 1986, Stoecker *et al.* 1989). Some selectivity may nevertheless take place (selective feeding - Rassoulzadegan *et al.* 1988). The plastids are most of the time observed enclosed with 3 membranes, two belonging to the plastid (whatever its origin), and the third named "periplastidial membrane", which should have a role of protection against digestion (see Laval-Peuto 1992). The plastids are associated with any organelles of the cells, but are located in a Vacuolated Endoplasmic Reticulum (VER - Laval-Peuto 1975, 1992). They have always been observed intact, and their degeneration or digestion (elimination) have rarely been observed, suggesting that they are eliminated before they become senescent (Laval-Peuto and Febvre 1986). Division of the plastids within the ciliates has neither been observed (Laval-Peuto and Febvre 1986). Observations of starved ciliates, which are then unable to renew their plastids, gave an indication of the time of survival of the plastids in the cell: 14 days in *Strombidium sp.* and from 6 days to 2 weeks in *Laboea strobila* (Stoecker *et al.* 1987, 1988). But Stoecker (1991) measured a turnover of less than a day in *S. capitatum.* The potential advantages of this association have been delineated (Laval-Peuto and Salvano 1991, Laval-Peuto 1992). These authors suggested that the photosynthates produced by the retained plastids could be stored as reserves in the polysaccharide plates, which are situated in the cortex of the plastid-retaining strombidiids (Fig. 5).

The nature of the photosynthetic pigments can be easily observed by fluorescence microscopy (Haas 1982, Laval-Peuto and Rassoulzadegan 1988, Bernard and Rassoulzadegan 1994). The fluorescence always appeared bright, confirming the general good shape of the plastids. McManus and Fuhrman (1986b) found chlorophyll a and phycoerythrin in *Laboea strobila*; the origin (prokaryotic or eukaryotic) of the phycoerythrin was not explained. The photosynthetic activity has been measured with ^{14}C incorporation by several authors with different protocols (Jonsson 1987, Stoecker *et al.* 1987, 1989, Putt 1990a), leading to a wide range of results (Table 3). Photosynthetic rates might be comparable to those of algae of the same size (Stoecker *et al.* 1987, 1989). Putt (1990b) estimated the contribution of ciliate chlorophyll as <15% of total chlorophyll in polar areas in summer. In the Mediterranean Sea, however, where plastid-retaining ciliates may at times represent 100% of the ciliate population, this estimation could be enhanced. Light might be a major factor influencing the photoautotrophic efficiency of the plastid-retaining ciliates. It has an influence on the vertical distribution of the plastid-retaining ciliates in the water column (McManus and Fuhrman 1986b, Dale 1987, Stoecker *et al.* 1987, 1989). In the Mediterranean Sea, a positive correlation has been found (Bernard and rassoulzadegan, 1994) between the seasonal occurrence of the plastid-retaining ciliates and light.

Since the plastid-retaining ciliates are primarily phagotrophs, they need to ingest prey mainly to satisfy some nutritional requirements: carbon, nitrogen, and perhaps phosphorus (Stoecker *et al.* 1988), as well as trace elements (vitamins, *etc* Provasoli 1977), but also to renew the plastids. The photosynthetic activity could from time to time play a significant role in the total carbon budget of the ciliates, but satisfy no more than the minimum energy requirements (no growth - Jonsson 1987). This might allow the ciliates to survive and compete efficiently in ecosystems with low food concentration or patchiness, but good light intensity (Laval-Peuto 1991, Bernard and Rassoulzadegan 1994).

Table 3. Plastid-retaining ciliates. Rates of photosynthesis assimilation in oligotrichs obtained by different authors using ^{14}C assimilation with different protocols.

Authors	Species	Rates of photosynthesis ($pg\ C\ h^{-1}\ ind^{-1}$)
Jonsson 1987	*Strombidium sp.1*	11
	Strombidium conicum	12
	Laboea strobila	41
Stoecker *et al.* 1987	*Laboea strobila*	586
Stoecker *et al.* 1988	*Laboea strobila*	925
Stoecker *et al.* 1989	*Strombium acutum*	35
	Strombium capitatum	78
	Strombium chlorophilum	98
	Strombium conicum	60

Ecological Implications of Mixotrophy

In the previous sections, we documented that mixotrophic flagellates and ciliates constitute a significant part of the microbial biomass in marine environments, and that a substantial amount of material flows through these protists. These findings have methodological, conceptual and ecological implications.

Concerning the methodology, there are no well-accepted routine procedures to determine mixotrophy in nano-sized protists. Ciliates are most often classified as mixotrophs, strict heterotrophs, and photoautotrophs by means of epifluorescence microscopy (Rassoulzadegan *et al.* 1988). This technique, however, is not adequate to determine the proportion of the pigmented nanoflagellates that are mixotrophs. At present, there is no alternative to microscopic examination of uptake of fluorescently labelled particles. The measured rates are reasonably precise, but the accuracy is unknown, and variations are presumably caused by species-specific selectivity regarding particles.

An important conceptual implication is that mixotrophic flagellates and ciliates represent an additional route in the flow of material between microbial compartments. The present understanding of the flow of material between microorganisms in the water column includes the classical food web and the microbial loop (*eg.* Azam *et al.* 1983, Fenchel 1988). In marine environments, it has hitherto been assumed that bacteria are grazed mostly by strictly heterotrophic protists (Fenchel 1982, Davies and Sieburth 1984). Alternatively, Bratbak *et al.* (1992) demonstrated that marine viruses infected 30% of the bacteria, and that this process played an important part in the carbon budget. Since mixotrophic flagellates also graze bacteria, rates of protist grazing on bacteria are underestimated when they are exclusively based on strict heterotrophs. In terms of ecological science, mixotrophy in natural microbial assemblages puts severe constraints on the measurement and quantification of competition for and limitation of nutrients. Measurements of nutrient uptake in relation to turnover rates cannot be used to illustrate microbial nutrient limitation when mixotrophs are present. Even though the rates measured of photoautotrophic nutrient uptake would indicate nutrient limitation, a significant proportion of the protist biomass might be mixotrophs, and their growth is not necessarily limited by nutrients, because they supply their carbon and nutrient requirements by means of phagotrophic feeding modes.

Mixotrophic flagellates include a gradient of algae that covers a range from almost entirely heterotrophs to almost complete photoautotrophs. Mixotrophic flagellates near the heterotrophic extreme as well as plastid-retaining flagellates and plastid-retaining ciliates can be considered mixotrophic protists that rely primarily on phagotrophy. Their photoautotrophic feeding mode contributes little to their nutrient requirements, but might ensure their survival during periods with low density of prey particles. Mixotrophic flagellates near the photoautotrophic extreme rely primarily on photosynthesis, and phagotrophy may enhance their survival during periods of low light intensity or low mineral nutrient concentration.

The limited knowledge of factors controlling succession and growth of natural populations of mixotrophs leaves only room for a few speculations on their role in the water column. On the basis of simple chemostat experiments, a number of important predictions can be made on the success of mixotrophic flagellates in the presence of bacteria, strictly heterotrophic flagellates, and specialized photoautotrophs. In addition, it seems obvious that mixotrophic protists can survive periods with nutrient depletion, low light, and low density of particles better than their specialized heterotrophic/photoautotrophic counterparts. However, the limited number of results from field experiments suggest that mixotrophic protists are important under a variety of environmental conditions. Knowledge of factors controlling the success of natural populations of mixotrophs is urgently needed.

Acknowledgements
We wish to thank Michèle Peuto-Moreau for comments to the ciliate section. This study was supported by MAST II contract MAST-CT92-0031-MEICE.

References

Andersson A, Falk S, Samuelsson G, Hagström, Å (1989) Nutritional Characteristics of a Mixotrophic Nanoflagellate, *Ochromonas* sp. *Microb Ecol* **17:** 251-262

Auf dem Venne H (1990) Distribution of auto-, mixo-, and heterotrophic planktonic ciliates in the Greenland Sea in late spring and fall 1988. *Comm Meet Int Cou Explor Sea C.M.-ICES/L*:**10**

Azam F, Fenchel T, Field J, Gray JS, Meyer-Reil LA and Thingstad F (1983) The ecological role of water column microbes in the sea. *Mar Ecol Prog Ser* **10:** 257-263

Azam F and Fuhrmann JA (1984) Measurements of bacterioplankton growth in the sea and its regulation by environmental conditions: . In: *Heterotrophic activity in the sea.* Hobbie JE and Williams PJLeB (eds). Plenum Press New York 179-1196

Bennett SJ, Sanders RW, Porter KG (1990) Heterotrophic, autotrophic and mixotrophic nanoflagellates: seasonal abundances and bacterivory in a eutrophic lake. *Limnol Oceanogr* **35:** 1821-1832

Bernard C and Rassoulzadegan F (1994) Bacteria or microflagellates as a major food source for marine ciliates: possible implications for the microzooplankton. *Mar Ecol Prog Ser* **64:** 147-155

Berninger U-G, Caron DA, Sanders RW (1992) Mixotrophic algae in three ice-covered lakes of the Pocono Mountains, U.S.A. *Freshwater Biol* **28:** 263-272

Biecheler B (1936) Observation de la capture et la digestion des proies chez un péridinien vert. *Comptes rendus des séances de la Société de Biologie* **122:** 1173-1175

Biecheler B (1952) Recherches sur les Péridiniens. *Bull Biol Fr Belg* **36 (suppl.)**: 1-149

Bird DF and Kalff J (1986) Bacterial Grazing by Planktonic Lake Algae. *Science* **231:** 493-495

Bird DJ and Kalff J (1987) Algal phagotrophy: Regulating factors and importance relative to photosynthesis in *Dinobryon* (Chrysophyceae). *Limnol Oceanogr* **32:** 277-284

Bird DF and Kalff J (1989) Phagotrophic sustenance of a metalimnetic phytoplankton peak. *Limnol Oceanogr* **34:** 155-162

Blackbourn DJ, Taylor FJR, Blackbourn J (1973) Foreign organelle retention by ciliates. *J Protozool* **20:** 286-288

Bockstahler KR and Coats DW (1993a) Spatial and Temporal Aspects of Mixotrophy in Chesapeake Bay Dinoflagellates. *J Euk Microbiol* **40(1):** 49-60

Bockstahler KR and Coats DW (1993b) Grazing of the mixotrophic dinoflagellate *Gymnodinium sanguineum* on ciliate populations of Chesapeake Bay. *Mar Biol* **116:** 477-487

Bratbak G, Heldal M, Thingstad TF, Riemann B, Haslund, OH (1992) Incorporation of viruses into the budget of microbial C-transfer. A first approach. *Mar Ecol Prog Ser* **83:** 273-280

Cachon M, Cachon J, Cosson J, Greuet C, Huitorel P (1991) Dinoflagellate flagella adopt various conformations in response to different needs.

Biol Cell **71:** 175-182

Caron AC, Porter KG, Sanders RW (1990) Carbon, nitrogen, and phosphorus budgets for the mixotrophic phytoflagellate *Poterioochromonas malhamensis* (Chrysophyceae) during bacterial ingestion. *Limnol Oceanogr* **35(2):** 433-443

Caron AC, Sanders RW, Lim EL, Marrasé C, Amaral LA, Whitney S, Aoki RB, Porter KG (1993) Light-Dependent Phagotrophy in the Freshwater Mixotrophic Chrysophyte *Dinobryon cylindricum. Microb Ecol* **25:** 93-111

Currie DJ and Kalff J (1984) The relative importance of bacterioplankton and phytoplankton in phosphorus uptake in seawater. *Limnol Oceanogr* **29:** 311-321

Davies PG and Sieburth JMcN (1984) Estuarine and oceanic microflagellate predation of actively growing bacteria: Estimation by frequency of dividing-divided bacteria. *Mar Ecol Prog Ser* **19:**237-246

Dale T (1987) Diel vertical distribution of planktonic ciliates in Lindaspollene, western Norway. *Mar Microb Food Webs* **2:** 15-28

Dodge JD and Crawford RM (1970) The morphology and fine structure of *Ceratium hirundinella* (Dinophyceae). *J Phycol* **6:** 137-149

Dolan JR (1992) Mixotrophy in ciliates: a review of *Chlorella* symbiosis and chloroplast retention. *Mar Microb Food Webs* **6:** 115-132

Droop MR (1963) Algae and invertebrates in symbiosis. *Symp Soc Gen Microbiol* **13:** 171-199

Estep KW, Davis PG, Keller MD, Sieburth JMcN (1986) How important are algal nanoflagellates in bacterivory? *Limnol Oceanogr* **31:** 646-50

Fenchel T (1988) Marine plankton food chains. *Ann Rev Ecol Syst* **19:** 19-38

Fenchel T (1991) Flagellate design and function. In: *The Biology of Free-living Heterotrophic Flagellates.* Patterson DJ and Larsen J:(eds) Syst Ass 45. Clarendon Press. Oxford 7-19

Fenchel T and Bernard C (1993) Endosymbiotic purple nonsulphur bacteria in an anaerobic ciliated protozoan. *FEMS Microbiol Lett* **110:** 21-25

Finlay BJ, Berninger U-G, Stewart LJ, Hindle RM, Davison W (1987) Some factors controlling the distribution of two pond-dwelling ciliates with algal symbionts (*Frontonia vernalis* and *Euplotes daidaleos*). *J Protozool* **34:** 349-356

Finlay BJ and Fenchel T (1989) Everlasting picnic for protozoa. *New Scientist.* **1671:** 66-69

Finlay BJ, Clarke KJ, Cowling AJ, Hindle RM, Rogerson A, Berninger U-G (1988) On the abundance and distribution of protozoa and their food in a productive freshwater pond. *Europ J Protistol* **23:** 205-217

Gourret P (1883) Sur les Péridiniens du golfe de Marseille. *Annales du musée d'histoire naturelle de Marseille. Zoologie.* T.1

Green JC (1991) Phagotrophy in prymnesiophyte flagellates. In: *The Biology of Free-living Heterotrophic Flagellates.* Patterson, DJ and Larsen, (eds) *J: Syst Ass* **45.** Clarendon Press. Oxford 21-38

Haas LW (1982) Improved epifluorescence microscopy for observing planktonic microorganisms. *Ann Inst Océanogr, Paris* **58:** 261-266

Hall JA, Barett DP, James MR (1993) The importance of phytoflagellate, heterotrophic flagellate and ciliate grazing on bacteria and picophytoplankton sized prey in a coastal marine environment. *J Plankt Res* **15(9):** 1075-1086

Hällfors G and Niemi Ä (1974) A *Chrysochromulina* (Haptophyceae) bloom under the ice in the Tvärminne archipelago, southern coast of Finland. *Memoranda Societatis pr Fauna et Flora fennica* 50: 89-10

Hansen PJ (1991) *Dinophysis* - a planktonic dinoflagellate genus which can act both as a prey and a predator of a ciliate. *Mar Ecol Prog Ser* **69:** 201-204

Havskum H and Riemann B (unpubl.) The ecological importance of mixotrophic flagellates in the Bay of Aarhus, Denmark.

Hobbie JE and Williams PJLeB (1984): *Heterotrophic activity in the sea.* Plenum Press. New York

Hofeneder H (1930) Über de animalische Ernährung von *Ceratium hirundinella* O.F. Müller und über die Rolle des Kernes bei dieser Zellfunktion. *Arch Protistenkunde* **71:** 1-32

Ishida Y and Kimura B (1986) Photosynthetic phagotrophy of Chrysophyceae: evolutionary aspects. *Microbiological Sciences* **3(5):** 132-135

Jacobson D and Anderson D (1986) Thecate heterotrophic dinoflagellates: feeding behavior and mechanisms. *J Phycol* **22:** 249-258

Jones HLJ, Leadbeater BSC, Green, JC (1993) Mixotrophy in marine species of *Chrysochromulina* (Prymnesiophyceae): Ingestion and digestion of a small green flagellate. *J Mar Biol Ass UK* **73:** 283-296

Jonsson PR (1987) Photosynthetic assimilation of inorganic carbon in marine oligotrich ciliates (Ciliophora, Oligotrichina). *Mar Microb Food Webs* **2:** 55-68

Jumars P, Penry DL, Baross JA, Perry MJ, Frost BW (1989) Closing the microbial loop: dissolved carbon pathway to heterotrophic bacteria from incomplete ingestion, digestion and absorption in animals. *Deep-Sea Res.* **36:** 483-495

Kawachi M, Inouye I, Maeda O, Chihara M (1991) The haptonema as a food-capturing device: observations on *Chrysochromulina hirta* (Prymnesiophyceae). *Phycologia* **30(6):** 563-573

Kawakami H (1991) An endosymbiotic *Chlorella*-bearing ciliate: *Platyophora chlorelligera* Kawakami 1989. *Eur J Protistol* **26:** 245-255

Kugrens P and Lee RE (1990) Ultrastructural Evidence for Bacterial Incorporation and Myxotrophy in the Photosynthetic Cryptomonad *Chroomonas pochmanni* Huber-Pestalozzi (Cryptomonadida). *J Protozool* **37(4):** 263-267

Larsen J (1988) An ultrastructural study of *Amphidiniumpoecilochroum* (Dinophyceae), a phagotrophic dinoflagellate feeding on small species of cryptophytes. *Phycologia* **27:** 366-377.

Laval-Peuto M (1975) Cortex, perilemme et réticulum vésiculeux de *Cyttarocylis brandti* (Cilié Tintinnide). Les cilies ê périlemme. *Protistologica* **11:** 83-98

Laval-Peuto M (1991) Symbiose plastidiale et Mixotrophie des Ciliés planctoniques marins oligotrichina (Ciliophora). Thèse Doctorat Etat, Université Nice-Sophia Antipolis. Fasc. 1: mémoire, 176 p. Fasc. 2: 13 articles, 191 p.

Laval-Peuto, M (1992) Plastidic protozoa. In:*Algae and symbioses*. Reisser, W (ed.) Biopress Ltd, Bristol, 471-499

Laval-Peuto M and Febvre M (1986) On plastid symbiosis in *Tontonia appendiculariformis* (Ciliophora, Oligotrichina). *BioSystems* **19:** 137-158

Laval-Peuto M and Rassoulzadegan F (1988) Autofluorescence of marine planktonic Oligotrichina and other ciliates. *Hydrobiologia* **159:** 99-110

Laval-Peuto M and Salvano P (1991) Cytochemistry of polysaccharides stored by ciliates in symbiosis with plastids (Ciliophora, Oligotrichina). *Symbiosis* (submitted)

Laval-Peuto M, Salvano P, Gayol P, Greuet C (1986) Mixotrophy in marine planktonic ciliates: ultrastructural study of *Tontonia appendiculariformis* (Ciliophora, Oligotrichina). *Mar Microb Food Webs* **1:** 81-104

Lindeman RL (1942) The trophic dynamic aspect of ecology. *Ecology* **23:** 399-418

Lindholm T (1985) *Mesodinium rubrum* - a unique photosynthetic ciliate. *Adv Aquat Microb* **3:** 1-48

Lindholm T, Lindroos P, Mürk A-C (1988) Ultrastructure of the photosynthetic ciliate *Mesodinium rubrum. BioSystems* **21:** 141-149

Lindholm T and Mürk, A-C (1989) Symbiotic algae and plastids in planktonic ciliates. *Memoranda Soc Fauna Flora Fennica* **65:** 17-22

Lopez E (1979) Algal chloroplasts in the protoplasm of three species of benthic foraminifera: taxonomic affinity, viability and persistence. *Mar Biol* **53:** 201-211

McManus GB and Fuhrmann JA (1986a) Bacterivory in seawater studied with the use of inert fluorescent particles. *Limnol Oceanogr* **31:** 420-426

McManus GB and Fuhrmann JA (1986b) Photosynthetic pigments in the ciliate *Laboea strobila* from Long Island Sound, USA. *J Plankt Res* **8:** 317-327

Moestrup, Ø and Andersen, RA (1991) Organization of heterotrophic heterokonts. In: *The Biology of Free-living Heterotrophic Flagellates*. Patterson DJ and Larsen J (eds) : *Syst Ass* **45**. Clarendon Press. Oxford: 333-360

Nielsen TG and Kiørboe T (1994) Regulation of zooplankton biomass and production in a temperate, coastal ecosystem. 2. Ciliates. *Limnol Oceanogr* **39:** 508-519

Nyggard K and Tobiesen A (1993) Bacterivory in algae: A survival strategy during nutrient limitation. *Limnol Oceanogr* **38(2):** 273-279

Olrik K and Nauwerck A (1993) Stress and disturbance in the phytoplankton community of a shallow, hypertrophic lake. *Hydrobiologia* **249:** 15-24

Parke M, Manton I, Clarke B (1956) Studies on marine flagellates. III. Three further species of *Chrysochromulina. J Mar Biol Ass UK* **35:** 387-414

Pascher A (1917) Flagellaten und Rhizopoden in ihren gegenseitigen Beziehungen. *Archiv für Protistenkunde* **38:** 1-87

Patterson DJ and Dürrschmidt M (1987) Selective retention of chloroplasts by algivorous heliozoa: Fortuitous chloroplast symbiosis? *Europ J Protistol* **23:** 51-55

Pengerud B, Skjoldal EF, Thingstad TF (1987) The reciprocal interaction between degradation of glucose and ecosystem structure. Studies in mixed chemostat cultures of marine bacteria, algae, and bacterivorous nanoflagellates. *Mar Ecol Prog Ser* **35:**111-117

Porter KG (1988) Phagotrophic phytoflagellates in microbial food webs. *Hydrobiologia* **159:** 89-97

Pringsheim EG (1952) On the nutrition of *Ochromonas*. *Quart J Microscopical Soc* **93 (1):** 71-96

Provasoli L (1977) Cultivation of animals. In: *Marine Ecology,* Vol. 3 Kinne, O (ed), Wiley, New York 1295-1320.

Putt M (1990a) Metabolism of photosynthate in the chloroplast-retaining ciliate *Laboea strobila. Mar Ecol Prog Ser* **60:** 271-282

Putt M (1990b) Abundance, chlorophyll content and photosynthetic rates of ciliates in the Nordic Seas during summer. *Deep Sea Res* **37:** 1713-1731

Rassoulzadegan F, Laval-Peuto M, Sheldon RW (1988) Partitioning of the food ration of marine ciliates between pico- and nanoplankton. *Hydrobiologia* **159:** 75-88

Reisser W (1986) Endosymbiontic associations of freshwater protozoa and algae. In: *Progress in Protistology.* Corliss JO and Patterson DJ (eds) Biopress Ltd. Bristol 195-214

Riemann, B and Søndergaard, M (1986) Regulation of bacterial secondary production in two eutrophic lakes and in experimental enclosures. *J Plankt Res* **8:** 519-536

Rogerson A, Finlay BJ, Berninger U-G (1989) Sequestered chloroplasts in the freshwater ciliate *Strombidium viride* (Ciliophora: Oligotrichida). *Trans Amer Micr Soc* **108:** 117-126

Rublee PA and Gallegos CL (1989) Use of fluorescently labelled algae (FLA) to estimate microzooplankton grazing. *Mar Ecol Prog Ser* **51:** 221-227

Salonen K and Jokinen S (1988) Flagellate grazing on bacteria in a small dystrophic lake. *Hydrobiologia* **161:** 203-209

Sanders RW (1991a) Trophic strategies among heterotrophic flagellates. In:*The Biology of Free-living Heterotrophic Flagellates* Patterson DJ and Larsen J (eds) . Syst Ass. 45 Clarendon Press Oxford: 21-38

Sanders RW (1991b) Mixotrophic protists in marine and freshwater ecosystems. *J Protozool* **38:** 76-81

Sanders RW, Porter KG, Bennett SJ, DeBiase AE (1989) Seasonal patterns of bacterivory by flagellates, ciliates, rotifers, and cladocerans in a freshwater planktonic community. *Limnol Oceanogr* **34:** 673-687

Sanders RW, Porter KG, Caron DA (1990) Relationship Between Phototrophy and Phagotrophy in the Mixotrophic Chrysophyte *Poterioochromonas malhamensis*. *Micro Ecol* **19:** 97-109

Schnepf E, Deichgräber G, Drebes G (1985) Food uptake and the fine structure of the dinophyte *Paulsenella* sp., an ectoparasite of marine diatoms. *Protoplasma* **124:** 188-204

Schnepf E, Winter S, Mollenhauer D (1989) *Gymnodinium aeruginosum* (Dinophyta): a blue-green dinoflagellate with a vestigial, anucleate, cryptophycean endosymbiont. *Pl Syst Evol* **164:** 75-91

Smetacek VS (1981) The annual cycle of protozooplankton in the Kiel Bight. *Mar Biol* **63:** 1-11

Spero HJ (1982) Phagotrophy in *Gymnodinium fungiforme* (Pyrrhophyta): the pedunkel as an organelle of ingestion. *J Phycol* **18:** 356-360

Stoecker DK (1991) Mixotrophy in marine planktonic ciliates: physiological and ecological aspects of plastid retention by oligotrichs. In: *Protozoa and their role in marine processes* Reid PC, Turley CM, Burkill PH (eds), G25. NATO ASI Series, Springer-Verlag, Berlin, Heidelberg 161-179

Stoecker DK, Buck KR, Putt M (1992) Changes in the sea-ice brine community during the spring-summer transition, McMurdo Sound, Antarctica. I. Photosynthetic protists. *Mar Ecol Prog Ser* **84:** 265-278

Stoecker-, DK and Michaels, AE (1991) Respiration, photosynthesis and carbon metabolism in planktonic ciliates. *Mar Biol* **108:** 441-447

Stoecker DK, Michaels AE, Davis LH (1987) Large proportion of marine planktonic ciliates found to contain functional chloroplasts. *Nature* **326:** 790-792

Stoecker DK and Silver MW (1987) Chloroplast retention by marine planktonic ciliates. In: *Endocytobiology III*, Vol **503** Lee, JJ and Fredericks, J (eds), Ann New York Acad Sci 562-565

Stoecker DK, Silver MW, Michaels AE, Davis LH (1988/1989) Enslavement of algal chloroplasts by four *Strombidium* spp. (Ciliophora, Oligotrichida). *Mar Microb Food Webs* **3:** 79-100

Stoecker DK, Taniguchi A, Michaels AE (1989) Abundance of autotrophic, mixotrophic and heterotrophic planktonic ciliates in shelf and slope waters. *Mar Ecol Prog Ser* **50:** 241-254

Suttle CA, Fuhrman JA, Capone DG (1990) Rapid ammonium cycling and concentration-dependent partitioning of ammonium and phosphate: Implications for carbon transfer in planktonic communities. *Limnol Oceanogr* **35:**424-433

Swale EMF (1969) A study of the nanoplankton flagellate *Pedinella hexacostata* Vysotskii by light and electron microscopy. *Br Phycol J* **4:** 65-86

Taylor FJR (1982) Symbioses in marine microplankton. *Ann. Inst. Océanogr., Paris* **58:** 6190

Thingstad TF and Pengerud B (1985) Fate and effect of allochthonous organic material in aquatic microbial ecosystems. An analysis based on chemostat theory. *Mar Ecol Prog Ser* **21:**47-62

Thingstad TF, Skjoldal EF, Bohne, RA (1993) Phosphorus cycling and algal-bacterial competition in Sandsfjord, western Norway. *Mar Ecol Prog Ser* **99:**239-259

Tranvik LJ, Porter KG, Sieburth, JMcN (1989) Occurrence of bacterivory in *Cryptomonas*, a common freshwater phytoplankter. *Oecologia* **78:** 473-476

Verity PG and Vernet M (1992): Microzooplankton grazing, pigments and composition of plankton communities during late spring in two Norwegians fjords. *Sarsia* **77:** 263-274

Wawrik F (1970) Mixotrophie bei *Cryptomonas borealis* Skuja. *Arch Protistenk* **112:** 312-313

Weis DS (1982) Protozoal Symbionts. In: *Experimental Microbial Ecology* Burns RG and Slater JH, (eds). pp. 320-341, Blackwell

Wehr JD, Brown LM, O'Grady K (1985) Physiological ecology of the bloom-forming alga *Chrysochromulina breviturrita* (Prymnesiophyceae) from lakes influenced by acid precipitation. *Can J Bot* **63:** 2231-2239

Wetherbee R and Andersen RA (1992) Flagella of a chrysophycean alga play an active role in prey capture and selection. Direct observations on *Epipyxis pulchra* using image enhanced video microscopy. *Protoplasma* **166:** 1-7

Wilcox LW and Wedemeyer GJ (1991) Phagotrophy in the Freshwater, Photosynthetic Dinoflagellate *Amphidinium cryophilum. J Phycol* **27:** 600-609

Wilson EO and Bossert WH (1971) *A primer of population biology.* Sinauer Associates Inc. Publishers. Stamford Connecticut

Successional Change in the Planktonic Vegetation: Species, Structures, Scales

Colin S. Reynolds
Freshwater Biological Association
NERC Institute of Freshwater Ecology
LA 22 0LP Ambleside, Cumbria
United Kingdom

Introduction

Plankton is the community of plants, animals and microbes which are adapted to live suspended in the pelagic of lakes and seas and which are liable to passive entrainment within the motion of the water. As is the case in terrestrial communities, such as forests or grasslands, planktonic communities, actually or potentially, comprise a considerable diversity of species with a variety of phylogenetic affinity, primary function, physiology, size and ecology. As our knowledge of pelagic organisms has steadily increased, however, it has become more difficult to make rigid distinctions between the phytoplankton, zooplankton and bacterioplankton and it has accordingly become less easy to distinguish the analogous community structuring of the communities. Moreover, debates have continued about the relative mechanistic roles of the autotrophs and heterotrophs in underpinning the biogeochemical processes upon which they impinge. In this paper, I attempt to show, primarily by reference to the freshwater pelagic photoautotrophs, how the adaptations of individual organisms influence the manner in which communities are assembled and altered through time and how, apart from the temporal and spatial scales at which they operate, these developments provide model examples of terrestrial vegetation processes. As the principal plant-life of open water, I seek to establish the propriety of referring it as pelagic vegetation.

Cell replication

The proper starting place for this consideration is at the point where the individual daughter unicellular alga is separated from its parent, often with a twin, the simultaneous product of mitotic division and binary fission of the mother. In order for it to reach the corresponding stage in its own life-cycle, when its mass is doubled and is ready to be apportioned between its own daughters, it will have to

NATO ASI Series, Vol. G 38
Molecular Ecology of Aquatic Microbes
Edited by I. Joint

have acquired the equivalent of its own mass of additional carbon, hydrogen, oxygen, nitrogen, phosphorus and approximately fifteen other elements, to have formed them into the correct proteins and lipids, then to have incorporated them into the proper cytological structures, destined to become those of the daughters and, finally, to have arranged them for apportionment among the new cells. All these processes occupy finite periods of time but some have been measured more precisely than others. Let us suppose that the active cell is offered all the nutrients it will require, in the most readily-assimilable forms and at concentrations saturating the maximum rates of uptake, under ideal conditions of continuous saturating light and constant, favourable temperature (say 20°C). To get round differences in the mode of cell-division among the various phylogenetic groupings represented among the planktonic species, let us also make certain further assumptions about the uniformity of the mass of the daughter cells, this being, in total, exactly twice that of the original parent. In spite of all this, we find that it still takes 9.05h for something the size of a *Chlorella* cell to double its biomass (Reynolds 1990).

The cell-specific growth rate, r, is derived from the doubling time, G:

$$\boldsymbol{r = \ln 2 / G}$$

or, in this case, $21.3 \times 10^{-6} s^{-1}$. Assuming equivalence with the carbon assimilation, that is 21.3μmol carbon incorporated per mol cell carbon every second.

It is functionally and ecologically important that the maximum rate of growth at this temperature varies among species and that the effect of altered temperature on the maximal rate of growth of each species differs too. For the moment, let us stay with our *Chlorella* and consider its means of gathering the raw materials to sustain even this rate of growth.

In our ideal environment, the usable carbon is fixed intracellularly from dissolved carbon dioxide, which is itself absorbed from the water bathing the cell. Given that the chlorophyll-specific photosynthetic rate is saturated by a photon flux of 10.9×10^{-3} mol photon g^{-1} chlorophyll *a* s^{-1} at $\sim 0.186 \times 10^{-3}$ mol C fixed g^{-1} chlorophyll *a* s^{-1} (Reynolds 1990) and that the chlorophyll content is a typical 1% of ash-free dry weight (or 240mg chlorophyll *a* mol^{-1} cell C), it may be calculated that the cell is capable of fixing 44.6×10^{-6} mol C (mol cell C)$^{-1}s^{-1}$ at a constant photon flux density of 125×10^{-6} mol photons $m^{-2}s^{-1}$. This is a rate sufficient to sustain the doubling of cell carbon in 4.3h. Indeed, if we assume even a 10% respiration and maintenance cost, it could be deduced that the same cell could have barely half the chlorophyll content or receive little more than half the saturating light intensity, and yet still maintain the maximal growth rate.

The need to acquire nutrients from solution may be subject to similar scrutiny. In the case of phosphorus, which the cell must acquire exclusively from the bathing medium, partly by molecular diffusion and in part, at least for cells larger than *Chlorella*, supplemented by active uptake (Reynolds 1994). Many workers have investigated these uptake rates into starved cells and found them consistently to conform to Monod-type kinetics, which are characterised in terms of S, the concentration of nutrient obtaining, K_v , the concentration required to sustain half the maximum rate of uptake (V_{max}). According to Nyholm (1977), V_{max} is equivalent to 13.5 x 10^{-18} mol P $(cell)^{-1}s^{-1}$, which, given a cell-carbon content of 0.61 x 10^{-12} mol C $(cell)^{-1}$, is 22.1 x 10^{-6} mol P (mol cell $C)^{-1}s^{-1}$; half this rate is achieved when the external concentration is (K_v =) 0.68 x 10^{-6} mol P l^{-1}, *ie* 21µg P l^{-1}).

The derived curves are often presented to support suppositions of the onset of phosphorus limitation but they do not do so. For if we assume that a replete, growing cell of *Chlorella* maintains a steady (Redfield-) ratio of intracellular phosphorus to carbon (1 mol P : 106 mol C), its uptake requirement is scaled to the growth rate of 21.3 x 10^{-6} s^{-1}, *ie*, it requires to take up 0.20 x 10^{-6} mol P s^{-1}, which the Monod model will allow down to concentrations of 6.3 x 10^{-8} mol P l^{-1} (2µg P l^{-1}) before growth rate is even slowed. Phosphorus has to be all but exhausted before rate-limitation could apply. On the other hand, the ability to obtain large amounts of phosphorus in short periods is another way of saying that it is collected opportunistically and stored intracellularly pending its demise from the water. This is why the Droop "internal stores" modification to the Monod model is a much better predictor of algal growth rate (Sommer 1991). Were the external concentration to be >1.5 x 10^{-6} mol P l^{-1} (>45µg P l^{-1}) and the maximum uptake rate to be maintained, the *Chlorella* cell could obtain, potentially, the requisite doubling of cell phosphorus within 430 s, or 7.1 minutes!

Similar reasoning underpinned Reynolds' (1990) view that, despite differences in affinity and kinetics of the various forms of dissolved inorganic nitrogen (DIN) species, an external concentration of 7 x 10^{-6} mol DIN l^{-1} (98µg N ml^{-1}) is probably sufficient to saturate the growth requirements of most species.

Population growth

Supposing that ecological success depended on superior competitive performances, of arriving first and reproducing fastest to the exclusion of other species, it would appear advantageous for phytoplankton species to adopt the dynamic capabilities of a *Chlorella,* or improve upon them. Yet there could be upwards of 4000 genera of freshwater phytoplankton, drawn from eight or (depending which authority is followed) more phyla of photoautotrophs and two kingdoms. Admittedly, most species are quite rare or sporadic in occurrence but some of them are sufficiently common and cosmopolitan to reveal consistent patterns of distribution, said to be indicative of particular water-body types, chemical composition, season, physical variability and community structure. How have these species survived at all or evolved in the first place? What features of the environment select for the periodic wax and wane in their abundance and, frequently, lead them to outcompete *Chlorella*?

I propose that the species have particular adaptations which suit them for growth in particular locations or at particular times: the fact that different kinds of algae are so associated either with given chemical or physical conditions is no more paradoxical than is the fact that scrub, grassland and forest exist simultaneously on the land surface; or that, given time, unmanaged temperate grassland will eventually be transformed into forest: time-scales apart, plankton environments are, to varying extents, subideal, are liable to successive colonisation and are subject to intermediate, allogenic disturbances which set back the succession to a less evolved state (Reynolds 1993a).

Adaptive criteria and selection

In earlier considerations of this problem (Reynolds 1988a, 1989a, 1991), I have used observational data from real seasonal sequences of successive dominants and from the sequences of the communities engineered in large limnetic enclosures (variously known as mesocosms, limnocorrals or Lund Tubes; see Lund and Reynolds, 1982) to demonstrate that the dynamic responses of phytoplankton populations are generally driven by the balance of the rates of biomass loss against that of cell anabolism. Moreover, this balance broadly conforms to one or other of three primary growth and survival strategies among phytoplankton species, each of which is consistently associated with defined morphological, physiological and behavioural properties of the organisms themselves. Similarly-adapted species, *ergo*, tend to respond analogously and simultaneously at the same, appropriate stages of the successional sequence. There is thus a satisfying circularity of adaptive cause and selective effect: the algae which are successful in certain types of water body or at certain times of the year share particular properties which are better suited to growth and survival under the conditions likely to obtain at those times of the year. Curiously, these properties largely, but not wholly, occur within all the main phylogenetic groupings of the algae, so that the functional identity of a species is often more ecologically relevant than is its taxonomic affinity.

Having once begun to diagnose the dynamic responses and to recognise that the morphological properties were not merely coincidental, I used alphanumeric symbols to refer to them (see Reynolds et al. 1983, Reynolds 1984), only later analogising them to Grime's (1979) primary *C, S* and *R* plant strategies. Even these remain the subject of controversy and the terms have scarcely been embraced by plankton biologists. Here, I attempt to re-affirm the validity of the original diagnosis by summarising the linkages between ecology, physiology and morphology of each basic category, as represented in Fig.1.

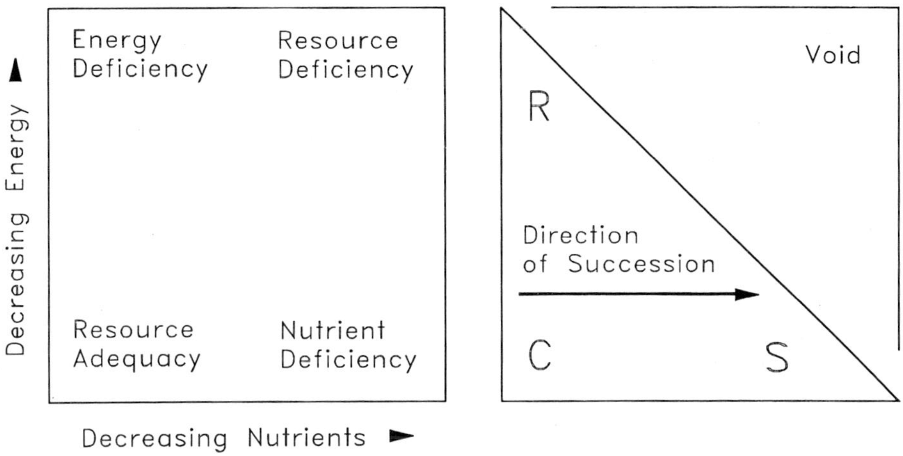

Fig 1 Environmental contingencies to be struck between adequacy or deficiency of nutrient resources and of photosynthetically-available energy, and the primary adaptive strategies (*C*, *S* or *R*)of algae to exploit them; note that no strategy copes with severe deficiency of both (cf. Grime 1979).

(i) INVASIVE (C) STRATEGISTS

Species which appear in new bodies of water, or soon after new hydrographic or chemical conditions are established, such as the onset of stable thermal stratification (direct or inverse). They are typically characterised by a relative ease of dispersal, which is assisted by small size (cells or coenobia generally $<10^3$ μm^3; diameter of sphere of same volume, $<12\mu m$); nutrient absorption, assimilation and rapidity of reproduction is assisted by favourable surface area/volume ratios ($>0.5\mu m^{-1}$; r at 20° C: $>10 \times 10^{-6} s^{-1}$;see Reynolds 1989a); grow successfully in cold water owing to a relative insensitivity to low temperature similarly favoured by high surface area/volume ratio (r at 0° C: $>1.0 \times 10^{-6} s^{-1}$); the area (of *eg* light-field) projected by cell is equivalent to at least 6.5 m^2 (mol cell C)$^{-1}$.

Typical representatives are picoplanktonic and nanoplanktonic algae such as *Chlorella, Ankyra, Koliella, Chlamydomonas, Synechococcus, Rhodomonas, Chrysochromulina, Monochrysis.* They may increase rapidly in numbers if they happen to be present when the nutrient resources saturate their requirements and the energy income exceeds their maintenance needs. The principal vulnerabilities of such cells is to their own relatively rapid basal metabolic rates ($\geq 1 \times 10^{-6}$ s^{-1} at 20°C) during prolonged darkness, to exhausting one or other nutrients supplied and to being directly grazed by planktonic ciliates, rotifers and, especially, larger filter-feeding Cladocerans such as *Daphnia*, populations of which can rapidly generate clearance rates exceeding those of algal growth.

(ii) ACQUISITIVE (S) STRATEGISTS

Species which dominate well-established, often deep, weakly-flushed bodies of water, but generally after substantial periods of direct thermal stratification. Poor dispersal is associated with relatively slow rates of growth and large size of unicells (among dinoflagellates) or colonial structures (cells or coenobia generally $>10^4$ μm^3; diameter of sphere of same volume, >30μm); nutrient absorption, assimilation and rapidity of reproduction is restricted by low surface area/volume ratio (<0.2μm^{-1} ; r at 20°C: $<8 \times 10^{-6}$s^{-1}); growth in cold water is weak owing to slow light-independent assimilation and transport in cells or colonies with low surface area/volume ratios (r at 0°C: $<0.4 \times 10^{-6}$s^{-1}); the areas projected by cells or by colonial units scarcely exceed 2.5 m^2 (mol cell C)$^{-1}$. We may deduce that most typical representatives (which include *Peridinium, Ceratium, Microcystis* and other colonial, bloom-forming Cyanobacteria and, arguably, such colony-forming algae as *Uroglena*) are mainly confined to warm, well-insolated shallow lakes and epilimnia. The organisms are, however, relatively robust; this, combined with a high affinity for nutrients in rarified environments, enables them not only to survive adverse conditions well but almost wholly to escape the attentions of filter-feeding crustaceans, except when they are still quite young. The power of controlled motility helps them to self-regulate position in the water column, to avoid excesses of irradiance and to locate and exploit deep reserves of nutrients, which they can take up and store in excess of immediate needs. Unlike the invasive species which invest in reproductive capacity (r-selected species of MacArthur and Wilson, 1967), these organisms allocate considerable resources to biomass preservation; almost all have well-developed means of perennation and build populations to the limit of the resource base (K-selection of MacArthur and Wilson, 1967).

(iii) ATTUNING (R) STRATEGISTS

These species are, variously, tolerant of, or are dependent upon, near-continuous entrainment within actively and deeply circulating mixed layers. They are usually pre-adapted to maximise their suspension opportunities, as well as to being able to maintain net growth in waters where the aggregate light dose is low and/or the photoperiods are severely truncated. Typical representatives of the group include most larger planktonic diatoms (*eg, Asterionella, Aulacoseira, Synedra*), planktonic desmids (*eg, Closterium, Staurastrum*) and Chlorococcalean genera (*eg, Pediastrum*) and solitary, filamentous members of the Oscillatoriacae (like

Planktothrix, Limnothrix and *Lyngbya* spp.). The principal trait common to these organisms is a high surface-to-volume ratio (generally >0.5μm^{-1}), attained not necessarily through small unit-size but by morphological complexity (filaments, attenuated or fenestrated coenobia, protuberances, *etc*) but offering the analogous consequences of rapid cell-specific uptake and growth rates (r at 20° C: >10 × $10^{-6} s^{-1}$), which mostly hold up well at low temperatures. The maximum mass-specific area projected at the appropriate orientation can be quite large (> 8 m^2 (mol cell C)$^{-1}$ in the case of 1-mm threads of *Planktothrix agardhii*; ~30 m^2 (mol cell C)$^{-1}$ in the case of an 8-celled colony of *Asterionella formosa*). This potential for light interception may be exploited in low-light environments for energy-harvesting advantage and where it is often considerably enhanced by photoadaptive measures. These include an exaggerated chlorophyll content and the production of a range of accessory pigments (xanthophylls, phycobilins), which gather light over a greater proportion of the spectrum than is absorbed by chlorophyll alone and to transfer the energy to the chlorophyll reaction centres; in this way, the energy harvested per mol cell carbon can be raised by a factor of 2-4 and nearly ten-fold in the case of *Planktothrix agardhii* (Reynolds 1994). Attuning species are most likely to dominate assemblages in cold or large, deep lakes which have consistently deep mixed layers (generally >10-20m), in continuously-mixed turbid, perhaps hypertrophic lakes and in a majority of other waters when thermal stratification weakens or is absent and turbulent mixing extends several meters down. They are well-represented in the spring-bloom periods of temperate lakes and seasonal convective episodes in the tropics (Lewis 1986). The species of phytoplankton thriving in larger rivers are drawn mainly from this group (Reynolds et al. 1994).

It should be noted that the group includes both *r*- and *K*- selected species. Thus, well- or frequently- mixed lakes can support a sequence of *R*-species which represents a true succession, even if only a plagioclimactic condition is ever reached (Reynolds 1993a). Those annual sequences culminating in dominant *Aulacoseira, Staurastrum* or *Planktothrix* provide good examples of progressions from an early phase of diatom dominance but with the eventual outcome influenced (respectively, in these instances) by physical variability and the reserve of soluble reactive silicon, by the continued clarity of a phosphorus-deficient epilimnion or by a continued light deficiency.

(iv) INTERMEDIATE STRATEGISTS

It must also be added that, unsurprisingly, not all phytoplankton species fit perfectly within one or other of the three categories but show properties intermediate between them. What is interesting, however, is that the intermediacy in morphological and physiological adaptation matches well the intermediacy in their ecologies. The 'space' between *C*- and *S*- species is occupied by genera such as *Dinobryon, Dictyosphaerium, Sphaerocystis, Gemellicystis,* arguably *Volvox* (see Reynolds 1983), *Eudorina, Aphanizomenon* and *Gloeotrichia* which diminish in surface-area/volume ratio, maximum growth rate and cold-tolerance but increase in their abilities to exploit and conserve nutrient resources. They are all organisms usually associated either with mid-successional or oligo- to meso-trophic summer assemblages (Reynolds 1993). By analogy, algae in the *C*- to *R*-axis include species of *Cyclotella, Scenedesmus,* and *Coelastrum* which could be said to be increasingly large, more convoluted and increasingly turbidity-tolerant algae. The *R-S* possibility is represented by relatively large, long-surviving acquisitive but self-regulating species which can persist for months to years on stable density gradients, notably *Planktothrix rubescens, P. mougeotii* and species of *Lyngbya* and *Phormidium* (Reynolds 1992). Finally, mention should be made of the genus *Cryptomonas* (note, not all Cryptomonads, as previously suggested; Reynolds 1984), which shows simultaneous characteristics close to those of *C*-, *S*- and *R*-strategists, in being (usually) unicellular with cells of moderate size (1 to $4 \times 10^3 \mu m^3$), of intermediate surface/volume 0.3 to $0.5 \mu m^{-1}$), in being capable of $r \sim 10 \times 10^{-6} s^{-1}$ at 20°C and $0.9 \times 10^{-6} s^{-1}$ at 0°C, and in projecting a maximum area of about $6 m^2$ (mol cell C)$^{-1}$. Moreover, cells are sufficiently motile to migrate or regulate position within a density gradient, are able to photoadapt to an extent but remain susceptible to grazing by crustaceans, some rotifers (*Brachionus, Polyarthra*) and large ciliates. Accordingly, *Cryptomonas* spp. are capable of dominating the spring bloom of small lakes (like *R*-species), being prominent in the (*C*-like) establishment of a post-stratification community and even to constitute dense, monspecific plates in meromictic lakes (Vicente and Miracle, 1988: a strong *S*-trait).

Assembly of communities

The species that happen to comprise the assemblage at the given moment that it is sampled for examination is best regarded as being fortuitous. The species represented will certainly have had to have been present and those which are relatively numerous can be assumed thitherto to have encountered conditions supportive of their growth. It cannot be assumed that those which are absent could not grow there, were they present, or that those which are present in small numbers are experiencing conditions which are perceived by them as being in any way adverse. Because the single sample does not even distinguish which populations are accreting and which are experiencing attrition, it cannot be apparent to which species the present conditions are favourable. Assuming resources (nutrients, energy, transport) to be adequate, the community remains a loose assembly of species each doing its best to increase its mass wherever and whenever the opportunity is presented.

These points may seem obvious but they are frequently ignored. They are a long way, for instance, from the implication advanced by some phytoplankton ecologists that the species are lined up awaiting the trigger of the right ratio of (nutrient) resources to make them grow.

Nevertheless, there *are* rules influencing the assembly of communities; these distinguish among the various "best" performances which, while the same given conditions persist, will bias the species structure of the assemblage. That is, of those present, algal species with the fastest *net* rates of increase or which field the greatest starting population are the ones most likely to characterise the structure of the community and, if not so already, to soon become dominant.

While the condition of resource adequacy holds, the ascendant species are most likely to be *r*-selected invasive *C*-species and fast-growing *R*-species. It is not easy to make a more robust prediction than that (as to what species will dominate, at what size of population and for how long), although the possibilities may reduce if the hindsight provided by records of the events of previous years is available.

However, it is no less certain that the "same given conditions" will not persist indefinitely. Change may come about for two main categories of reasons: those which occur in consequence of the ascendancy ("autogenic" causes) and those which emanate quite independently through external agencies ("allogenic" forcing).

Autogenic effects

Changes in the first category include the biomass-increase itself, which progressively imposes a self-shading effect, as the penetration of light into the water is reduced by the presence of more algae. In the case of an epilimnion or shallow lake of near-constant mixed depth, the species which will be able to maintain growth longest will be those best equipped to maintain residence in the layer and to adapt to a reducing biomass-specific energy quota, thus affording an advantage to photoadaptive *K*-selected *R*-species; if the mixing intensity is variable at the scale of hours-to-days, it may well be self-regulating *S*-species which are the more strongly favoured, through their ability to adjust to the sharp diel fluctuations in the underwater irradiance it is possible to receive (Reynolds 1989b).

Autogenic changes imposed by a growing population also include the progressive removal of nutrients from solution and into algal cells, leaving it increasingly difficult for cells recruited subsequently, of whatever species, to grow at all. Then, acquisitive (*S*-) species which, supposedly, maintain themselves for longer in the absence of the nutrient, have more opportunities to locate alternative sources deeper in the water column and can then store, intracellularly, any excess over the immediate needs. In short, they prove to be better competitors for the shrinking extracellular resource and, only then, are they likely to gain the dynamic advantage. Even this selective benefit might not express itself in terms of a significantly increased share of the community, were it not for the additional and simultaneous attrition of the dominant population. Standing populations of *C*-species may go into increasingly rapid decline once the rates of respirational loss or of their ingestion by grazers exceed the rate of cell replication, while the competing *S*-species experience fewer losses and can perhaps still continue net increase at their own modest rates. The processes, dynamics and resource-partitioning along the progression from *Ankyra*- to *Sphaerocystis*- and then to *Microcystis*-dominance, catalogued in Reynolds (1988b), or the steps implicit in the *Cryptomonas*→ *Eudorina*→ *Anabaena*→ *Ceratium* sequence observed in small eutrophic lakes (Reynolds 1984) provide good examples of autogenic community changes.

In a stratified lake, the combination of growth-depleted nutrient conditions near the surface, and the presence of a stable density gradient which suppresses the vertical hydraulic exchange of water is itself paralleled by a gradient of diminishing irradiance. There is also a series of biogenic exports of intracellular resources to depth (the sedimentation of live and dead cells, particulate products therefrom, as well as the faeces of planktonic grazers) which leads to an increasingly pronounced segregation of the lake into a nutrient-deficient epilimnion and an energy-deficient and, often, electrochemically-reducing hypolimnion, which layers are separated by a narrow continuum of differentiated metalimnetic microhabitats. Indeed, it may frequently transpire that this metalimnetic region is effectively the only productive part of the entire water column and becomes so distinguished by the presence of a "deep chlorophyll maximum" of living algae (Moll and Stoermer 1982). Finer resolution shows that the continuous scale of interacting growth-limiting factors complies readily with Tilman's (*eg* 1982) theory of selection through resource-based competition and, that given time (Reynolds 1992), also allows for competitive exclusion to operate finally and uniquely in favour of different species of algae, cyanobacteria and bacteria at the various available combinations of limiting light, nutrient and redox levels.

Planktonic successions

Such kinds of community restructuring broadly conform to many of the criteria said to be diagnostic of ecological succession (Odum 1969; Reynolds 1988b, 1993a). These include the progression of dominance from smaller to larger species, of short-lived species to slow-growing or slow-maturing ones, of species increasingly more complex in their life histories, physiology and perennation strategies, and to species better adapted to the environmental conditions as altered in consequence of the activities of those species dominating previously. It is also evident that the structure of the assemblage becomes decreasingly the outcome of the dynamics of any one species but that of an increasingly complex array of interactions among them and with an increasing order and organisation that these mutual relationships impart to the assembling community. The general course of succession is summarised in Fig.2.

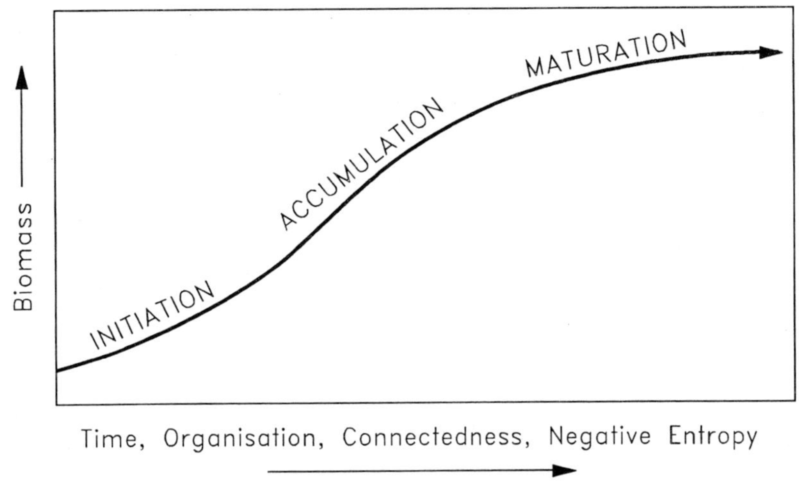

Fig 2 The course of ecological succession, from the initial low-biomass pioneer stages through to the maturation of a high-biomass climax with a single, competitively-excluded dominant. Organisation, connectedness and negative entropy are acquired through this series.

The real test of a plant succession, however, is the attainment of a recognisable, steady-state climax vegetation. So far as the pelagic environment of a lake is concerned, few plant ecologists can properly be satisfied by anything less than its complete extinction through the accumulation of biogenic sediment and centripetal hydroseral encroachment by macrophytes, reedswamp, fen carr and, ultimately, high forest. Yet, on its own, far less remote timescales, the planktonic vegetation can be shown to be capable of achieving the criteria of the climactic steady-state. In this context, the remarkable similarity among the planktonic communities of meromictic monimolimnia, of near-amictic permanently ice-covered polar lakes and of the metalimnia of sheltered, tropical forest lakes which remain stably stratified for over eleven months of every twelve, points to an achievable, enduring steady state, of global fidelity (Reynolds

1992). In each case, production and decomposition are balanced, the capacity-limiting nutrient is largely intrabiotic and its cycling is closed. It is of no matter that, within a few months or years, the community may have disappeared or have been destroyed by mixing, for when measured in terms of the generation times of the dominating organisms, the 15-25 required to establish a *Quercus*- or *Fagus*-dominated temperate woodland through much of the Devensian postglacial (*ie* the last 10,000 years), is comparable with those required to produce a monocultural, competitively-excluded *Microcystis*-bloom at the same latitude (Reynolds 1989) in half a summer! Year-to-year endurance of algal populations is akin to trees surviving an ice age, or returning after one.

Other enduring monocultural, competitively-excluded planktonic associations maturing about continuous environmental constraints are not unknown. Continuous rapid flushing by nutrient-rich ground water (*vide* the uninterrupted dominance of nanoplankton in Montezuma Well; Boucher *et al.*, 1984), the gradient of cooling hot-spring water across Lake Rotowhero, New Zealand (with its persistent population of *Chlorella:* Jolly and Brown 1975) and the relatively more frequently-recognised situation of a continuously-mixed, turbid, hypertrophic, shallow lake, being distinctively dominated by large populations of *Planktothrix* or *Limnothrix*; see Berger 1975) provide cases in point. In each of these instances, the steady-state selection depends not on a biogenic outcome, but is one maintained by an externally imposed forcing; the condition fits our understanding of a plagioclimax.

Allogenic effects: interrupted succession

The imposition of physically disruptive forcing is rarely so continuous but, then, the autogenic achievement of a climactic steady state is scarcely common. This final section attempts to reconcile the great majority of weakly-organised but species-diverse assemblages of phytoplankton to successional theory, as applied to plants with relatively very short generation times. Once it is appreciated that, in such accumulating, mid-successional assemblages, most of the specific populations then present are in a state of flux, either increasing or decreasing, then it is not difficult to accept that most communities are, at any point in time, in some stage between being dismantled and reassembled.

As has already been suggested, the cues for such structural reorganisations are set by alterations to the physical environment, imposed by such arbitrary factors as stochastic weather variations or quite sharp non-random factors like the onset or breakdown of seasonal stratification. Assuming that the newly-imposed conditions now favour different species and, simultaneously, target a new, steady-state outcome, it follows that the structure of the community will also alter, not immediately but progressively, as the dynamically advantaged species increase while those simultaneously disadvantaged may steady in numbers or even decline as they begin to die off.

Many examples of this supplanting of one successional sequence by external forcing have been given, for instance, the restoration of diatom dominance following wind-mixed episodes in small, stratifying lakes, or the onset of low biomass in lakes and rivers following vigorous flushing (for summaries, see Reynolds, 1984, 1993). Without prejudice to subsequent events, the immediate effects may include a loss of existing cells and a weakening of community structure, as well as a rejuvenation of productivity and nutrient availability.

The extent of the disruption is related to the power of the external force relative to the structural resilience of the community. Reduction to a more minimal, supportable structure, is taken as evidence of a disturbance to the original successional sequence (Pickett et al. 1989). Note, however, that the stimulus is not quantified in this definition. The new level of organisation may be permitted to "re-group" and restore a sequence not greatly dissimilar from that which had existed prior to the disturbance. Further disturbances of this nature could produce analogous reactions and, given the appropriate frequency, could prevent the community from ever advancing very much further towards any potential steady-state outcome. Instead, it is kept substantially short of any maturation by frequent intermediate disturbances.

That many ecosystems are characterised by the absence of steady-state dynamics was hypothesised by Connell (1978) to be thus attributable to such intermediate disturbances. Taking advantage of the short temporal scales at which pelagic communities operate, Padisák *et al.* (1993) presented a series of studies on plankton dynamics as a valid test of Connell's hypothesis. In particular, it may be invoked to explain the often high and, once, "paradoxical" (Hutchinson 1961) diversity of phytoplankton and the failure of biotic processes to bring about competitive exclusion. In addition, such biodiverse, incompletely evolved structures remain productive, organisationally weak but still some way from thermodynamic equilibrium owing to their frequent resetting at a more primitive stage than they had reached (Reynolds 1995). Ultimately, it *is* the frequency of abiotically-imposed environmental variability which decides the progress permitted in a *C-* to *S-* succession, and the extent to which it is either recapitulated or replaced by the tendency of R-species to maintain some kind of ecological plagioclimax (Reynolds 1993). These possibilities are represented in Fig. 3.

This elegant and useful theory can be invoked to explain the wide diversity of species characterising most natural ecosystems, as well as the fact that their species compositions are manifestly not subject to competitive exclusion.

However, the theory does have several practical flaws (Juhász-Nagy 1993) which need to be overcome if its potential value is to be realised. The most glaring of these is that disturbance is recognised by and, indeed, is measured exclusively by the biotic response; in this way, it is possible for the same magnitude of stimulus to have quite different effects on a poorly organised, early-successional stage from those on one already well-advanced; alternatively, the increased resistance of advancing successions may require a much stronger impetus to bring about an equivalent level of disruption, or, indeed, any at all (Reynolds et al., 1993). This is logically unsatisfactory and it is downgrading important information about the community organisation of the systems concerned.

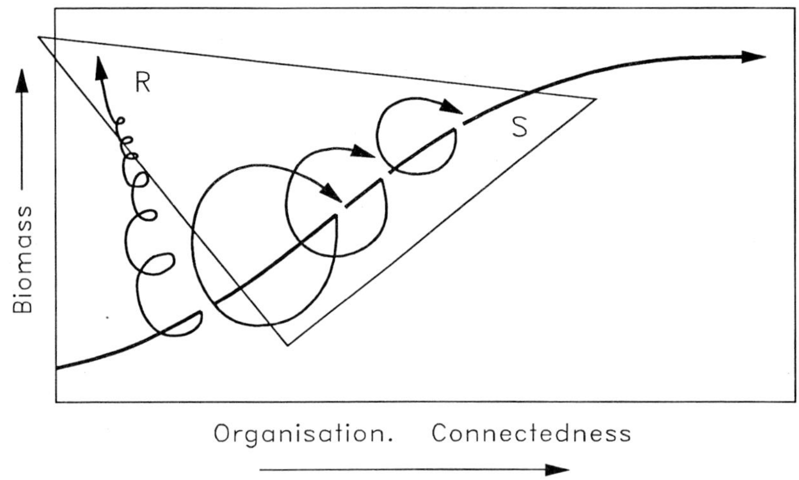

Fig 3 The superimposition of the *C-S-R* triangle from Fig 1 on the successional course from Fig 2 to show how species are selected from within the *C*- or *S*-groupings, how disturbances of varying intensity set back the succession and again select for more tolerant or faster re-colonising species, and how frequent or constant energy-deficient episodes select increasingly for *R* -species.

Thermodynamic perspective of community development

Nevertheless, a likely way forward may be founded on the structural-dynamic modelling approach of Jørgensen (1986) and the quantification of the extent to which the advancing succession has removed itself from the abiotic, thermodynamic equilibrium, wherein the radiant energy flux, E, is balanced exclusively by the dissipation of the energy as heat (Q_O). The intervention of a functional ecosystem, based on photoautotrophs, has the effect of siphoning off a part or, potentially, all the radiant energy flux into the production of biomass *(P)*, net of that proportion which is consumed in maintaining the biomass (M, mainly community respiration). The abiotic heat loss is now reduced to

$$Q_B = \mathrm{E} - [\,P - M\,];$$

the difference, Q_B - Q_o , is equivalent to the energetic distance by which the system is removed from thermodynamic equilibrium, which Jørgensen (1986) refers to as its `exergy' or, roughly, its *negative entropy flux*, here symbolised as *N*.

In the abiotic system, the influx of radiant energy (*E*) is balanced by the totality of heat loss, Q_O , such that:

$$N = 0\,;$$

In the presence of an early successional ecosystem,

$$N = Q_B - Q_O = E - [P - M] - E$$
$$= M - P$$

Note the negative value. In later successional stages, the value of *M* approaches that (albeit now rather greater) of *P*, until:

$$N = M - P = 0$$

While the ecosystem builds, *N* remains negative. Once the energetic cost and entropy of maintenance balances the biotically harvested energy income, the climactic condition recognised by Odum (1969) is properly achieved and defined thermodynamically.

We now incorporate allogenic forcing, *F*, either as a cumulative subtractive term in respect of (say) the energy of wind stress, or the effect of the reduction in energy income through the extension of the surface mixed layer and its impingement upon the access of planktonic autotrophs to light. Alternatively, seasonal changes in daylength and cloud cover can be effective by reducing P but not M. If we call that quantity *F,* we can then say that if:

$$N = (M + F) - P < 0$$

then the ecosystem can continue to assemble towards a climactic condition and no disturbance from *F* need have been evident. Once the value of *N* turns positive, some resetting to a more minimal state is inevitable. Of particular interest, however, is the fact that, in the early stage of succession, the values of *P*, *M* and *N* are all absolutely small and are so relative to the potential scale of *F*: early successional stages are vulnerable to external disturbance. As the ecosystem builds, however, the value of *F* required to overcome the increasing absolute value of (*M* - *P*) is also raised. This then offers a measure of the added resilience that successional advance imparts to ecosystem structure. In late succession, however, when $M \rightarrow P$, the resilience is accordingly reduced. Provided *N* remains negative, even late successions advance slowly to the steady state. If not, or the productive capacity is restricted by the supply of a basic nutrient, by the onset or collapse of stratification, or the succession is simply truncated or reversed (disturbed), the structure of the ecosystem must reorganise itself down to a level sustainable under the conditions obtaining.

Synthesis: scales of successional processes

These ideas are, as yet, fairly primitive and a good deal of refinement is necessary before they can be elevated to the level of a testable hypothesis. For the present, they do provide a framework for recognising that few ecosystems are fully evolved but are only so advanced as the time since the last time that the structure was effectively disturbed. The point that is made is usefully illustrated by comparison with the difference, in terms of floristic organisation and productivity, between scrub and forest, or between the relative ease of harvesting, every year or less, from the same area of land, a herbaceous crop compared to cropping timber and the time it takes to restore the croppable biomass. The species diversity associated with the area of land is greatest not when left alone but when the succession is disturbed at intervals measured in generation times of the key species(a year or so for pioneer species, decades for shrubs, centuries for trees). One has only to consider the recolonisation of the canopy gap left by a fallen forest tree or the regeneration of woodland after traditional coppicing to appreciate the mechanisms of disturbance and successional rejuvenation.

Applying these ideas to phytoplankton, where the key plants have generation times of hours to days, and where a calendar year can circumscribe numbers of generations comparable to those of forest-tree species since the last glaciation, the analogues abound. Among the numerous published studies of plankton periodicity, the several which have distinguished community-driven changes from responses to external forcing all take the starting point of a new succession after a large disturbance - a flood, or a storm, or the onset of thermal stratification (Trimbee and Harris 1983, Ashton 1985, Reynolds 1993a). The pattern of colonisation by *r*-selected nanoplankton and its replacement by increasingly *K*-selected *S*-strategists is well-established, even though the transitions may be very rapid (see, for instance, sequences presented by Viner and Kemp 1983 and Lewis 1986) or are very protracted (Boucher et al. 1984),or are frequently recapitulated, as is probably the case in many temperate lakes (Reynolds 1984). Elsewhere, it may require some 30-60 days to accommodate the 16-or-so generations required to establish a stable, dominant population, and many undisturbed months or years to develop vertical arrays of niche-differentiated, competitively-excluded monospecific layers (Reynolds 1992). Equally, where the successional development has to respond to (say) continuous mixing, the trend is from *r*-selected, attuning species to *K*-selected ones (*R*-strategists). Moreover, besides flood, the critical disturbing factor could be the onset of thermal stratification. The actual proportionate displacement or the absolute depths of mixing may determine the extent of the disturbance but the loss of spring diatom crops from rivers may be due to elevated discharge while, in lakes, it may be often ultimately be attributable to sinking out of truncated mixed layers (Reynolds et al. 1994). On the other hand, persistence of the initial conditions may eventually bring about competitive exclusion by *Planktothrix.*

In a majority of smaller lakes, there is discernable, stochastic weather-related oscillation in habitat structure and, hence, in community organisation, with a frequency perhaps sufficient to promote and select against species alternately in a way which keeps the maximum diversity. Sommer *et al.* (1993) estimated that the time taken for the newly-selected species to make an impact on the existing species structure (albeit probably in decline is about four generation times. Certainly, effective disturbances coming at intervals of 5-15 days achieve the highest diversity indices (Holzmann 1993, Sommer 1993). These, then, are truncated successions - the aquatic equivalent of coppiced woodland or, indeed, almost all terrestrial vegetation outside of tundra, desert and subpolar forest.

References

Ashton PJ (1985) Seasonality in southern hemisphere freshwater phytoplankton assemblages. *Hydrobiologia* **125:** 179-190

Berger C (1975) Occurrence of *Oscillatoria agardhii* Gomont in some shallow eutrophic lakes. *Verh int Verein Limnol* **19:** 2687-2697

Boucher P, Blinn DW, Johnson DB (1984) Phytoplankton ecology in an unusually stable environment (Montezuma Well, Arizona, USA). *Hydrobiologia* **119:**149-160

Connell J (1978) Diversity in tropical rain forests and coral reefs. *Science* **199:** 1304-1310

Grime JP (1979) *Plant strategies and vegetation processes.* Wiley-Interscience, Chichester

Holzmann R (1993) Seasonal fluctuations in the diversity and compositional stability of phytoplankton communities in small lakes in Upper Bavaria. *Hydrobiologia* **249:** 101-109

Hutchinson, GE (1961) The paradox of plankton. *Am Nat* **95:** 137-147

Jolly VH and Brown JMA (1975) *New Zealand Lakes.* Auckland University Press, Auckland

Jørgensen S E (1992) Structural dynamic eutrophication models. In *Eutrophication, research and application to water supply,* Sutcliffe D W, Jones J G (eds) Freshwater Biological Association, Ambleside 59-72

Juhász-Nagy P (1993) Notes on compositional diversity. *Hydrobiologia* **249:**173-182.

Lewis WM (1986) Phytoplankton succession in Lake Valencia, Venezuela. *Hydrobiologia* **138:** 189-203

Lund JWG and Reynolds CS (1982) The development and operation of large limnetic enclosures in Blelham Tarn, English Lake District and their contribution to phytoplankton ecology. In *Progress in phycological research* Vol 1, Round FE, Chapman DJ (eds) Elsevier, Amsterdam 1-65.

Moll RA and Stoermer EF (1982) A hypothesis relating trophic status and subsurface chlorophyll maxima in lakes. *Arch Hydrobiol* **94:** 425-440

Odum EP (1969) The strategy of ecosystem development. *Science* **164:** 262-270

Padisák J, Reynolds CS, Sommer U (eds) (1993) *Intermediate disturbance hypothesis in phytoplankton ecology*. Kluwer, Dordrecht

Reynolds CS (1983) Growth-rate responses of *Volvox aureus* Ehrenb. Chlorophyta, Volvocales) to variability in the physical environment. *Br phycol J* **18:** 433-442

Reynolds CS (1984) Phytoplankton periodicty: the interactions of form, function and environmental variability. *Freshwat Biol* **14:** 111-142

Reynolds CS (1988a) Functional morphology and the adaptive strategies of freshwater phytoplankton. In *Growth and reproductive strategies of freshwater phytoplankton,* Sandgren CD (ed) Cambridge, New York 388-433

Reynolds CS (1988b) The concept of ecological succession applied to the seasonal periodicity of phytoplankton. *Verh int Verein Limnol* **23:** 683-691

Reynolds C S (1989a) Physical determinants of phytoplankton succession. In *Plankton Ecology* Sommer U (ed), Brock-Springer, Madison 1-56

Reynolds CS (1989b) Relationships among the biological properties, distribution and regulation of production by planktonic Cyanobacteria. *Toxicity Ass* **4:** 229-255

Reynolds CS (1990) Temporal scales of variability in pelagic environments and the responses of phytoplankton. *Freshwat Biol* **23:** 25-53

Reynolds CS (1991) Lake communities: an approach to their management for conservation. In *The scientific management of temperate communities for conservation,* Spellerberg IF, Goldsmith FB, Morris MG (eds) Blackwell, Oxford 199-225.

Reynolds CS (1992) Dynamics, selection and composition of phytoplankton in relation to vertical structure in lakes. *Ergebn Limnol* **35:** 13-31

Reynolds CS (1993) Scales of disturbance and their role in plankton ecology. *Hydrobiologia* **249:** 157-171

Reynolds CS (1994) The role of fluid motion in the dynamics of phytoplankton in lakes and rivers. In *Aquatic Ecology - scale, pattern, processes* Giller PS, Hildrew AG, Raffaelli D (eds), pp.141-187. Blackwell, Oxford

Reynolds CS (1995) Successional development, energetics, and diversity in plankton communities. In *Ecological perspective of biodiversity*, Abe T (ed) Center for Ecolocical Research, Kyoto (in press)

Reynolds CS, Descy J-P, Padisák J (1994) Are phytoplankton dynamics in rivers so different from those in shallow lakes? *Hydrobiologia* **289:** 1-7

Reynolds CS, Wiseman SW, Godfrey BM, Butterwick C (1983) Some effects of artificial mixing on the dynamics of phytoplankton in large limnetic enclosures. *J Plankt Res* **5:** 203-234

Sommer U (1991) A comparison of the Droop- and the Monod-model of nutrient-limited growth applied to natural populations of phytoplankton. *Functional Ecol* **5:** 535-544

Sommer U (1993) Disturbance-diversity relationships in two lakes of similar nutrient chemistry but contrasted disturbance regimes. *Hydrobiologia* **249:** 59-65

Sommer U, Padisák J, Reynolds CS, Juhász-Nagy P (1993) Hutchinson's heritage: the diversity-disturbance relationship in phytoplankton. *Hydrobiologia* **249:**1-7

Tilman D (1982) *Resource competition and community structure.* Princeton University, Princeton

Trimbee AM and Harris GP (1983) Use of time-series analysis to demonstrate advection rates of variables in a small lake. *J Plankt Res* **5:** 819-833

Viner AB and Kemp L (1983) The effect of vertical mixing on the phytoplankton of Lake Rotongaio (July 1979 - January 1981). *New Zealand J mar freshwat Sci* **17:** 407-422.

Can Molecular Techniques Change Our Ideas About the Species Concept?

Linda K. Medlin[1], Martin Lange[1], Gary L. A. Barker[2], and Paul K. Hayes[2]

[1] Alfred-Wegener-Institute
Am Handelshafen, 12
D-27515 Bremerhaven
Germany

[2] School of Biological Sciences
University of Bristol
Bristol BS8 1UG
UK

Abstract

Molecular techniques can now be used to address many important questions concerning taxonomic affinity, genetic diversity, gene flow and dispersal. These data (protein or nucleic acid sequence information) can augment our understanding of a species in the marine environment. There is growing evidence that speciation and dispersal mechanisms in the marine environment are very different from terrestrial systems and occur at different rates. We have examined species/genetic diversity in three ecologically important members of the marine phytoplankton: the prymnesiophytes *Phaeocystis* and *Emiliania huxleyi* and the diatom *Skeletonema costatum*. All are high dispersal taxa. Differences among *Phaeocystis* species as measured by 18S rDNA sequence comparison indicate that extant *Phaeocystis* species probably arose from a cosmopolitan warm-water ancestor. Speciation events in this genus appear to have responded to major global cooling events. Two species complexes are apparent: one corresponds to taxa from polar regions and the other to taxa from temperate to tropical regions. *Emiliania huxleyi*, a much younger species, has dispersed across many oceanographic barriers during much colder climatic conditions. Sequence data from coding and non-coding regions confirm that it is a single taxon, but RAPD techniques reveal extensive genetic diversity with both spatial and short-term temporal resolution. Sequence comparison of isolates of *Skeletonema costatum*, a cosmopolitan neritic form, reveals at least one cryptic species, which can also be differentiated by certain morphological features, and a cluster of isolates that may be sibling species. It may be that in many planktonic forms molecular speciation has proceeded, whereas morphological divergence has not. The abundance of sibling species in the marine environment is only now being revealed (Knowlton

NATO ASI Series, Vol. G 38
Molecular Ecology of Aquatic Microbes
Edited by I. Joint

1993). The fitness of form resulting in similar if not identical morphotypes has analogies at all taxonomic levels in the sea (Knowlton 1993, Sournia 1988).

Introduction

The recognition of a species is one of the most complex and controversial topics in biology. Many definitions or concepts of a species are known (see reviews in Gosling 1994; Manhart and McCourt 1992), such as the morphological, biological, phylogenetic and palaeontological species. Of these, the morphological species is the most widely used. The biological species is often the most difficult to document especially in organisms such as marine phytoplankton where little is known of life histories and where most reproduction is asexual. The biological concept of a species is further complicated by the observation that for many marine plants, *eg* the red algae, broad hybridisation zones, albeit sometimes only unidirectional, can be documented between accepted morphological species. The phylogenetic species concept, defined as "an irreducible (basal) cluster of organisms, diagnosably distinct from other such clusters, and within which there is a parental pattern of ancestry and descent" (Cracraft 1989) represents monophyletic groupings and introduces an element of time into the species concept (Gosling 1994). Molecular data can be most easily incorporated into this form of the species concept, although they can be used to augment any of the other concepts. Notably, species may be differentiated based on molecular divergence but may not be reproductively isolated (syngens). It is also possible that apparent conflicts between phylogenetic/molecular speciation and morphological speciation may result from the different rates at which molecular and morphological changes accumulate.

In our work with marine phytoplankton species, we have used molecular markers from both coding and non-coding regions of the nuclear and plastid genomes to interpret species level differentiation based primarily on morphological and physiological characters. In order to evaluate how much variation in molecular markers can be tolerated within a species, we have compared variation in molecular markers with species level markers from other concepts. In the first instance, values obtained from DNA/DNA hybridisation studies were compared with corresponding variation in the small subunit (ssu) ribosomal RNA (rRNA) gene (16S and 18S rDNA) (Table 1). DNA/DNA hybridisation values of less than 80% similarity are generally taken to indicate that separate species are involved. Such species (strains within a single species) are usually separated by more than 4 nucleotides (99.8% similar) when sequences of their ssu rRNA genes are compared. Thus, DNA/DNA hybridisation can reveal cryptic species within morphological similar but physiologically distinct strains of unicellular algae and bacteria, although rDNA sequence comparisons may be more sensitive.

Table 1. Comparison of DNA/DNA hybridisation values and 16S - 18S RNA similarity

Species x Species	DNA/DNA hybridisation	rRNA similarity	Reference
Chlorella sorokiniana x *Chl. vulgaris*	12-15%	99.6%	Huss *et al.* 1986, 1989, 1991, unpubl
Chl.fusca v. *vac* x *Chl. kessleri*	7-14%	95.2%	"
Chl. vulgaris x *Chl. kessleri*	11-12%	98.4%	"
Chl. sorokiniana groups of strains			
1st group (3 strains)	87-98%	100%	"
2nd group (3 strains)	47-54%	not det.	"
3rd group (3 endo strains)	93-102%[a]	99.8-99.9%[b]	"
4th group	15-16%	99.5-99.8%	"
[a] endosymbionts from *Spongilla*		[b] endosymbionts from *Hydra*	
Cladophoropsis membranacea			
Canary Islands x Caribbean	80%	99.9%	Kooistra 1993
Fibrobacter succinogenes (15 strains)			
Group 1	95-100%	99.8-100%	Amann *et al.* 1992
Group 2	40-65%	99.1-99.8%	
Groups 3-4	<40%	<99%	

In the second instance, variation in the D2 region of the large subunit (lsu) rRNA gene between sibling species of two protists were compared (Preparata *et al.* 1992). Both the dinoflagellate *Crypthecodinium cohnii* and the protozoan *Tetrahymena pyriformis* were initially considered to be monospecific; however extensive breeding compatibility and isoenzyme studies have revealed that each can be divided into many sibling species. The variable D2 region of the lsu rRNA gene has now been sequenced from over 60 strains of both taxa (Beam *et al.* 1993) (Figs. 1-2). *C. cohnii*, composed of 46 sibling species, is represented by 9 different ribosets. The ribosets contain from one up to eleven different sibling species and are separated from one another by 1 - 22 nucleotide substitutions.

Notably, many sibling species share identical sequences for this region of the lsu rRNA gene and are placed in the same riboset. In contrast, each of the 25 sibling species comprising *T. pyriformis* has a unique sequence for the variable D2 region with 1 - 45 nucleotide differences separating the sibling species (Nanney *et al.* 1989).

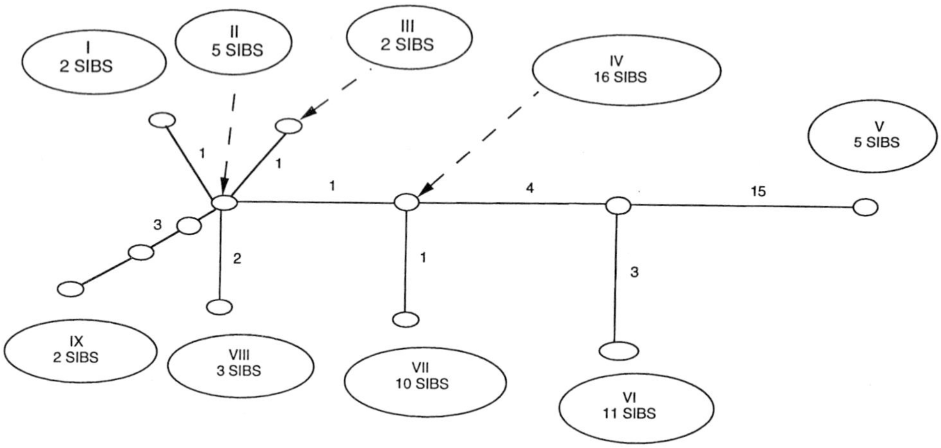

Fig. 1. Sequence differences in lsu rDNA's between major sibling species sets of *Crypthecodinium cohnii*, redrawn from Preparata *et al.* 1992. Numbers on lines indicate number of nucleotide substitutions from nearest branch point.

The relative age of the two sibling species swarms has been used to explain the differences in the accumulated nucleotide differences separating the sibling species. Thus, if the reproductive isolation of the sibling species is too recent, then the sibling species will not have coexisted long enough to accumulate nucleotide substitutions in either their ssu or lsu rDNA's. However, if polymorphisms in coding regions are observable within strains of a morphologically well-defined species, then the possibility of a species complex (sibling species) should be investigated.

We also know that local variation between individuals and populations does exist in both coding and non-coding regions of the genome. In comparisons of 18S rDNA sequences Bird *et al.* (1994) found up to 6 nucleotide differences separating isolates of the red algal species *Gracilaria verrucosa*; two other known species of *Gracilaria* [*G. pacifica* and *Gracilaria* sp. (North Carolina)] lie within the complex of *Gracilaria verrucosa* isolates in both the parsimony and maximum likelihood trees presented by Bird *et al.* 1994 (see her fig. 5),

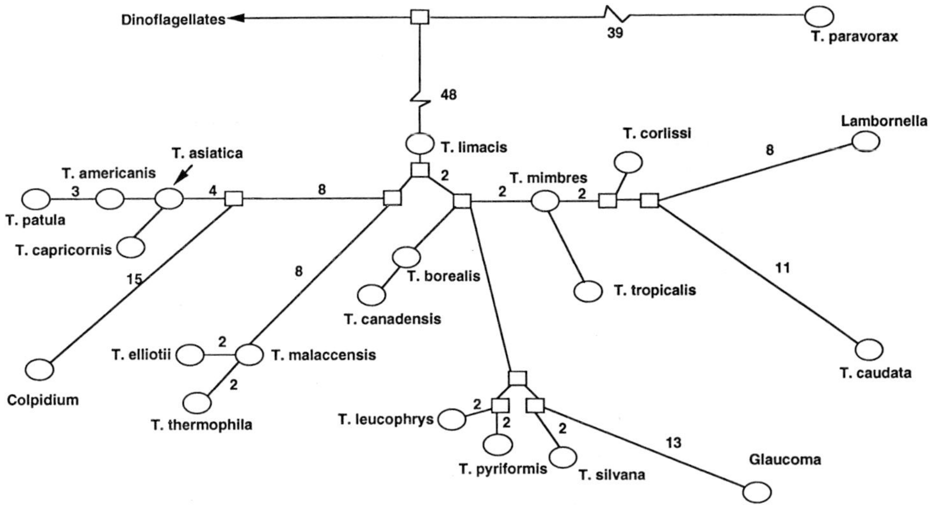

Fig. 2. Sequence differences in lsu rDNA's between major sibling species sets of *Tetrahymena pyriformis,* redrawn from Preparata *et al.* 1993. Numbers on lines indicate number of nucleotide substitutions from nearest branch point.

suggesting that the observed sequence variation within *Gracilaria verrucosa* isolates may reflect a species complex. Isolates of *Gracilariopsis lamaneiformis* differed at at least 49 positions in their 18S rDNA sequences; these differences can been used to separate the isolates into at least two cryptic species (Bird *et al.* 1994). Goff *et al.* (1994) compared non-coding regions within the rDNA and rbc gene clusters at several taxonomic levels within the red algal genera, *Gracilaria* and *Gracilariopsis,* to evaluate the usefulness of these regions in delineating species (Table 2). The two regions did not provide the same level of resolution in the two genera and Goff *et al.* (1994) concluded that one of the regions must be evolving at a different rate. Restriction maps of the 18S rDNA or its group I intron could not differentiate between isolates of *Porphyra nereocystis* ranging from Oregon, USA to British Columbia, Canada (Stiller and Waaland 1993), suggesting a single species within this geographical range (Fig. 3), whereas one cryptic species within *Porphyra thuretii* was found from a similar geographical range using the same technique. Species complexes within European *Gelidium* species showing local population variation were also detected using rbcL sequence data (Freshwater and Rueness 1994).

Table 2 Comparison of % variation in ITS and RUBISCO spacer regions between *Gracilaria* and *Gracilariopsis* at different taxonomic levels. Taken from Goff *et al.* 1994.

Taxa compared	ITS	RUBISCO
Gracilariopsis		
Interindividual	0-3	0
Interpopulation	0.2-4	0
Interspecies	7-18	3-6
Gracilaria		
Interpopulation	0.4	0-1
Interspecies - *verrucosa* type	6-17	3-7
Interspecies - *tunuistipata* type	9-31	2-8
Gracilariopsis x *Gracilaria*	not alignable	18-25

These examples serve to reinforce the point that the detection of variation in local populations using molecular techniques will depend on the technique used, probably the age of the group investigated, and perhaps the rate of evolution in the region selected.

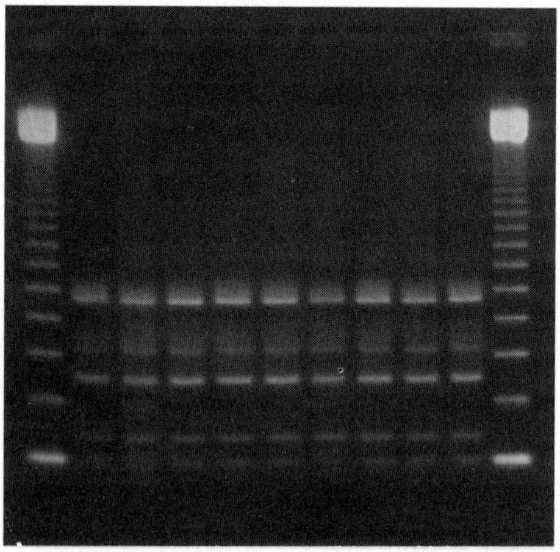

Fig. 3. Agarose gel showing amplified 18S rRNA gene including a group 1 intron from *Porphyra nereocystis* sampled from Oregon to British Columbia (left to right) and digested with *Msp*I (Stiller and Waaland 1994).

From the above studies, it can be concluded that sequences of rRNA genes are conserved at the species level; polymorphisms usually indicate either a species complex or local variation. More information regarding species delineation and population level variation can be gathered when non-coding regions and/or data from other species level measurements are also included. Given that sufficient care (such as sampling of interpopulation variation, direct sequencing of PCR products, pooling of multiple clones, construction of secondary structures, *etc*) is used to produce sequence data free of methodological error for species level enquiries, then these data can be reliably used to delineate species. The incorporation of molecular techniques with more traditional means of delineating species provides a powerful tool to help in our assessment of a species.

The combination of several methods to assess species level differences is especially needed in the marine environment (Wood and Leatham 1992) because virtually all species concepts have been based on terrestrial systems where species typically have low dispersal and small population sizes (Gosling 1994, Palumbi 1992). Such is not the case in marine systems, where benthic taxa with planktonic larvae and zoo- and phytoplankton taxa have large geographical ranges and population sizes (Gosling 1994, Palumbi 1992). In addition, planktonic organisms are subjected to ever changing environmental conditions (physical and chemical) so any individual's adaptive advantage will vary through time.

We have used sequence data from both coding and non-coding regions to assess species/genetic diversity in three ecologically important members of the marine phytoplankton: the prymnesiophytes *Phaeocystis* and *Emiliania huxleyi* and the diatom *Skeletonema costatum*. All are high dispersal taxa and the information from these studies has provided an insight into important questions concerning taxonomic affinity, genetic diversity, and gene flow and dispersal; all of which have an impact on the assessment of biodiversity and production in marine ecosystems.

Phaeocystis

Members of the genus *Phaeocystis* are recognised both as nuisance algae and as ecologically important phytoplankton species (see review in Lancelot *et al.* 1987). Their polymorphic life cycle with both colonial and flagellated cells (Kornmann 1955) has caused many taxonomic problems. *Phaeocystis* was erected by Langerheim (1893) to accommodate the colonial state of the alga originally described as *Tetraspora poucheti* by Hariot in Pouchet (1892). *Phaeocystis pouchetii* occurs in cold waters and forms globular, cloud-like colonies with cells arranged in packets of four (see Jahnke and Baumann 1987, for illustrations). *Phaeocystis globosa* (Scherffel 1900), a warm water species, forms spherical colonies with cells arranged homogeneously within the gelatinous matrix (Jahnke and Baumann 1987). Early workers separated both taxa based on their different distributions and different colonial morphologies until Kornmann (1955) demonstrated, in his life cycle studies on *P. globosa*, that *P. globosa* cell types were juvenile forms of *P. pouchetii*. Since that time, colony morphology has been judged an unreliable specific character.

Sournia (1988) reviewed the diagnostic features of *Phaeocystis* and recognised only two of its nine species as valid: *Phaeocystis scrobiculata* Moestrup known only from the flagellated state (Moestrup 1979) and *P. pouchetii,* which included *P. globosa* as a later synonym. Upon Sournia's (1988) recommendation, most marine ecologists report colonies as *P. pouchetii* (the older name) or as *Phaeocystis* sp. to avoid confusion. Thus, colonial *Phaeocystis* is recorded as cosmopolitan forming massive blooms in Antarctic and Arctic waters and extending into north temperate regions where areas of high nutrient enrichment occur (Lancelot *et al.* 1991, Davidson and Marchant, 1992).

Table 3 Comparison of DNA content in *Phaeocystis* strains determined by flow cytometry with strains whose 18S rRNA gene sequence has been determined (Vaulot *et al.*, 1994 and Vaulot unpubl)

Strain	**DNA content** pg cell^{-1}	**Ploidy**	**DNA cluster**
Naples (new species)	0.12 (0.24)	1	1
P. antarctica SK 22	0.37	2	2
P. globosa			
Gulf of Mexico	0.34	2	3
South Africa	0.32	2	3
Thailand	0.35	2	3
P. globosa			
Surinam	0.38	2	4
Galapagos	0.38	2	4
P. globosa			
North Sea	0.21(0.42)	1	5
Eastern Atlantic	0.44	2	5

Recent studies dispute this over-simplification of the genus (Baumann and Jahnke 1986, Jahnke and Baumann 1986, 1987, Jahnke 1989, Larsen and Moestrup 1989, Moestrup and Larsen 1992, Baumann *et al.* 1994b). We have investigated the validity of the separation of the colonial forms of *Phaeocystis* into three distinct species, as suggested by Moestrup and Larsen (1989) and Baumann *et al.* (1994b), using sequence data from the nuclear-encoded ssu rRNA gene. Our analysis supports the separation into three species (Fig. 4, Medlin *et al.* 1994 a, b) and we recognise two species complexes. The cold water complex contains two species: the Antarctic *Phaeocystis antarctica* described by Karsten (1905) and the Arctic *Phaeocystis pouchetii*. The warm-water complex contains isolates of *Phaeocystis globosa*. This separation is further supported by differences in DNA content, pigment spectra and morphology of the colonial stage (Medlin *et al.* 1994a, b, Vaulot *et al.* 1994). Although we cannot resolve differences in the 18S rRNA gene in the warm water complex, Vaulot *et al.* (1994) can detect three clusters of isolates having different DNA content as measured by flow cytometry (Table 3).

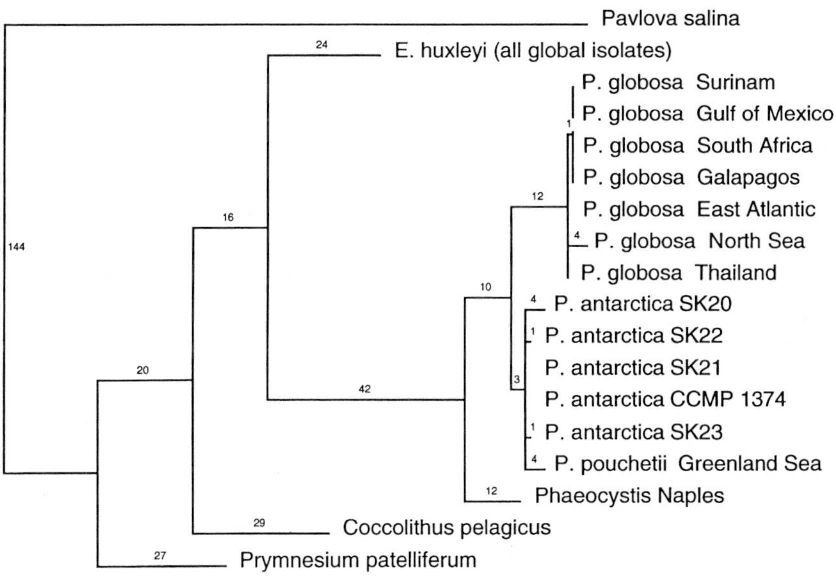

Fig. 4. Phylogenetic representation of the genetic distance among strains of *Emiliania huxleyi* and *Phaeocystis* using 18S rDNA sequence data. All isolates of *E. huxleyi* are identical, whereas the isolates of *Phaeocystis* fall into two distinct species complexes found either in warm or cold water.

Similarity values and absolute number of base substitutions (Medlin *et al.* 1994 a, b) between these three colonial *Phaeocystis* species are comparable to species differences within genera of diatoms (Medlin *et al.* 1991), protozoa (Sogin *et al.* 1986) and green algae (Huss and Sogin 1991, Friedl and Zeltner

1994). A maximum parsimony tree inferred from these data show that *Phaeocystis* appears to have originated as a unicellular genus (Fig. 4). The first isolate to diverge is the unicellular *Phaeocystis* isolate from Naples, a new species with a distinctive scale morphology (Zingone unpubl.). This is followed by a separation of the warm and cold water complexes, then a later divergence within these complexes (Fig. 4).

Using a 1% divergence in the nucleotide sequence of the ssu rRNA gene per 25 million years as documented for higher plants (Ochmann and Wilson 1987), Dasycladales (Olsen *et al.* 1994) and diatoms (Medlin *et al.* unpubl.), we have plotted the divergence of the three colony forming *Phaeocystis* species against surface water changes in the Antarctic since the Cretaceous (Fig. 5).

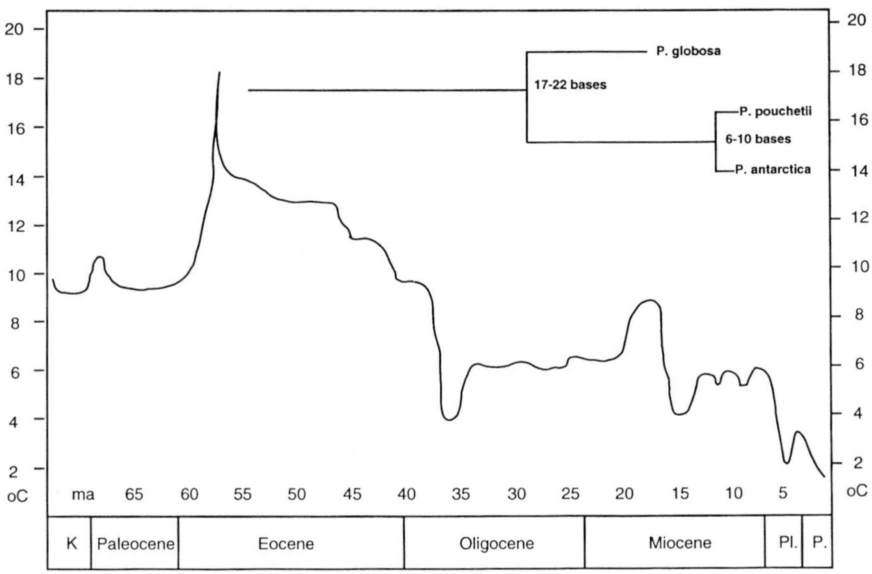

Fig.5. Reconstruction of phylogenetic changes in *Phaeocystis* species corresponding to a 1% divergence per 25 million years and correlated with changes in Antarctic surface water changes since the Cretaceous. Adapted from Crame 1993.

Based on these temperature changes and other tectonic events, we believe that the divergence of the warm and cold water forms of *Phaeocystis* corresponded to a major cooling event at the end of the Eocene following a warmer period when *Phaeocystis* would have been present throughout the world's oceans (Medlin *et al.* 1994b). Relative rate tests using *Emiliania huxleyi* as outgroup indicate that both complexes are evolving at nearly the same rate; both have identical generation times (Medlin *et al.* 1994b). A scenario involving vicariant and dispersal events must be invoked to explain the recent divergence of the two cold water species. Transtropical migrations from the Antarctic northwards are probably more common than previously thought and occurred for longer periods resulting in many cases of bipolarity (Crame 1993).

Emiliania huxleyi

Emiliania huxleyi (Lohm.) Hay et Mohl., the most abundant representative of the Prymnesiophyta, can be found in oceanic and neritic waters from sub-polar to tropical latitudes. Annual north Atlantic blooms may cover up to half a million square kilometres, with cell densities of up to 10^5 ml^{-1} (see review in Aiken *et al.* 1992) and have a major impact on both carbon and sulphur cycles.

E. huxleyi's ability to exploit a wide range of habitats and to produce blooms under a multitude of environmental conditions implies extensive phenotypic plasticity or genetic variation within the taxon, either at the species level or below. Morphometric, physiological, biochemical, and immunological studies support the sub-division of *E. huxleyi.* Two morphotypes, designated A and B, exist (Young and Westbroek 1991, van Bleijswijk *et al.* 1991, Medlin *et al.* 1994b): B morphotype cells are generally larger with more coccoliths per cell than A morphotype cells. Not only are growth rates of A and B morphotypes significantly different (van Bleijswijk *et al.* 1991) but also individual clones of the A morphotype of *E. huxleyi* taken from a single sampling bottle exhibit different acclimatised growth rates, suggesting that the A morphotype is also genetically diverse (Brand 1982). Antisera raised against coccolith associated polysaccharide from either A or B morphotypes are morphotype specific, a feature retained by cells kept in culture for 20 years and implying a stable genetic basis (Young and Westbroek 1991), but antisera raised against naked cells reacts with both morphotypes (Campbell *et al.* 1989). Coastal and oceanic strains also exhibit different key biomarker compounds and different amounts of fucoxanthin and 19'-hexanoyloxyfucoxanthin, but these isolates do not segregate into A and B morphotypes and tend to exhibit high within strain variation (Conte *et al.* 1994 and Mantoura *et al.* unpubl.).

The molecular techniques initially used to investigate the genetic relatedness of geographically isolated clones of *E. huxleyi* involved a comparison of the nuclear-encoded ssu rDNA from three geographically isolated type A clones to that previously published (Bhattacharya *et al.* 1992) for a single type B clone, 92D. Only a single ambiguous nucleotide position separated clone 92D and M181 (A morphotype) from the remaining clones (A morphotypes). Because ambiguous positions are ignored in phylogenetic analyses, all clones are considered to be identical (Fig. 4).

Our analysis of species level diversity in *E. huxleyi* extended to the functionally non-conserved spacer region separating plastid-borne genes encoding the large and small subunits of ribulose 1,5-bisphosphate carboxylase (RUBISCO spacer). The applicability of this region for the discrimination of taxa at the species level or below has been evaluated by Goff *et al.* (1994). The RUBISCO spacer region from each of the A and B morphotypes examined to date is identical with the exception of one possible ambiguous position in the 92D sequence.

Our interpretation of the lack of variation in coding and non-coding regions of *E. huxleyi* isolates is that *E. huxleyi* is a cosmopolitan species. Its distribution into all of the major ocean basins has occurred in recent geological time across major oceanographic barriers during a period of cooler temperatures than that of the purported cosmopolitan ancestor of extant *Phaeocystis* species.

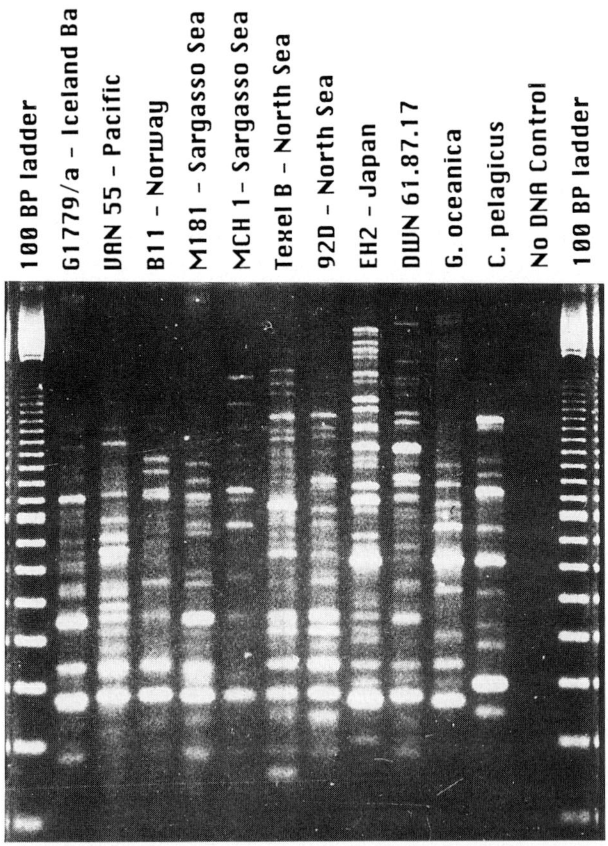

Fig. 6. Agarose gel showing amplification fragment length polymorphisms (AFLPs) produced from globally distributed *E. huxleyi* isolates with the RAPD-PCR technique using the 10-mer oligonucleotide primer OPE-02 .

We also recognise that if the two morphotypes are distinct species, then their separation may have been too recent in geological time to have allowed nucleotide substitutions to accumulate in the regions we have investigated; other regions of the genome may reveal differences between the two morphotypes. At present recognition of the two morphotypes from field material can be recorded at the varietal level (Young and Westbroek 1991).

Given the lack of variation found in both coding and non-coding regions of the genome, we have implemented the RAPD-PCR analysis to distinguish the clones and to estimate the amount of genetic variation present within and between populations. Preliminary evidence suggests that all isolates of *E. huxleyi* are different but shared bands can be found between *E. huxleyi* isolates and *Gephyrocapsa oceanica*, its purported ancestor (Fig. 6), which are not found in *Coccolithus pelagicus*, another coccolithophorid species. Also, bloom events are not clonal but are highly diverse within both temporal and spatial scales (Fig. 7). Split-decomposition analysis of distance matrix data from bands generated by the RAPD technique indicates isolates taken from close geographical proximity share more bands (data not shown).

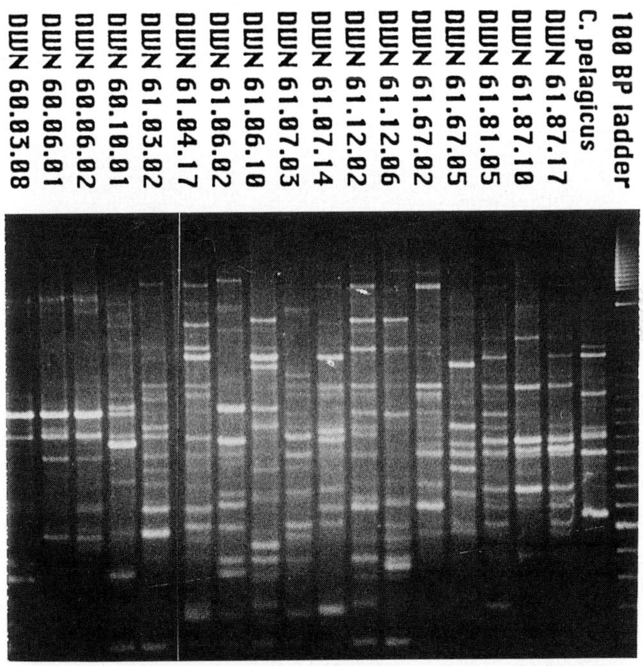

Fig. 7. Agarose gel showing amplification fragment length polymorphisms (AFLPs) produced from *E. huxleyi* cells isolated from a bloom in the North Atlantic with the RAPD-PCR technique using the 10-mer oligonucleotide primer OPE-02 .

We are continuing an in-depth study of the development of two bloom events, one from the North Atlantic and a second from near-shore mesocosm experiments to investigate the amount of genetic diversity in bloom-forming phytoplankton taxa and to determine if we can map gene flow in phytoplankton communities using the RAPD analysis. Certainly these preliminary results indicate that bloom events are seeded by many cells.

Skeletonema costatum

Skeletonema costatum (Grev.) Cleve is a major bloom-forming diatom species in neritic waters. It is well-known for its extreme variation in size and shape as well as its ability to survive different environmental conditions (Hasle 1973). Like *E. huxleyi*, its ability to form major blooms under a variety of water conditions implies extensive genetic variation within the taxon. Gallagher (1980) first demonstrated that seasonal populations of *S. costatum* from Narragansett Bay, Rhode Island were as genetically diverse as sibling species of terrestrial plants and animals. Superimposed upon this temporal diversity were several distinct but spatially separated populations (Gallagher 1990a). Gallagher has subsequently reinterpreted her data and believes that all of these genetically distinct populations represent a single interbreeding cline (Gallagher 1990b).

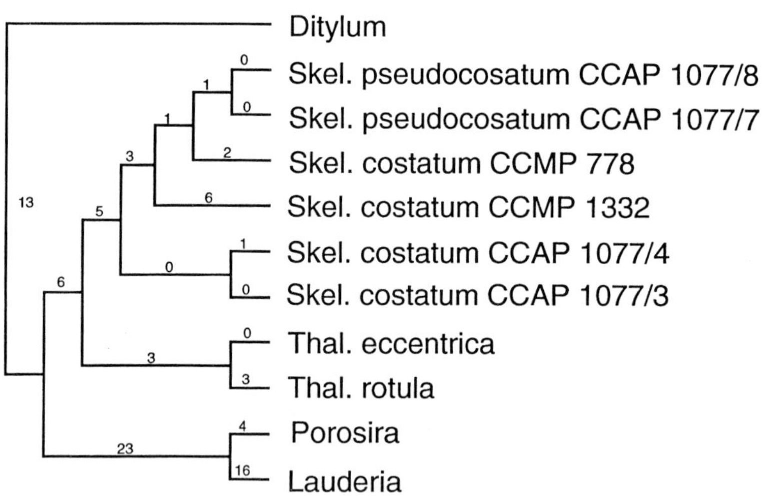

Fig. 8. Differences within a 106 bp hypervariable region of the 18S rRNA gene from *Skeletonema*. One cryptic species (*Skeletonema pseudocostatum*) can be distinguished morphologically from the normal morphology assignable to *Skeletonema costatum*.

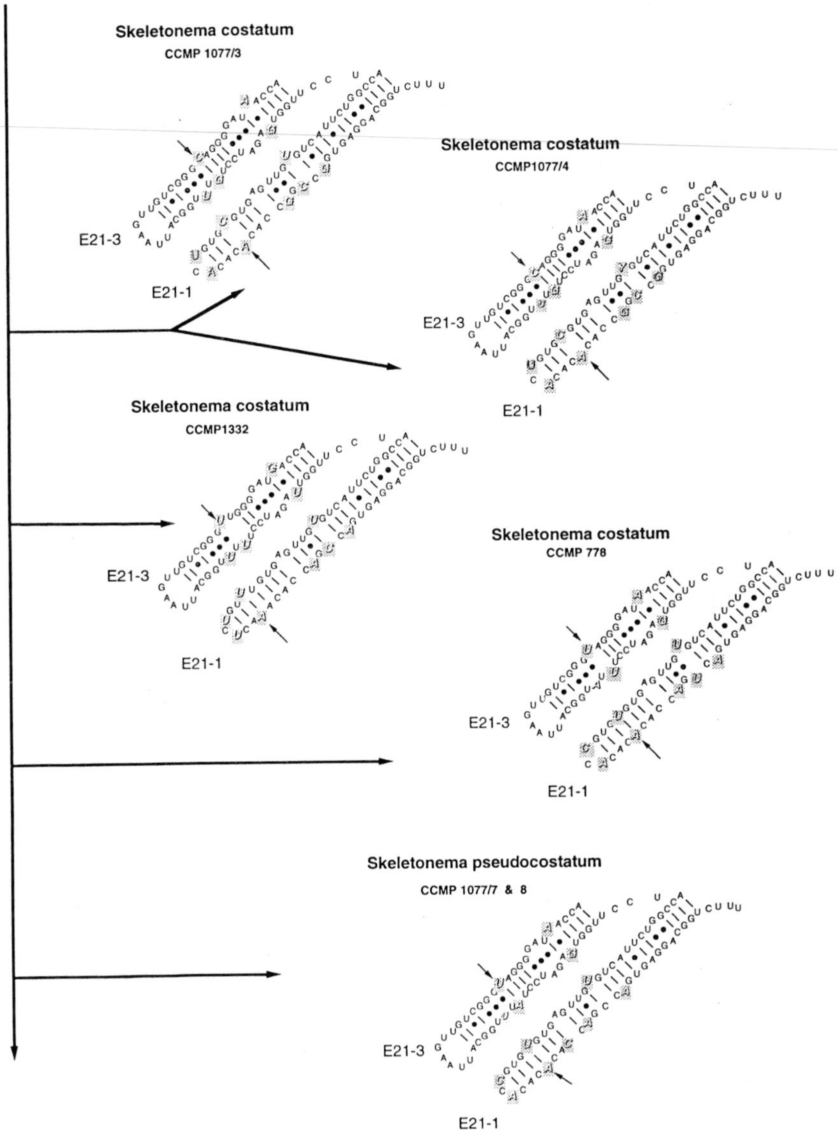

Fig. 9. Secondary structure model of two helices in the 18S rRNA gene from isolates of *Skeletonema*. Base changes are shadowed. Small arrows indicate compensatory base changes. Large arrows indicate the direction of change inferred from Fig. 8.

However, sequence analysis of a 106 bp variable region within the 18S rRNA gene has revealed extreme polymorphisms among clones of *S. costatum*, which strongly suggests a species complex (Fig. 8). Each of the base changes shown in this maximum parsiomny tree are autapomorphic for each clone.

Mapping of these changes to a secondary structure model of this region indicates several changes involve compensatory base changes (Fig. 9). Sufficient morphological differences from the normal morphotype of *S. costatum* have been found in two of these clones to warrant separation at the species level. These morphological differences, albeit small, can be detected in field material to document succession of four *Skeletonema* species on an annual basis in a Mexican bay (Hernandez unpubl.). *Skeletonema costatum* is an economically important member of the phytoplankton, being used extensively in mariculture. Physiological differences among strains of *Skeletonema* that may reflect species level differences may have an impact on their use under various mariculture conditions.

Conclusions

Molecular biology provides tools to 1) identify multi-species complexes and define a species limits, 2) interpret the taxonomic level to which physiological and morphological differences should be applied, 3) interpret gene flow and dispersal mechanisms, and 4) depict the phylogenetic history of a group and interpret its biogeographic distribution. The molecular technique and the region of the genome to be used must be carefully selected in order to provide answers to any questions being asked. If the separation of sibling species is too recent in geological time, then a variety of molecular techniques may be needed to resolve the problem.

The examples presented here illustrate how molecular techniques have been specifically used to enhance our understanding of a "species" in certain ecologically important marine phytoplankton taxa. The potential for recognising genetic individuality at the DNA sequence level is only just being realised and its use in clustering individuals into biologically meaningful groups reflecting their overall relatedness will probably increase the diversity in the marine environment by an order of magnitude (Hedgecock 1994).

Acknowledgements

This work was supported in part by grants from the NERC (GST/02/223, GR3/8139 and GR9/715), the DFG (Schwerpunkt "Antarktisforschung" SM 22/5-1) and the EC (MAST-PL92-0058). The technical support of T. Colbourn, M. Conte, J. C. Green, F. Hinz, R. Jordon, L. Madureira, T. Ruttkowski, A. Thompson, and U. Wellbrock is gratefully acknowledged. This is Contribution No. 815 from the Alfred-Wegener-Institut.

References

Amann RI, Lin C, Key R, Montgomery L, Stahl DA (1992) Diversity among *Fibrobacter* isolates: Towards a phylogenetic classification. *System Appl Microbiol* **15:** 23-31

Aiken J, Moore GF, Holligan PM (1992) Remote sensing of oceanic biology in relation to global climate change. *J Phycol* **28:** 579-590

Baumann MEM and Jahnke J (1986) Marine Planktonalgen der Arktis. I. Die Haptophycee *Phaeocystis pouchetii. Mikro* **75:** 262-5

Baumann MEM, Brandini FP, Staubes R (1994a) The influence of light and temperature on carbon specific DMS-release by cultures of *Phaeocystis antarctica* and three antarctic diatoms. *Mar Chem* **45:** 129-136

Baumann MEM, Lancelot C, Brandini FP, Sakshaug E, John DM (1994b) The taxonomic identity of the cosmopolitan prymnesiophyte *Phaeocystis*: a morphological and ecophysiological approach. *J Mar Sys* **5:** 23-39

Beam CA, Preparata R-M, Himes M, Nanney DL (1993) Ribosomal RNA sequencing of members of the *Crypthecodinium cohnii* (Dinophyceae) species complex: comparison with soluble enzyme studies. *J Euk Microbio* **40:** 660-667

Bhattacharya D, Medlin L, Wainwright PO, Arizitia EV, Bibeau C, Stickel SK, Sogin ML (1992) Algae containing chlorophylls *a* + *c* are paraphyletic: molecular evolutionary analysis of the Chromophyta. *Evol* **46:** 1801-1817

Bird CJ, Ragan MA, Critchley AT, Rice EL, Gutell RR (1994) Molecular relationships in the Gracilariaceae (Rhodophyta): further observations on some undetermined species. *Eur J Phycol* **29:** 195-202

Brand LE (1982) Genetic variability and spatial patterns of genetic differentiation in the reproductive rates of the marine coccolithophores *Emiliania huxleyi* and *Gephyrocapsa oceanica. Limnol Oceanogr* **27:** 236-245

Campbell L, Shapiro LP, Haugen EM, Morris L (1989) Immunochemical approaches to the identification of the ultraplankton: assets and limitations. In *Novel Phytoplankton Blooms* EM Cosper, VM Bricelj, EJ Carpenter (eds) Springer-Verlag, Berlin 39-56

Conte MH, Volkman JK, Eglinton G (1994) Lipid biomarkers of the Prymnesiophyceae. In *The Haptophyte Algae* JC Green and BSC Leadbeater (eds) Clarendon Press, Oxford 351-378

Cracraft J (1989) Speciation and its ontology: the empirical consequences of alternative species concepts for understanding patterns and process of differentiation. In *Speciation and Its Consequences* D Otte and JA Endler (eds) Sinauer Assoc, Sunderland 28-59

Crame JA (1993) Latitudinal range fluctuations in the marine realm through geological time. *Tree* **3:** 162-166

Davidson AT and Marchant H (1992) The biology and ecology of *Phaeocystis* (Prymnesiophyceae). In *Progress in Phycological Research* Vol 8 (FE Round and DJ Chapman eds) Biopress, Bristol 1-45

Freshwater DW and Rueness J (1994) Phylogenetic relationships of some European *Gelidium* (Gelidiales, Rhodophyta) species, based on *rbc*L nucleotide sequence analysis. *Phycologia* **33:** 187-194

Friedl T and Zeltner C (1994) Assessing the relationships of some coccoid green lichen algae and the Microthamniales (Chlorophyta) with 18S ribosomal RNA gene sequence comparisons. *J Phycol* **30:** 500-506

Gallagher JC (1980) Population genetics of *Skeletonema costatum* (Bacillariophyceae) in Narragansett Bay. *J Phycol* **16:** 464-474

Gallagher JC (1990a) The relative roles of space and time in controlling genetic differentiation in populations of the marine diatom *Skeletonema costatum*: preliminary data and a comparison of methods of analysis. *J Phycol* **26** (suppl):17

Gallagher JC (1990b) Comparisons of sampling methods and data analysis of molecular evolutionary studies of diatoms. Abstract 11th International Symposium on Living and Fossil Diatoms, San Francisco

Goff LJ, Moon DA, Coleman AW (1994) Molecular delineation of species and species relationships in the red algal agarophytes *Gracilariopsis* and *Gracilaria* (Gracilariales). *J Phycol* **30:** 521-537

Gosling EM (1994) Speciation and species concepts in the marine environment. In *Genetics and Evolution of Aquatic Organisms* (AR Beaumont ed) Chapman and Hall, London 1-15

Hasle GR (1973) Morphology and taxonomy of *Skeletonema costatum* (Bacillariphyceae). *Norw J Bot* **20:** 109-137

Hedgecock D (1994) Population genetics of marine organisms. *US GLOBEC News* **6:** 1-3, 11

Huss VAR, Huss G, Kessler E (1989) Deoxyribonucleic acid reassociation and interspecies relationships of the genus *Chlorella* (Chlorophyceae). *Pl Syst Evol* **168:** 1-82

Huss VAR, Dörr R, Grossman U, Kessler E (1986) Deoxyribonucleic acid reassociation in the taxonomy of the genus *Chlorella. Arch Microbiol* **145:**329-333

Huss VAR and Sogin ML (1991) Phylogenetic position of some *Chlorella* species within the Chlorococcales based upon complete small-subunit ribosomal RNA sequences. *J Mol Evol* **31:**432-442

Jahnke J (1989) The light and temperature dependence of growth rate and elemental composition of *Phaeocystis_globosa* Scherffel and *P. pouchetii* (Har.) Lagerh. in batch cultures. *Neth J Sea Res* **23:** 15-21

Jahnke J and Baumann MEM (1986) Die marine planktonalge *Phaeocystis globosa*: eine Massenform unserer Küstengewässer. *Mikro* **75:** 357-359

Jahnke J, Baumann M (1987) Differentiation between *Phaeocystis pouchetii* (Har.) Lagerheim and *Phaeocystis globosa* Scherffel. I. Colony shapes and temperature tolerances. *Hydro Bull* **21:** 141-147

Knowlton N (1993) Sibling species in the sea. *Ann Rev Ecol Sys* **24:** 189-216

Karsten G (1905) Das Phytoplankton des Antarktischen Meeres nach dem Material der Deutschen Tiefsee-Expedition 1898-1899. *Wiss Erg Deut Tiefsee-Exp 'Valdivia' 1898-1899.* **Band II Teil 2** 1-136

Kooistra WHCF (1993) Historical biogeography in tropical Atlantic populations of *Cladophoropsis membranacea* and related species. Ph D Dissertation. University of Groningen, pp 111

Kornmann P (1955) Beobachtungen an *Phaeocystis*-Kulturen. *Hel Wiss Meeres* **5:** 218-233

Lagerheim G 1893) *Phaeocystis* nov. gen. grundadt på *Tetraspora poucheti Har. Bot Not* **1:** 32-33

Lancelot C, Billen G, Sournia A, Weisse T, Colijn F, Veldhuis MJW, Davies A, Wassmann P (1987) *Phaeocystis* blooms and nutrient enrichment in the continental coastal zones of the North Sea. Ambio **16:** 38-46

Lancelot C, Billen G, Barth H (1991) The dynamics of *Phaeocystis* blooms in nutrients enriched coastal zones. *Wat Poll Res Rep* **23:** 1-106

Larsen J and Moestrup Ø (1989) *Guide to Toxic and Potentially Toxic Marine Algae.* Fish Inspection Service, Minister of Fisheries, Copenhagen, 49-51

Manhart JR and McCourt RM (1992) Molecular data and species concepts in the algae. *J Phycol* **28:** 730-737

Medlin LK, Elwood HJ, Stickel S, Sogin ML (1991) Morphological and genetic variation within the diatom *Skeletonema costatum*_(Bacillariophyta): evidence for a new species *Skeletonema pseudocostatum. J Phycol* **27:** 514-524

Medlin LK, Lange M, Baumann MEM (1994a) Genetic differentiation among three colony-forming species of *Phaeocystis*: further evidence for the phylogeny of the Prymnesiophyta. *Phycologia* **33:** 199-212

Medlin LK, Barker GLA, Baumann, MEM, Hayes PK, Lange M (1994b) Molecular biology and systematics. In *The Haptophyte Algae* (JC Green and BSC Leadbeater eds), Clarendon Press, Oxford 393-412

Moestrup Ø (1979) Identification by electron microscopy of marine nanoplankton from New Zealand including the description of four new species. *N Zea J Bot* **17:** 61-95

Moestrup Ø and Larsen J (1992) Potenially toxic Phytoplankton. 1 Haptophyceae (Prymnesiophyceae). In *ICES Identification Leaflets for Plankton*, Leaflet No 179 (JS Lindley ed),. Natural Environmental Research Council, Plymouth 1-11

Nanney DL, Meyer EB, Simon EM, Preparata R-M (1989) Comparison of ribosomal and isozymic phylogenies of Tetrahymenine ciliates. *J Prot* **36:** 1-8

Ochman H and Wilson AC (1987) Evolution in bacteria: evidence for a universal substitution rate in cellular genomes. *J Mol Evol* **26:** 74-86

Olsen JL, Stam WT, Berger S, Menzel D (1994) 18S rDNA and evolution in the Dasycladales (Chlorophyta): modern living fossils. *J Phycol* in press

Palumbi SR (1992) Marine speciation on a small planet. *Tree* **7:** 114-118

Pouchet G (1892) Sur une algue pélagique nouvelle. *Compte Rendues séance à 16 Janvier*

Preparata R-M, Beam CA, Himes M, Nanney DL, Meyer EB, Simon EM (1992) *Crypthecodinium* and *Tetrahymena*: an exercise in comparative evolution. *J Mol Evol* **32:** 209-218

Scherffel A (1900) *Phaeocystis globosa* nov. spec. nebst einiger Betrachtungen über die Phylogenie niederer ins besonderer brauner Organismen. *Wiss Meeres Ab Hel* **4:** 1-29

Sogin ML, Ingold A, Karlok M, Nielsen H, Engberg J (1986) Phylogenetic evidence for the acquisition of ribosomal RNA introns subsequent to the divergence of some of the major *Tetrahymena* groups. *EMBO* **5:** 3625-3630

Sournia A (1988) *Phaeocystis* (Prymnesiophyceae): how many species? *Nova Hedw* **47:** 211-217

Stiller JW and Waaland JR (1993) Molecular analysis reveals cryptic diversity in *Porphyra* (Rhodophyta). *J Phycol* **29:** 506-517

van Bleijswijk J, van der Wal P, Kempers R, Veldhuis M, Young JR, Muyzer G, deVrind-deJong E, Westbroek P (1991) Distribution of two types of *Emiliania huxleyi* Prymnesiophyceae in the northeast Atlantic region as determined by immunofluorescence and coccolith morphology. *J Phycol* **27:** 566-570

Vaulot D, Birrien J-L, Marie D, Casotti R, Veldhuis M, Kraay G , Chrétiennot-Dinet M-J (1994) Morphology, ploidy, pigment composition and genome size of cultured strains of *Phaeocystis* (Prymnesiophyceae). *J Phycol* in press

Wood AM and Leatham T (1992) The species concept in phytoplankton ecology. *J Phycol* **28:** 723-729

Young JR and Westbroek P (1991) Phenotypic variation in the coccolithophorid species *Emiliania huxleyi. Mar Micropaleont* **18:**5-23

How Do Cyanobacteria Perceive and Adjust to Their Environment?

Jean Houmard
Physiologie Microbienne (CNRS URA 1129)
Département B.G.M., Institut Pasteur
28 rue du Dr Roux
75724 Paris Cedex 15
France

Abstract
Cyanobacteria have been recognized in almost all of the ecosystems that have been examined heretofore, and they probably represent the most highly diverse group of prokaryotes. They are quite remarkable in their abilities to adapt to their environment, whether the changes concern nutrient availability, temperature, intensity or spectral quality of the light. The environmental parameters have to be sensed before cells can answer and adapt to the various changes. A series of mechanisms are involved in the sensing and transduction of signals. Both the mechanisms and the connections that may exist between them have recently started to be elucidated. A few examples are described and summarized which have been studied at a molecular level in cyanobacteria. Particular emphasis is given to photoregulation of gene expression since light plays a major role in the metabolism of these photoautotrophic organisms.

Introduction
Cyanobacteria appeared some 3 billion years ago. They have since colonized the Earth and cyanobacterial species have been recognized in almost every ecosystem that has been studied, whether aquatic or terrestrial. Like all living organisms, cyanobacteria must constantly monitor the external ambient conditions and accordingly adjust their metabolism, physiology and behaviour. This implies, first, the perception of the environmental conditions and secondly, the transfer of the signals, which can imply their amplification, and thirdly the interaction with the intracellular target(s) where the required modification(s) will finally be initiated. Only recently have some of the molecular effectors started to be identified. Obviously, many different kind of receptors have to be discovered since the signals can be, in nature, physical as well as chemical. Signals can be variations of pressure, temperature, intensity or wavelengths of the incident light,

NATO ASI Series, Vol. G 38
Molecular Ecology of Aquatic Microbes
Edited by I. Joint

as well as changes in heavy metal concentrations or nutrient availability, for example. In addition, since cyanobacteria probably represent one of the most diverse groups of micro-organisms, it is difficult to envisage that a single and common answer will be given by every species to similar modifications of the environment. Finally, one could expect that the elucidation of the underlying control mechanisms will often be complicated by the interconnections and cross-talks that exist between the various transduction pathways.

In general, two major means of regulation have been recognized: i) the control of enzyme activities via oligomerisation, post-translational modifications, and/or the binding of ligands which can sometimes lead to allosteric responses; ii) transcriptional regulations under specific growth conditions which could either turn on (positive) or off (negative control) the expression of a gene or a set of genes. Although regulatory adjustments in the expression of some genes may also occur at the translational level, such control is often superimposed on changes of the steady-state mRNA levels over longer periods.

Most of the molecular studies have so far been performed with only a limited number of cyanobacterial species: *eg* the unicellular *Synechococcus* sp. PCC6301 and PCC7942, *Synechocystis* sp. PCC6714 and PCC6803, the filamentous heterocystous *Anabaena* sp. PCC7120, related *Nostoc* sp. strains, as well as *Calothrix* sp. PCC7504 and PCC7601. For each of these strains, the known mechanisms of adaptation will be presented, as well as the connections between them, if any.

The unicellular strains

Synechococcus sp. PCC6301 and PCC7942 were isolated from fresh water in Texas and California (USA), respectively. They appear as unicellular rods of about 3μm length and 1μm width. These two *Synechococcus* strains, also known as *Anacystis nidulans* or *Synechococcus leopoliensis,* are very closely related (Golden *et al.*, 1989; Rippka and Herdman, 1993). *Synechococcus* sp. PCC7942 is often considered as the cyanobacterial counterpart of the *Escherichia coli* paradigm. Indeed, this unicellular cyanobacterium is a fast-growing species that is transformable by DNA without any pretreatment of the cells; it has a genome size of about 2×10^9 daltons, recombines very easily, and many genetic tools like shuttle and cosmid vectors have been constructed for it (Kuhlemeier *et al.*, 1981 and 1983, Tandeau de Marsac *et al.*, 1982), as well as a platform system that allows insertion of any DNA fragment which is carried by a pBR322 derivative into a precise location on the chromosome (van der Plas *et al.*, 1990). These *Synechococcus* strains are strict photoautotrophs, unable to fix molecular nitrogen or differentiate any particular cell type. From studies with these species, the mechanisms that are common to most of the cyanobacteria may be worked out without too many interferences from other more specific regulatory pathways. The ability of *Synechococcus* PCC7942 to sense and adapt to a wide range of ambient CO_2 concentrations involves many changes which are discussed in detail by Ronen-Tarazi *et al.* (1995).

Synechocystis PCC6714 and PCC6803 (*Aphanocapsa* 5-3A and N-1) were both isolated from fresh water in California. These euryhaline facultative photoheterotrophic cyanobacteria have a diameter of about 2.5μm. Like the *Synechococcus* strains, their genome size is about 2 x 10^9 daltons (Rippka and Herdman, 1993). *Synechocystis* PCC6803 takes up DNA by transformation without requiring any specific pretreatment of the cells. Because of its photoheterotrophy, it has been regarded as the cyanobacterium of choice for genetic studies of photosynthesis.

PHOTOACCLIMATION

The photosynthetic apparatus of the cyanobacteria and its functioning are very similar to those of higher plants and algae: two photosystems (PSI and PSII) are located within the thylakoid membrane; electron transfer between the two PS is mediated by the cytochrome *b6-f* complex and the plastoquinones (PQ). The major difference resides in the lack of chlorophyll *b*-protein complexes, the antenna function for light harvesting being assured by the phycobilisomes (PBsomes), multimolecular structures usually hemidiscoidal, which are regularly arrayed on the surface of the photosynthetic membrane. The PBsomes transfer the excitons mostly to the chlorophyll *a*-protein complexes that form the reaction centre of the PSII (Glazer, 1989). At present, we do not really know how cyanobacteria perceive changes in the available photosynthetic photon flux density (PPFD, often referred to as light intensity), but they adapt well to it. Standard growth irradiance is about 100μmol m^{-2} s^{-1} for the *Synechococcus* sp. strains. They suffer from chronic inhibition of PSII at irradiances above 300μmol $m^{-2}s^{-1}$ (direct sunlight is about 1,700μmol m^{-2} s^{-1}). In *Synechococcus* sp., PBsomes are associated almost exclusively with PSII, whereas most of the chlorophyll *a* is associated with PSI. The qualitative distribution of pigments between the photosystems is not strongly affected by growth irradiance (Khanna *et al.*, 1983).

It has long been known that pigment concentration and thylakoid content vary inversely with photon fluence rate (Allen, 1968). Upon light shift-downs, phycobilisome rods elongate before the number of PBsomes per unit area of membrane increases with the consequence that any available surrounding photons can be collected (Lönneborg *et al.*, 1985). However, although the phycobiliprotein content decreases in response to light shift-ups, an unexpected increase in phycocyanin (*cpcBA*) mRNA concentration with PPFD has been observed, implying post-transcriptional controls (Kalla *et al.*, 1993). Whether these controls occur at the level of translation, of chromophore attachment to the apoprotein, of mRNA stability or result from an increased protein turnover is not known. In addition, when exposed to a PPFD higher than that required for optimal growth, cells respond by decreasing CO_2 fixation and O_2 evolution, the phenomenon known as photoinhibition.

The light effects on the expression of genes that encode the main components of the PSII reaction centre, namely *psbA* (D1 protein) and *psbD* (D2 protein), have been studied in detail. The former protein, D1, is supposed to represent the initial site of photoinhibition and could represent a sensor for light acclimation (Koenig, 1990; Clarke *et al.*, 1993). In *Synechococcus* sp. PCC7942, there are three copies of *psbA*: *psbAI* specifies FormI of D1, while *psbAII* and *psbAIII* encode an identical Form II of the D1 protein. Form II differs from Form I by only 25 amino acids, half of these differences being located in the 16 N-terminal residues of the protein (Golden *et al.*, 1986). Schaefer and Golden (1989) have demonstrated that, depending upon the incident PPFD, either Form I under low light, or Form II under high light, predominates in the thylakoid membrane. It was also shown that at high PPFD, the half-life of Form I of the D1 protein is 36min and that of Form II is 90min, Form II replacing FormI within the D1 pool. In parallel, gene inactivation and measurements of steady-state levels of *psbA* mRNAs, as well as of the levels of expression of reporter genes fused to the various P_{psbA} promoters, have been performed under varying PPFD (Bustos *et al.*, 1990; Kulkarni and Golden, 1994). Shifting cells to a high PPFD (500µmol m^{-2} s^{-1}) results in a rapid and large increase in the level of *psbAII* (20-fold) and *psbAIII* mRNAs (8-fold). If cells are maintained under these conditions, the *psbAII* and *psbAIII* mRNA levels remain high for at least a few generations. Following a major decrease within the first 15min, the concentration of *psbAI* transcripts remained low up to 1h but reached a 1.5-fold increase by 6h after the light shift. Altogether, there is a 3- to 4-fold increase in *psbA* mRNA concentration. This modulation in the expression of the *psbA* gene family in response to shifts in the available PPFD differs from standard stress response, but follows circadian rhythmicity (Kulkarni and Golden, 1994). The existence of circadian rhythms was demonstrated by monitoring luciferase activity after having fused promoterless *luxAB* genes to P_{psbAI}. Following growth for 10 days under standard PPFD (46µmol m^{-2} s^{-1}), a 12h period of darkness is sufficient to reset the clock. Rhythmicity then decreases with time but is still visible after 8 days (Kondo and Ishiura, 1994). For the D2 protein, acclimation also occurs through differential expression of the members of the *psbD* gene family, *psbDI* and *psbDII*. These two genes encode an identical polypeptide, but *psbDII* is a monocistronic unit while *psbDI* is part of the *psbDIC* operon. In addition to D2, this operon encodes CP43, another protein of PSII. Upon shift to high light, a 5-fold increase in *psbDII* mRNA can be observed. The induced synthesis of D2 compensates for a faster turnover of the gene product (Bustos and Golden, 1992). Finally, from gel shift experiments, Bustos and Golden (1991) have concluded that the last step in the transduction of the light signal could be the binding of specific proteins to the promoter regions of *psbAII* and *psbDII,* but the corresponding proteins have not yet been characterized and the occurrence of a specific regulon remains to be substantiated.

Like the *Synechococcus* strains, *Synechocystis* PCC6714 and PCC6803 possess three copies of *psbA*, but only the two which encode an identical gene product seem to be expressed (Ajlani *et al.*, 1989; Mohamed *et al.*, 1993). One copy is much more expressed than the other, transcription of the latter being up-regulated under high-light conditions. Using *psbA* mutations which result in D1 proteins that confer resistance to herbicides, the D1 composition of the thylakoid membrane was shown to correlate with the level of expression of each of the two *psbA* genes (Bouyoub *et al.*, 1993). To encode the D2 protein, *Synechocystis* PCC6803 has two *psbD* genes that are arranged as in *Synechococcus* PCC7942 and, as with the latter, a greater accumulation of *psbD* mRNA was observed under high-light relative to low-light conditions (Mohamed and Jansson, 1989). In *Synechocystis* PCC6803, the *psaA* and *psaB* genes which encode the two P700 apoproteins of PSI that participate in electron transfer have also been characterized and, as in all cyanobacteria examined so far, they are cotranscribed. When *Synechocystis* PCC6803 cells are shifted from light-saturating to light-limiting conditions, depending on the genes, either the transcription and/or the mRNA stability of *psbA*, *psbD*, *psaA*, *psaB* and *rbcL* (gene encoding the ribulose bis-phosphate carboxylase) are regulated to control gene expression (Smart and McIntosh, 1991; Mohamed and Jansson, 1991).

Diurnal and seasonal variations affect both the PPFD and the spectral quality of the light. It has been shown that in cyanobacteria, most of the orange light is absorbed *via* the PBsomes and under these conditions PSI content becomes larger than PSII. When switched to red light, which is absorbed by chlorophyll *a*, the PSI/PSII ratio is about 1. The PSI/PSII composition is regulated by spectral light quality (inverse chromatic adaptation) so as to maintain the right balance between the electron influx which results from PSII functioning and the electron efflux which is due to PSI action (Fujita *et al.*, 1987). According to Murakami and Fujita (1991a,b), the response of *Synechocystis* PCC6714 cells to shifts in light intensiy depends on the preexisting PSI/PSII ratio. The redox state of the cytochrome *b6-f* complex and not that of the PQ pool would be closely related to the signal that regulates the PSI/PSII ratio in response to both light quality and light intensity. These authors have also mentioned that protein phosphorylation could be involved in the signal transduction for PSI/PSII regulation. Recently, by using a mutant of *Synechocystis* PCC6803 that lacks PSII (deletion of *psbDIC* and *psbDII*), Mullineaux (1994) concluded that the presence of PSII is required for state transitions to occur, but not for the mediation of an efficient energy transfer from PBsomes to PSI.

Synechococcus sp. could also recognize changes in the spectral quality of the light, and respond to it by the so-called light-state transitions. According to Allen (1992), state transitions would be the result of a redox control at the level of the plastoquinone. Phosphorylation reactions would be involved in this regulation, but the specific kinase/phosphatase(s) involved await isolation. A recent study performed with a *sodB* (iron superoxide dismutase) mutant of *Synechococcus* PCC7942 led to the conclusion that, during acclimation to growth irradiance, the amount of phycocyanin is regulated independently of photosystem stoichiometry (Samson *et al.*, 1994). The adjustment of the latter could be signaled by an imbalance in the activity of one or both of the photosystems which would act as a light sensor and self-regulate. At this point, it is worth mentioning that the harvesting of light energy and its conversion to chemical energy must be properly coordinated with the metabolism of carbon and nitrogen which requires ATP and NADPH. The number and the chemical nature of the light sensor(s) remain to be established but recently some regulatory elements have started to be elucidated (Ronen-Tarazi *et al.1995;* Mann 1995; and see below).

INTERMEDIARY METABOLISM

On Earth, molecular nitrogen is almost the only available source of N atoms. Its chemical combination with H and O atoms only marginally results from natural phenomena like thunderstorms. The biological process of nitrogen fixation is, by far, the most important reductive pathway, the production of fertilizers being very recent on the geological time scale. At present, diazotrophy seems to be restricted to some prokaryotes which start the biological chain. No mine of nitrogenous compounds exists on our planet, except for nitrate in South America for example, but these 'mines' correspond in fact to guano which is an accumulation of bird excrements. Depending on availability, the nitrogen source for cyanobacteria can be either ammonium, nitrite or nitrate in that order of preference. This order parallels the energy requirement for their assimilation if more than one N source is accessible. Only a few unicellular cyanobacteria can reduce molecular nitrogen to ammonia, this property being much more widespread within the group of the filamentous cyanobacteria (see below). One of the mechanisms used to solve the paradox of expressing the highly O_2-

sensitive nitrogenase in a single cell that also has to perform oxygenic photosynthesis is to temporarly space the reduction of N_2 and the photosynthetic reactions (Fay, 1992).

In enteric bacteria, the control of nitrogen assimilation is mediated through the Ntr regulon which involves a specific sigma factor (σ^{54} = NtrA = RpoN), and a transcriptional activator (NtrC) which becomes active after its phosphorylation by an histidine-kinase (NtrB). NtrB and NtrC belong to the family of the two-component systems (Magasanik, 1993; Simon, 1995). No major difference has been recognized in the pathways used by cyanobacteria to metabolize the inorganic forms of nitrogen when compared with other bacteria (Guerrero and Lara, 1987). An uptake system common to nitrite and nitrate is built up by *Synechococcus* PCC6301 and PCC7942 upon ammonium starvation (Luque *et al.*, 1994a). However, none of the proteins that form the enteric Ntr regulon have yet been recognized in cyanobacteria. Instead two proteins, NtcA and PII (GlnB) play an important role in sensing the nitrogen status of the cell (Luque *et al.*, 1994b; Forchhammer and Tandeau de Marsac, 1994).

NtcA belongs to the CRP-FNR family of transcriptional activators, (Mann, 1995). On the other hand, the cyanobacterial PII is a trimer that, like its enteric counterpart, is modified post-translationally to various extents depending on nitrogen availability, ammonium preventing the modification (Forchhammer and Tandeau de Marsac, 1994; Vasudevan *et al.*, 1994). However, in contrast to the enteric PII which is uridylylated (Adler *et al.*, 1975), the *Synechococcus* protein is phosphorylated on a seryl residue (Forchhammer and Tandeau de Marsac, 1994). O-phosphorylations are more typical of eukaryotic sensing systems but protein kinases and a tyrosine/serine protein phosphatase of the eukaryotic-type have been found recently in filamentous cyanobacteria (Potts *et al.*, 1993; Zhang, 1993). Following studies using a $GlnB^-$ *Synechococcus* mutant and various metabolic inhibitors of GS (glutamine synthetase), GOGAT (glutamate synthase) or of CO_2 fixation, Forchhammer and Tandeau de Marsac (personal communication) propose that the PII modifying system is controlled by the glutamine/α-ketoglutarate ratio, that is by both the GS-GOGAT and CO_2 fixation pathways. Their conclusion is that PII seems not necessary for ammonium repression *per se* but is required for the coordination of carbon and nitrogen assimilation.

The role of GS in the control of ammonium assimilation remains an open question. In contrast to many bacterial GS, adenylylation has never been demonstrated for the cyanobacterial enzymes, but another still obscure activation/deactivation process has been reported for *Synechocystis* PCC6803 (Mérida *et al.*, 1991a,b). In addition, a second GS (GlnN) has been characterized from this strain that may account for about 20% of total GS activity under N stress. The *glnN* expression and/or GlnN activity are sensitive to ammonium and are thus regulated by the N status of the cell. A *glnN* homolog has been detected by Southern hybridization experiments in all the unicellular and filamentous non-heterocystous cyanobacteria tested (Reyes and Florencio, 1994).

OTHER RESPONSES

Bacteria often sense and respond to environmental changes by means of the so-called two-component systems: a sensor protein perceives some physiologically relevant signals and phosphorylates a response regulator. Then, depending on its phosphorylation state, the latter protein differentially affects some physiological response (Simon, 1995). The function of the response regulator is often but not always to modulate the transcription of a given set of genes (Parkinson, 1993). Such mechanisms also operate in cyanobacteria and for *Synechococcus* sp. strains they constitute part of the response to phosphate limitation, for example (Aiba *et al.*, 1993; Mann,1995; Nagaya *et al.*, 1993 and 1994). As already mentioned, transcriptional activators of the CRP-FNR family can also be part of signal transduction pathways: NtcA for nitrogen and CysR for sulphur sensing. The effects of these various nutrient limitations and the fact that cell responses might vary depending of the endogeneous intracellular reserves are reviewed by Mann (1995; see also Tandeau de Marsac and Houmard, 1993).

Mann (1995) has mentioned adaptation mechanisms linked to shortage in trace elements. Opposite situations might occur in polluted environments and *Synechococcus* sp. strains exhibit an unexpected response mechanism. Upon exposure of the cells to increased Cd concentrations, *Synechococcus* PCC6301 and PCC7942 amplify and rearrange the *smt* chromosomal locus (Huckle *et al.*, 1993). DNA amplification coupled to rearrangement of chromosomal genes were up to recently considered as a property restricted to metal-tolerant eukaryotes. The *Synechococcus smtA* gene encode a class-II metallothionein, a protein that efficiently complexes metal ions. Transcription of *smtA* is controlled by the trans-acting SmtB repressor. The *smtB* gene is adjacent to *smtA* on the chromosome and divergently transcribed. Its deletion in Cd-tolerant *Synechococcus* sp. strains involves a highly iterated palindrome and leads to amplification of *smtA via* genome rearrangements (Gupta *et al.*, 1993).

The filamentous strains

Only a few cyanobacteria that all belong to the group of the filamentous heterocystous strains have been studied at a molecular level; *Anabaena* PCC7120 and PCC7937, as well as *Nostoc* spp. are members of the family *Nostocacae* while *Calothrix* PCC7504 and PCC7601 are members of the family

Rivulariaceae (Rippka and Herdman, 1993). These cyanobacteria display one or more of the three differentiation processes that can be regarded as responses to changes in the environment: N-starvation induces the development of heterocysts, the cells specialized in N_2 fixation; the differentiation of hormogonia — short filaments of small, often gas-vacuolated and/or motile, cells that allow the dispersion of the organism in new habitats — can be stimulated by a stress plus red light wavelengths or by a plant partner: unfavourable growth conditions may induce the formation of akinetes which can be compared to dormant spores. Only heterocysts cannot regenerate vegetative filaments; the hormogonium stage is always transient and akinetes can germinate when returned to a replete environment (Rippka and Herdman, 1985).

Anabaena PCC7120 (*Nostoc muscorum* ISU) is an obligate photoautotroph of unknown origin which under low PPFD synthesizes phycoerythrocyanin; its genome size is around 3.9 x 10^9 daltons. This strain is particularly used for studies dealing with heterocyst differentiation. *Anabaena* PCC7937 (*Anabaena variabilis*) was isolated from fresh water in Mississipi (USA); besides heterocysts it can differentiate gas-vacuolated hormogonia. *Nostoc* PCC73102 comes from a root section of an Australian *Macrozamia* sp. A symbiotic association can be reconstituted from this free-living *Nostoc* sp. and a bryophyte, *Anthoceros punctatus* (Meeks, 1988); vegetative cells (5-6μm width) are facultative photo- and chemoheterotrophs able to form gas-vacuolated hormogonia. This strain belongs to group II of the complementary chromatic adaptors, being able to modulate its phycoerythrin (PE) but not its phycocyanin (PC) content in response to changes in the incident light wavelengths; its genome size has been estimated to be 4.9 x 10^9 daltons. For all these strains, DNA transfer can be achieved *via* conjugation although with variable efficiency, most of the genetic tools having been developed in Wolk's laboratory for *Anabaena* PCC7120 (Elhai and Wolk, 1988).

Calothrix PCC7504 and PCC7601 (*Fremyella diplosiphon* UTEX481) are fresh water strains isolated from Sweden and Connecticut (USA), respectively. They are facultative photo- and chemoheterotrophs able to form motile gas-vacuolated hormogonia; they belong to the group III for complementary chromatic adaptation, modulating both PE and PC2 synthesis (see below). By analogy with a relative strain, *Tolypothrix* PCC7101, their genome size can be estimated to about 7.8 x 10^9 daltons. *Calothrix* PCC7504 is a true heterocystous diazotrophic strain while PCC7601, which can only differentiate proheterocysts, requires anaerobic conditions to express *nif* genes and reduce acetylene (Rippka and Herdman, 1993). Although DNA transfer by conjugation has been reported, these manipulations are not yet as easy to perform as with *Anabaena/Nostoc* spp. (Chiang *et al.*, 1992; Cobley *et al.*, 1993).

PHOTOACCLIMATION

For the filamentous strains, sensing of and adaptation to changes in PPFD are likely to imply mechanisms similar to those used by unicellular strains (see above). Multiple copies of *psbA* genes have indeed been found in *Anabaena* PCC7120 and *Calothrix* PCC7601 (Mulligan *et al.*, 1984; Vrba *et al.*, 1989), but more emphasis has been given to the study of their peripheral photosynthetic antennae, the PBsomes. One of the main reasons is that most of the visible changes that occur during complementary chromatic adaptation affect the distal part of the PBsomes. This adaptative phenomenon concerns some of the cyanobacteria that can synthesize PE: strains of group I regulate neither PE nor PC while those of group II only modulate PE synthesis; strains of group III make PE under green light but synthesize a second PC (PC2) under red light (Tandeau de Marsac and Houmard, 1993).

From the physiological, biochemical and biophysical studies that have been performed, it is generally accepted: i) the light sensing *per se* is a purely photochemical process which can occur even at low temperature; ii) action maxima of the photoreceptor(s) are 540nm (green) and 640nm (red); iii) green light promotes the synthesis of PE and red light that of the second PC, PC2; iv) in cells adapted to darkness, the pigment content resembles that of red-light grown cells; v) if alternating flashes of chromatic light are given, the last one determines the type of phycobiliprotein that will be synthesized (Tandeau de Marsac, 1991). In contrast, and thanks to the new techniques of molecular biology, some of the elements of the signal transduction pathway have been identified. Using cloned genes, it was first demonstrated that, in *Calothrix* PCC 7601, the regulation of complementary chromatic adaptation occurs at least in part at the level of transcription and concerns the three operons that encode the distal part of the PBsome, PE and its associated linker polypeptides (β^{PE}, α^{PE} and three $L_R{}^{PE}$), as well as PC2 and its linkers. Levels of mRNAs corresponding to the *cpeBA* (β^{PE} and α^{PE}) and *cpeCDE* ($L_R{}^{PE}$) operons are maximum under green light, mRNAs corresponding to *cpcB2A2H2I2D2* (β^{PC2}, α^{PC2} and $L_R{}^{PC2}$, *cpc2* operon) being only made under red light (Conley *et al.*, 1985; Tandeau de Marsac *et al.*, 1988). Because the transcription of the *cpe* operons is on when that of *cpc2* is off and *vice-versa*, the existence of activator(s) and/or repressor(s) was postulated (see Grossman *et al.*, 1993; Tandeau de Marsac and Houmard, 1993). Proteins that specifically bind to the promoter region (P) of the *cpeBA* and *cpc2* operons have indeed been identified (Schmidt-Goff and Federspiel, 1993; Sobczyk *et al.*, 1993 and 1994). By using gel shift and footprinting experiments, the targets for the DNA-binding proteins on the promoter regions (P_{cpeBA} and P_{cpc2}) have been located. Fig.1 presents our current model for the last steps in this regulatory pathway.

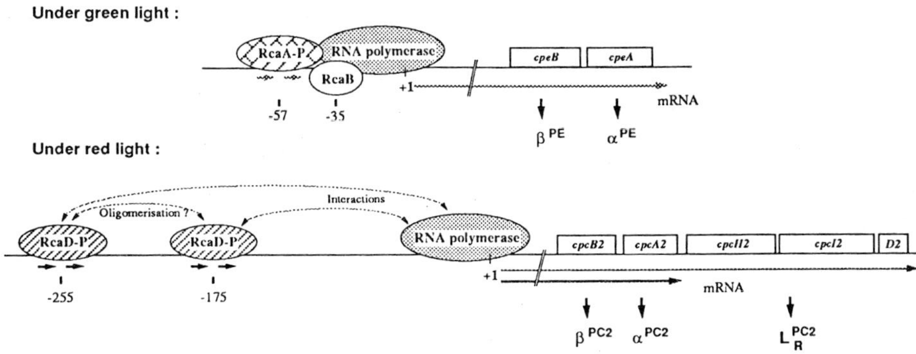

Fig.1: Model for the regulation of the expression of the *cpeBA* and *cpc2* operons during complementary chromatic adaptation in *Calothrix* PCC 7601. Tandem arrowheads represent the binding sites for the DNA-binding proteins. Nomenclature for the polypeptides is according to Glazer (1989).

Under green light, *Calothrix* PCC7601 cells possess two proteins able to bind to P*cpeBA*, RcaA and RcaB, and under red light they have RcaD which can bind to P*cpc2*. The working hypothesis, mostly because of the location of its binding site and of the poor footprints obtained with DNaseI, is that RcaB would be a specific sigma factor. The affinity of RcaA and RcaD for their cognate sequences is lost upon treatment with alkaline phosphatase (Sobczyk *et al.*, 1993 and 1994), and the post-translational modification of RcaA and RcaD could thus be an O-phosphorylation, a kind of modification already observed for the PII protein in *Synechococcus* PCC7942 (see above). Only one binding site was recognized for RcaA-P while two were found for RcaD-P that are situated much further upstream than the RcaA-P site with regards to the transcriptional start sites. At a molecular level, the regulatory mechanisms that allow to inversely switch on or off the *cpeBA* and *cpc2* operons thus differ. The identification of the photoreceptor(s) that can sense the prevalent light wavelengths received by the cells during growth will represent an important breakthrough, since no progress have been made for more than ten years.

DIFFERENTIATION PROCESSES

It is well known that upon nitrogen starvation many filamentous cyanobacteria can differentiate some of their vegetative cells into a new cell type, the heterocyst, which is specialized in the reduction of molecular nitrogen. However, a few other filamentous strains like the marine *Trichodesmium* sp. can fix nitrogen under aerobic conditions, although no particular cell type can be recognized (Zehr, 1995). Heterocyst differentiation probably requires more than a hundred genetic loci which have to be orderly expressed. Thanks to the development of genetic tools and to mutant analysis, a few hints concerning this process have recently been reported. Some of the characterized gene products are members of the two-component systems or of ATP-dependent transport devices while for others the molecular modes of action have yet to be established (for reviews, see Buikema and Haselkorn, 1993; Wolk *et al.*, 1994). Since it is now possible to follow the expression of a given gene within the individual cells of a filament (Elhai and Wolk, 1990) and even to amplify the signal *in vivo* (Wolk *et al.*, 1993), responses to variations in the environment can be precisely monitored for each cell type. In some but not all of the filamentous strains, akinete formation occurs adjacent to heterocysts (Rippka and Herdman, 1985). Very little is known about this differentiation process but progress was recently made on the ontogenesis of akinetes and heterocysts that do share common envelope polysaccharide and glycolipid components. Leganés *et al.* (1994) have reported that mutation in *hetR* renders *Nostoc ellipsosporum* defective in both heterocyst and akinete differentiation while a *hetP* mutation prevents heterocyst but not akinete formation. The location within the filament as well as the number of heterocysts vary with strains, but heterocyst differentiation resembles the formation of bacteroids within the nodules of the *Rhizobium*-legume symbiosis since neither bacteroids nor heterocysts can regenerate to vegetative cells.

In contrast, hormogonium differentiation may affect all the cells of a filament and is always transient since every cell regenerates into a vegetative form (Tandeau de Marsac, 1994). Various environmental factors have been reported to either induce or inhibit hormogonium differentiation. Usually, hormogonia are abundantly formed upon transfer of stationary phase cultures to fresh medium and a dialyzable inhibitor yet to be identified is produced by the vegetative cells during growth (Herdman and Rippka, 1988). Hormogonium differentiation tends to occur each time cells are stressed by environmental changes which could be: dilution to fresh medium, altered nitrogen metabolism, a cold shock or an increase in light intensity, for example. The light effect has been studied in some detail for the *Calothrix* sp. strains. It was first demonstrated that red light is required for full induction to occur and that green light can prevent and sometimes stop hormogonium differentiation (Damerval *et al.*, 1991). Since the same wavelengths are implicated in the control of complementary chromatic adaptation, the simplest hypothesis was that a single photoreceptor could govern both processes. However, using inhibitors of the photosynthetic electron transport chain, conditions have been found under which dissociation of the two phenomena was observed. For the control of hormogonium differentiation light directly operates through differential excitation of PSI and PSII and during the time hormogonia form, heterocyst differentiation is prevented (Campbell *et al.*,

1993). Profound intracellular modifications occur during hormogonium differentiation, the transcription of a number of genes being mutually exclusive throughout the process (Damerval *et al.*, 1991).

With *Nostoc* PCC73102, hormogonium differentiation can be induced by specific factors, produced by the plant partner with which it will establish a symbiotic association. In the absence of ammonia, *Anthoceros punctatus* produces an extracellular compound which triggers the differentiation into hormogonia of *Nostoc* sp. filaments, the formation of motile hormogonia representing a developmental stage required for the establishment of the symbiosis (Campbell and Meeks, 1989). Work is in progress towards the isolation of this compound and the identification of the corresponding cyanobacterial receptor, in particular *via* the use of interposon mutagenesis and complementation studies (JC Meeks, personal communication). Another interesting property of symbiotic cyanobacterial trichomes is that they usually contain a much higher percentage of heterocysts than free-living filaments. This is reminiscent of the differentiation of the rhizobial cells into bacteroids within the nodules of the leguminous symbiotic partner. In cyanobacterial symbiosis, the signal as well as the transduction pathway that govern heterocyst differentiation remain to be identified.

Conclusions and prospects

Environmental changes in natural habitats are most often more gradual than in laboratories where extreme conditions can be applied to obtain clear-cut responses. As pointed out by Wyman and Fay (1987), illumination exclusively in the red waveband, for example, remains an exotic condition in the context of the natural environments. In addition, the other members of the ecosystem certainly do not stay inactive and competition for essential nutrients as well as syntrophy do occur. Homeostasis requires an intricate network of regulation between the essential molecular mechanisms that ensure light harvesting, CO_2 fixation and nitrogen assimilation. In filamentous cyanobacteria, the existence of differentiated cell types fulfills the requirements for mutually exclusive transcription of certain subsets of genes. In *Anabaena* and *Nostoc* spp. strains *nif* gene expression, for example, only occurs following precise DNA rearrangements within the heterocysts and the switching off of some operons (*rbcLS, cpcBA* and *apcAB*) (Wolk *et al.*, 1994). In *Calothrix* sp., during hormogonium differentiation, there is a general repression of the transcription of the phycobiliprotein genes as long as the genes that encode the components of the gas vesicles (*gvp*) are on, the *rbcLS* operon being unaffected (Damerval *et al.*, 1991).

Sequential and/or temporarly exclusive expression of genes immediately suggest that many transcription factors come into play, and very few of the actors have yet been identified (see above). Three and four sigma factors, respectively, have been found in *Anabaena* PCC7120 and *Synechococcus* PCC7942 but, except for the absolute requirement for SigA (RpoD1), the physiological role of the other members of the family awaits characterization (Brahamsha and Haselkorn, 1991 and 1992; Tanaka *et al.*, 1992). Other sigma factors, anti-sigmas and DNA-binding proteins are likely to be involved in these regulations, as well as RNA-binding proteins such as that recently shown to be expressed by *Anabaena variabilis* upon a cold shock (Sato, 1994). Multifunctional proteins which, like the *R. meliloti* FixL, have an enzyme function (kinase) and a transmembrane sensory domain (oxygen-binding haem moiety) represent good examples of what we could expect as receptors for the environmental stimuli (Gilles-Gonzalez *et al.*, 1991). The identification of the molecular constituents of the networks is the goal of an increasing number of laboratories and many elements as well as the connections that link them may bring interesting new concepts.

Acknowledgements
I am very grateful to Dr. N. Tandeau de Marsac for her constant support and fruitful collaboration. I also wish to thank all my colleagues, and in particular Dr. M. Herdman for valuable comments and a careful reading of the manuscript.

References

Aiba H, Nagaya M, Mizuno T (1993) Sensor and regulator proteins from the cyanobacterium *Synechococcus* species PCC7942 that belong to the bacterial signal-transduction protein families: implication in the adaptive response to phosphate limitation. *Mol Microbiol* **8:** 81-91

Ajlani G, Kirilovsky D, Picaud M, Astier C (1989) Molecular analysis of *psbA* mutations responsible for various herbicide resistance phenotypes in *Synechocystis* 6714. *Plant Mol Biol* **13:** 469-479

Allen JF (1992) Protein phosphorylation in regulation of photosynthesis. *Biochim Biophys Acta* **1098:** 275-335

Allen MM (1968) Photosynthetic membrane system in *Anacystis nidulans*. *J Bacteriol* **96:** 836-841

Bouyoub A, Vernotte C, Astier C (1993) Functional analysis of the two homologous *psbA* gene copies in *Synechocystis* PCC 6714 and PCC 6803. *Plant Mol Biol.* **21:** 249-258

Brahamsha B, and Haselkorn R (1991) Isolation and characterization of the gene encoding the principal sigma factor of the vegetative cell RNA polymerase from the cyanobacterium *Anabaena* sp. strain PCC 7120. *J Bacteriol* **173:**2442-2450.

Brahamsha B and Haselkorn R (1992) Identification of multiple RNA polymerase sigma factor homologs in the cyanobacterium *Anabaena* sp. strain PCC 7120: cloning, expression, and inactivation of the *sigB* and *sigC* genes. *J Bacteriol* **174:** 7273-7282

Buikema WJ and Haselkorn R (1993) Molecular genetics of cyanobacterial development. *Annu Rev Plant Physiol Plant Mol Biol* **44:** 33-52

Bustos SA and Golden SS (1991) Expression of the *psbDII* gene in *Synechococcus* sp. strain PCC 7942 requires sequences downstream of the transcription start site. *J Bacteriol* **173:** 7525-7533

Bustos SA and Golden SS (1992) Light-regulated expression of the *psbD* gene family in *Synechococcus* sp. strain PCC 7942: evidence for the role of duplicated *psbD* genes in cyanobacteria. *Mol Gen Genet* **232:** 221-230

Campbell D, Houmard J, Tandeau de Marsac N (1993) Electron transport regulates cellular differentiation in the filamentous cyanobacterium *Calothrix. Plant Cell* **5:** 451-463

Campbell EL and Meeks JC (1989) Characteristics of hormogonia formation by symbiotic *Nostoc* spp. in response to the presence of *Anthoceros punctatus* or its extracellular products. *Appl Environ Microbiol* **55:** 125-131

Chiang GG, Schaefer MR, Grossman AR (1992) Complementation of a red-light-indifferent cyanobacterial mutant. *Proc Natl Acad Sci USA* **89:** 9415-9419

Clarke AK, Hurry VM, Gustafsson P, Öquist G (1993) Two functionally distinct forms of the photosystem II reaction-center protein D1 in the cyanobacterium *Synechococcus* sp. PCC 7942. *Proc Natl Acad Sci USA* **90:** 11985-11989

Cobley JG, Zerweck E, Reyes R, Mody A, Seludo-Unson JR, Jaeger H, Weerasuriya S, Navankasattusas S (1993) Construction of shuttle plasmids which can be efficiently mobilized from *Escherichia coli* into the chromatically adapting cyanobacterium, *Fremyella diplosiphon. Plasmid* **30:** 90-105

Conley PB, Lemaux PG, Grossman AR (1985) Cyanobacterial light-harvesting complex subunits encoded in two red light-induced transcripts. *Science* **230:** 550-553

Damerval T, Guglielmi G, Houmard J, Tandeau de Marsac N (1991) Hormogonium differentiation in the cyanobacterium *Calothrix*: a photoregulated developmental process. *Plant Cell* **3:** 191-201

Elhai J and Wolk CP (1988) Conjugal transfer of DNA to cyanobacteria. *Methods Enzymol.* **167:** 747-754.

Elhai J and Wolk CP (1990) Developmental regulation and spatial pattern of expression of the structural genes for nitrogenase in the cyanobacterium *Anabaena. EMBO J* **9:** 3379-3388.

Fay P (1992) Oxygen relations of nitrogen fixation in cyanobacteria. *Microbiol Rev* **56:** 340-373.

Forchhammer K and Tandeau de Marsac N (1994) The PII protein in the cyanobacterium *Synechococcus* sp. strain PCC 7942 is modified by serine phosphorylation and signals the cellular N-status. *J Bacteriol* **176:**84-91

Fujita Y, Murakami A, Ohki K (1987) Regulation of photosystem composition in the cyanobacterial photosynthetic system: the regulation occurs in response to the redox state of the electron pool located between the two photosystems. *Plant Cell Physiol* **28:** 283-292

Gilles-Gonzalez MA, Ditta GS, Helinski DR (1991) A haemoprotein with kinase activity encoded by the oxygen sensor of *Rhizobium meliloti. Nature* **350:** 170-172

Glazer AN (1989) Light guides. Directional energy transfer in a photosynthetic antenna. *J Biol Chem* **264:** 1-4

Golden SS, Brusslan J, Haselkorn R (1986) Expression of a family of *psbA* genes encoding a photosystem II polypeptide in the cyanobacterium *Anacystis nidulans* R2. *EMBO J* **5:** 2789-2798

Golden SS, Nalty MS, Cho DSC (1989) Genetic relationship of two highly studied *Synechococcus* strains designated *Anacystis nidulans*. *J Bacteriol* **171:** 24-29

Grossman AR, Schaefer MR, Chiang GG, Collier JL (1993) The phycobilisome, a light-harvesting complex responsive to environmental conditions. *Microbiol Rev* **57:** 725-749

Guerrero MG and Lara C (1987) Assimilation of inorganic nitrogen. In *The Cyanobacteria.* Fay P, Van Baalen C (eds) Elsevier, Amsterdam, 163-186

Gupta A, Morby AP, Turner JS, Whitton BA, Robinson NJ (1993) Deletion within the metallothionein locus of cadmium-tolerant *Synechococcus* PCC 6301 involving a highly iterated palindrome (HIP1). *Mol Microbiol* **7:** 189-195

Herdman M and Rippka R (1988) Cellular differentiation: hormogonia and baeocytes. *Methods Enzymol* **167:** 232-242 and 851-853 (Addendum)

Huckle JW, Morby AP, Turner JS, Robinson NJ (1993) Isolation of a prokaryotic metallothionein locus and analysis of transcriptional control by trace metal ions. *Mol Microbiol* **7:** 77-187

Kalla R, Bhalerao RP, Gustafsson P (1993) Regulation of phycobilisome rod proteins and mRNA at different light intensities in the cyanobacterium *Synechococcus* 6301. *Gene* **126:** 77-83

Khanna R, Graham JR, Myers J, Gantt E (1983) Phycobilisome composition and possible relationship to reaction centers. *Arch Biochem Biophys* **224:** 534-542

Koenig F (1990) Shade adaptation in cyanobacteria. Further characterization of *Anacystis* shade phenotype as induced by sublethal concentrations of DCMU-type inhibitors in strong light. *Photosynth Res* **26:** 29-37

Kondo T and Ishiura M (1994) Circadian rhythms of cyanobacteria: monitoring the biological clocks of individual colonies by bioluminescence. *J Bacteriol* **176:** 1881-1885

Kuhlemeier CJ, Borrias WE, van den Hondel CAMJJ, van Arkel GA (1981) Vectors for cloning in cyanobacteria: construction and characterization of two recombinant plasmids capable of transformation to *Escherichia coli* K12 and *Anacystis nidulans* R2. *Mol Gen Genet* **184:** 49-254

Kuhlemeier CJ, Thomas AAM, van der Ende A, van Leen RW, Borrias WE, van den Hondel CAMJJ, van Arkel GA (1983) A host-vector system for gene cloning in the cyanobacterium *Anacystis nidulans* R2. *Plasmid* **10:** 156-163

Kulkarni RD and Golden SS (1994) Adaptation to high light intensity in *Synechococcus* sp. strain PCC 7942: Regulation of three *psbA* genes and two forms of the D1 protein. *J Bacteriol* **176:** 959-965

Leganés F, Fernández-Piñas F, Wolk CP (1994) Two mutations that block heterocyst differentiation have different effects on akinete differentiation in *Nostoc ellipsosporum*. *Mol Microbiol* **12:** 679-684

Lönneborg A, Lind LK, Kalla SR, Gustafsson P, Öquist G (1985) Acclimation processes in the light-harvesting system of the cyanobacterium *Anacystis nidulans* following a light shift from white to red light. *Plant Physiol* **78:** 110-114

Luque I, Flores E, Herrero A (1994a) Nitrate and nitrite transport in the cyanobacterium *Synechococcus* sp PCC 7942 are mediated by the same permease. *Biochim Biophys Acta* **1184:** 296-298

Luque I, Flores E, Herrero A (1994b) Molecular mechanism for the operation of nitrogen control in cyanobacteria. *EMBO J* **13:** 2862-2869

Magasanik B (1993) The regulation of nitrogen utilization in enteric bacteria. J Cell Biochem **51:** 34-40

Mann NH (1995) How do cells express nutrient limitation at the molecular level? *This volume* 171-190

Meeks JC (1988) Symbiotic associations. *Methods Enzymol* **167:** 113-121

Mérida A, Candau P, Florencio FJ (1991a) In vitro reactivation of in vivo ammonium-inactivated glutamine synthetase from *Synechocystis* sp. PCC 6803. *Biochem Biophys Res Commun* **181:** 780-786

Mérida A, Candau P, Florencio FJ (1991b) Regulation of glutamine synthetase activity in the unicellular cyanobacterium *Synechocystis* sp. strain PCC 6803 by the nitrogen source: effect of ammonium. *J Bacteriol* **173:** 4095-4100

Mohamed A and Jansson C (1989) Influence of light on accumulation of photosynthesis-specific transcripts in the cyanobacterium *Synechocystis* 6803. *Plant Mol Biol* **13:** 693-700

Mohamed A and Jansson C (1991) Photosynthetic electron transport controls degradation but not production of *psbA* transcripts in the cyanobacterium *Synechocystis* 6803. *Plant Mol Biol* **16:** 891-897

Mohamed A, Eriksson J, Osiewacz HD, Jansson C (1993) Differential expression of thc *psbA* genes in the cyanobacterium *Synechocystis* 6803. *Mol Gen Genet* **238:** 161-168

Mulligan B, Schultes N, Chen L, Bogorad L (1984) Nucleotide sequence of a multiple-copy gene for the B protein of photosystem II of a cyanobacterium. *Proc Natl Acad Sci USA* **81:** 2693-2697

Mullineaux CW (1994) Excitation energy transfer from phycobilisomes to photosystem I in a cyanobacterial mutant lacking photosystem II. *Biochim Biophys Acta* **1184:** 71-77

Murakami A and Fujita Y (1991a) Regulation of photosystem stoichiometry in the photosynthetic system of the cyanophyte *Synechocystis* PCC 6714 in response to light-intensity. *Plant Cell Physiol* **32:** 223-230

Murakami A and Fujita Y (1991b) Steady state of photosynthetic electron transport in cells of the cyanophyte *Synechocystis* PCC 6714 having different stoichiometry between PS I and PS II: analysis of flash-induced oxidation-reduction of cytochrome f and P700 under steady state of photosynthesis. *Plant Cell Physiol* **32:** 213-222

Nagaya M, Aiba H, Mizuno T (1993) Cloning of a sensory-kinase-encoding gene that belongs to the two-component regulatory family from the cyanobacterium *Synechococcus* sp. PCC7942. *Gene* **131:** 119-124

Nagaya M, Aiba H, Mizuno T (1994) The sphR product, a two-component system response regulator protein, regulates phosphate assimilation in *Synechococcus* sp. strain PCC 7942 by binding to two sites upstream from the *phoA* promoter. *J Bacteriol* **176:** 2210-2215

Parkinson JS (1993) Signal transduction schemes of bacteria. *Cell* **73:** 857-871

Potts M, Sun H, Mockaitis K, Kennelly PJ, Reed D, Tonks NK (1993) A protein-tyrosine/serine phosphatase encoded by the genome of the cyanobacterium *Nostoc commune* UTEX 584. *J Biol Chem* **268:** 7632-7635

Reyes JC and Florencio FJ (1994) A new type of glutamine synthetase in cyanobacteria: the protein encoded by the *glnN* gene supports nitrogen assimilation in *Synechocystis* sp. strain PCC 6803. *J Bacteriol* **176:** 1260-1267

Rippka R and Herdman M (1985) Division patterns and cellular differentiation in cyanobacteria. *Ann Microbiol (Inst. Pasteur)* **136A:** 33-39

Rippka R and Herdman M (1993) Pasteur Culture Collection of Cyanobacterial Strains in Axenic Culture. In *Catalogue & Taxonomic Handbook. Institut Pasteur, Paris, vol I: Catalogue of Strains*, 1-103

Ronen-Tarazi M, Schwarz R, Bouevitch A, Lieman-Hurwitz J, Erez J, Kaplan A (1995) Response of photosynthetic microorganisms to changing ambient concentrations of CO_2. *This volume* 323-334

Samson G, Herbert SK, Fork DC, Laudenbach DE (1994) Acclimation of the photosynthetic apparatus to growth irradiance in a mutant strain of *Synechococcus* lacking iron superoxide dismutase. *Plant Physiol* **105:** 287-294

Sato N (1994) A cold-regulated cyanobacterial gene cluster encodes RNA-binding protein and ribosomal protein S21. *Plant Mol Biol* **24:** 819-823

Schaefer MR and Golden S.S (1989) Light availability influences the ratio of two forms of D1 in cyanobacterial thylakoids. *J Biol Chem* **264:** 7412-7417

Schmidt-Goff CM and Federspiel NA (1993) In vivo and in vitro footprinting of a light-regulated promoter in the cyanobacterium *Fremyella diplosiphon*. *J Bacteriol* **175:** 1806-1813

Simon, M (1995) Signal transduction in microorganisms. *This volume* 205-215

Smart LB and McIntosh L (1991) Expression of photosynthesis genes in the cyanobacterium *Synechocystis* sp. PCC 6803: *psaA-psaB* and *psbA* transcripts accumulate in dark-grown cells. *Plant Mol Biol* **17:** 959-971

Sobczyk A, Schyns G, Tandeau de Marsac N, Houmard J (1993) Transduction of the light signal during complementary chromatic adaptation in the cyanobacterium *Calothrix* sp. PCC 7601: DNA-binding proteins and modulation by phosphorylation. *EMBO J* **12:** 997-1004

Sobczyk A, Bely A, Tandeau de Marsac N, Houmard J (1994) A phosphorylated DNA-binding protein is specific for the red light signal during complementary chromatic adaptation in cyanobacteria. *Mol Microbiol* **13:** 875-885

Tanaka K, Masuda S, Takahashi H (1992) Multiple *rpoD*-related genes of cyanobacteria. *Biosci Biotech Biochem* **56:**1113-1117

Tandeau de Marsac N (1991) Chromatic adaptation by cyanobacteria. In *Cell Culture and Somatic Cell Genetics of Plants.* Bogorad L, Vasil I (eds). Academic Press Inc., New York, vol. 7B - The Photosynthetic Apparatus: Molecular Biology and Operation, 417-446

Tandeau de Marsac N (1994) Differentiation of hormogonia and relationships with other biological processes. In *The Molecular Biology of Cyanobacteria.* Bryant DA (ed). Kluwer Academic Publishers, Dordrecht, Ch 28, In press

Tandeau de Marsac N and Houmard J (1993) Adaptation of cyanobacteria to environmental stimuli: new steps towards molecular mechanisms. *FEMS Microbiol Rev* **104:** 119-190

Tandeau de Marsac N, Borrias WE, Kuhlemeier CJ, Castets AM, van Arkel GA, van den Hondel CAMJJ (1982) A new approach for molecular cloning in cyanobacteria: cloning of an *Anacystis nidulans met* gene using a Tn*901*-induced mutant. *Gene* **20:** 111-119

Tandeau de Marsac N, Mazel D, Damerval T, Guglielmi G, Capuano V, Houmard J (1988) Photoregulation of gene expression in the filamentous cyanobacterium *Calothrix* sp. PCC 7601: light-harvesting complexes and cell differentiation. *Photosynth Res* **18:** 99-132

van der Plas J, Hegeman H, de Vrieze G, Tuyl M, Borrias M, Weisbeek P (1990) Genomic integration system based on pBR322 sequences for the cyanobacterium *Synechococcus* sp. PCC7942: transfer of genes encoding plastocyanin and ferredoxin. *Gene* **95:** 39-48

Vasudevan SG, Gedye C, Dixon NE, Cheah E, Carr PD, Suffolk PM, Jeffrey PD, Ollis DL (1994) *Escherichia coli* PII protein: purification, crystallization and oligomeric structure. *FEBS Lett* **337:** 255-258

Vrba JM and Curtis SE (1989) Characterization of a four-member *psbA* gene family from the cyanobacterium *Anabaena* PCC 7120. *Plant Mol Biol* **14:** 81-92

Wolk CP, Elhai J, Kuritz T, Holland D (1993) Amplified expression of a transcriptional pattern formed during development of *Anabaena. Mol Microbiol* **7:** 441-445

Wolk CP, Ernst A, Elhai J (1994) Heterocyst metabolism and development. In *The Molecular Biology of Cyanobacteria.* Bryant DA (ed). Kluwer Academic Publishers, Dordrecht, Ch 27, In press

Wyman M and Fay P (1987) Acclimation to the natural light climate. In *The Cyanobacteria.* Fay P and van Baalen C (eds). Elsevier, Amsterdam, 347-376

Zehr J (1995) Nitrogen fixation in the sea: why only *Trichodesmium*? *This volume* 335-364

Zhang CC (1993) A gene encoding a protein related to eukaryotic protein kinases from the filamentous heterocystous cyanobacterium *Anabaena* PCC 7120. Proc Natl Acad Sci USA **90:** 11840-11844

How Do Cells Express Nutrient Limitation at the Molecular Level?

Nicholas H. Mann
Department of Biological Sciences
University of Warwick
Coventry CV4 7AL
UK.

Introduction

Microbial growth in both marine and freshwater environments is frequently restricted by nutrient availability. The process of nutrient limitation is not a sudden transition from an actively growing to a non-growing state, but rather involves sequential changes that first enable the organism to utilize appropriate reserves, scavenge trace amounts of the limiting nutrient, or to utilize alternative less preferred sources of the nutrient and then to undergo the various physiological and ultrastructural changes that allow prolonged survival in the non-growing state. The nature of the limiting nutrient will vary in response to a range of factors (geographical, seasonal, anthropogenic *etc*), but will most commonly be for one of the quantitatively major nutrients, namely carbon, nitrogen, phosphorus and sulphur. Though there is evidence that trace elements such as iron and zinc may limit productivity in some environments. Furthermore, in the case of photoautotrophic micro-organisms it is reasonable to consider light as a nutrient.

This chapter concentrates primarily on how molecular approaches have facilitated analysis of the responses to nutrient limitation of the microorganisms contributing substantially to primary production in both the marine and freshwater environments. Consequently, the planktonic cyanobacteria will be the major group of organisms to be considered, but where necessary the information derived from the heterotrophic bacterioplankton and model organisms such as *Escherichia coli* will be used. Although C_i is major inorganic nutrient for cyanobacteria responses to limited C_i availability are dealt with elsewhere in this volume (Ronen-Tarazi *et al.* 1995).

NATO ASI Series, Vol. G 38
Molecular Ecology of Aquatic Microbes
Edited by I. Joint

Overview

Once the environmental concentration of a major nutrient falls below a certain level the cell will respond by inducing high affinity transport systems and pathways which can exploit alternative sources of the nutrient. In many cases these responses are regulated by two component sensory systems which are considered in detail elsewhere in this volume. The transport of all inorganic nutrients into microbial cells involves highly specific membrane transport systems. Typically, for non-starved cells the majority of inorganic nutrients enter the cell by active transport systems (symports, antiports and uniports) driven by the proton motive force or by another ion gradient, itself formed at the expense of the proton motive force. This situation has become more complicated with respect to ionic homeostasis as a result of the discovery in prokaryotes of P-type ATPases which drive the uptake or efflux of mono- and divalent cations (see Fagan and Saier, 1994). For certain major inorganic nutrients there exists a range of inducible high affinity transport systems that become operative when the relevant nutrient is limiting. As will become apparent below, when considering the responses of microorganisms in aquatic ecosystems to nutrient limitation a group of transport systems known as the ABC (ATP-binding cassette) proteins (Hyde *et al.*, 1990) or traffic ATPases (Ames and Lear, 1992) are of central importance (for recent reviews see Higgins, 1992; Doige and Ames, 1993). This superfamily of membrane transporters includes not only prokaryotic systems, but also eukaryotic examples, including the medically important multidrug resistance protein and the cystic fibrosis transmembrane regulator. Members of the family exhibit considerable primary sequence and secondary structure similarity. The bacterial permease members accumulate substrates against very large concentration gradients (as much as 1:10,000) and with apparent K_ts in the range 0.01 - 1μmol l^{-1}. The key features of the bacterial permease traffic ATPases are that they include a periplasmic substrate-binding protein and a cytoplasmic membrane-bound complex of four subunits, consisting of two hydrophobic proteins and two subunits of a hydrophilic protein. The receptor-substrate complex interacts with the membrane complex and translocation of the substrate occurs via a series of conformational changes concomitant with ATP hydrolysis. The membrane complex is typically composed of two hydrophobic proteins with multiple membrane-spanning domains, which may form a channel or binding site for the substrate, and two subunits of the hydrophilic protein which is tightly membrane-associated and has the ATP hydrolysing activity.

A second feature which appears as a common theme linked with the traffic ATPases in cyanobacteria is their association in cyanobacteria with regulatory proteins which belong to CRP - FNR family of transcriptional activators (Shaw *et al.*, 1983). CRP (the cAMP receptor protein) stimulates transcription of the genes of the *lac* operon, and other operons subject to catabolite repression, by about 50 fold when it is complexed with cAMP and bound to the CRP site upstream from promoter. Upon binding to DNA, CRP induces a substantial bend of the order of 100° - 130° (see Zinkel and Crothers, 1991). Two models have been proposed for CRP one involving protein-protein interactions and the other involving changes in DNA structure. The present evidence suggests that the DNA wraps around the CRP protein in a way that puts the promoter at the terminal loop of an interwound negative superhelical domain (Zinkel and Crothers, 1991). Formation of this special structure may lead to transduction of the superhelical stress to unwinding of the promoter during formation of the open complex with the σ^{70} RNA polymerase. FNR is a transcriptional activator for a number of operons involved in anaerobic metabolism in *E. coli.* The sequence similarity between CRP and FNR is most pronounced in the C-terminal region which includes the 'helix-turn-helix' motif involved in DNA binding. FNR, however, lacks the nucleotide binding site of CRP and instead has a cysteine rich N-terminal extension which is thought to form part of a metal-binding redox sensing domain (see Guest, 1992). The consensus DNA-binding sites for CRP and FNR are very similar and in fact can be functionally interconverted. Other proteins now thought to be members of this family include FixK (*Rhizobium meliloti*), HlyX (*Actinobacillus pleuropneumoniae*), FnrN (*Rhizobium leguminosarum*), ANR (*Pseudomonas aeruginosa*) (see Vega-Palas *et al.*, 1992).

Light

Cyanobacteria, like other components of the phytoplankton, possess an oxygenic photosynthesis apparatus, utilizing two distinct photosystems and organized in the thylakoid membranes, which themselves are usually distinct from the cytoplasmic membrane. Chlorophyll *a* is involved in the reaction centres of both photosystems, but cyanobacteria lack light-harvesting complex II (LHCII) and instead phycobiliproteins, organized into macromolecular complexes known as phycobilisomes, represent the primary light harvesting antennae (for review see Grossman *et al.*, 1993). Cyanobacteria are capable of responding to changes in both the intensity and spectral quality of light by both short and long term adjustments to the photosynthetic apparatus. Among the responses to changes in spectral quality are state transitions, which involve a redistribution of excitation energy between PSI (photosystem I) and PSII (photosystem II), and chromatic adaptation in which the complement of light-harvesting phycobiliproteins becomes modified (see elsewhere in this volume). A potentially lethal result of their phototrophic lifestyle is the risk of photoinhibition, when the rate at which light energy is harvested exceeds the rate at which it can be converted into useful metabolic energy. This danger is dramatically enhanced by nutrient limitation, which greatly increases the likelihood of non-productive excitation of photosynthetic reaction centres and under such conditions the phycobilisomes may be rapidly degraded (see below). Furthermore, under conditions where C_i availability is restricted, photosynthetic electron transport may lead to the extensive production of superoxide radicals with the potential for extensive oxidative damage to the cell. These factors emphasize the critical importance to the physiology of cyanobacteria of integrating the central metabolic process of light harvesting with inorganic nutrient acquisition.

The regulation of light-induced state transitions in cyanobacteria has accumulated a very considerable and somewhat contradictory literature, which could not be dealt with adequately in this short chapter. Fortunately this topic has been recently reviewed by Allen (1992). The debate concerns the models which have been put forward to explain two central aspects of state transitions, to wit control of the process and the mechanism of energy redistribution. It is variously proposed that the control of state transitions is determined by (i) the redox state of components of the intersystem chain, (ii) localized electron gradients around PSI and PSII, and (iii) cyclic electron flow around PSI. The opposing ideas concerning energy redistribution centre round spill-over from PSII to PSI or redirection of energy harvested by the phycobilisome directly to PSI. Correlative evidence implicates thylakoid protein phosphorylation in the state transition process. It is important in the context of considering responses to nutrient limitation, particularly given the potential for photoinhibition, to recognize that state transitions, and/or closely related phenomena, may be caused by nutritional factors such as nitrogen metabolism and C_i limitation.

Phosphate limitation

Phosphate is commonly recognized as the limiting nutrient in freshwater environments, and there is increasing evidence that it may limit productivity in some oceanic systems. Recently, molecular biological techniques have begun to be employed to study phosphate acquisition in cyanobacteria. In this laboratory we wished to develop molecular probes that would be capable of defining the nutrient status of individual oceanic cyanobacterial picoplankton. Using cultures of *Synechococcus* sp. WH7803 we compared the protein profiles of cells grown under phosphate-replete or deplete conditions and detected the increased synthesis of several polypeptides under the limiting conditions (Scanlan *et al.*, 1993). Amino-terminal sequence information was obtained for one of the polypeptides approximately 32 kDa in size and was used to design an oligonucleotide probe to screen for the cloned gene. Sequence analysis of the gene revealed it to encode a polypeptide which exhibited 35% identity with the phosphate starvation inducible periplasmic phosphate-binding protein (PstS) of *E.*

coli. Furthermore, the key residues implicated in phosphate binding by PstS were conserved in the cyanobacterial protein. This observations taken together with its location in the cell envelope and induction by P_i limitation led us to propose that the protein was a cyanobacterial PstS homologue. The *pstS* was transcribed as a 1.2 kb monocistronic RNA under P_i replete conditions and as a 1.4 kb RNA of greatly increased abundance under P_i limiting conditions. Both Southern and western blotting confirmed the presence of the gene and protein in other oceanic *Synechococcus* strains and remarkably in a *Prochlorococcus marinus* isolate, but did not detect homologues in freshwater cyanobacteria (this laboratory, DJ Scanlan - unpublished data). However, Aiba and Mizuno (1994) recently described the induction by phosphate limitation of a 33 kDa protein in the freshwater cyanobacterium *Synechococcus* sp. PCC 7942. Sequence analysis of the cloned gene *sphX* revealed an open reading frame (ORF) of 337 amino acids which had a putative bacterial signal-like sequence at the N-terminus; however, no similarities could be detected with entries in the Swiss Prot database. Furthermore, a mutant in which the *sphX* gene was disrupted was reported to grow in phosphate-free medium. At approximately the same time, Wagner *et al.* (1994) reported the induction by phosphate limited growth of a polypeptide with an apparent size of 28-35 kDa in the freshwater cyanobacterium *Anacystis nidulans* (*Synechococcus* species) strain 1402-1. The protein which was described as a high affinity phosphate binding protein could be functionally reconstituted into liposomes. Amino-terminal sequence analyses of tryptic fragments of the protein were carried out, but no significant similarities to other proteins were found. In fact the two sequences reported are identical to two regions of SphX and Mann and Scanlan (1994) have proposed that, in fact, they are members of a phosphate-binding protein family. An alignment of SphX, for which there is complete sequence information, with proteins previously identified as phosphate-binding proteins revealed that, apart from overall similarities between SphX and the phosphate-binding proteins, there is apparent conservation of several residues which have been implicated in interactions with phosphate. Thus it seems that a phosphate-binding protein in the cell envelope is common to both marine and freshwater cyanobacteria as well as a marine prochlorophyte. It is somewhat surprising that the freshwater SphX protein exhibits greatest homology (22.7% identity, 48.5% similarity) to the mycobacterial phosphate-binding protein, rather than to its marine cyanobacterial counterpart.

The fact that an oceanic cyanobacterial phytoplankton should have a phosphate binding protein homologous to that of *E. coli* prompted speculation as to whether there were further similarities to the extremely well characterized enterobacterial Pho regulon which includes several genes and operons encoding proteins involved in the uptake and utilization of extracellular phosphorus sources (for review see Rao and Torriani, 1990). In *E. coli* there is a low affinity transport system (Pit) and when the external Pi concentration falls below about 20μM the high affinity transport system (Pst), which includes the phosphate binding protein PstS, is induced. Expression of the Pho regulon is under the control of the two component sensory system pair PhoR (histidine protein kinase) and PhoB (response regulator) and consequently we attempted to detect the homologous genes in *Synechococcus* sp. WH7803. Comparison of the phosphate response regulator PhoB protein from *E. coli* and *B. subtilis* revealed regions of considerable sequence conservation which we used to design two oligonucleotide probes for screening *Synechococcus* gene libraries. Sequence analysis of a positive clone revealed two large ORFs and the codon usage analysis very strongly suggested that they were genuine coding regions. One ORF exhibited 41% identity with the PhoB homologue of *B. subtilis* and a second ORF exhibited 26% identity with PhoR of *E. coli.* The two ORFs contained all the diagnostic residues and domains characteristic of a response regulator and histidine protein kinase respectively. A northern blotting experiments with RNA extracted from a cultures growing either in normal medium or in P_i-limited medium confirmed the greatly increased expression of these genes under P_i limitation and indicated a transcript size of about 4.4 kb (GMF Watson, unpublished results - this laboratory). Aiba *et al.* (1993) have cloned two component sensory system genes *sphS* and *sphR* from the freshwater cyanobacterium *Synechococcus* sp. PCC 7942 that are also involved in the regulation of phosphate acquisition. Mutants lacking these genes were unable to induce alkaline phosphatase and other polypeptide upon P_i starvation. The *E. coli* PhoB binds to specific sites known as Pho-boxes upstream from genes of the Pho regulon. Nagaya *et al.* (1994) have shown that SphR is phosphorylated by phosphotransfer from SphS and binds to two distinct sites upstream from the *phoA* (alkaline phosphatase) promoter. One of these sites contains within it a sequence identified as a possible cyanobacterial Pho-box by Ray *et al.* (1991). In the case of the oceanic picoplankton *Synechococcus* sp. WH7803, sequences bearing reasonable similarities to the *E. coli* Pho-box are found upstream from *pstS* and *phoBR.* A third possible Pho-box is found downstream of PstS immediately upstream of an ORF designated *ptrA.* These marked similarities between the situation in cyanobacteria and enterobacteria suggest that cyanobacteria can also be considered to have a Pho regulon and consequently one would predict that the actual high affinity P_i transport system would involve members of the highly conserved super-family of traffic ATPases. Sequence analysis of the *ptrA* ORF has revealed it to potentially encode a protein of the CRP - FNR family of transcriptional activators (J Bourne, unpublished research - this laboratory). The strongest sequence similarity occurs in the carboxy-terminal part of the protein which contains the helix-turn-helix motif involved in DNA binding. As yet there is no evidence to establish the nature of the signal recognized by the N-terminal domain. In the event that PtrA is involved in the response to P_i limitation it will represent the first significant difference between the cyanobacterial and enterobacterial Pho regulons.

Nitrogen limitation

Nitrogen is frequently the limiting nutrient in many oceanic environments and occasionally in freshwater systems. Most microorganisms can use nitrate, nitrite or ammonium as their N-source and to a greater or lesser extent will utilize organic N-sources. Furthermore, when all such fixed forms of nitrogen become exhausted some organisms, including many species of cyanobacteria, become capable of fixing di-nitrogen. It is commonly accepted that the phycobiliproteins act as a nitrogen-reserve, as well as functioning as light-harvesting pigments and starvation for N in non-dinitrogen fixing cyanobacteria causes chlorosis due to a dramatic loss of phycobiliproteins (see below). Where both ammonium and nitrate are available cyanobacteria tend to preferentially utilize ammonium and there is a strict negative regulation of the proteins involved in nitrate transport and the enzymes involved in its assimilation *eg* nitrate reductase by ammonium. In the enterobacteria the global adaptive response to ammonium starvation involves induction of the Ntr regulon and a central role in the process is played by the *glnA ntrBC* operon. The genes *ntrB* and *ntrC* respectively encode the histidine protein kinase and response regulator proteins of a typical two component sensory system involved in controlling expression of the Ntr regulon (see Stock *et al.*, 1989). Despite extensive searching no homologues of the *ntrB* and *C* genes have been found in cyanobacteria. However, a gene designated *ntcA* gene was found to be required for the full expression of proteins subject to ammonium repression (Vega-Palas *et al.*, 1990). Using an *ntcA* mutant, impaired in the expression of nitrate reductase, for complementation studies Vega-Palas *et al.* (1992) were able to characterize the *ntcA* gene which encodes a protein of 24.8 kDa which belongs to CRP-FNR family of transcriptional activators (Shaw *et al.*, 1983) and is closely related to the CysR protein involved in cyanobacterial sulphur regulation (see below). PtrA protein putatively involved in the response to Pi limitation (see above). NtcA has now been shown (Luque *et al.*, 1994) to bind to promoters of nitrogen regulated genes at a site involving the palindromic DNA sequence $GTAN_8TAC$, which substitutes for the canonical -35 sequence. These promoters have a -10 region in the form TAN_3T. As yet the mechanism by which the cell controls the concentration of active NtcA has not been established.

Starvation for all forms of fixed nitrogen leads many strains of cyanobacteria to develop the ability to fix dinitrogen. In filamentous strains terminally differentiated, specialized cells termed heterocysts are the sites of nitrogen fixation and protect the enzyme from oxygen inactivation. However, nitrogen fixation is not confined to filamentous strains and occurs in several unicellular strains. A great deal is now known about the molecular biology of heterocyst development and has been the subject of a recent review (Buikema and Haselkorn, 1993). Genomic re-arrangements involving the excision of substantial regions of DNA take place during heterocyst differentiation and one of these excisions involves the site-specific recombinase XisA. A protein BifA (VF1) was identified by virtue of its interaction with sequences immediately upstream from xisA (Chastain *et al.*, 1990) and was shown to interact with both vegetative cell- and heterocyst-specific promoters with a consensus recognition sequence TGT N_{9-10}ACA. It is now recognized that BifA is the *Anabaena* PCC 7120 homologue of NtcA (Wei *et al.*, 1993; Frías *et al.*, 1993). Cassette mutagenesis of the *Anabaena* sp. PCC 7120 *ntcA* gene yielded a null mutant which was unable to grow on nitrate or dinitrogen as the N-source, was unable to form heterocysts confirming the central role of NtcA in the process of adaptation to ammonium limitation (Wei *et al.*, 1994).

There is also now information regarding the molecular nature of the high affinity nitrate transport system induced by ammonium limitation. When grown with nitrate as the N-source *Synechococcus* sp. PCC 7942 expresses a 45 kDa protein as a major protein of the cytoplasmic membrane (Omata and Ogawa, 1987); however sequence analysis of the *nrtA* gene shows the NrtA protein to have only two hydrophobic amino acid segments that could span a membrane, and is therefore, unlikely to be directly involved in the actual translocation of nitrate across the membrane. A cluster of three genes (*nrtB*, *nrtC* and *nrtD*) has been characterized downstream from *nrtA* (Omata *et al.*, 1993). NrtB is a hydrophobic protein, which sequence similarity indicates is an integral membrane component of highly conserved super-family of traffic ATPases. Similarly, NtrC and NtrD exhibit sequence similarity with the ATP-binding components of the family. Studies with mutants confirmed the roles of these genes in nitrate transport.

Sulphur limitation

Given the high concentration of sulphate (approx. 27mM) in marine environments, it is highly unlikely to be a limiting nutrient. However, in freshwater environments, sulphate concentrations are much lower; in the freshwater strains such as *Synechococcus* sp. PCC 6301 and PCC 7942, starvation for a sulphur source leads to chlorosis (see below), a loss of photosynthetic membranes, and accumulation of storage compounds including glycogen, polyphosphate and β-hydroxybutyrate (see Tandeau de Marsac and Houmard, 1993). *Synechococcus* sp. PCC 7942 exhibited a 10-20 fold increase in the V_{max} of sulphate transport following sulphur deprivation but there was no change in the K_t suggesting that this response represented further induction of already functional transport system (Green and Grossman, 1988). These changes were accompanied by characteristic changes in the polypeptide composition of the cytoplasmic membrane. A region of the *Synechococcus* sp. PCC 7942 genome required for sulphate transport was cloned and partial sequence analysis revealed a gene (*cysA*) which encoded a protein with marked sequence similarity to the ATP-binding protein component of the traffic ATPase family. The involvement of such a transport system was confirmed by further sequence analysis (Laudenbach and Grossman, 1991) of the cloned fragment which revealed a gene (*sbpA*) adjacent to *cysA*, but transcribed in the opposite direction, encoding a protein exhibiting 40% identity with the *E. coli* periplasmic sulphate-binding protein. Two other genes cysT and cysW encode proteins that may form the channel for transport of sulphate across the cytoplasmic membrane. A fourth gene (*cysR*), located between *cysT* and *cysW* encodes a protein belonging to the CRP-FNR family of transcriptional activators. As was the case for NtcA, the mechanism by which CysR becomes functional as an activator has not yet been established.

Recently, Nicholson and Laudenbach (1993) have implicated the large (approx. 20 kb) cryptic plasmid pANL of *Synechococcus* sp. PCC 7942 in the acclimation to sulphur stress. Inactivation of the *cysR* gene, amongst other effects, prevents the synthesis of two polypeptides (24 kDa and 36 kDa) which are normally induced by sulphur deprivation. The gene encoding the 36 kDa polypeptide was localized to plasmid pANL and sequence analysis revealed the protein to have weak homology to the catalase family of enzyme. Additional open reading frames downstream of the gene showed significant similarities to transport proteins from other bacterial species and Northern analysis revealed that a large region of the plasmid was both sulphur and CysR regulated.

A further adaptive response to sulphur limitation has been characterized in the cyanobacterium *Calothrix* sp. PCC 7601 (Mazel and Marlière, 1989). This organisms encodes sulphur-depleted versions of its most abundant proteins, phycocyanin and associated auxiliary polypeptides, which are expressed under conditions of sulphur limitation.

Trace elements

Several trace elements, including iron, copper, zinc, manganese and cobalt are required as nutrients and function as cofactors for many enzymes. There is considerable theoretical and experimental evidence to suggest that iron may be a limiting nutrient in sea water and freshwater, and there is growing interest in the acquisition of manganese, zinc and copper. Indeed, it has recently been proposed that Zn^{2+} availability may limit primary production by marine phytoplankton (Morel *et al*., 1994). In this context it is interesting to note that potentially toxic Cd^{2+} can substitute for Zn^{2+} in Zn^{2+}-limited cultures of the marine diatom *Thalassiosira weissflogii* (Price and Morel, 1990) probably by replacement of Zn^{2+} in carbonic anhydrase. Marine microorganisms have transport systems capable of accumulating essential trace elements at low oceanic concentration; 0.001 to 1 nmol l^{-1}, and in marine phytoplankton Fe, Mn, Zn and Ni transport has been shown to involve complexation by membrane carriers (Hudson and Morel, 1993). There is only limited information on the detailed molecular mechanisms by which bacteria transport Mn, Ni and Zn (see Silver and Walderhaug, 1992). All bacteria so far studied seem to have high affinity, high specificity Mn^{2+} transport systems. Although Ni^{2+} may be transported by the broad specificity Mg^{2+} transport system they may also be specific high affinity transport systems. Zn^{2+} transport has so far received very little attention.

Copper is required for the activity of several redox proteins and in some cyanobacteria and eukaryotic algae copper-containing plastocyanin and cytochrome *c*-553 function as alternative electron carriers between the cytochrome b_6/f complex and photosystem I. There is contrasting evidence regarding the nature of the control by copper availability of the expression of plastocyanin and cytochrome c-553. In the green alga *Pediastrum boryanum* synthesis is regulated at the pre-translational level (Nakamura *et al*., 1992). In *Synechocystis* sp. PCC 6803 Briggs *et al*. (1990) reported that the control of expression occurred at the translational or post-translational level; however, Zhang et al. (1992) provide evidence for pre-translational control. In *Anabaena* sp. PCC 7937 (Bovy *et al*., 1992) and *Anabaena* sp. PCC 7120 (Ghassemian *et al*., 1994) control appears to occur at the level of the initiation of transcription. Furthermore, two genes encoding P-type ATPases have been cloned from *Synechococcus* sp. PCC 7942 (Kanamaru *et al*., 1993) and one of these genes *pacS* has been shown to a copper-transporting ATPase associated with the thylakoid membranes (Kanamaru *et al*., 1994). The expression of PacS is regulated in response to increased copper concentrations and pacS-inactivated mutants are hypersensitive to elevated concentrations of copper. However, it is still not certain whether PacS functions as a copper accumulating system in the thylakoid lumen or as a copper-efflux system from the lumen to the thylakoid compartment.

The responses of cyanobacteria to iron limitation have received a comparatively detailed examination (see Tandeau de Marsac and Houmard, 1993) and amongst the most characteristic changes are the synthesis of novel polypeptides, reduction in the number of thylakoids, phycobilisomes and carboxysomes and increases in glycogen reserves. Many microorganisms respond to iron starvation by synthesizing and exporting iron-chelating compounds (siderophores) and transport systems capable of taking up the siderophore-iron complexes (for review see Briat, 1992). Several species of cyanobacteria have been shown to produce siderophores (Kerry *et al.*, 1988). A well characterized response of some cyanobacteria to iron limitation is the replacement of ferredoxin by the FMN-containing flavodoxin. There is evidence from *Synechococcus* sp. PCC 7942 that iron protects the ferredoxin I transcript via *cis*-regulatory sequences in the 5' part of the mRNA and thus iron-starvation will cause degradation of ferredoxin mRNA (Bovy *et al.*, 1993). In *Synechococcus* sp. PCC 7942 the *isiB* (flavodoxin) gene is co-transcribed with *isiA* and upstream from the transcription start point of the operon are three 17 bp sequences which resemble 'iron boxes' (Laudenbach *et al.*, 1988). In *E coli* the operons involved in iron acquisition are regulated by the Fur protein which binds to 'iron box' sequences in the promoter regions of these operons (Bagg and Neilands, 1987) A similar *isiBA* operon has been characterized in *Synechococcus* sp. PCC 7002 (Leonhardt and Straus, 1992). The polypeptide encoded by the *isiA* from both *Synechococcus* strains exhibits a marked similarity to the chlorophyll binding protein CP-43 and confirmation that the *isiA* gene encodes a component of the iron-limitation inducible CPVI-4 complex was provided by Burnap *et al.* (1993). A gene, *irpA,* which is thought to encode a protein involved in iron transport in *Synechococcus* sp. PCC 7942 has been characterized and also has an upstream sequence similar to an 'iron box' (Reddy *et al.*, 1988).

Utilizing reserves

Cyanobacteria, in common with other microorganisms, can accumulate significant reserves of carbon, nitrogen, phosphorus and iron (see Carr and Wyman, 1986). Carbon is stored in form of glycogen granules and whilst these tend to accumulate during starvation for N or Fe, glycogen is utilized during conditions of light limitation or carbon dioxide starvation. Phosphorus is stored in polyphosphate granules and tends to increase during sulphur starvation. Cyanophycin is a high molecular weight polymer of arginine and aspartate, synthesized by a non-ribosomal process, which has a widespread distribution in cyanobacteria, but is absent from all other microorganisms. Cyanophycin is rapidly degraded in response to nitrogen starvation confirming its role as a nitrogen reserve. Iron storage presents problems in the presence of oxygen, both in terms of solubility and toxicity. Cyanobacteria, at least in the case of *Synechocystis* sp. PCC 6803, possess the iron storage protein ferritin though this probably does not represent the bulk of the iron store which is associated with low molecular weight (<10 kDa) uncharacterized compounds (Laulhère *et al.*, 1992).

The phycobilisomes which contain the bulk of the soluble protein of the cyanobacterial cell act not only as the light-harvesting structures, but also as reserves of nitrogen. Both the structure and abundance of phycobilisomes in the cell are responsive to a range of environmental parameters including the intensity and spectral quality of the light, but are particularly sensitive to nutrient availability. Following nutrient limitation cyanobacterial cells may lose their typical distinct pigmentation and become chlorotic or bleached. The nature and extent of this process appears to be determined by the specific limiting nutrient. The most detailed studies on this process have been carried out by Collier and Grossman (1992) using *Synechococcus* sp. PCC 7942. They found that the response of the cells to N or S deprivation was similar with chlorophyll synthesis stopping almost immediately and cell division stopping after one (N) or two (S) cycles. The phycobiliproteins were degraded rapidly reflecting an ordered breakdown of the phycobilisomes. In contrast phosphate limitation permitted cell division to carry on for four cycles and chlorophyll continued to accumulate for several hours. Phycobilisomes were not extensively degraded in response to P-limitation although phycobiliprotein accumulation ceased. These differences may reflect the extensive stores of polyphosphate. It is also worth considering that this process of phycobilisome degradation may also serve to protect the cell from damage by oxygen radicals produced by non-productive stimulation of PSII which is enhanced under nutrient limited conditions. There is evidence that in the early phases of nutrient stress, before the phycobilisome is completely degraded that energy transfer to PSII is largely uncoupled (Grossman *et al.*, 1993)

The mechanism of phycobilisome degradation has been investigated using mutants of *Synechococcus* sp. PCC 7942 defective in the process (Collier and Grossman, 1994). Complementation studies with these 'non-bleaching' mutants revealed two genes both of which are regulated by N and S limitation. involved in the process. *nblA* encodes a polypeptide of 59 amino acids which is required for phycobilisome degradation and which might directly interact with the phycobilisome, or a protease, or might regulate gene expression. A second gene *txlA* encoding a protein of 190 amino acids is not involved in the bleaching response, but may be involved in energy transfer from the phycobilisome to PSII (Collier and Grossman, 1993).

Signalling starvation

Once the high affinity transport systems have scavenged any remaining nutrient and cellular reserves are exhausted the cell will enter stationary phase. A major question about the entry into stationary phase concerns the nature of the signal. Does starvation for C, N, or P generate the same signal, such as an alteration in the energy charge, or are there different signals associated with starvation for each particular nutrient. At present there are no definitive answers, but most evidence regarding possible signals has come from studies with *E. coli.* One proposed signal is the observed methylation of elongation factor Tu in response to nutrient deprivation, which would provide a direct link with the translation machinery (Young and Bernlohr, 1991). The involvement of the nucleotide ppGpp (guanosine 3',5' bispyrophosphate) has also attracted interest. The intracellular of ppGpp has been found to increase in *E. coli* in response to amino acid, energy source and nitrogen starvation, though not significantly to phosphate limitation (see Gentry *et al.*, 1993). Further evidence supporting the role of ppGpp as a signal of incipient starvation is provided by the fact that strains of *E. coli* lacking the ability to make ppGpp fail to induce significant amounts of the stationary phase positive regulator σ^S (see below) in response to glucose, phosphate, or amino acid starvation and artificially elevated levels of ppGpp enhanced σ^S synthesis (Gentry et al., 1993). The nucleotide ppGpp has been detected in cyanobacteria and its synthesis responds very rapidly to a growth shift down obtained by a drastic reduction in light intensity (Mann *et al.*, 1975). Further evidence for the role of ppGpp has been obtained with the ultramicrocell-forming marine *Vibrio* strains (Kjellenberg *et al.*, 1993). A particularly interesting recent development implicates a homoserine lactone in sensing starvation (Huisman and Kolter, 1994). An *E. coli* chromosomal locus was found which, when expressed at high copy number, would block starvation-induced expression of σ^S. Sequence analysis revealed two ORFs designated *rspA* and *rspB.* The amino acid sequence of RspA is similar to that of Spa2 a protein from *Streptomyces ambofaciens* involved in starvation-induced pigmentation and also to two catabolic enzymes from *Pseudomonas putida*; mandelate racemase and chloromuconate cycloisomerase. These similarities, particularly that with the lactonizing enzyme chloromuconate cycloisomerase, led to the hypothesis that a lactone might be the starvation signal. Homoserine lactone (HSL) derivatives act as inducers for several cell density-dependent systems including bioluminescence, Ti plasmid transfer and synthesis of exoenzymes and antibiotics (for review see Swift *et al.*, 1994). The biosynthetic pathway of HSL is unknown, but it was speculated that homoserine (HS) or homoserine phosphate (HSP), both intermediates in the threonine biosynthetic pathway might be precursors. A consequence of this model would be that mutants blocked in HS or HSP synthesis should be defective in σ^S induction. Screening of mutants blocked at every step of threonine biosynthesis revealed that strains blocked before HS did not induce σ^S, but induction did occur if HS or HSL was included in the medium. Further evidence that RspA was involved in HSL derivative metabolism was provided by the observation that overexpressed RspA could interfere with a second HSL-related pathway, namely bioluminescence. Huisman and Kolter (1994) proposed the following model to explain HSL accumulation at the onset of stationary phase. Imbalances in amino acid biosynthesis occur as cells begin to starve, which lead to ribosome stalling and concomitant ppGpp synthesis and stimulation of expression of the threonine biosynthetic pathway. Increased threonine levels cause feedback inhibition and HS and HSP accumulate. These intermediates could bound by tRNA synthetases and the 'proof-reading' function of the enzyme would produce HSL.

Other lines of evidence point to the metabolite acetyl phosphate playing a key role in signalling starvation (McCleary *et al.*, 1993). Acetyl phosphate is a high energy compound produced from ATP and acetate by acetate kinase or from acetyl-CoA and P_i by phosphotransacetylase. Acetyl phosphate can directly phosphorylated some two component system response regulator without the participation of the relevant kinase (Lukat *et al.*, 1992) and has been implicated in nitrogen and phosphorus regulation as well as to carbon source availability and motility (see McCleary *et al.*, 1993). It may be that acetyl phosphate under certain circumstances may eliminate the need for a histidine kinase. Indeed, Wanner and Wilmes-Riesenberg (1992) have proposed that acetyl phosphate is involved in a regulatory coupling between the PHO regulon and genes for enzymes in central metabolism. McCleary *et al.* (1993) proposed that acetyl phosphate may represent a global signal that there is an imbalance between carbon flowing through glycolysis and into the TCA cycle. More recently, Nyström (1994) has shown that the ability to synthesize acetyl phosphate is essential to starvation survival following glucose limitation.

Surviving starvation

In response to starvation, bacteria from many genera are able to undergo cellular differentiation to form dormant spores, cysts or multicellular aggregates. However, bacteria which are not able to form these resting structures are also able to undergo a developmental programme which leads to the formation of cell which is metabolically less active, capable of surviving starvation and resistant to a number of other stresses. There is an extensive series of morphological and physiological changes accompanying the development of the starved cell and these have been extensively reviewed (Siegele and Kolter, 1992; Kolter *et al.*, 1993; Kjelleberg *et al.*, 1993). It is commonly the case that bacterial cells decrease in size with decreasing growth rate and the onset of starvation. This is particularly true for marine *Vibrio* spp which form ultramicrocells. The changes to the cell envelope may include changes in membrane lipid composition, structural changes to the peptidoglycan, increase in surface hydrophobicity and synthesis of fimbriae-like structures. Changes in the negative superhelical density of DNA occur and the nucleoid may become highly condensed. The energy charge of the cell decreases, though the proton motive force remains approximately constant. Stable RNAs (rRNA, tRNA) are reduced in abundance and protein turnover increases markedly. There is a development of resistance to other stresses including salt shock, heat shock and oxidative stress. These are in part due to the induction of enzymes such as catalase and the accumulation of trehalose.

Central to starvation survival is the synthesis of a specific set of proteins during entry into stationary phase. The patterns of proteins synthesize during this transition have been extensively studied in organism from several genera. In *E. coli* the pattern of proteins synthesized is determined by the nutrient which limits the cell; however, there is common core set of 15-30 proteins which are designated the Pex (post-exponential phase) proteins (Matin *et al*., 1989). In *E. coli* expression of many of the starvation-induced genes is under the control of an alternative σ-factor; the stationary phase-specific sigma factor (σ^S or σ^{38}) encoded by the *rpoS* gene. More than 30 genes require σ^S for their expression and about one third of protein products have been identified (see Kolter *et al*., 1993). These protein include catalase, exonuclease III, trehalose synthesis enzymes, a cell shape determination protein, Dps (a DNA-binding protein with protective and regulatory roles), and a lipoprotein. In spite of the characterization of a number of σ^S dependent genes no consensus sequence has yet been established for σ^S promoters. A variety of explanations have been put forward to explain this (see Kolter *et al*., 1993). It has been suggested that this may be due to the fact that it is not known which of these promoters are directly recognised by σ^S; also it may be that some promoters recognized by σ^S can be transcribed by σ^{70}, but only in combination with an activator. Furthermore, sequences outside the -10 and -35 may be important in promoter discrimination and, finally, post-translational modification of σ^S may regulate its activity. There are other regulators of stationary phase gene expression in *E. coli*; for example, during carbon starvation three 'heat shock' proteins DnaK, GroEL and HtpG are dependent on the heat shock σ-factor, σ^{32}. Such alternative σ-factors have been identified in cyanobacteria. Brahamsha (1993) has reported two σ-factors SigB and SigC whose expression is regulated by nutrient starvation and alternative σ-factors which respond to nutrient limitation have been characterized in *Synechococcus* sp. PCC 7002 (Caslake and Bryant, 1993).

References

Aiba H and Mizuno T (1994) A novel gene whose expression is regulated by the response, SphR, in response to phosphate limitation in *Synechococcus* species PCC7942. *Mol Microbiol* **13:** 25-34

Aiba H, Nagaya M, Mizuno T (1993) Sensor and regulator proteins from the cyanobacterium *Synechococcus* species PCC7942 that belong to the bacterial signal-transduction protein families: implication in the adaptive response to phosphate limitation. *Mol Microbiol* **8:** 81-91

Allen JF (1992) Protein phosphorylation in regulation of photosynthesis. *Biochim Biophys Acta* **1098:** 275-335

Bagg A and Neilands JB (1987) Molecular mechanism of regulation of siderophore-mediated iron assimilation. *Microbiol Rev* **51:** 509-518.

Bovy A, de Kruif J, de Vrieze G, Borrias M, Weisbeek P (1993) Iron-dependent protection of the *Synechococcus* ferredoxin I transcript against nucleolytic degradation requires *cis*-regulatory sequences in the 5' part of the messenger RNA. *Plant Mol Biol* **22:** 1047-1065

Bovy A, de Vrieze G, Borrias M, Weisbeek P (1992) Transcriptional regulation of the plastocyanin and cytochrome c_{553} genes from the cyanobacterium *Anabaena* species PCC 7937. *Mol Microbiol* **6:** 1507-1513

Brahamsha B (1993) Identification of multiple RNA polymerase sigma factor homologs in cyanobacteria: group 2 sigmas in *Anabaena* sp. strain PCC 7120. *Cyanobacterial Workshop* Abstracts p22

Briat J-F (1992) Iron assimilation and storage in prokaryotes. *J Gen Microbiol* **138:** 2475-2483

Briggs LM, Pecoraro VL, McIntosh L (1990) Copper-induced expression , cloning, and regulatory studies of the plastocyanin gene from the cyanobacterium *Synechocystis* sp. PCC 6803. *Plant Mol Biol* **15:** 633-642

Buikema WJ and Haselkorn R (1993) Molecular genetics of cyanobacterial development. *Ann Rev Plant Physiol Plant Mol Biol* **44:** 33-52

Burnap RL, Troyan T, Sherman LA (1993) The highly abundant chlorophyll-protein complex of iron-deficient *Synechococcus* sp. PCC7942 (CP43') is encoded by the *isiA* gene. *Plant Physiol* **103:** 893-902

Carr NG and Wyman M (1986) Cyanobacteria: their biology in relation to the oceanic phytoplankton. In *Photosynthetic Picoplankton.* Platt T, Li WKW (eds) *Can Bull Fish Aquat Sci* **214:** 121-158

Caslake LF and Bryant DA (1993) Characterization of sigma factors associated with the RNA polymerase of the unicellular cyanobacterium *Synechococcus* sp. PCC 7002. *Cyanobacterial Workshop* Abstracts p30

Chastain CJ, Brusca JS, Ramasubramanian TS, Golden JW (1990) A sequence specific DNA-binding factor (VF1) from *Anabaena* sp. strain PCC 7120 vegetative cells binds to three adjacent sites in the *xisA* upstream region. *J Bacteriol* **172:** 5044-5051

Collier JL and Grossman AR (1992) Chlorosis induced by nutrient deprivation in *Synechococcus* sp. strain PCC 7942: Not all bleaching is the same. *J Bacteriol* **174:** 4718-4726

Collier JL and Grossman AR (1993) Two novel genes involved in the biosynthesis of the photosynthetic apparatus in *Synechococcus* sp. strain PCC 7942. *Cyanobacterial Workshop* Abstracts p34

Collier JL and Grossman AR (1994) A small polypeptide triggers complete degradation of light-harvesting phycobiliproteins in nutrient-deprived cyanobacteria. *EMBO Journal* **13:** 1039-1047

Doige CA and Ames GF (1993) ATP-dependent transport systems in bacteria and humans: relevance to cystic fibrosis and multidrug resistance. *Annu Rev Microbiol* **47:** 291-319

Fagan MJ and Saier MH Jr (1994) P-type ATPases of eukaryotes and bacteria: Sequence analyses and construction of phylogenetic trees. *J Mol Evol* **38:** 57-99

Frías JE, Mérida A, Herrero A, Martín-Nieto J, Flores E (1993) General distribution of the nitrogen control gene *ntcA* in cyanobacteria. *J Bacteriol* **175:** 5710-5713

Gentry DR, Hernandez VJ, Nguyen LH, Jensen DB, Cashel M (1993) Synthesis of the stationary phase sigma factor σ^S is positively regulated by ppGpp. *J Bacteriol* **175:** 7982-7989

Ghassemian M, Wong B, Ferreira F, Markley JL, Straus NA (1994) : Cloning, sequencing and transcriptional studies of the genes for cytochrome c-555 and plastocyanin from *Anabaena* sp. PCC 7120. *Microbiology* **140:** 1151-1159

Grossman AR, Schaefer MR, Chiang GG, Collier JL (1993) The phycobilisome, a light-harvesting complex responsive to environmental conditions. *Microbiol Rev* **57:** 725-749

Green LS and Grossman AR (1988) Changes in sulfate transport characteristics and protein composition of *Anacystis nidulans* during sulfur deprivation. *J Bacteriol* **170:** 583-587

Green LS, Laudenbach DE, Grossman AR (1989) A region of a cyanobacterial genome required for sulfate transport. *Proc Natl Acad Sci USA* **86:** 1949-1953

Higgins CF (1992) ABC transporters: from microorganisms to man. *Annu Rev Cell Biol* **8:** 67-113

Grossman AR, Schaefer MR, Chiang GG, Collier JL (1993) The phycobilisome, a light-harvesting complex responsive to environmental conditions. *Microbiol Rev* **57:** 725-749

Guest JR (1992) Oxygen regulated gene expression in *Escherichia coli*. *J Gen Microbiol* **138:** 2253-2263

Hudson, RJM and Morel FMM (1993) Trace metal transport by marine microorganisms: implications of metal coordination kinetics. *Deep-Sea Res* **40:** 129-150

Huisman GW and Kolter R (1994) Sensing starvation: a homoserine lactone-dependent signaling pathway in *Escherichia coli*. *Science* **265:** 537-539.

Hyde SC, Emsley P, Hartshorn MJ, Mimmack MM, Gileadi U (1990) Structural model of ATP-binding proteins associated with cystic fibrosis, multidrug resistance and bacterial transport. *Nature* **346:** 362-365

Kanamaru K, Kashiwagi S, Mizuno T (1993) The cyanobacterium, *Synechococcus* sp. PCC7942, possesses two distinct genes encoding cation-transporting P-type ATPases. *FEBS Lett* **330:** 99-104

Kanamaru K, Kashiwagi S, Mizuno T (1994) A copper-transporting P-type ATPase found in the thylakoid of the cyanobacterium *Synechococcus* species PCC7942. *Mol Microbiol* **13:**369-377

Kerry A, Laudenbach DE, Trick CG (1988) Influence of iron limitation and nitrogen source on growth and siderophore production by cyanobacteria. *J Phycol* **24:** 566-571.

Kjellenberg S, Albertson N, Flärd K, Holmquist L, Jouper-Jaan Å, Marouga R, Östling J, Svenblad B, Weichart D (1993) How do non-differentiating bacteria adapt to starvation. *Antonie van Leewenhoek*: **63:** 333-341

Kolter R, Siegele DA, Tormo A (1993) The stationary phase of the bacterial life cycle. *Annu Rev Microbiol* **47:** 855-874

Laudenbach DE (1993) The involvement of a plasmid in the acclimation of cyanobacteria to sulfur stress. *Cyanobacterial Workshop* Abstracts p69

Laudenbach DE and Grossman AR (1991) Characterization and mutagenesis of sulfur-regulated genes in a cyanobacterium: Evidence for function in sulfate transport. *J Bacteriol* **173:** 2739-2750

Laudenbach DE, Reith ME, Straus NA (1988) Isolation, sequence analysis and transcriptional studies of the flavodoxin gene from *Anacystis nidulans* R2. *J Bacteriol* **170:** 258-265

Laulhère JP, Labouré AM, van Wuytswinkel O, Gagnon J and Briat JF (1992) Purification, characterization and function of bacterioferritin from the cyanobacterium *Synechocystis* PCC 6803. *Biochem J* **281:** 785-793

Leonhardt K and Straus NA (1992) An iron stress operon involved in photosynthetic electron transport in the marine cyanobacterium Synechococcus sp. PCC 7002. *J Gen Microbiol* **138:** 1613-1621

Lukat GS, McCleary WR, Stock AM, Stock JB Phosphorylation of bacterial response regulator proteins by low molecular weight phospho-donors. *Proc Natl Acad Sci USA* **89:** 718-722.

Luque I, Flores E, Herrero A (1994) Molecular mechanism for the operation of nitrogen control in cyanobacteria. *EMBO J* **13:** 2862-2869

Mann N and Carr NG (1974) Control of macromolecular composition and cell division in the blue-green alga *Anacystis nidulans*. *J Gen Microbiol* **83:** 399-405.

Mann N, Carr NG, Midgley JEM (1975) RNA synthesis and the accumulation of guanine nucleotides during growth shift down in the blue-green alga *Anacystis nidulans*. *Biochim Biophys Acta* **402:** 41-50

Mann NH, Scanlan DJ (1994) The SphX protein of *Synechococcus* species PCC 7942 belongs to a family of phosphate binding proteins. *Mol Microbiol.* In press

Mazel D and Marlière P (1989) Adaptive eradication of methionine and cysteine from cyanobacterial light harvesting proteins. *Nature* **341:** 245-248

Matin A, Auger EA, Blum PH, Schultz JE (1989) Genetic basis of starvation survival in nondifferentiating bacteria. *Annu Rev Microbiol* **43:** 293-316

McCleary WR, Stock JB, Ninfa AJ (1993) Is acetyl phosphate a global signal in Escherichia coli? *J Bacteriol* **175:** 2793-2798

Morel FMM, Reinfelder JR, Roberts SB, Chamberlain CP, Lee JG, Yee D (1994) Zinc and carbon c0-limitation of marine phytoplankton. *Nature* **369:** 740-742.

Nagaya M, Aiba H, Mizuno T (1994) The *sphR* product, a two-component system response regulator protein, regulates phosphate assimilation in *Synechococcus* sp. strain PCC 7942 by binding to two sites upstream from the *phoA* promoter. *J Bacteriol* **176:** 2210-2215

Nakamura M, Yamagishi M, Yoshizaki F, Sugimura Y (1992) The syntheses of plastocyanin and cytochrome c-553 are regulated by copper at the pre-translational level in a green alga, *Pediastrum boryanum*. *J Biochem* **111:** 219-224

Nyström T (1994) The glucose-starvation stimulon of Escherichia coli: induced and repressed synthesis of enzymes of central metabolic pathways and role of acetyl phosphate in gene expression and starvation survival. *Mol Microbiol* **12:** 833-843

Omata T (1991) Cloning and characterization of the *nrtA* gene that encodes a 45 kDa protein involved in nitrate transport in the cyanobacterium *Synechococcus* PCC 7942. *Plant Cell Physiol* **32:** 151-157

Omata T, Andriesse X, Hirano A (1993) Identification and characterization of a gene cluster involved in nitrate transport in the cyanobacterium *Synechococcus* sp. PCC7942. *Mol Gen Genet* **236:** 193-202

Omata T and Ogawa T (1987) Immunochemical studies on the major proteins in cytoplasmic membranes of cyanobacteria. In *Progress in photosynthesis research*, Biggins J (ed) Vol 4. Martinus Nijhoff, Dordrecht, 309-302

Price NM and Morel FMM (1990) Cadmium and cobalt substitution for zinc in a marine diatom. *Nature* **344:** 658-660

Rao NN and Torriani A (1990) Molecular aspects of phosphate transport in *Escherichia coli Mol Microbiol* **4:** 1083-1090

Ray JM, Bhaya D, Block MA and Grossman AR (1991) Isolation, transcription, and inactivation of the gene for an atypical alkaline phosphatase of *Synechococcus* sp. strain PCC 7942. *J Bacteriol* **173:** 4297-4309

Reddy KJ, Bullerjahn GS, Sherman DM and Sherman LA (1988) Cloning, nucleotide sequence, and mutagenesis of a gene (*irpA*) involved in iron-deficient growth of the cyanobacterium *Synechococcus* sp. strain PCC7942. *J Bacteriol* **170:** 4466-4476

Ronen-Tarazi M, Schwarz R, Bouevitch A, Lieman-Hurwitz J, Erez J, Kaplan A (1995) Response of photosynthetic microorganisms to changing ambient concentrations of CO_2. *This volume* 323-334

Scanlan DJ, Mann NH, Carr NG (1993) The response of the picoplanktonic marine cyanobacterium *Synechococcus* species WH7803 to phosphate starvation involves a protein homologous to the periplasmic phosphate-binding protein of *Escherichia coli*. *Mol Microbiol* **10:** 181-191

Shaw DJ, Rice DW, Guest JR (1983) Homology between CAP and Fnr, a regulator of anaerobic respiration in *Escherichia coli*. *J Mol Biol* **166:** 241-247

Siegele DA and Kolter R (1992) Life after log. *J Bacteriol* **174:** 345-348

Stock JB, Ninfa AJ, Stock AM (1989) Protein phosphorylation and regulation of adaptive responses in bacteria. *Microbiol Rev* **53:** 450-490

Swift S, Bainton NJ, Winson MK (1994) Gram-negative bacterial communication by N-acyl homoserine lactones: a universal language? *Trends Microbiol* **2:** 193-198

Tandeau de Marsac N, Houmard J (1993) Adaptation of cyanobacteria to environmental stimuli: new steps towards molecular mechanisms. *FEMS Microbiol Rev* **104:** 119-190.

Vega-Pala MA, Flores, E, Herrero A (1992) NtcA, a global nitrogen regulator from the cyanobacterium Synechococcus that belongs to the Crp family of bacterial regulators. *Mol Microbiol* **6:** 1853-1859

Vega-Palas MA, Madueño F, Herrero A, Flores E (1990) Identification and cloning of a regulatory gene for nitrogen assimilation in the cyanobacterium *Synechococcus* sp. strain PCC 7942. *J Bacteriol* **172:** 643-647

Wagner F, Gimona M, Ahorn H, Peschek GA, Falkner G (1994) Isolation and functional reconstitution of a phosphate binding protein of the cyanobacterium *Anacystis nidulans* induced during phosphate-limited growth. *J Biol Chem* **269:** 5509-5511

Wanner BL and Wilmes-Riesenberg MR (1992) Involvement of phosphotransacetylase, acetate kinase, and acetyl phosphate synthesis in control of the phosphate regulon in *Escherichia coli*. *J Bacteriol* **174:** 2124-2130

Wei TF, Ramasubramanian TS, Pu F, Golden JW (1993) *Anabaena* sp. strain PCC 7120 *bifA* gene encoding a sequence-specific DNA-binding protein cloned by *in vivo* transcriptional interference selection. *J Bacteriol* **175:** 4025-4035

Wei TF, Ramasubramanian TS, Golden JW (1994) *Anabaena* sp. strain PCC 7120 *ntcA* gene required for growth on nitrate and heterocyst development. *J Bacteriol* **176:** 4473-4482

Young CC and Bernlohr RW (1991) Elongation factor Tu is methylated in response to nutrient deprivation in *Escherichia coli*. *J Bacteriol* **173:** 3096-3100.

Zhang L, McSpadden B, Pakrasi HB, Whitmarsh J (1992) Copper-mediated regulation of cytochrome c_{553} and plastocyanin in the cyanobacterium Synechocystis 6803. *J Biol Chem* **267:** 19054-19059

Zinkel SS and Crothers DM (1991) Catabolite activator protein-induced DNA bending in transcriptional initiation. *J Mol Biol* **219:** 201-205.

The Problem of Excess and/or Limitation of the Habitat Conditions: Do Natural Assemblages Exist?

Ricardo Guerrero and Jordi Mas-Castellà
Department of Microbiology,
University of Barcelona,
Spain

Abstract
The structure of the communities present in both Lake Cisó and the microbial mats at the Ebro Delta, in Catalonia, Spain, as well as their changes with time and depth, were studied following a variety of approaches including taxonomic and physiological analyses, trophic state, population dynamics and structure of the food web, and competition among bacteria. Habitats with either an excess or a limitation of some natural condition harbour different kinds of microbial assemblages, which can be regarded as a means for self-regulation of the participating populations. Assemblages of phototrophic bacteria (cyanobacteria and purple and green sulphur bacteria) are among the most stable and balanced microbial communities. The population structure of a phototrophic community in Lake Cisó is determined by light penetration and by the abundance of sulphide. These two environmental factors are rarely found together in present day aquatic systems, except in habitats such as the metalimnion of some lakes (as in the case of Lake Cisó) and shallow water marine sediments (microbial mats). Simultaneous presence of green and purple sulphur bacteria has been observed in many occasions in karstic lakes. However, few studies have analysed which environmental factors are important in determining whether purple or green sulphur bacteria should be more abundant at a given time of the year. Light quantity and light quality have been implicated in the selection of purple or green sulphur bacteria in different lakes. This phenomenon has been studied in detail in Lake Cisó, where both groups of bacteria are present at different times of the year, thus providing a natural experiment where changes in environmental conditions change the relative abundance of the green and purple bacteria. Major factors determining the abundance of these two groups of bacteria are: solar radiation, temperature, accumulated rainfall, and the vertical distribution of both oxygen and sulphide.

NATO ASI Series, Vol. G 38
Molecular Ecology of Aquatic Microbes
Edited by I. Joint

Introduction

Microbial populations rarely occur alone in nature but rather interact with each other forming complex communities. Traditionally, increasing complexity of ecological organisation are considered at the individual, population, community and ecosystem levels. A community can be regarded as an assemblage of microorganisms living together at a given place or habitat. The community is the highest biological unit in an ecological hierarchy (Atlas and Bartha, 1993).

Community structure can be more or less stable depending on its degree of development. This usually involves the succession of populations in an orderly sequential change. The concept of natural assemblage lies between that of community and that of interacting populations. Natural assemblages can be regarded as biological self-regulating communities. A natural assemblage may become stable and persistent after many years of Darwinian evolution.

Habitats where an excess or a limitation of a natural condition occurs are clearly related with microbial assemblages. Almost every environmental variable becomes limiting at both edges of its range of permissiveness. For instance, resistance of a single cell to high temperatures depends only on its cellular adaptation mechanisms. However, the functioning of an ecosystem at high temperatures (*eg* hydrothermal vents) depends on the assemblage of several forms of life, either prokaryotic or eukaryotic or a combination of both. Comparatively, bacterial populations depending on sulphide for performing anoxygenic photosynthesis (*eg* purple and green sulphur bacteria in karstic lakes) are limited in their growth by either low or high concentrations of the compound. But when they form an assemblage, they can expand the limits which allow their living together (Madigan, 1989).

Another case of microbial assemblages may be seen in chemically polluted soils. Frequently, the toxic and complex chemicals present are not degradable by a single species. A combination of actions of several species performs the complete degradation of the toxic compounds. The assemblage confers a new function to the previous capabilities of the species present (Weightman and Slater, 1988).

Lake Cisó: planktonic natural assemblage of phototrophic prokaryotes

Lake Cisó is a small, holomictic anaerobic, sulphurous water body, located 200 m west of Lake Banyoles, 175m above sea level. Its shape is roughly circular, with an area of 599 m^2 and a maximum depth of 8m. It is situated in the karstic area of Banyoles (Catalonia, NE of Spain, 42° 07' 35"N, 2° 45' 05" E). The lake has been described in detail elsewhere (Guerrero *et al.*, 1985; Gasol, 1988; Pedrós-Alió and Guerrero, 1993). Water enters the lake mostly by seepage; there is only one small surface outlet, which dries up in summer. Underground water percolates through gypsum-rich layers before entering the lake and this causes high sulphate concentration (around 10mM) and, as a consequence of bacterial reduction, high sulphide concentrations in the water (from 0.4 to 2mM at the bottom on occasions). Lake Cisó is holomictic, with one mixing season lasting from mid-October to mid-March. At the beginning of April, stratification separates a shallow epilimnion from the hypolimnion. Summer epilimnetic temperatures are close to 24°C, and the hypolimnion is always at <14°C. The epilimnion is not very transparent (the extinction coefficient is 3-4 m^{-1}) and the water is deeply coloured due to dissolved organic matter. The depth at which 1% of surface light reaches is only 1-1.5m. Alkalinity is high (8.7 meq l^{-1}) especially in the hypolimnion; the pH oscillates around 7.5 in the whole lake, while nitrate and nitrite appear in high concentrations (up to 5μM for both ions) only in the epilimnion. But they can be found at all depths when the lake has just mixed. Ammonium is present throughout the lake, with hypolimnetic concentrations up to 1mM, and 10-50μM in the epilimnion. For most of the year, soluble reactive phosphorus was below detection limits, but appeared at given times during the studied period, usually below 10μg PO_4-P l^{-1} (Pedrós-Alió and Guerrero, 1993).

Different groups of organisms are present in Lake Cisó: phototrophic bacteria (Guerrero *et al.*, 1985), algae (Gasol *et al.*, 1991), ciliates (Dyer *et al.*, 1986; Gasol *et al.*, 1992), and rotifers and crustaceans (Alfonso and Miracle, 1987). The relative contribution of these organisms to total biomass has been established (Pedrós-Alió *et al.*, 1987; Gasol *et al.*, 1991). Among phototrophs, for example, 90% of the biomass is formed by three species: the purple phototrophic bacteria *Chromatium minus* and *Amoebobacter purpureus* and the flagellate *Cryptomonas phaseolus*. Purple phototrophic bacteria in Lake Cisó have the highest biomass value among other microbial groups. In general, bacterial biomass is typically 4 times greater in anoxic than in oxic waters of lakes (Cole *et al.*, 1993). Cell size is not related to total phosphorus or to bacterial abundance. Bacteria from anoxic hypolimnia are typically 2–10 times larger that those from oxic water. Cell size is independent of temperature under either oxic or anoxic conditions. Diatoms, dinoflagellates and chlorophytes are present only in small amounts and only at certain times of the year. Phototrophic biomass is dominated by purple and green sulphur bacteria.

Among the eukaryotes, rotifers are the most important, followed by ciliates and copepods. Rotifers are mostly represented by *Anuraeopsis fissa*, ciliates by *Coleps hirtus* and *Prorodon* sp. and the crustaceans by *Thermocyclops dybowskii*.

Life at the Z-axis

The simultaneous presence of purple (Chromatiaceae) and green (Chlorobiaceae) sulphur phototrophic bacteria has been observed on several occasions in lakes. However, few studies have analysed which environmental factors result in the dominance of Chromatiaceae or Chlorobiaceae at any time of the year. Light quantity and light quality have been implicated in selection of Chromatiaceae or Chlorobiaceae in different lakes. This phenomenon has been studied in detail in Lake Cisó where both types of bacteria are present at different times of the year, thus providing a natural experiment where changes in environmental conditions result in changes in relative abundances of green and purple bacteria. Factors important for determining the abundance of these two types of bacteria are: solar radiation, temperature, accumulated rainfall, and oxygen and sulphide vertical distribution.

The amount of solar radiation follows a pattern of annual change and is always the same — high in June and July and low in the winter. As light passes through the water column of a lake, it is attenuated in both intensity and spectral quality. The effect of biotic or abiotic light filtering on the species composition of phototrophic bacteria in lakes has been studied in Lake Cisó. Purple sulphur bacteria are always positioned in a layer on top of the green sulphur bacteria. Purple bacteria need higher light intensities to grow at the same rate as green bacteria due to the better location of light harvesting pigments in the chlorosomes of green bacteria. Due to absorption of light by the *Chromatium* layer, light that passes through to the *Chlorobium* zone is of a spectral quality that only supports growth of green-coloured *Chlorobium* species: the latter containing carotenoids that absorb maximally at around 450nm, the window of light passed by the *Chromatium* layer. Light and other physicochemical conditions in Lake Cisó favour the growth of purple bacteria over that of green bacteria. Perhaps the fact that the photic zone in Lake Cisó is limited to the top 3m favours the purple bacteria; green bacteria are more common in deeper lakes where both light quality and intensity are greatly affected by the size of the water column (Montesinos *et al.*, 1983). Seasonal distribution shows that biomass of purple bacteria increases with light intensity in the spring and decreases in the fall. Light intensity explains 89% of the variability in the biomass of Chromatiaceae. All of the evidence, thus, points to a light limitation of the phototrophic bacteria in Lake Cisó (Pedrós-Alió *et al.*, 1983, 1984).

Lake Cisó is a seepage lake like all the others in the Banyoles area. Using fluorescein as tracer,it was shown that water disappearing from river beds 20km north of the area takes about 13 days to travel underground and reach the waters of Lake Banyoles. Rain has two different effects on the lake; first, an immediate one increasing the water level through direct rainfall and surface runoff. An increase of 4.5cm has been observed in the water level after 1 hour of thunderstorm. Secondly, a delayed effect through increased seepage and, thus, increased water outflow. Sulphide concentrations in Lake Cisó were very high compared to other lakes with phototrophic bacteria; *eg* 0.02 to 0.05mM in freshwater Japanese lakes (Takahashi and Ichimura, 1968), or 0.01 to 0.15mM in several Wisconsin lakes (Parkin and Brock, 1980). Sulphide was more or less uniform throughout the water column for most of the year, the only exception being a shallow aerobic epilimnion from March to September. This distribution of sulphide meant that phototrophic bacteria could bloom at the very surface at several times during the year, colouring the entire lake red. Since the inputs to Lake Cisó are direct rainfall and seepage, and the outputs are evaporation and outflow, accumulated rainfall can be used as an estimate of the inputs to the lake. Subtracting evaporation from this variable gives an estimate of the water flux through the lake. Evaporation can be a major factor; we have observed the water level lower 1cm per day in the absence of outflow during the month of July (Pedrós-Alió et al, 1984; Van Gemerden *et al.*, 1985).

The problem of excess or limitation of the habitat conditions

Each of the phototrophic bacterial strains found in the lake presents a set of characteristics which separates it from the others. The purple bacteria differ from the green bacteria in being larger (about 20–50μm^3) instead of small (0.5–1μm^3), motile instead of nonmotile, accumulating sulphur intracellularly instead of depositing it outside the cells, and being more tolerant of microaerophilic conditions. The green bacteria are more tolerant of high sulphide concentrations and are able to use lower light intensities. Thus the purple bacteria are able to position themselves farther up in the water column than the green bacteria and the latter, in turn , are better adapted to the deeper zones of the lake.

The main genera found at the time of the present study were *Chromatium* and *Chlorobium.* Since both niches overlap and resource limitation exist, competition should take place. It was apparent that there was an inverse relationship between the biomass of these two genera. Relationships between biomass and rainfall or solar radiation which could only be guessed from the data were also confirmed by correlation coefficients (Pedrós-Alió *et al.*, 1984). Biomass of *Chromatium* was positively correlated with solar radiation and temperature, while biomass of *Chlorobium* was negatively correlated with these two variables. Rainfall and sulphide did not show any correlations with either organism. On the other hand, cumulative rainfall and rainfall minus evaporation showed negative correlations with the biomass of *Chlorobium.*

These results can be interpreted in the following way. There are two main types of variables influencing phototrophic bacteria in lakes. First, some variables, such as solar radiation and temperature, change smoothly and continuously both temporally (during the year) and spatially (with depth). Thus, their effect on biomass can be reflected by correlation coefficients which will show both the strength and the direction (positive or negative) of the relationship. Secondly, other variables do not have an effect proportional to their magnitude. Instead, bacteria are unable to survive if these variables cross upper and/or lower threshold levels, but the bacteria are not strongly affected by intermediate values of such variables. Thus, sulphide can be limiting if below a certain level or inhibitory if above another level specific for each species. Likewise, outflow can vary from zero, when it does not have any influence on the growth of the populations, to very high values, which are able to greatly reduce the population densities. In fact, phototrophic bacteria were completely eliminated from the lake after a period of heavy rainfall in 1982, with flow rates of 30 to 50litre min^{-1}. As for dilution rate in a chemostat, outflow affects growth rate but not biomass in the lake, except at high outflow (dilution) rates, when bacteria are eliminated from the system.

The populations actually found are controlled by the first type of factors as long as the second type variables present values within threshold levels. But when one of the later crosses either the upper or the lower threshold for one species it becomes the single most important factor.

Phototrophic bacteria in Lake Cisó follow opposite strategies in this respect. *Chromatium* could be considered a stenoecic organism, specialised to live under stratification conditions, when flow rate is low, and it can find sulphide levels within its relatively narrow tolerance range by swimming to the thermocline. By growing at this depth it can shade the nonmotile *Chlorobium* and outcompete it. On the other hand, *Chlorobium* adapts passively by being euryoecic with relation to light and sulphide, pushing the tolerance limits beyond those of *Chromatium.* Thus, when conditions are favourable for *Chromatium, Chlorobium* survives with low numbers, but when any of the *Chromatium* threshold levels are crossed, *Chlorobium* is able to grow and become the dominant organism in the population (Pedrós-Alió *et al.*, 1983).

In summary, the presence or absence of the purple sulphur bacteria is determined by certain variables (outflow and sulphide), which have to be maintained within certain permissive limits. However, purple sulphur bacteria biomass is proportional to variables which change continuously with space and time (light and temperature). On the other hand, total biomass of green sulphur bacteria is mostly determined by the presence or absence of competing (shading) populations, *ie* the purple bacteria.

In general, growth of each individual population can be expressed by a relationship in the field analogous to the Monod's equation for the chemostat:

$$dN / dt = (m - w) N$$

in which **N** is the population density in a given time, **m** is the specific growth rate, and **w** is the addition of the factors' constants which reduce the number of cells in the population (outflow, if any, predation, lysis, sedimentation, *etc* [Guerrero and Mas, 1989]).

Microbial mats of the Ebro Delta: benthic natural assemblages

Microbial mats are laminated benthic communities composed mainly of phototrophic prokaryotes together with chemotrophic organisms, that can develop in a variety of microgradients established between water and sediments. Microbial mats are present in a variety of biotopes. The most extensively studied include hot springs and marine intertidal or hypersaline environments. Today, they are confined to a restricted range of habitats, including hypersaline and coastal marine environments, hot springs and alkaline lakes. In the last few years, these systems have been the subject of a major study and several books have been published (Cohen *et al.*, 1984; Cohen and Rosenberg, 1989; Stal and Caumette, 1994), which extensively review the structure, physiology, genetics and evolution of microbial mats.

A preliminary investigation of these microbial ecosystems in Spain, distributed along the west Mediterranean coast, from Almería (SE limit of the Iberian Peninsula) to the Ebro Delta has been done (Esteve *et al.*, 1992). More detailed research have been done about microbial mats of La Banya spit, the southern peninsula of the Ebro Delta (Mir *et al.*, 1991; Rampone *et al.*, 1993; Guerrero *et al.*, 1993). At the Alfacs peninsula (La Banya spit) sand flats, extensive microbial mats thrive in several areas that may be flooded during the occasional rises in the sea level due to tides, waves and storms (Esteve *et al.*, 1994).

The studied microbial mats are located at the Ebro Delta (40°35' N, 0°40' E), 50 km south from the city of Tarragona (Catalonia, Spain). They occur all along the coast in the narrow ephemeral ponds of the backshore, on the flanks of the storm inlets and, most commonly, on the sand flats and channels of La Banya spit. In the last few years, ten sites along La Banya spit, selected in order to study spatial variations, have been studied. All of them lie 1 to 7cm below the water surface during the flooded periods. The temperature of water covering these microbial mats ranged from 12 to 30°C, salinity from 40 to 75‰, conductivity from 59 to 105mS.cm^{-1}, and pH from 7.5 to 9.0. Changes in oxygen microgradients as well as temperature, salinity and overlying water were the most important factors controlling microbial mat communities development (Rampone *et al.*, 1993).

These ecosystems are shallow benthic communities of phototrophic microorganisms composed of a surface layer of oxygenic cyanobacteria which overlay one or more layers of anoxygenic phototrophic bacteria. However, mats containing exclusively anoxygenic phototrophic bacteria occur in hot springs, and deep-sea mats composed of chemolithoautotrophic bacteria living in hydrothermal vents have also been described.

These microbial mats in the Ebro Delta have an average thickness of 4 to 5mm (the phototrophic populations), and they usually comprise three distinctively coloured layers. These colours are due to the pigment composition of the different phototrophic species present in each layer. Vertical stratification of these communities is not only the consequence of intensity and spectral quality of penetrating light but it is also a consequence of the physiological requirements and sensitivities of the mat-organisms to the vertical physicochemical gradients, such as O_2, H_2S, and pH.

This laminated community consisted of surface gold layer of diatoms mixed with cyanobacteria and a lower dark green layer dominated by filamentous cyanobacteria. Usually, under the green layer there was a orange or purple layer of anoxygenic phototrophic bacteria composed primarily of purple sulphur bacteria (Chromatiaceae: *Cromatium* spp. and *Thiocapsa* spp.), but in which bacteriochlorophylls *c*, *d* or *e* (corresponding to the Chlorobiaceae) could be detected.

Assemblages at intimate contact: the earliest biospheres

Well-developed mats were vertically stratified into three distinct pigmented layers. In the uppermost yellow-brown layer, diatoms and coccoid cyanobacteria were dominant. The second layer was composed mostly of filamentous (frequently sheathed) and coccoid cyanobacteria. The principal pigment in both layers was chlorophyll *a*. Bacteriochlorophyll *a*, predominant in Chromatiaceae, but also present in green sulphur bacteria (Chlorobiaceae), increased with depth. The deepest layer, purple or orange in colour depending on sampling site, was composed basically of anoxygenic phototrophic bacteria - Chromatiaceae and some Chlorobiaceae.

Several species of non phototrophic bacteria — spirochetes and other unidentified small rods and cocci — could be observed in these communities. Although it is not common to find eukaryotes, with the exception of diatoms, a copepod and an unidentified nematode were present in these communities. At the Ebro Delta microbial mats, the largest free-living spirochete known has been identified and described for the first time. It has been named *Spirosymplokos deltaiberi*; it can be up to 100μm long and 0.4μm to 3.0μm wide, and it presents a composite structure (Guerrero *et al.*, 1993).

The top layer was yellowish-brown and it was from 0.68 to 1.2mm thick. The dominant microorganisms in this layer were different species of diatoms, *Nitzschia* sp., *Navicula* sp., and *Amphora* sp, and coccoid cyanobacteria usually of the *Gloeocapsa* group. These coccoid cyanobacteria appeared mainly single or in pairs, but sometimes aggregates of cells surrounded by slime material could be seen.

The second layer was dark green in colour, leathery in texture and showed a thickness of 0.96 to 1.89mm. The dominant microorganism in this layer was the filamentous sheathed cyanobacterium *Microcoleus* sp. Other species of filamentous cyanobacteria could also be found such as *Lyngbya* sp., as well as different coccoid cyanobacteria but, at this depth, the cell aggregates were dominant over the single cells. This layer was dense and very cohesive.

The same genera were usually present in all the sites studied, but at different ratios. The filamentous cyanobacteria *Microcoleus* sp. and *Lyngbya* sp. dominated the coccoid forms in most of the mats. *Spirulina* sp., another filamentous cyanobacteria, was also identified, but it was only predominant in one of the selected sites. A well-developed purple layer was seen at different sites, which were submerged when sampled, whilst the presence of a black layer of sulphate reducing bacteria was ubiquitous and it ranged from 3mm to several cm (Rampone *et al.*, 1993).

The third layer was orange in some sites and could reach 1.28mm in thickness. This colour was due to anoxygenic phototrophic bacteria, *Chromatium gracile* and *Thiocapsa roseopersicina*, which had pigments of the spirilloxantina series as the main pigment. This third layer was purple in other sites, and had a thickness of 1.36mm. Carotenoids of anoxygenic phototrophic bacteria were responsible for this colour. Purple sulphur bacteria mostly belonging to the genus *Chromatium* sp. were arranged in tight clusters. Small amounts of carotenoids belonging to Chlorobiaceae were also detected. Below this last pigmented layer, there was a thick (1 to 3cm) black sediment; the colour was due to the production of H_2S by *Desulfovibrio*-like bacteria and later combination of the gas with the abundant available iron.

The different mat populations occurred within a few millimeters where close contacts could be established among them. The extracellular polymer production in several microorganisms allowed their attachment to sediment particles. External polymers as well as extracellular surface structures are involved in the establishment of cellular clusters. The presence of many cells in close proximity favours the establishment of relationships requiring physical contact.

The establishment of benthic microbial communities requires attachment mechanisms to surfaces by microorganisms. External polymer production and the development of outer polysaccharide layers in several marine microorganisms will determine their ability to colonise solid/water interfaces. Consequently, the cellular morphology and the growth characteristics of the different mat organisms affect the microbial assemblage, showing a different pattern of cohesion in each mat layer. Thus, in several sites the upper layer was not very cohesive, although extracellular polymers were produced, probably due largely to diatoms. The green layer was the most cohesive lamination because of binding of sediment particles by numerous filamentous microorganisms held together with their extracellular polymers. Finally, in the bottom pigmented layer, tight globular clusters of coccoid cells were dominant (Rampone *et al.*, 1993; Guerrero *et al.*, 1993; Esteve *et al.*, 1994).

The mats from Ebro Delta are found in an environment where salinity can change depending on climatic conditions and water level, just as in salterns. Observations made through the year indicate that the water level covering the mats in the sampling sites fluctuated considerably. In winter no water covered these microbial mats, but in spring water began inundating this area, reaching the maximum water level in summer. Some sites were inundated for longer periods of time than the other sites; this fact could partially explain the different mat development in the different sampling sites. Oxygen diffusion is greater in the air than in water, and consequently, during desiccation, oxygen can penetrate deeper in the mat and it is more difficult for anoxygenic communities to develop.

The absence of green sulphur bacteria can be correlated with their lesser tolerance to oxygen than purple sulphur bacteria. They are very sensitive to oxygen traces and in fluctuating conditions they can not establish themselves.

Concluding remarks: From individuals to community

The introduction of molecular techniques based on the analysis of RNA promises to allow comparisons of specific microbial diversity similar to those already established among groups of animals and plants. But the question of existing diversity among microbial assemblages will remain open until a more clear analysis of trophic relationships and biological interspecies modification of the habitat conditions will be available (Moyer *et al.*, 1994, Pedrós-Alió and Guerrero, 1994).

Let us take the example of a given bacterial genus: the purple sulphur bacteria *Chromatium*. In the overall community of Lake Cisó, the population of *Chromatium*, not the individual, is absolutely necessary. Any individual is entirely dispensable. The population of bacteria in the lake has certain characteristics exactly analogous to those of any animal, for instance, a cow. A cow does not reproduce using all its cells. The reproducing part of a cow is limited to eggs in the ovaries. Only very few cells can reproduce and thus maintain the species through time. The cow has a shape, just as the lake has a shape, that of geometry of its basin. The cow's shape is somewhat more fixed; in the case of the phototrophic bacteria, the shape of the population is the shape of the lake. The lake is like a "liquid animal" in shape. The bacterial community has "layers", and each of them is a population. One part of this community is composed of cells which divide very actively and therefore produce more cells. *Chromatium* cells fall towards the bottom of the lake; their remains feed other bacteria, the sulphate-reducers of the sediment. The sulphate-reducers produce the sulphide necessary for the photosynthesis of *Chromatium*. All of this activity maintains the "liquid organism's" physiology, its nutrient cycling. So bacteria are not acting as individuals but intrinsically as a community of interacting members, which are the different populations. Some of the cells reproduce single bacteria, but the entity of interest that makes sense is the community. An "individual" is intrinsically complex and intrinsically divisible. It sounds paradoxical, but in the prokaryotic world (*ie* all life until about 1.5 billion years ago) nothing smaller than a community can be an individual. The integrating mechanisms of the members of the community make the individual, and the permanence of the community modifies the habitat conditions to allow the evolution of new communities.

Acknowledgements
We thank the many persons who have participated in tasks of many kinds, field sampling, laboratory analyses, preparation of articles and fruitful discussions on Lake Cisó and Ebro Delta studies through the years. The latest part of the research and field work has been supported by project EV 4V-719 of the 4th Environmental R and D Programme of the UE, and by project MAR91-0874 of the Spanish Agency for Science and Technology (CICYT).

References

Alfonso MT and Miracle MR (1987) Temporal variation of zooplanktonic populations in the anoxic Lake Cisó (Girona). *Limnetica* **3:** 167-177. [In Spanish.]

Atlas RM and Bartha R (1993) *Microbial ecology. Fundamentals and applications.* 3rd edition. The Benjamin/Cummings Publishing Co.

Cohen Y, Castenholz RW, Halvorson HO (eds.). 1984. *Microbial Mats: Stromatolites.* Alan R. Liss, New York.

Cohen Y and Rosenberg E (eds.). (1989) *Microbial Mats: Physiological Ecology of Benthic Microbial Communities.* American Society for Microbiology, Washington.

Cole JJ, Pace ML, Caraco NF, Steinhart GS (1993) Bacterial biomass and cell size distributions in lakes: more and larger cells in anoxic waters. *Limnol. Oceanogr.* **38**: 1627-1632.

Dyer B, Gaju DN, Pedrós-Alió C, Esteve I, Guerrero R (1986) Ciliates from a freshwater sulfuretum. *BioSystems* **19**: 127-135.

Esteve I, Martínez-Alonso M, Mir J, Guerrero R (1992) Distribution, typology and structure of microbial mats communities in Spain: Preliminary studies. *Limnetica* **8:** 185-195.

Esteve I, Ceballos D, Martínez-Alonso M, Gaju N, Guerrero R (1994) Development of versicolored microbial mats: Succession of microbial communities. In: *Microbial Mats. Structure, Development and Environmental Significance*, Stal LJ and Caumette P (eds.). Springer-Verlag. Berlin, Heidelberg. *NATO ASI Series*, **G 35:**. 415-420

Gasol JM (1988) *Microbial Ecology of Lake Cisó.* Ph. D. Thesis. Autonomous University of Barcelona. [In Catalan.]

Gasol JM, Guerrero R, Pedrós-Alió C (1991) Seasonal variations in size structure and prokaryotic dominance in sulfureous Lake Cisó. *Limnol. Oceanog.* **36:** 860-872.

Gasol JM, Peters F, Guerrero R, Pedrós-Alió C (1992) Community structure in Lake Cisó: Biomass allocation to trophic groups and differing patterns of seasonal succession in the meta- and epilimnion. *Arch. Hydrobiol.* **123:** 275-303.

Guerrero R, Montesinos E, Pedrós-Alió C, Esteve I, Mas J, van Gemerden H, Hofman PAG, Bakker JF (1985) Phototrophic sulphur bacteria in two Spanish lakes: Vertical distribution and limiting factors. *Limnol. Oceanogr.* **30:** 919-931.

Guerrero R and Mas J (1989) Multilayered microbial communities in aquatic ecosystems: Growth and loss factors. In *Microbial Mats: Ecological Physiology of Benthic Microbial Communities,* Cohen Y and Rosenberg E (eds.),. American Society for Microbiology, Washington 37-51

Guerrero R, Ashen J, Solé M, Margulis L (1993) *Spirosymplokos deltaeiberi* nov. gen. nov. sp. variable-diameter composite spirochete from microbial mats. *Arch. Microbiol.* **160:** 461–470.

Guerrero R, Urmeneta J, Rampone G (1993) Distribution of types of microbial mats at the Ebro Delta, Spain. *BioSystems* **31:** 135-144.

Madigan MT (1989) Microbiology, physiology, and ecology of phototrophic bacteria. In *Biology of anaerobic microorganisms* Zehnder AJB (ed)

Mir J, Martínez-Alonso M, Esteve I, Guerrero R (1991) Vertical stratification and microbial assemblage of a microbial mat in the Ebro Delta (Spain). *FEMS Microbiol. Ecol.* **86:** 59-68.

Montesinos E, Guerrero R, Abellà C, Esteve I (1983) Ecology and physiology of the competition for light between *Chlorobium limicola* and *Chlorobium phaeobacteroides* in natural habitats. *Appl. Environ. Microbiol.* **46:** 1007-1016.

Moyer CL, Dobbs FC, Karl DM (1994) Estimation of diversity and community structure through restriction fragment length polymorphism distribution analysis of bacterial 16S rRNA genes from a microbial mat at an active hydrothermal vent system, Loihi Seamount, Hawaii. *Appl. Environ. Microbiol.* **60:** 871-879.

Parkin TB and Brock TD (1980) The effects of light quality on the growth of phototrophic bacteria in lakes. *Arch. Microbiol.* **125:** 19–27.

Pedrós-Alió C, Montesinos E, Guerrero R (1983) Factors affecting phototrophic bacterial pigments in holomictic Lake Cisó, Spain. *Appl. Environ. Microbiol.* **46:** 999–1006.

Pedrós-Alió C, Abellà C, Guerrero R (1984) Influence of solar radiation, water flux and competition on biomass of phototrophic bacteria in Lake Cisó, Spain. *Verh. Internat. Verein. Limnol.* **22:** 1097–1101.

Pedrós-Alió C, Gasol JM, Guerrero R (1987) On the ecology of a *Cryptomonas phaseolus* population forming a metalimnetic bloom in Lake Cisó, Spain: Annual distribution and loss factors. *Limnol. Oceanogr.* **32:** 285-298.

Pedrós-Alió C and Guerrero R (1993) Microbial ecology in Lake Cisó. *Adv. Microbial Ecol.* **13:**155–209.

Pedrós-Alió C and Guerrero R (1994) Prokaryotology for the limnologist. In *Limnology Now: A Paradigm of Planetary Problems,* Margalef R (ed) Elsevier, Amsterdam 37-57

Rampone G, Urmeneta J, Puigdefàbregas C, Guerrero R (1993) Geographical distribution of the microbial mats in the Ebro Delta: Structural diversity and role in the stabilization of the sediments. *Verh. Internat. Verein. Limnol.* **25:** 1014-1019.

Takahashi M and Ichimura I (1968) Vertical distribution and organic matter production of photosynthetic sulphur bacteria of Japanese lakes. Limnol. Oceanog. **13**: 644–655.

Stal L and Caumette P (1994). *Microbial Mats. Structure, Development and Environmental Significance*. NATO ASI Series, Vol G 35. Springer-Verlag. Berlin, Heidelberg.

Van Gemerden H, Montesinos E, Mas J, Guerrero R (1985) Diel cycle of metabolism of phototrophic purple sulphur bacteria in lake Cisó (Spain). *Limnol. Oceanogr.* **30:** 932-943.

Weightman AJ and Slater JH (1988) The problem of xenobiotics and recalcitrance. In *Micro-organisms in action: Concepts and Applications in Microbial Ecology*, Lynch JM and Hobbie JE (eds) Blackwell, Oxford 322-347

Signal Transduction in Microorganisms

Melvin I. Simon
Division of Biology 147-75
California Institute of Technology
Pasadena, California 91125
USA

Introduction

In addition to the flow of metabolites and energy that drives the distribution of populations of a wide variety of micro-organisms. These cells respond to specific signals from their environment and from other organisms. The responses allow them to adapt to changing conditions and to generate stable ecological interactions. The stimuli vary enormously and include changes in light, temperature, osmolarity, nutrient availability and changes in the presence of adherable surfaces. The organisms generally respond to these changes by modulating their metabolic potential and gene expression. They have mechanisms that allow them to sense the density of members of their own species or in the case of parasites and symbionts of appropriate host organisms. They can adapt by inducing the expression of previously repressed genes and gaining new catalytic capacity, *eg* the adaptation to anaerobiosis and nitrate utilization (Stewart *et al.*, 1989). In addition, micro-organisms can undergo extensive morphological and physiological change, *eg* sporulation or lateral flagella formation and they can modify their own environment, for example, by growing in microbial mats to allow for efficient nutrient utilization. How do micro-organisms detect environmental change? How do they integrate environmental information to generate an appropriate response? And how do these mechanisms evolve so that they can be adapted to the ecological strategy of specific organisms?

Bacterial Chemotaxis

The work in our laboratory has focused on the mechanisms of bacterial chemotaxis. This system which is present in a large variety of flagellated bacteria involves a sensing apparatus that can detect changes in the presence of small amounts of 20 to 30 different chemicals and nutrients. It allows the cells to modulate their swimming behaviour so that they can accumulate in regions of high nutrient and move away from potentially toxic locations. The bacterial chemotaxis system provides insight into the paradigms that cells use to process

NATO ASI Series, Vol. G 38
Molecular Ecology of Aquatic Microbes
Edited by I. Joint

information, and many of the characteristic features of chemotaxis are found in other systems. Even relatively simple organisms develop circuits that function through specific protein-protein interaction to convert the binding of a ligand to a receptor at the surface of the cell into two intracellular signals; an excitation signal and an adaptation signal. These in turn control a cellular process. In the case of bacterial chemotaxis, they modulate the direction of flagellar filament rotation and thus control the length of swimming in a fixed direction as opposed to the frequency of changes in swimming direction. The control of flagellar rotational bias (defined as the ratio of the time that bacterial flagella rotate in the counter clockwise direction to the time that they rotate in the clockwise direction) results in the regulation of the random walk that the organism normally executes in an isotropic environment. The cell behaves as if it perceives a change in concentration of "attractant" substances as a function of time and lengthens its runs in the direction of increasing attractant concentrations while decreasing the length of a run in the direction of the gradient where attractant concentrations are lower (for reviews of bacterial chemotaxis see Stock *et al.*, 1989; Bourret *et al.*, 1991; Parkinson *et al.*, 1992; Alex and Simon, 1994). The basis of this sensing system lies in a series of proteins that were first identified as the products of *Che* genes, *ie* genes wherein mutations resulted in the absence of chemotaxis. Fig. 1 and Table 1 lists these genes and the activities that have been associated with them. Fig. 1 also shows in a schematic fashion the role that each of the gene products plays in regulating the flow of information from the cell surface receptors to the development of an integrated response by modulating the mechanism that controls the bias of flagellar rotation.

The receptors all have the same general structure; there is a region that is found in the periplasmic space that functions as the ligand binding region. It is different in each of the four receptors and is preceded and followed by a transmembrane region. The ligand binding region of the aspartate receptor has been crystallized and its atomic structure determined (Milburn *et al.*, 1991). The polypeptide crystallizes as a dimer with one molecule of aspartate shared by a binding site made up of elements contributed by each of the components of the homodimer. The structure of the transmembrane domains has been deduced (Pakula and Simon, 1992). The molecule on the cytoplasmic side of the membrane can also be subdivided into functional domains. One domain is required for generating excitation signals and two regions are required in order for the cell to adapt to fixed levels of attractant. The excitation signaling region was identified by clusters of mutations resulting in a phenotype that fixed the cell in a constant tumbling or constant unidirectional swimming mode. The adaptation mutants are confined to sites on the molecule that are modified during chemotaxis by the addition of methyl groups to specific glutamic acid residues.

Table 1. Chemotaxis Components

	Excitation Response	
1.	CheA -	An autophosphorylating histidine kinase
2.	CheW -	Couples CheA to Receptor
3	CheY -	a) Substrate for CheA b) Phosphorylated on Asp-57 c) Autodephosphorylates d) Interacts with flagellar switch
4.	CheZ -	Acts as a phosphatase to accelerate the dephosphorylation of CheY
	Adaptation Response	
1.	Receptor -	Activates the kinase Activity is modulated by reversible methylation of specific glutamyl residue Binds specific subset of ligands
2.	CheB -	a) Receptor methylesterase b) Regulated by aspartyl phosphorylation c) Substrate for CheA-phosphorylation increases methylesterase activity
3.	CheR -	Receptor methyltransferase transfers methyl groups to glutamyl residues

The theme of a large protein with functional subdomains that can be defined by mutagenesis and that controls specific functions is found in the other elements of the chemotaxis system. Thus, for example, the histidine kinase, *CheA* is composed of at least four distinguishable domains. The C-terminal 80-90 amino acids can be deleted and the molecule loses little or no kinase activity but it does lose its ability to couple to the receptor (Oosawa *et al.*, 1988; Bourret *et al.*, 1993). Ordinarily, mixing molecules of purified activated receptor (*eg* TAR receptor in the absence of ligand) with the CheW protein and CheA protein induces a greater than 300-fold increase in the rate of CheA autophosphorylation. CheA uses ATP to autophosphorylate on histidine residue-48. It can then transfer this phosphate to an aspartate residue on the CheY protein or to a domain of the CheB protein. Phosphorylated CheY and CheB then go on to generate the excitation signal and the adaptation signal. CheY phosphate interacts with the base of the flagellar motor inducing a change in the direction of flagellar rotation. While the phosphorylation of a segment of CheB that is homologous in structure to CheY increases CheB activity by 30-fold (Stock and Koshland, 1978; Lupus and Stock, 1989), this results in rapid demethylation of the receptor and subsequent decrease in the receptors signaling functions (Fig. 1).

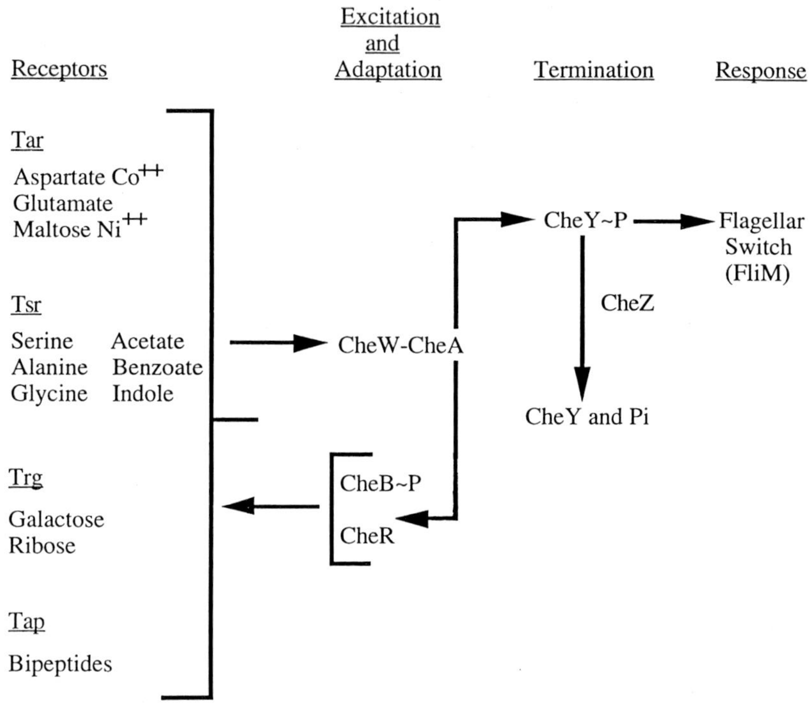

Fig 1. Components involved in chemotaxis. Receptors are transmembrane proteins that respond to a variety of substrates by binding them specifically. Ligand binding modulates kinase (CheA) activity. It phosphorylates CheY initiating the excitation response, *ie* modification of flagellar rotation, and adaptation by phosphorylating CheB which can then efficiently modify the receptor.

The ability of CheA to couple to receptor and to be activated by receptor is confined to the C-terminus of CheA. The central portion of the molecule (including amino acid residues 260-537) is required for kinase activity and includes a nucleotide binding site. Two other domains can be distinguished on the N-terminus of the molecule; the region from amino acid 1-134 (the P-1 fragment) is a segment involved in phosphate transfer both to CheB and CheY (see Fig. 2). The region from amino acid 150 to 233 has been called the P-2 region (Swanson *et al.*, 1993a). This portion of the molecule acts as a binding site for CheY and for CheB (Swanson *et al.*, 1993a). The modular nature of the CheA molecule has been further demonstrated by actually synthesizing and purifying protein fragments corresponding to these domains (Swanson *et al.*, 1993b). The domains were found to be inactive by themselves but when mixed together they function. Thus, the kinase fragment will deliver phosphate to a specific residue (His-48) on either the phosphate donating domain (P-1) or on the phosphate donor-substrate binding fragment (P1-P2). When these are phosphorylated they can transfer phosphate to aspartate residue 57 (Sanders *et al.*, 1989; Bourret *et al.*, 1990) on the CheY molecule or to the CheY-like domain on the CheB-methylesterase. Fig. 2 provides a cartoon that illustrates the various domains of the receptor kinase complex. Ligand binding in the periplasmic space

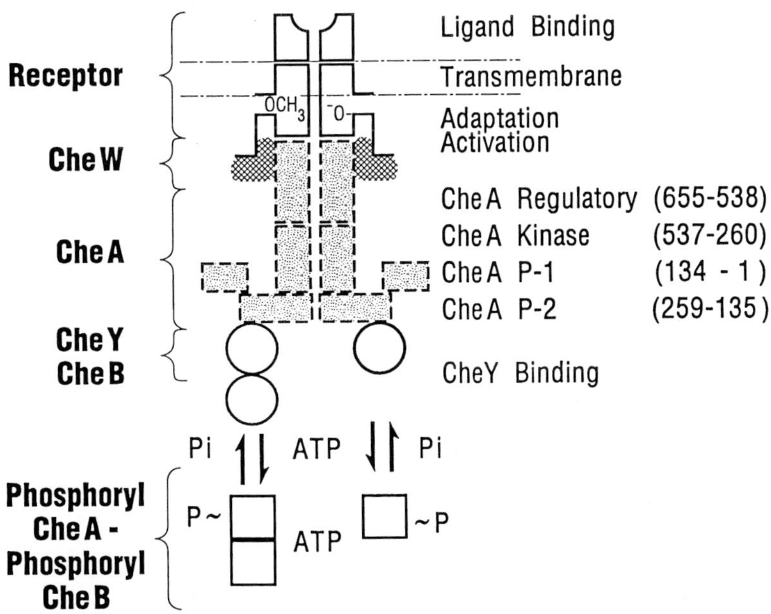

Fig. 2 The domains of the receptor kinase complex

presumably transmits a conformational change which induces the inhibition of CheA kinase activity (Borkovich and Simon, 1990). Alternatively, the absence of bound ligand or the binding of a repellent molecule to the receptor results in the marked stimulation of kinase activity and the subsequent accumulation of CheY-phosphate and the activated form of the CheB methyl esterase.

The existence of the complex of receptor-CheW-CheA and CheY was demonstrated directly by measuring ambient interactions using plasmon resonance (Schuster *et al.*, 1993). Complexes of the various proteins were formed and CheY was tethered to a dextran-matrix. When a preformed complex of receptor-kinase and CheW was passed over the matrix the complex bound to CheY when ATP was added and the complex was proficient at phosphorylating CheY because of the absence of attractant, the complex was released from the matrix. These results were all consistent with previous measurements of phosphorylation (Borkovich and Simon, 1990; Schuster *et al.*, 1993) by complexes of these proteins. They suggested that the complex, receptor-CheW-CheA has high affinity for CheY but much lower affinity for CheY~P. Recent evidence (Swanson, Dahlquist *et al.*, manuscript in preparation) involving measurement of interactions between the P1-P2 fragment and CheY suggest that a specific face of the CheY molecule is involved in interaction with CheA. When CheY is phosphorylated a change in its structure is induced. This change rearranges the surface of the molecule that interacts with CheA and releases it from CheA. In a similar way CheB competes with CheA for the P-2 site and when CheB is activated it can modify the receptor to decrease the efficiency with which the receptor activates the kinase. Thus, the scheme for information processing that emerges from this analysis includes two important general mechanisms: (1) That the signal transducing machinery, *ie* the receptor and the kinase are composed of separable functional domains. The receptor includes a ligand binding domain, a transmembrane domain, an activation domain, and a region that controls the activation domain and is involved in adaptation. The kinase has a domain that interacts with receptor and with the coupling protein CheW. This domain regulates kinase activity. It is joined to the catalytic kinase region which is in turn linked to a substrate binding and a phosphoryl domain.
(2) Signal transmission probably occurs through a change induced by ligand binding or ligand release that affects the structure of a supermolecular complex that includes the three chemotaxis proteins, receptor, CheW, and CheA. This results in ATP hydrolysis and phosphoryl transfer to CheY, which in turn stabilizes CheY in a novel configuration. Thus, the ligand binding or release event is converted into a stabilized conformational change in the target molecule CheY. A single ligand interaction event can result in the conversion of hundreds of CheY molecules thus amplifying the initial interaction.

The phosphorylated form of CheY can be viewed as a second messenger which has a limited lifetime dependent upon the lifetime of its phosphorylated form. During this time it can bind to the flagellar switch and presumably induce a change in flagellar rotation. Multiple molecules of phosphorylated CheY are thought to act (Kuo *et al.*, 1987) cooperatively to induce changes in flagellar rotation. The lifetime of the phosphorylated form is critical and it is controlled by two factors. One, an endogenous mechanism for autodephosphorylation which is an intrinsic part of the CheY molecule and second, CheZ another component of the chemotaxis pathway which can accelerate CheY dephosphorylation (Hess *et al.*, 1988). CheZ acts as if it were a CheY phosphatase and thus controls the relative concentration of CheYP.

The chemotaxis pathway is an integrative informational processing circuit. There are multiple receptors, at least four in *E. coli,* that each represent a different channel of information (see Fig. 1). The relative numbers of these receptors, their degree of coupling to CheA and their occupancy all contribute to the regulation of the levels of CheY phosphate at any instant. Thus, this system is capable of adding or subtracting, in a scaled fashion, inputs regarding the extracellular concentration of multiple (30-40) different ligands and conditions. Furthermore, the cellular response to changes in conditions is very rapid. The latent time (Ishihara *et al.*, 1983), i.e. the time from binding of attractant to receptor until the time when an effect on flagellar rotation can be discerned is approximately 200 milliseconds. Thus, the system is capable of rapid integration. Intracellular factors may also affect flagellar rotation providing yet another input into the system.

OTHER INFORMATION PROCESSING SYSTEMS

The chemotaxis system represents only one kind of information processing system. It deals primarily with multiple inputs and integrates over short time periods affecting cellular behaviour in the short term by affecting flagellar rotation. The same paradigm is used in a large variety of systems that control the genetic response of cells to changes in environmental conditions. We find the same modules as in the chemotaxis system, however, they are rearranged and modified by mutation to be tailored to the individual requirements of different information processing systems. Parkinson and Kofoid (1992), in a recent review, surveyed many of these systems. Some systems, for example, the NtrB-NtrC system which regulates responses to intracellular nitrogen metabolites involve a soluble histidine kinase, NtrB which is controlled by the PII protein a soluble protein which senses intracellular levels of a ketoglutarate and glutamate (Ninfa and Magasanik, 1986; Keener and Kustu, 1988). NtrB activates NtrC by phosphorylation. It (NtrC) is a complex protein with a multimerization site and a DNA binding site that is specific for binding to a region adjacent to the glutamine synthetase promoter. The phosphorylated form of NtrC with bound nucleotide functions as an activator of RNA polymerase and acts to modulate transcription of the glutamine synthetase gene. NtrB has marked homology to CheA and the regulatory region of NtC has all the hallmarks of the CheY module.

There are over 50 "two component" systems that have been characterized in a wide variety of bacterial systems, including Cyanobacteria (Nagaya *et al.*, 1994), Rhizobia, Bordetella, Agrobacterium, Bacillus and Vibrio (see Parkinson and Kofoid, 1992). It is probably fair to conclude that they will be found in all prokaryotes. Recent evidence indicates that they also function in eukaryotic cells including microorganisms, *eg* yeast, *Neurospora* and *Dictyostelium* as well as plants, *eg Arabidopsis* and probably in animals as well (Alex and Simon, 1994). This paradigm has succeeded because the separable functional domains that make up the kinase and the regulator can be recombined in a variety of different ways providing the organism with tremendous flexibility and yet the general mechanism of information transfer is probably conserved. Thus, it has been possible to generate chimeric systems that still retain function. For example in *E. coli* one of the systems involved in regulating osmolarity is the EnvZ-OmpR system. EnvZ is a receptor-kinase, *ie* it has a periplasmic region capable of sensing changes in osmolarity and transmitting this information to the cytoplasmic portion of the molecule which act as a kinase-phosphatase. It is able to phosphorylate OmpR, which includes a CheY-like domain and a DNA binding domain (Igo and Silhavy, 1988). Phosphorylation enhances DNA binding to a positive regulatory site adjacent to genes that encode porins, molecules that regulate entry of solutes into the periplasmic space and that modulate osmolarity (Aiba *et al.*, 1989). It has been possible to generate chimeric molecules that have the aspartate binding region of the chemotaxis receptor replacing the periplasmic osmolarity sensor. The chimeras now respond to changes in amino acid concentration by regulating porin gene transcription (Jin and Inouye, 1993). These results suggest that the mechanism involved in transmembrane transmission and cytoplasmic kinase activation may be similar in the two systems.

Evidence for substrate similarity has also been found. This is usually referred to as "crosstalk". Thus the overproduction of the components of one system can lead to the spill over of signal to another system. It has been possible to suppress mutants that are deficient in EnvZ by overproducing related kinases which are part of different sensory systems (Nagasawa *et al.*, 1993). It is not clear if such crosstalk is physiologically important or if it represents recognition mechanisms that are the result of a common ancestral evolutionary origin of members of these systems.

In addition to sensory systems that are integrative at the level of the cell surface receptor such as the chemotaxis system, where a variety of sensors activate the same kinase, there are many compound systems where integration results from a number of kinase regulator systems all controlling a cellular function and distributive systems where multiple responses are coordinately regulated. Compound systems involve a cascade of kinases that function either in parallel or in series to regulate a complex property of the cell. Thus, for example, recent evidence in *Vibrio harveyi* (Bassler *et al.*, 1994) suggests that there are multiple inputs that are required to control the operon that is transcribed to endow the cell with bioluminescence. Two separate sensory kinases and two response regulators affect the repressor-activator system that is required for the regulation of transcription of the luminescence operon. Indeed, at the DNA level the operator region that controls transcription may also act as a site for integrating signals mediated by the "two component" systems and other signals such as those regulated by catabolite concentration or by other intracellular and extracellular regulators. Sporulation control in *B. subtilis* represents another compound system where a phospho-relay involving at least two kinase-response regulator systems acting in tandem is required for the initiation of sporulation. In addition, a variety of other signaling systems contribute information that is required before the cell commits irreversibly to spore formation.

Distributive circuits are also found to be widespread. These are sensory systems generally controlled by complex kinases that can activate response regulators that switch entire metabolic pathways. Thus, for example, the ArcA-ArcB (Iuchi and Lin, 1992) system controls the expression of numerous target operons all involved in regulating the expression of genes required for aerobic respiration in *E. coli*. ArcA is the response regulator that binds at a variety of promoter sites when phosphorylated. The NarX, NarQ, NarL system is a compound-distributive pathway (Schroder *et al.*, 1994) where the expression of anaerobic respiratory genes for nitrate reductase, dimethyl sulfoxide reductase and fumarate reductase in response to nitrate concentration is mediated by a two component system that may have two kinases that activate or regulate the level of phosphorylation of NarL the response regulator which can then modulate the expression of a variety of genes.

The kinases involved in many of these systems are also complex structures often including in addition to a cytoplasmic domain, transmembrane region and kinase domain an integrated response regulator domain which is phosphorylated by the kinase domain on an aspartyl residue (Iuchi and Lin, 1992; Alex and Simon, 1994). It is thought that this region might act as a pseudo-substrate-inhibitor and thus modulate kinase function.

The widespread distribution of the two component regulatory systems is well established. In a large number of systems the autophosphorylating nature of the kinase component and the phosphorylation of the response regulator have been demonstrated. However, finding these motifs is not enough to tell us about the mechanisms by which these systems function. The variety of combinations of sensory kinase and regulator modules, the complex circuitry that suggests that these systems can be "wired" together to generate both global regulatory responses to single or multiple inputs and single outputs to complex environmental changes indicates the malleability of these motifs. It is necessary to analyze each of these systems individually and study how they interact to regulate various responses. The complexity of these systems suggests that we will need in addition to the usual tools of biochemistry, genetics and physiology, novel approaches that have not been extensively applied to regulatory systems. It is important to generate models that will allow us to simulate the effects of changed in the environment on gene regulation and microbial behaviour. By combining data on the biochemical properties of some of these information processing systems we may be able to predict the behaviour of populations of microorganisms in response to changes in nutrients, energy or other parameters. Bacteria provide an excellent starting system. While their genomes are relatively small, *ie* 1000-5000 genes per cell they clearly have sufficient sophistication in their ability to regulate individual cellular responses as well as population responses to environmental change.

References

Aiba H, Mizuno T, Mizushima S (1989) Transfer of a phosphoryl group between two regulatory proteins is involved in osmoregulatory expression of the Omp7 and OmpC genes. *J Biol Chem* **264**:8563-8567

Alex LA and Simon MI (1994) Protein histidine kinases and signal transduction in prokaryotes and eukaryotes. *Trends Genet* **10**:133-138

Bassler B, Wright M, Silverman MR (1994) Multiple signaling systems controlling expression of luminescence in *Vibrio harveyi*. *Molecular Microbiol* **13**:273-286

Borkovich KA, Alex LA, Simon MI (1992) Attenuation of sensory receptor signalling by covalent modification. *Proc Natl Acad Sci USA* **89**:6756-6760

Borkovich KA and Simon MI (1990) The dynamics of protein phosphorylation in bacterial chemotaxis. *Cell* **63**:1339-1348

Bourret RB, Davagnino J, Simon MI (1993) The carboxy-terminal portion of the CheA kinase mediates regulation of autophosphorylation by transducer and CheW. *J Bact* **175**:2097-2101

Bourret RB, Borkovich KA, Simon MI (1991) Signal transduction pathways involving protein phosphorylation in prokaryotes. *Ann Rev Biochem* **60**:401-441

Bourret RB, Hess JF, Simon MI (1990) Conserved aspartate residues and phosphorylation in signal transduction by the chemotaxis protein CheY. *Proc Natl Acad Sci USA* **87**:41-45

Hess JF, Oosawa K, Kaplan N, Simon MI (1988) Phosphorylation of three proteins in the signaling pathway of bacterial chemotaxis. *Cell* **53**:79-87

Igo MM and Silhavy TJ (1988) EnvZ, a transmembrane environmental sensor of *E. coli* is phosphorylated *in vitro*. *J Bacteriol* **170**:5971-5973

Ishihara A, Segal JE, Block SM, Berg HC (1983) Coordination of flagella on filamentous cells of *Escherichia coli*. *J Bacteriol* **155**:228-237

Iuchi S and Lin ECC (1992) Purification and phosphorylation of the Arc regulatory components. *J Bacteriol* **174**:5617-5623

Jin T and Inouye M (1993) Ligand binding to the receptor domain regulates the ratio of kinase to phosphatase activities in the signaling domain of the hybrid *E. coli* transmembrane receptor Taz1. *J Mol Biol* **232**:484-492

Keener J and Kustu S (1988) Protein kinase and phosphoprotein phosphatase activities of nitrogen regulatory proteins NtrB and NtrC of enteric bacteria. *Proc Natl Acad Sci USA* **85**:4976-4980

Kuo SC and Koshland DE (1987) Roles of CheY and CheZ gene products in controlling flagellar rotation in bacterial chemotaxis of *Escherichia coli. J Bacteriol* **169**:1307-1314.

Lupus A and Stock JB (1989) Phosphorylation of an N-terminal regulation domain activates the CheB methylesterase. *J Biol Chem* **264**:17337-17342

Milburn M, Prive G, Milligan D, Scott W, Yeh J, Jancarik J, Koshland DE, Kim S-H (1991) *Science* **254**:1342-1347

Nagasawa S, Ishige, K Mizuno T (1993) Novel members of the two component signal transduction genes in *E. coli. J Biochem* **114**:350-357

Nagaya M, Aiba H, Mizuno T (1994) The *Sph*R product, a two-component system response regulator protein regulates phosphate assimilation in Synechococcus sp. strain PCC 7942. *J Bacteriol* **176**:2210-2215

Ninfa AJ and Magasanik B (1986) Covalent modification of the *gln*G product Nr1 by the *gln*L product NRII, regulates transcription of the *gln*ALG operon. *Proc Natl Acad Sci USA* **83**:5909-5913

Oosawa K, Hess JF, Simon MI (1988) Mutants defective in bacterial chemotaxis show modified protein phosphorylation. *Cell* **53**:89-96

Pakula AA and Simon MI (1992) Determination of transmembrane protein structure by disulfide cross-linking: The *Escherichia coli* Tar receptor. *Proc Natl Acad Sci. USA* **89**:4144-4148

Parkinson JS and Kofoid EC (1992) Communication modules in bacterial signaling proteins. *Ann Rev Genet* **26**:71-112

Sanders DA, Gillece-Castro BL, Stock AM, Burlingame AL, Koshland DE Jr. (1989) Identification of the site of phosphorylation of the chemotaxis response regulator protein CheY. *J Biol Chem* **264**:21770-21778

Schroder I, Wolin CD, Cavicchioli R, Gunsalus RP (1994) Phosphorylation and dephosphorylation of the NarQ, NarX and NarL proteins. *J Bacteriol* **176**:4985-4992

Schuster SC, Swanson RV, Alex LA, Bourret RB, Simon MI (1993) Assembly and function of a quaternary signal transduction complex monitored by surface plasmon resonance. *Nature* **365**:343-347

Stewart V, Parales J Jr, Merkel SM (1989) Structure of genes *narL* and *narX* of the *nar* locus in *Escherichia coli* K-12. *J. Bacteriol.* **171**:2229-2234.

Stock JB, Ninfa AJ, Stock AM (1989) Protein phosphorylation and regulation of adaptive responses in bacteria. *Microbiol Rev* **53**:450-490

Stock JB and Koshland DE Jr (1978) A protein methylesterase involved in bacterial sensing. *J Biol Chem* **256**:10826-10833

Swanson RV, Schuster SC, Simon MI (1993a) Expression of CheA fragments which define domains encoding kinase, phosphotransfer, and CheY binding activities. *Biochem* **32**:7623-7629.

Swanson RV, Bourret RB, Simon MI (1993b) Intermolecular complementation of the kinase activity of CheA. *Molecular Microbiology* **8**:435-441

Microbial Diversity in Oceanic Systems: rRNA Approaches to the Study of Unculturable Microbes

Stephen J. Giovannoni, Thomas D. Mullins, and Katharine G. Field
Department of Microbiology, 220 Nash Hall
Oregon State University
Corvallis, OR 97331-3804
USA

Introduction: The Great Plate Count Anomaly

A standard method for enumerating marine bacteria from a water sample is to grow them on nutrient plates and count colonies. In low nutrient aquatic habitats, however, the number of bacterial colonies that can be grown using standard plating methods is frequently from one to four orders of magnitude fewer than the number of cells observed by direct microscopic count (Jannasch and Jones 1959). This well-documented observation, dubbed the "great plate count anomaly" (Staley and Konopka 1985), has led to much speculation about the identity, physiological ability and metabolic status of the observed but apparently uncultivatible microbes. These are well-defined questions that have nevertheless gone unanswered since the late 1800's (Roszak and Colwell 1987).

Most of what is known about marine bacteria has been synthesized from studies of cells in cultivation; it is not known how much of this can be extended to uncultivated cells. The subject has been reviewed by a number of writers (Brock 1987; Roszak and Colwell 1987; Staley and Konopka 1985).

The following questions have been debated: 1. Do the cells than can be grown on plates represent an unbiased sample of the original community? Brock (1987) stresses that they do not. For example, plates are selective for unicellular organisms, organisms that produce many propagules, and organisms which grow well on the surface of a medium. Dormant cells, not active in the natural community, may grow on media plates. Jannasch and Jones (1959) speculate that part of the difference between direct and plate counts of bacterial cells could be due to the presence in water samples of a large number of aggregated cells, which, when plated, only produce one colony and thus are counted as one. In addition, they speculate that certain forms they observed directly, such as spirilli, are difficult to grow.

NATO ASI Series, Vol. G 38
Molecular Ecology of Aquatic Microbes
Edited by I. Joint

2. Are the observed but unculturable cells dead, dormant or actively growing? What is their contribution to the microbial community? In 1983, it was first reported that unculturable cells were physiologically active (Kurath and Morita 1983). As measured by a variety of techniques, most (85%) of the total direct count of bacterial cells from aquatic and terrestrial habitats are metabolically active (Staley and Konopka 1985). Staley and Konopka hypothesize that the bacteria that don't form visible colonies may be obligate oligotrophs, which, although difficult to study using conventional equipment and procedures, are numerically and physiologically important members of the community.

Some marine bacteria show a very long doubling time (Carlucci and Williams 1978; Jannasch 1969). In addition, when bacteria are grown under conditions simulating oligotrophy, they may develop into small, rounded cells; these can, however, grow to "normal" size upon provision of excess substrate (Casida 1965; Conn 1948; Ensign 1970; Hoppe 1976). These small cells have been interpreted as starved bacteria (Sieburth 1979). Growth studies have indicated that populations of bacteria in the death or decline phase reach a phase of "cryptic growth" (Postgate 1976), a steady state in which 2 to 4% of the cells remain viable (Ribbons and Dawes 1963). The small cell size may also result from an adaptive response to starvation: size reduction through a series of divisions during which cell numbers increase without an increase in biomass (Morita 1982; Novitsky and Morita 1978).

Kjellberg and colleagues (1982, 1983) hypothesized that most marine bacteria require high concentrations of nutrients for growth, and therefore exist in a predominantly dormant stage, adapted for high-nutrient zones found in rapidly-changing environments. These are the so-called "copiotrophs". It has been proposed that marine bacteria are of two types: either oligotrophic, adapted to low nutrient concentrations, or copiotrophic, needing high nutrient concentrations, but possessing various adaptations to survive periods of low nutrients (Martin and Macleod 1984).

To summarize, the original observation, that most marine bacteria seen by direct counts cannot be (or are not usually) grown *in vitro*, still remains to be explained. Although studies of marine species in cultivation have provided much useful information, it is not known to what extent knowledge of cultivated marine bacteria is applicable to the majority of marine bacteria which remain uncultivated.

Phylogenetic Analyses of Bacterioplankton Populations by Ribosomal RNA Sequence Analysis

SUMMARY OF PREVIOUS STUDIES

The use of ribosomal RNAs as markers for species diversity is providing a fresh approach for investigating the structure of microbial communitites and the identities and physiological status of the uncultivated microbes observed in them (Ward *et al.* 1992). Studies of rDNA diversity have dramatically improved our understanding of bacterioplankton communities.

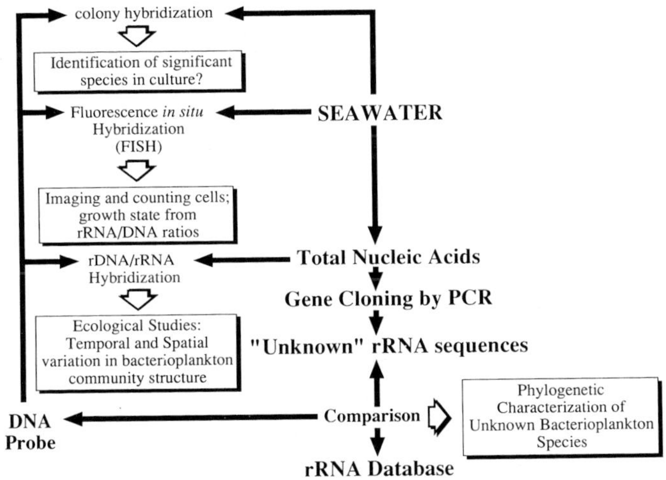

Fig. 1: Schematic outline of the strategies used for genetic investigations of bacterioplankton diversity. These approaches do not rely on cultivation, but instead focus on natural populations of cells obtained by filtration. (Reprinted with permission from Giovannoni and Cary, 1993).

The general approach is to collect total nucleic acids (DNA and RNA) from natural populations, clone ribosomal RNA genes by gene amplification or shotgun cloning, and then compare the recovered RNA gene sequences with the large collection of sequences available in data bases. The sequences are then used to design oligonucleotide probes which are specific to different constituents of the community; these probes can used to quantify distributions and identify cell types. In Fig. 1, the strategies employed in this kind of investigation are charted.

Any conserved gene can be used for this kind of study (*eg nifH,* Ben-Porath and Zehr 1994; *rbcL*, Paul *et al.* 1990; *petB/D,* Urbach 1994) but because of the large ribosomal RNA sequence data bank, rRNA genes confer an advantage when phylogenetic identification of organisms is a goal. Biases in the filtration and cloning procedures used in these studies are not yet well understood, but they appear to be less limiting than the biases associated with microbial cultivation. These methods for investigating the structure of microbial communities are now in wide use in a variety of habitats; here we concern ourselves with marine bacterioplankton communities only.

16S rRNAs isolated thus far from natural bacterioplankton samples are diverse, but predominantly fall into several distinct phylogenetic groups (Fig. 2). A majority of the clones belong to the cyanobacteria and proteobacteria, although Gram-positive bacteria, flavobacteria and novel Archaea genes have also been found. Most marine samples studied so far have come from the surface of subtropical regions of the ocean, although samples from 100m, 500m, and marine snow have been examined (DeLong 1992; DeLong *et al.* 1993; Fuhrman *et al.* 1993; Fuhrman *et al.* 1992), and there are recent coastal studies (Michael Rappe', personal communication).

We have used these methods to study the identities and the temporal and spatial distributions of bacterioplankton from an oligotrophic habitat, the Sargasso Sea (Britschgi and Giovannoni 1991; Giovannoni *et al.* 1990; Giovannoni *et al.* 1990; Mullins *et al.* 1995). The DNA amplification primers used in this study were specific for the domain Bacteria; therefore Archaea and Eukarya rRNA genes were not recovered. We cloned and analyzed 60 bacterial 16S rRNA genes amplified from mixed population DNA.

The genes cloned were identified as members of the α-proteobacteria subdivision, γ-proteobacteria subdivision, and cyanobacteria. Twelve major bacterial lineages were represented in the SAR collection of genes. Only four of the 60 clones could be identified with cultured isolates (96% to 98% identity). The four identified genes include the marine *Synechococcus* group, *Prochlorococcus marinus,* and the cultured species *Photobacterium phosphoreum* and *Alteromonas haloplanktis*.

In a recent study we compared the phylogenetic diversity of bacterioplankton communities from Atlantic and Pacific Oceans (Mullins *et al.* 1995). The sixty 16S rDNA clones from the SAR gene library were compared to partial sequences from 37 Bacterial 16S rDNA clones from a surface picoplankton sample collected at ALOHA station in the North Pacific gyre (ALO; Schmidt *et al.* 1991) and to the partial sequences of twenty-nine Bacterial 16S rRNA clones obtained from sites near Bermuda (BDA; Fuhrman *et al.* 1993) and the Western California Current (NH; Fuhrman *et al.* 1993). The major bacterioplankton groups whose genes were recovered in these studies are listed in Table 1. An important finding of this study was that closely-related lineages

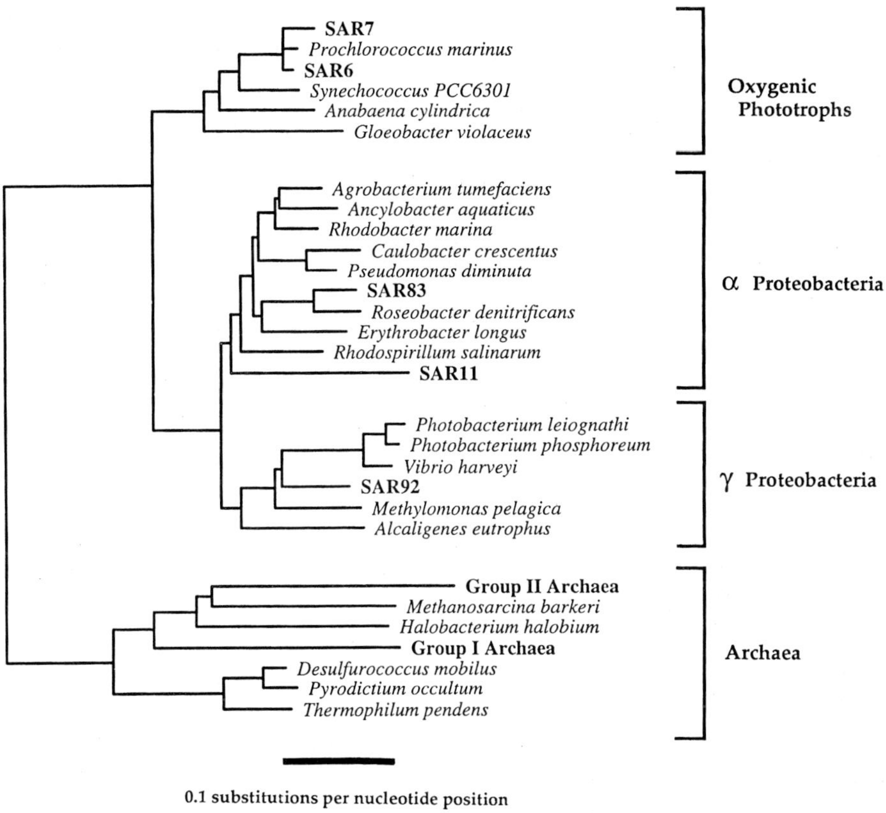

Fig. 2: Evolutionary comparisons of picoplankton 16S rDNAs to genes from cultured organisms. Genes with the prefix "SAR" were cloned directly from a surface water bacterioplankton population collected at Hydrostation S in the Sargasso Sea. Otherwise, the names refer to cultured species. The genes from Group I and Group II Archaea shown here were cloned from coastal water samples (DeLong, 1992), but similar genes were cloned from mesopelagic water samples in the open Pacific by Fuhrman *et al.* (1992; 1993). This phylogenetic tree was inferred by a distance-matrix method. (Reprinted with permission from Giovannoni and Cary 1993).

occurred in clone libraries that had been prepared from different oceans by entirely different cloning methods. In the SAR and ALO libraries, fourteen gene lineages were found altogether, of which six were found in both libraries. Several genes cloned from the Sargasso Sea were nearly identical to genes cloned from the Pacific populations, suggesting that these previously unrecognized and uncultivated bacteria groups (SAR11, SAR122, SAR86) are distributed widely in the surface waters of subtropical oceans. This result has important implications

Table 1. Major bacterioplankton lineages obtained from rDNA libraries: SAR (Sargasso Sea, north central Atlantic: Britschgi and Giovannoni 1991; Giovannoni *et al.* 1990; Mullins *et al.* 1995); ALOHA (north central Pacific: Schmidt *et al.* 1991): BDA/NH (north central Atlantic and Pacific: Fuhrman *et al.* 1993); OM (north coastal Atlantic off Cape Hatteras, North Carolina: unpublished data, M. Rappe'); OCS (north coastal Pacific off Newport, Oregon: unpublished data, M. Rappe'). Boxed areas indicate overlap between the libraries. "*" indicates lineages represented by more than one clone. "**" indicate lineages which match or are closely related to cultivated bacteria.

	SAR	ALOHA	BDA/NH	OM	OCS
Alpha Proteobacteria	SAR11*	ALO21*	BDA/NH*	OM136*	
	SAR83*	ALO17	NH16-3	OM42*	OCS126*
	SAR116*	ALO24			
	SAR121				
Beta Proteobacteria				OM43*	
				OM180*	
					OCS7*
Gamma Proteobacteria	SAR86*	ALO4*	BDA/NH*	OM10*	OCS5*
	SAR92				
	SAR95				
	SAR125				
	SAR134**				
	SAR145				
	SAR160**				
		ALO18			
		ALO40			
Cyanobacteria	SAR6**	ALO7	NH16-15*		
	SAR7**	ALO11*			
Green non-sulphur bacteria *Herpetosiphon*	SAR202				
Marine Group A	SAR406		NH6-12*		
Gram positive	SAR432		BDA1-5	OM1	OCS155

for molecular studies of bacterioplankton communities: it suggests that probes for a limited number of taxa might account for a majority of cells in ocean surface samples from different sites. In addition, it suggests that whatever the biases of cloning methods, they are not sufficient to obscure the most basic elements of microbial diversity in marine systems, which appear repeatedly despite differences in cloning methods.

One of the remarkable discoveries to result from investigations of marine microbial diversity using rRNA sequences has been the observation of Archaea as important members of bacterioplankton communities (DeLong 1992; DeLong *et al.* 1994; Fuhrman *et al.* 1993; Fuhrman *et al.* 1992). Genes of two major archaeal groups have been recovered from marine systems. The first, Group I Archaea, are members of the Crenarchaeota. Their physiology is unknown. The second, Group II Archaea, are members of the Euryarchaeota, and are related to methanogenic bacteria. Group I crenarchaeote genes have been recovered from a variety of samples from both the Atlantic and Pacific oceans (DeLong 1992; Fuhrman *et al.* 1993; Fuhrman *et al.* 1992), and more recently from Antarctica (DeLong *et al.* 1994). Originally concerns were expressed that these clones might have originated from bacteria displaced from hydrothermal vent systems. The discovery of these organisms in Antarctic marine samples suggests, on the contrary, that they are widespread inhabitants of cold pelagic waters. In our work at the Bermuda-Atlantic Time Series Station in the Sargasso Sea, we found that Group I Archaea genes were common in the lower mixed layer throughout much of the year. We rarely observed archaeal genes in surface waters, except in winter, when mixing brought deeper surface layer water to the surface. The remarkable aspect of the Group I Archaea is their relatively deep phylogenetic position in evolutionary trees. This discovery is now contributing towards a reevaluation of relationships in the domain Archaea (Barns *et al.*, 1994).

Rehnstam *et al.* (1993) suggested that sequence-specific blooms of culturable marine bacteria might dominate seawater samples collected from inshore areas. Conclusions from these data differ from those from similar experiments in offshore areas. It is possible that inshore and offshore areas are significantly different with respect to the cultivatability of species. Evidence supporting differences in community composition between inshore and offshore areas was provided by the DNA:DNA hybridization studies of Lee and co-workers (Lee and Fuhrman 1990).

In summary, the use of rRNA markers to identify bacterioplankton from natural populations without cultivating them has tended to confirm the "great plate count anomaly" in two important ways. First, only a few genes have been recovered which match previously-cultured marine bacteria. Second, a number of previously-unknown groups identified by this methodology (SAR11, SAR122,

SAR86, various Archaea) now appear to be important constituents of marine bacterioplankton communities worldwide.

ASSUMPTIONS, BIASES AND CAVEATS

Biases in cloning: how does the choice of cloning methods affect the distribution of genes in clone libraries?

Since an objective of these studies was to overcome biases imposed by bacterial cultivation, concerns soon arose that the application of molecular techniques was simply substituting novel biases for the ones already known. This is almost certainly the case. Nevertheless, it is highly unlikely that the biases imposed by gene cloning methods influence rDNA distributions as severely as cultivation requirements influence bacterial distributions in culture. Microbial cultivation techniques are remarkably diverse and specific, depending upon the microorganisms being sought. In contrast, quite general methods are used for the isolation of rDNA genes.

A number of methodological approaches have been employed for the direct isolation of genes from natural assemblages of microorganisms. Some studies have employed libraries prepared by shotgun cloning into bacteriophage lambda. In this method, the total DNA from a bacterioplankton population is cut into pieces of a suitable size for insertion into bacteriophage lambda. Hybridization probes are used to search for the gene of interest among the resulting clones, which may represent entire genomes. Here, the "screening" step (picking out specific genes, such as rRNA genes, from the entire nucleic acids sample) occurs after total DNA is cloned. In order for the rRNA genes recovered to reflect the distribution of bacterioplankton lineages in the original population, the following assumptions must be true: all cells contributed their nucleic acids proportionally to the nucleic acid preparation; all cells had a similar genome size with an equal number of rRNA genes. In addition, bias could be caused by the physiological impact of unknown genes which lie close to ribosomal RNAs. For example, some gene fragments are unclonable because they interfere with cellular functions of organisms used as hosts for cloning vectors. When deleterious genes are present on fragments cloned into lambda bacteriophage, the packaged virus may be unviable. Perhaps for these reasons, shotgun clone libraries are not used in molecular biology to reconstruct gene frequencies.

The most widely used method for the isolation of genes from natural populations of microorganisms has been the polymerase chain reaction, PCR (Saiki *et al.* 1988). In PCR, a specific gene is amplified from bulk nucleic acids by use of "primers", oligonucleotides designed to complement known conserved sequences at either end of the gene to be amplified. Amplification products are then cloned for further study. The "screening" takes place at the gene amplification step. The amplification primers themselves act as probes; for the resulting clone library to be unbiased, there must be no primer mismatch. Additionally, in order for the clone library to exactly reflect the distribution of bacterioplankton lineages in the original population, the following assumptions must be true: all cells contributed their nucleic acids proportionally to the nucleic acid preparation; all cells had a similar genome size with an equal number of rRNA genes; no biases in amplification (such as those that would be caused by differences in genomic guanine plus cytosine content or primer mismatch) occurred between different templates; and all amplification products were cloned equally efficiently.

It is worthwhile to ask what the utility of rDNA distributions from natural systems would be, if, in fact, there were no biases in cloning. It would probably make little difference to our subsequent understanding of natural systems, for several reasons. In the first place, different microorganisms have different numbers of rDNA genes, ranging from a single copy to as many as 10. Without knowing the rDNA copy numbers and genome sizes of the constituent organisms, and, in some cases, the distribution of their genes on chromosomes and tandem repeats, it would be impossible to infer actual cell counts from rDNA gene abundances, even if these abundances were known with complete precision. One might argue that while rDNA distributions don't accurately reflect cell abundances, they reflect something about the growth potential of different phylogenetic groups, since organisms with the capacity for more rapid growth have higher gene copy numbers (Stahl and Urbance 1990). However, this requires a number of assumptions, and probably does not represent ecologically useful information. Even if rDNA gene abundances were known with precision, we would still need other methods to address important ecological questions.

A second consideration which is of even greater concern is the issue of sample bias. Natural systems are highly variable with time, as many previous studies of bacterial distributions, as well as our studies at the Bermuda/Atlantic Time Series Station, have shown us. A sampling matrix to produce a suitable environmental average would be quite complex. Most rDNA libraries are obtained from a single water sample. If absolute gene abundances were known for those samples, we would still know little about the average abundances of organisms in systems under study.

The relative abundances of gene clones in clone libraries may be influenced by biases introduced during gene amplification by PCR. Such systematic biases might produce a skewed view of microbial distributions, particularly if PCR is used as a quantitative measure of gene abundances. An example of bias in PCR was reported by Reysenbach and co-workers (Reysenbach *et al.* 1992). They found that rDNA genes could not be readily amplified from the genomic DNA of extremely thermophilic Archaea, presumably because the high mole percentage of guanine and cytosine residues prevented denaturation of the genomic DNA at temperatures routinely used to promote denaturation in PCR. In natural systems such a bias could cause microorganisms with high mole percentages of guanine and cytosine in their genomic DNA to be underrepresented in clone libraries. This problem was corrected by the addition of 5% acetamide to PCR reactions (Reysenbach *et al.* 1992). Acetamide has the same effect on double-stranded DNA as the more familiar compound formamide; both lower the melting point of double stranded DNA molecules by forming hydrogen bonds with the bases of single stranded DNA molecules in solution. Acetamide does not seem to influence the yield of PCR reactions, although it might be expected to influence specificity of DNA priming. Other potential sources of biases in PCR include variations in gene sequences at priming sites and the formation of secondary structures which inhibit primer extension.

The selection of specific rDNA genes by primers would occur if a hypothetical organism in a natural system had very unusual base substitutions at the priming sites used to obtain rDNAs by amplification. This cannot be entirely ruled out; however, sequence data bases, prepared largely by the analysis of organisms in culture, provide a degree of confidence that such mismatches are rare events. The priming sites routinely used for the preparation of rDNA clone libraries are extraordinarily conserved: in some cases, universal; in others, bacterial or archaeal. Given the level of conservation of these priming sites, it is somewhat surprising how frequently this concern surfaces. Straightforward probability, calculated from sequence data bases, provides a strong argument that flaws in clone libraries produced by priming mismatches are extremely rare. Detection of organisms which don't match well known priming sites may possibly be the challenge of future efforts.

A number of primers currently in use are degenerate; that is, the primers are mixtures of several slightly different versions, each one exactly complementing a particular sequence found at the priming site. This is a potential source of bias in PCR clone libraries which we consider to be a matter of considerable concern. The different versions of a primer corresponding to variations in sequence at priming sites may adequately represent sequence diversity in a natural system but still fail to translate that diversity into PCR products in the correct proportion. First, different versions of a primer could have different free energies of binding, and, at a given temperature, one template-primer hybrid might be more stable than another. This could result in the differential amplification of templates. Second, two different templates which correspond to two different primer sequences might amplify equally well, but nonetheless be misrepresented in the final products because of primer exhaustion.

In our studies we have encountered evidence that this occurs. Imagine that one template is ten times more abundant than another in a natural sample, and that the two templates correspond to two versions of a primer, present in an amplification reaction at equimolar concentrations. In early rounds of PCR, the two templates produce PCR products in the same proportion as genes in the original template mixture. In later rounds of replication the primer corresponding to the more abundant template is exhausted. At this point, if the abundant template hybridizes poorly to the remaining primer (which contains a mismatch), then amplification of that template ceases, while the less abundant template continues to be amplified. The final products could therefore contain equal proportions of both template genes, although their original proportions were one to ten. The net result of this mechanism would be for certain rare genes to be over-represented in libraries, skewing clone abundances.

For these reasons, libraries of environmental rDNA clones are most useful when viewed in terms of their potential as markers for other ecological studies. This is not to say that the abundance of clones in libraries is not useful information; it has been our experience in most cases that abundant clones do correspond to abundant organisms, and thus clone abundance can reveal fruitful directions for experimentation.

Sequence errors introduced by PCR: how do they affect analyses?

PCR rapidly became the method of choice for analyzing microbial diversity in natural populations following its first use for the purpose in 1990. PCR problems are greatly outweighed by its utility, despite two well-known limitations: misincorporation of bases during PCR and formation of chimeric genes.

Rates of misincorporation of nucleotide bases by different thermostable DNA polymerases range from 10^{-4} to 10^{-7} in optimized reactions (Ling *et al.* 1991). *Taq (Thermus aquaticus)* polymerase lacks a 5'->3' exonuclease proofreading activity. Lower error rates are found when other thermostable DNA polymerases, such as *Pfu (Pyrococcus furiosus)* DNA polymerase, are used because these enzymes possess proofreading activity. Slightly higher error rates have been reported in routine applications of PCR (reviewed in Giovannoni 1991). Nevertheless, an error rate as high as one half of one percent would still be too low to influence the results of phylogenetic sequence comparisons, and thus would not affect assessments of microbial diversity. However, polymerase errors could cause problems in two other areas. First, a misincorporated base might be chosen as part of the target site for a probe designed from a cloned sequence. In this case the probe would not work as intended. The odds of such an occurrence are low, and can be considerably reduced by the use of multiple sequences for probe design. Secondly, there are concerns when very closely related sequences are compared to identify separate lineages within clusters of highly related genes. The unique identity of lineages within clusters must be supported by analyses of secondary structures, which reveal covariation in base pairs. Although this approach does not entirely rule out error contributing to gene clusters, it does place a lower boundary on the number of unique gene sequences which can comprise a cluster.

The formation of chimeric genes during PCR (Shuldiner *et al.* 1989) was known before the first use of the method for the investigation of natural populations; thus the occurrence of chimeras in rDNA clone libraries was anticipated (Giovannoni *et al.* 1990; Liesack *et al.* 1991). Chimeric genes are formed when a template molecule is not completely copied during a round of replication in PCR. In a subsequent round of replication, if that partial copy forms a hybrid with a homologous rDNA which then serves as the template to complete the synthesis of the gene, the result will be a gene copied from two different templates. Such chimeric genes will behave anomalously in phylogenetic analyses, often appearing to be "fast clock".

Most chimeric genes can be readily detected. Congruence between phylogenetic trees from two ends of the same gene is the most sensitive test for a unique origin for an amplified gene. We routinely sequence the 5' and 3' ends of all genes which are the subject of sequencing efforts. These gene fragments are phylogenetically analyzed in separate data sets. Using this method, we have estimated the rate of chimera formation to be approximately 10% in clone libraries constructed with either *Taq* or *Pfu* polymerase. Chimeric genes can also be detected using a program available from the ribosomal RNA data base project (Larsen *et al.* 1993), CHECK_CHIMERA. Finally, secondary structures are a useful means of verifying the presence of chimeric genes.

Some classes of chimeric genes may be impossible to detect; for example, when chimeras between very closely related gene sequences have break points which occur outside of stem domains. If there is concern that a particular gene is chimeric, it is best to sequence related genes until duplicate copies are obtained. The probability of cloning and sequencing two identical chimeric genes is negligible.

One aspect of concern regarding chimeric genes is their contribution to estimates of microbial diversity when measures such as RFLP (restriction fragment length polymorphisms) are used to sort clones into groups. The mechanism of chimeric gene formation dictates that a certain chimeric gene will appear only once in any clone library (with exceptions that are very, very rare); consequently, chimeric genes may lead to over-estimates of microbial diversity. This is particularly the case when RFLP data are used to estimate coverage. Coverage estimates based on sequence information (see below) are likely to be much more accurate, as are rarefaction curves (a method of assessing the completeness of sampling for genetic diversity; see Hurlbert 1971), which are impacted less by chimeric genes.

Coverage of clone libraries: how good is the sample?

The portion of a clone library of infinite size that would be sampled by an actual clone library can be calculated. This value, termed coverage (C) by Good (1953), is derived using the equation:

$$\mathbf{C = 1\text{-}(n_1/N)}$$

Here, n_1 is the number of clone types which occurred only once in the actual library (frequency = 1), and N is the total number of clones examined. For this calculation we assume that clones which are >97% similar are identical. This value is conservative; it would lump together closely related genera such as *Salmonella* and *Escherichia*, but exclude variation introduced by PCR artefacts and microheterogeneities in rDNA gene families.

For our Sargasso Sea library, coverage was 81% (Mullins *et al.* 1995). Thus, this library accounted for about 81% of the clones that would be found in a similar clone library of infinite size. The available information does not permit us to estimate the diversity of the remaining 19% of rDNAs which are unaccounted for; they may be of few or many rDNA types. If we assume that rDNA copy numbers do not vary greatly for marine bacteria, and that PCR and cloning provide a reasonable representation of rDNA frequency, then this analysis suggests that a majority of the total bacterial rDNA genes from Sargasso samples were identified in the SAR clone library. This result is primarily of interest for studies of carbon flux, since taxa contributing significantly to carbon flux are also likely to be abundant. Focusing on these abundant taxa is an obvious choice for addressing community dynamics. However, many bacteria important to other marine processes may not be accounted for, since species present in relatively low numbers may participate substantially in biogeochemical or pathogenic processes.

THE LIMITED OVERLAP BETWEEN CULTIVATED MARINE BACTERIOPLANKTON AND SEQUENCES RECOVERED FROM NATURAL POPULATIONS

Why do rDNA sequences obtained directly from natural populations fail to match rDNA sequence data from cultured microorganisms? There are several plausible explanations. The simplest possibility is that most marine bacterioplankton are, in fact, species which cannot be cultivated using standard bacteriological methods. But there are other explanations; the most obvious is that bacterioplankton are culturable but have been studied insufficiently by taxonomic methods based on cultures. If so, then one would expect that the study of bacterioplankton by routine methods would readily lead to the discovery of new taxonomic groups. A third explanation for the discrepancy is that sequence data bases incompletely reflect the known taxonomy of cultured bacterioplankton. Each of these explanations is valid and to some extent probably correct. However, the latter two explanations, which focus on the incompleteness of bacterial taxonomy and sequence data bases, cannot possibly explain all the data.

How complete are ribosomal RNA sequence data bases from cultured species? These data bases currently contain over 2,500 sequences. Cultured bacteria number about 2,600. An examination of cultured marine bacteria reveals that 16S rRNA sequences for most of them are available (Table 2). These facts alone attest to the power of sequence data bases as a means of identifying unknown genes. However, the situation is not quite so simple as these figures make it sound. Bacterial species are poorly defined. At best they are largely grouped according to constellations of phenotypic traits. There is no doubt that such constellations of phenotypic traits occur in nature. For example, with minor effort, anyone can isolate *Vibrio harveyi* from temperate and tropical sea waters. If one does so and compares 100 or so strains, it will be found that they are not all identical phenotypically. Nevertheless, they can be recognized as belonging to a particular genus and species. In this sense, one can say bacterial species certainly exist from a pragmatic standpoint: "bacterial species form discrete taxonomic units in both genotypic and phenotypic characters, rather than a broad continuum as anticipated in light of gene exchange" (Ochman and Wilson 1987, p1650). However, the number of bacterial species that have been defined based on large numbers of strains is relatively few, and patterns of phenotypic variation within species, or groups of strains, are not well defined. Although there are some exceptions, in most cases molecular phylogeneticists have tended to pick a single example of each taxonomic group for sequencing studies. Seldom did anyone deliberately sequence many examples of the same microorganism. In the early days of sequencing by reverse transcriptase, such comparisons would have required unusually precise sequencing; even today, they would be demanding. As a consequence, for most bacterial species, the actual extent of rDNA sequence variation is unknown and generally considered to be too slight to be of scientific interest. Where the extent of rRNA genetic variability has been examined among strains of the same species, it varies from 0 to 4.7% (Amann *et al.* 1992; Fox *et al.* 1992). Among congeneric species, there is no "expected" amount of genetic divergence in rRNA gene sequences; divergences as low as 0.2% and as high as 17.3% were observed among *Bacillus* species that had been defined based on phenotypic traits (Fox *et al.* 1992); two species of *Fibrobacter* differed by 7 to 9% (Amann *et al.* 1992).

Table 2. A list of known marine bacteria in culture, compiled from *Bergey's Manual of Systematic Bacteriology* (1989), *Bergey's Manual of Determinative Bacteriology* (9th Edition) and Biological Abstract citations (1991-1994), and listed in the order they appear in *Bergey's Manual.* "y" indicates isolate has reference sequence in the Ribosomal Database Project (RDP). "*" indicates sequence is available for a congeneric species, but not for listed species. "**" indicates isolate has counterpart in genes cloned from Sargasso Sea bacterioplankton samples. Bacteria with no source of isolation indicated were obtained from marine environments.

Marine Bacteria	Source	RDP Sequence
Spirochaeta halophila	marine mud	y
S. litoralis	H_2S containing mud of high salinity pond	y
S. plicatilis	brackish water, marine mud	
Strain MA-2	obligately anaerobic marine isolate	y *(S. isovalerica)*
Cristispira pectinis	(has molluscan host) no culture	
Leptospira biflexa		y
Bdellovibrio (marine species)		*
Marinomonas communis		y
M. vaga		y
Microvibrio marinus roscoffensis		*
Oceanospirillum beijerinckii		y
O. hiroshimense	mussels	
O. japonicum	mussels	y
O. linum	mussels	y
O. maris subsp. *maris*		y
O. maris subsp. *williamsae*		
O. minutulum	mussels	y
O. multiglobuliferum	mussels	y
O. pelagicum	mussels	
O. pusillum	mussels	y
Runella		y *(R. slithyformis)*
Flectobacillus marinis		y
Acinetobacter		*
Agrobacterium ferrugineum		
A. atlanticum		
A. meteori		
A. gelatinovorum		
A. stellulatum		
Alcaligenes aestus		*
A. aquamarinus		*

Table 2 (cont)

Marine Bacteria	Source	RDP Sequence
A. cupidus		*
A. pacificus		*
A. venustus		*
Alteromonas aurantia		
A. citrea		
A. espejiana		y
*A. haloplanktis***		y
A. luteoviolacea		
A. nigrifaciens		y
A. piscicida		
A. rubra		
A. tetraodonis		y
A. sp. 2ML 26 (KMM156)	mussel	
A. sp. 2MM 6 (KMM158)	mussel	
A. sp. 2MC 12 (KMM155)	mussel	
Deleya marina		y
D. aquamarina		
Flavobacterium halmophilum	brine seawater	y
F. marinotypicum		y
F. oceanosedimentum	ocean sediment	
F. okeanokoites		
F. uliginosum		
LTR	Sargasso Sea water	y
Moraxella		y
Pseudomonas aeruginosa		y
P. atlantica		
P. beaufortensis		
P. corallina		
P. doudoroffii		
P. droebachense		
P. elongata		
P. floridana		
P. fluorescens		y
P. gelidicola		
P. iridescens		
P. nautica		y
P. piscicida	marine fish pathogen	
P. roseola		
P. stutzeri		y
P. vesicularis	Suyeong Bay	
Roseobacter litoralis		y
R. denitrificans		
Escherichia coli		y
Aeromonas caviae		y

Table 2 (cont)

Marine Bacteria	Source	RDP Sequence
A. hydrophila subsp. *anaerogenes*		y
A. media		y
A. salmonicida subsp. *masoucida*		y
A. veronii		y
Listonella anguillara		y
L. damsela		y
L. pelagia		y
Plesiomonas shigelloides		y
Vibrio aestuarianus		y
V. alginolyticus	coastal waters	y
V. anguillarum	diseased fish	y
V. campbellii		y
V. carchariae		y
V. cholerae		*
V. costicola		y
V. diazotrophicus		y
V. fischeri	water column	y
V. fluvialis	coastal waters	y
V. furnissii		
V. gazogenes	salt water marsh	y
V. harveyi	water column	y
V. hollisae		y
V. logei		y
V. marinus	seawater	y
V. metschnikovii		
V. mimicus		y
V. natriegens		y
V. ordalii		y
V. orientalis		y
V. parahaemolyticus		y
V. pelagius		y
V. proteolyticus		y
V. nereis		
V. nigripulchritudo		y
V. splendidus biovar. I,II and V		y
V. tubiashi		y
Photobacterium angustum		y
P. leiognathi		y
*P. phosphoreum***		y
Chromobacterium marinum		*
Shewanella putrefaciens		
S. alga		

Table 2 (cont)

Marine Bacteria	Source	RDP Sequence
Desulfuromonas acetoxidans		y
D. acetexigens		
D. africanus		y
D. desulfuricans subsp. *desulfuricans*		*
D. d. subsp. *aestuarii*		*
Desulfovibrio gigas		y
D. salexigens	sea water, muds	y
Desulfobacter postgatei		y
Desulfobulbus propionicus		y
Desulfococcus multivorans		y
Desulfosarcina variabilis		y
Micrococcus halobius	saline environment	
Planococcus citreus		
P. halophilus		
Bacillus badius	marine sources	
B. marinus	marine sediment	y *(Flectobacillus marinus)*
B. methanolicus		y
B. pacificus		
B. stearothermophilus	deep sea sediments	y
B. subtilus	Suyeong Bay	y
Clostridium baratii	Puget Sound	
C. bifermentans	marine sediment	y
C. fallax		
C. magenotii	marine sediment	y
C. oceanicum	marine sediment	y
C. perfringens	marine sediment	y
C. sporogenes	marine sediment	y
C. subterminale	marine sediment	y
Renibacterium salmoninarum	salmon	y
Acetobacterium woodii	marine anaerobic sediment	
Brevibacterium stationis	sea water	
Mycobacterium marinum	fish aquarium	y
Rhodococcus marinonascens	marine sediment	
Chromatium purpuratum	marine sediment	
C. vinosum	seawater and hydrogen sulfide	y
C. volutan		
Thiocystis violacea	seawater hydrogen sulfide	
Ectothiorhodospira		y *(E. halophila, E. halochloris)*
Rhodobacter sulfidophilus	seawater	y
Rhodomicrobium		y
Rhodopseudomonas marina		y
Rhodospirillum salexigens	seawater	y

Table 2 (cont)

Marine Bacteria	Source	RDP Sequence
R. salinarum	seawater	y
Chlorobium phaeovibrioides	anoxic marine	
C. vibrioforme	anoxic marinc	y
Choroherpeton thalassium	marine littoral sediments	
Erythrobacter longus	seawater	y
Synechococcus group** marinc cluster A	coastal and opcn watcr	
marine cluster B	coastal samples	
marine cluster C	coastal waters, sediment	
Cyanothece group - BG43511	coastal marine algal water	
BG 43522	coastal marine algal water	
Synechocystis group - marine cluster		
Synechocystis sp. strains		*
PCC7330		
ATCC 29108		
ATCC 27266		
ATCC 27153		
Dermocarpa	marine aquarium	y
Dermocarpella	intertidal marine snail	
Myxosarcina	intertidal marine snail	y
Pleurocapsa	intertidal marine snail	y
Xenococcus	marine aquarium	
Oscillatoria trichodesmium		
Lyngbya (Phormidium) PCC 7419	mat in shallow marine	y
Microcoleus chthonoplastes	intertidal	*
Spirulina sp.		y
Anabaena	brackish marine	y
Aphanizomenon	Baltic sea	
Calothrix parietina	marine shores	*
Gloeotrichia		
Nodularia spumigena	Baltic sea	*
Prochloron didemni	marine shores extracellular symbiont	y
*Prochlorococcus marinus***	open ocean	y
Arthrobacter sp. EK1		y (*Arthrobacter* sp. H1)
Nitrospina gracilis	seawater	
N. mobilis	seawater	
N. marina	seawater	
Thiobacillus denitrificans	marine sediment	
T. neapolitanus		y
T. versutus	salt marsh	y
Thiomicrospira pelophila	marine hydrothermal vents	*
T. denitrificans	tidal flats	

Table 2 (cont)

Marine Bacteria	Source	RDP Sequence
T. crunogena	hydrothermal vent	
Thiospira bipunctata	sulfurous marine brackish water	
Thiovulum majus	sulfur-containing H_2O mud	
Bilophococcus magnetotacticus	marine sediment	
Nitrosococcus oceanus		y
Nitrosomonas europaea	ocean	y
N. oceanus	ocean	
N. marina		y
N. mobilis	brackish water	
Caulobacter halobacteroides		
Caulobacter maris		
Hyphomicrobium indicum		
Hyphomonas hirschiana	hydrothermal vents	
H. jannaschiana	hydrothermal vents	y
H. neptunium	seawater	
H. oceanitis	Baltic sea	
Prosthecomicrobium litoralum	marine	
Pirellula marina	Baltic Sea	y
Planctomyces maris	Baltic Sea	y
Cytophaga agarovorans	marine mud	y
C. aprica	marine mud	
C. fermentans	marine mud	y
C. haloflava var. *nonreductans*	fishing nets	
C. latercula	marine aquarium	y
C. krzemieniewskae	coastal	
C. lytica	coastal	y
C. marina	blackened sea bream	
C. marinoflava	North Sea	y
C. rosea	sea water	
C. salmonicolor	marine mud	y
C. sensitiva	sea water	
C. uliginosa	marine bottom sediment	y
Flexithrix dorotheae		
Flexibacter elegans	grows on sea water (not regarded as marine)	y
Microscilla aggregans	marine sediment	y
M. arenaria	marine sediment	y
M. furvescens	marine sediment	
M. litoralis	marine aquarium	
M. marina	marine aquarium	y
M. polymorphus	Beggiatoa	
M. sericea	marine aquarium	y
M. tractuosa	coastal sand	
Saprospira gigantea	coastal	

Table 2 (cont)

Marine Bacteria	Source	RDP Sequence
S. grandis	marine sand and mud	
S. nana	marine sand and mud	
S. toviformis	tide pool	
Achromatium oxaliferum		
*Beggiatoa (*strain MS-81-6)		
Thioploca araucae		
T. chileae		
T. marina		
Desulfonema limicola	anoxic sediment	
D. magnum	anoxic sediment	
GSB isolate		
Leucothrix mucor		
Nannochloris sp. strain GSBn		
Herpetosiphon cohaerens	sea water for growth	
H. nigricans	sea water for growth	
H. persicus	sea water for growth	
Myxococcus coralloides	coastal seawater	y
M. fulvus	littoral seawater	
M. stipitatus	coastal seawater	
M. xanthus	coastal seawater	y
Rhodococcus marinonascens	northeast Atlantic	
Micromonospora	beach; deep sediment; White Sea; Black Sea	
Streptomyces albidoflavus	marine sediment	
Streptosporangium corrugatum	beach sand	
Archaea		
Methanococcus halophilus		
M. jannaschii	hydrothermal vent	y
M. maripaludis	salt-marsh sediment	y
M. thermolithotrophicus	geothermally heated sediment	y
M. vannielii	San Francisco Bay	y
M. voltae	marine sediment	y
Methanomicrobium paynteri	marine sediment	
Methanospirillum hungatei	salt tolerant strain from marine sediment	y
Methanogenium	anaerobic marine	y
Methanoplanus endosymbiosus	marine ciliate	y *(M. limicola)*
Methanococcoides methylutens	marine trench	y
Methanolobus siciliae	sea sediment	y
M. tindarius	sea sediment	y
M. vulcani	sea sediment	y
Methanosarcina acetivorans	marine sediment	y

Table 2 (cont)

Marine Bacteria	Source	RDP Sequence
Methanosarcina barkeri	marine mud (other sources)	y
Haloarcula	Dead Sea and saltern multi species	*y* *(H. marismortui)*
Halobacterium denitrificans	San Francisco Bay salterns	*
H. saccharovorum	marine salterns	y
H. salinarium		*
H. sodomense	Dead Sea	*
H. volcanii		*
Halococcus morrhuae	codfish	y
Haloferax	Dead Sea and salterns multi species	y
Thermotoga maritima		y
Thermococcus celer		y
Pyrococcus woesei		
Staphylothermus marinus	black smoker	
Pyrodictium	hot sea floor	y *(P. occultum)*
Sulfolobus acidianus	heated acidic marine environment	*

Because most bacterial species are defined in data bases by a sequence from a single strain, when a sequence fails to match any from the data base, it is difficult to say with certainty that the unknown sequence in fact does not correspond to any known species. On the other hand it is unlikely that sequence variation within most known bacterial species is greater than a few percent, though there may be exceptions. Although we do not know the significance of minor sequence variation between environmental gene clones and sequences from cultivated species, it is probably safe to assume that when an environmental clone is *very* dissimilar to any known gene sequence, it is likely to represent a sequence from a species which is not described taxonomically from cultured strains. To be certain of this conclusion, it is necessary to compare sequence data bases with taxonomies for the organisms of interest. If organisms within the group are well described by systematic monographs and also well represented in sequence data bases, then it becomes unlikely that a unique new sequence belongs to that group.

How likely is it, then, that the unique and novel genes obtained from natural picoplankton assemblages are in fact derived from species which could readily be grown? To answer this, one must consider to what length microbiologists have gone in order to cultivate bacterioplankton species. It is useful to examine occasions in recent literature in which new and important microbial species have been isolated from bacterioplankton, and to ask how this came about. Two significant examples from the past 20 years are the isolation of marine *Synechococcus* and *Prochlorococcus* species from sea water.

Although it is an almost ubiquitous, widely distributed species in waters of the ocean surface, marine *Synechococcus* was not discovered until 1977 (Waterbury *et al.* 1979). Its discovery came about because epifluorescent microscopes were used to count bacteria from sea water, using the stain acridine orange. Early users of this method noticed that there were cells which were autofluorescent in unstained control preparations. It was later discovered that these cells were cyanobacteria, and that they constituted 20 to 30 percent of the overall community (Waterbury *et al.* 1987) A short time thereafter marine *Synechococcus* came under close study using culture methods, and strains representing several different phenotypic groups became widely available (Waterbury and Rippka 1989). Why was marine *Synechococcus* not cultivated and studied sooner? The answer is that it is not easy to cultivate.

Another example of the same pattern of discovery followed by cultivation occurred in the case of *Prochlorococcus*, an abundant and remarkably significant oxygenic phototroph. Because of its weak autofluorescence, *Prochlorococcus* was not as easy to observe by epifluorescent microscopy as was marine *Synechococcus*. However, when flow cytometers were first applied to the problem of counting phytoplankton in the 1980's, cells were observed that were unique in their combined properties of autofluorescence and light scattering, suggesting that they were prokaryotes which contained chlorophyll but lacked phycoerythrin. The chlorophylls were identified as divinyl chlorophylls *a* and *b*, not the chlorophylls *a* and *b* typical of green algae and plants (Chisholm *et al.* 1988). These cells were phenotypically highly unique and very abundant. Despite the years of effort that had gone into culturing marine *Synechococcus* species, no cultures were available of *Prochlorococcus.* This situation lasted until *Prochlorocccus* was cultured in 1992 (Chisholm *et al.* 1992). If *Prochlorococcus* was so ubiquitous, and accomplished scientists were using specialized methods to isolate other oxygenic phototrophic prokaryotes from sea water, then why was *Prochlorococcus* not discovered first as a culture? The answer here is clear: *Prochlorococcus* is extremely difficult to culture, though once one has a culture which grows readily it can be propagated in any laboratory.

There is a clear pattern of discovery for novel marine bacterial oxygenic phototrophs. First the existence of the organism is revealed by a physical method, and its significance established. With those facts in hand, a conscientious effort by experienced microbiologists results in cultures being obtained. Once those cultures are obtained, they are spread from laboratory to laboratory, and additional cultures from new sources typically soon follow.

If one extrapolates from the results of studies with *Prochlorococcus* and marine *Synechococcus*, then one should expect that there are other marine bacteria which are ubiquitously distributed and significant in marine ecosystems, but uncultured. These organisms are not necessarily unculturable species. They may simply be uncultured because we do not know the conditions required for their propagation in the laboratory. New and important species of marine bacteria are not usually discovered serendipitously by culturing. Rather, new culturing methods are developed in a focused effort to acquire new species in culture, when the defining characteristics of the target organism are known. It is to be expected, therefore, that some of the unknown rDNA types defined by early sequencing studies will be obtained in culture after significant effort. rDNA probes provide a means by which culturing efforts can be focused on a single taxonomic group, much as flow cytophotometry and epifluorescent microscopy serve that purpose for the oxygenic phototrophs.

To summarize, there is probably more than one explanation for the observed mismatch between clone libraries prepared from natural bacterial assemblages and rDNA sequence data bases. It should be possible to isolate new bacterial species simply by applying creative and novel approaches to the cultivation of bacteria from seawater. A number of new bacterial strains have been described in recent years by the application of just such methods. It is unlikely, however, that the remarkably abundant organisms of the SAR11 cluster or Group I Archaea will be isolated accidentally. Future studies are being undertaken to culture these groups by systematic methods based on ribosomal RNA probes.

Ribosomal RNA Genes as Genetic Markers: Population studies using oligonucleotide probes

WHAT KINDS OF INFORMATION CAN BE OBTAINED WITH THESE PROBES?

The first investigations of microbial diversity in natural ecosystems using rDNA genes cloned directly from natural samples were criticized because they were "fishing expeditions." In retrospect, however, this approach seems justified by the results, which have been a better understanding of the evolutionary phylogenetic range of microbial species, and a better perspective on the diversity of cultivatable species in the light of a more accurate measure of real diversity. An example of the power and importance of sequencing studies is the recent description of novel Archaea from Yellowstone Hot Springs (Barns *et al.* 1994), resulting in a better understanding of archaeal diversity, and possibly the discovery of a new kingdom.

From the outset, however, sequencing studies had a second goal, quite apart from the cataloguing of microbial diversity. This goal was the identification of marker sequences which could be used to design oligonucleotide probes for the specific detection of microorganisms in natural ecosystems. In many respects this second goal is independent of the first. Whereas surveys of diversity depend heavily upon the unbiased nature of clone libraries, investigations of microbial ecology simply require markers. The representation of clones and libraries may be skewed without changing the utility of the markers, so long as they are valid markers for real species from the ecosystems in question. Viewed from this perspective, clone libraries become a means to an end, the end being the design of specific oligonucleotide probes.

The applications of oligonucleotide probes to studies of microbial ecology are numerous. Probes may be labelled with radioactive or chemical markers and hybridized to ribosomal RNAs extracted directly from natural systems. The same probes can also be hybridized to mixtures of rDNAs amplified by PCR. Probes may be labelled with fluorochromes and hybridized to fixed cells for the microscopic visualization of organisms. Each of these approaches provides different information and is subject to different sources of errors and problems with interpretation; each is valid and useful for a particular purpose.

The hybridization of rDNA probes to bulk nucleic acids prepared from natural samples was one of the first applications of oligonucleotides to microbial ecology (Stahl *et al.* 1988). Despite uncertainties about metabolism of unknown microorganisms, it is known that all cells synthesize ribosomal RNAs for the purpose of protein synthesis, and that rRNA synthesis is tightly regulated by cellular growth rates (Kemp *et al.* 1993; Kerkhof and Ward 1993). The ribosomal RNA content in cells is also related to the size of those cells. The ribosomal RNA abundance of an unknown species in a natural system is probably best represented as a function of the protein biosynthetic capacity of that species, and does not necessarily bear a linear relationship to population numbers or biomass. For example, when a particular species goes through a period of rapid increase in cellular numbers as a consequence of rapid growth, and a subsequent period of decline as a consequence of the exhaustion of substrates, one would expect its rRNA abundance to peak in time, out of phase with the subsequent peak in cell numbers. In fact, it is possible for rRNA abundance to be very low while a very large population of viable cells is maintained; rRNA abundance is not a proxy for cell numbers or biomass.

Hybridizations of oligonucleotide probes to ribosomal RNAs extracted from natural marine systems are limited by the lack of cultivatable organisms from which control rDNAs can be obtained. The efficiencies with which different oligonucleotide probes hybridize to their targets vary widely (Hill *et al.* 1990). Hence, it is valid to compare probe hybridization between samples, but the results cannot be expressed as absolute numbers. A possible way of avoiding this difficulty would be the synthesis of synthetic rRNAs from cloned rDNA genes. In our experience *in vitro* transcription systems for the synthesis of RNA are not efficient for molecules over a few hundred bases in length, thus limiting their utility for this application.

The measurement of rDNA abundance in amplified rDNA preparations is attractive for many reasons, perhaps foremost that an accurate measure of rDNA abundance is not as sensitive to short-term fluctuations and ecosystem dynamics as are measures of rRNA abundance. For example, in the oligotrophic ocean, rRNA may be in very low abundance at times when cell numbers are significant, reflecting a low metabolic rate for the microbial populations at the time of sampling. However, the rDNA copy number for cells is not known for uncultivated species, nor for most cultured species. Thus any attempt to interpret rDNA abundance in absolute terms is probably futile. In addition, as noted above, there are poorly understood potential problems related to the bias in selection of templates during the polymerase chain reaction.

Despite the above caveats, clear-cut and valid uses of bulk hybridization measurements are obvious. Spatial and temporal measurements of rDNA abundance are primarily concerned with changes in cellular activity and abundance, rather than the absolute values of these terms. For example, not much could be concluded from the knowledge that Group I Archaea are 20% of a particular population as opposed to 40% of that population; in both cases they are clearly significant members of the community. However, the *relative* distribution of genes in a vertical profile or in time series may provide important insights into the ecology of the organisms. For example, we have observed that Group I Archaea are very abundant in the lower surface layer at the Sargasso Sea, but absent or much less abundant in the mixed layer during stratified periods. In this case, it is differences in the relative abundance of genes as a function of depth that are most interesting. The consistently higher abundances of archaeal genes observed in the lower surface layer suggest that these organisms are best adapted to the conditions of the upper mesopelagic. Group I Archaea are rarely found at the surface of the open ocean, probably because they are not adapted to the conditions found there. In this example, it is not the absolute numbers of Archaea that are important, but rather the clear trend in their distribution.

Another example of the approach described above is provided in Fig. 3, which shows the distribution of prochlorophytes (SAR6) and SAR11 16S rDNAs in the upper 250 meters of the water column in the Sargasso Sea during August 1991 when the water column was highly stratified. The data indicate that the SAR11 cluster genes are most abundant in the upper euphotic zone, whereas at the time of this sampling prochlorophytes were most abundant near the deep chlorophyll maximum.

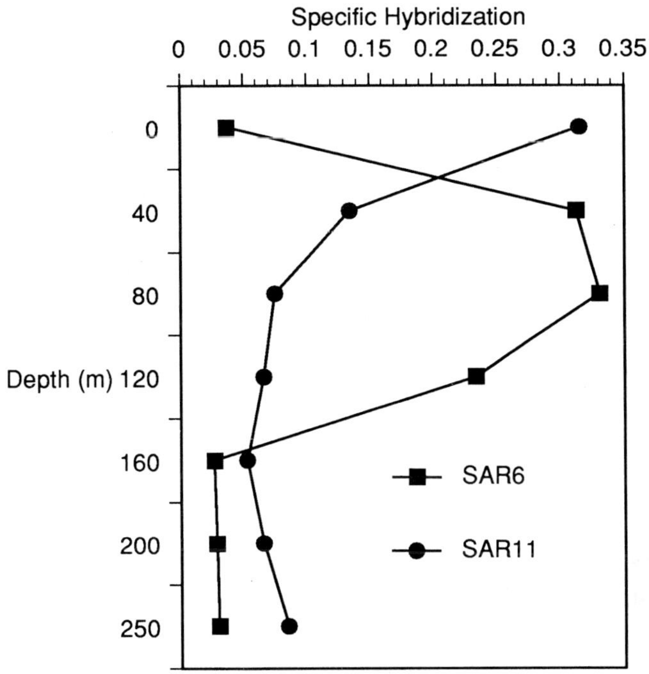

Fig. 3. Ribosomal DNA profiles at the Bermuda Atlantic Time Series Station in the Sargasso Sea (Aug, 1991). Shown here are the distributions of two gene clusters, SAR11 and prochlorophytes (SAR6). To produce these profiles 16S rDNAs were amplified from plankton genomic DNA by the polymerase chain reaction using general Bacterial primers, blotted onto nylon membranes, and hybridized to radioactively-labeled probes. (Reprinted with permission from Giovannoni and Cary 1993).

Combinations of rDNA and rRNA bulk hybridization data may be particularly useful because the two types of data are complementary, and provide insight into dynamic changes in populations that occur over time and space. For example, we have noted that Group I Archaea are mixed to the surface during strong seasonal forcing events, and that their rRNA abundance at the surface may be very low while DNA abundance is high. It is possible that this observation results from the inhibition of Group I Archaea following their displacement from the lower mixed layer to surface waters. These observations provide an example of the type of community dynamics that may be observed when measures of rRNA abundance and rDNA abundance are combined.

FLUORESCENT PROBES

Fluorescent oligonucleotide probes provide another important application of genetic markers to the study of uncultured bacterioplankton. One obvious advantage of this approach is that it makes it possible to microscopically visualize the unknown microorganisms. Fluorescent rRNA probes have received widespread attention, but the success of applications to the marine system has been mixed. The most significant follow-up to early observations has come from Paul Kemp's laboratory. These studies showed that with multiple probes, 75% of cells from coastal systems could be detected using a photometer (Lee *et al.* 1993). Kemp's work also verified earlier studies that had shown that cellular ribosome content is a nearly linear function of growth rate, and extended these studies to environmentally realistic doubling times (Kemp *et al.* 1993; Lee *et al.* 1993). Not surprisingly, it was found that the slope of ribosomes/growth rate curves varied between experimental organisms. A substantial component of this variation could be explained by differences in cell volume. Kemp and co-workers concluded that ribosomal RNA probes could be used to study growth rates, as was suggested by previous studies, but that only probes for specific taxa could yield such data, and that absolute values for growth rates were unobtainable unless the organisms were available in culture for the calibration of curves (Kemp *et al.* 1993; Lee *et al.* 1993).

The above results suggest that future applications of ribosomal RNA probes to marine systems will focus on the use of probes for particular microbial taxa rather than measurements of cellular ribosome content using universal probes. Studies such as these are tied closely to rDNA sequencing studies which provide the data bases necessary for the design of highly specific probes. Obvious applications of fluorescent probes may include the detection of specific microorganisms in enrichments obtained by amending sea water with metabolic substrates. This subject is reviewed more thoroughly elsewhere in this volume.

Acknowledgements

We are grateful to the Bermuda Biological Station BATS group (Bermuda Atlantic Time Series), the crew of the RV *Weatherbird II* , and Nanci Adair for support. We would also like to thank Ena Urbach and Michael Rappe' for useful comments. This work was supported by National Science Foundation grant OCE 9016373 and Department of Energy grant FG0693ER61697.

References

Amann RI, Lin C, Key R, Montgomery L, Stahl DA (1992) Diversity among *Fibrobacter* isolates: toward a phylogenetic classification. *System Appl Microbiol* **15:** 23-31

Barns SM, Fundyga RE, Jeffries MW, Pace NR (1994) Remarkable archaeal diversity detected in a Yellowstone National Park hot spring environment. *Proc Natl Acad Sci USA* **91:** 1609-1613

Ben-Porath J and Zehr JP (1994) Detection and characterization of cyanobacterial *nifH* genes. *Appl Environ Microbiol* **60:** 880-887

Britschgi TB and Giovannoni SJ (1991) Phylogenetic analysis of a natural marine bacterioplankton population by rRNA gene cloning and sequencing. *Appl Environ Microbiol* **57:** 1313-1318

Brock TD (1987) The study of microorganisms *in situ*: progress and problems, In Symp Soc Gen Microbiol **41** Fletcher M, Gray TR & Jones JG (eds.) Cambridge University Press, Cambridge, UK 1-17

Carlucci AF and Williams PM (1978) Simulated *in situ* growth rates of pelagic marine bacteria. *Naturwissenschaften* **65:** 541

Casida LE (1965) Abundant microorganisms in soil. *Appl Microbiol* **13:** 327-334

Chisholm SW, Frankel SL, Goericke R, Olson RJ, Palenik B, Waterbury JB, West-Johnsrud L, Zettler ER (1992) *Prochlorococcus marinus nov.* gen. *nov.* sp.: An oxyphototrophic marine prokaryote containing divinyl chlorophyll *a* and *b*. *Arch Microbiol* **157:** 297-300

Chisholm SW, Olsen RJ, Zehler ER, Goericke R, Waterbury JB, Welschmeyer NA (1988) A novel free-living prochorophyte abundant in oceanic euphotic zone. *Nature* **334:** 340-343

Conn HJ (1948) The most abundant groups of bacteria in soil. *Bacteriol Rev* **12:** 257-273

DeLong EF, Wickham GS, Pace NR (1989) Phylogenetic stains: ribosomal RNA-based probes for the identification of single cells. *Science* **243:** 1360-1363

DeLong EF (1992) Archaea in coastal marine bacterioplankton. *Proc Natl Acad Sci* **89:** 5685-5689

DeLong EF, Franks DG, Alldredge AL (1993) Phylogenetic diversity of aggregate-attached vs. free-living marine bacterial assemblages. *Limnol Oceanogr* **38:** 924-934

DeLong EF, Wu KY, Prezellin BB, Jovine RVM (1994) High abundance of Archaea in Antarctic marine picoplankton. *Nature* **371:** 695-697

Ensign JC (1970) Long term starvation of rod and spherical cells of *Arthrobacter crystallopoietes*. *J Bacteriol* **103:** 569-577

Fox GE, Wisotzkey JD, Jurtshuk P (1992) How close is close: 16S rRNA sequence identity may not be sufficient to guarantee species identity. *Int J Syst Bacteriol* **42:** 166-170

Fuhrman JA, McCallum K, Davis AA (1992) Novel major archaebacterial group from marine plankton. *Nature* **356:** 148-149

Fuhrman JA, K. McCallum, Davis AA (1993) Phylogenetic diversity of subsurface marine microbial communities from the Atlantic and Pacific oceans. *Appl Environ Microbiol* **59:** 1294-1302

Giovannoni SJ, DeLong E, Olsen GJ, Pace NR (1988) Phylogenetic group-specific oligodeoxynucleotide probes for *in situ* microbial identification. *J Bacteriol* **170:** 720-726

Giovannoni SJ (1991) The polymerase chain reaction, in *Sequencing and hybridization techniques in bacterial systematics*, (John Wiley & Sons, New York 177-201

Giovannoni SJ, Britschgi TB, Moyer CL, Field KG (1990) Genetic diversity in Sargasso Sea bacterioplankton. *Nature* **345:** 60-63

Giovannoni SJ and Cary SC (1993) Probing marine systems with ribosomal RNAs. *Oceanography* **6:** 95-104

Giovannoni SJ, DeLong EF, Schmidt TM, Pace NR (1990) Tangential flow filtration and preliminary phylogenetic analysis of marine picoplankton. *Appl Environ Microbiol* **56:** 2572-2575

Good IJ (1953) The population frequencies of species and the estimation of the population parameters. *Biometrika* **40:** 237-264

Hill WE, Weller J, Gluick T, Merryman C, Marconi RT, Tassanakajohn A, Tapprich W (1990) Probing ribosome structure and function by using short complementary DNA oligomers, In *The ribosome: structure, function and evolution*, Hill WE, Dahlberg A, Garrett RA, Moore PB, Schlessinger D, & Warner JR (eds) Am Soc Microbiol, Washington, DC 253-264

Hoppe HG (1976) Determination and properties of actively metabolizing heterotrophic bacteria, investigated by means of microautoradiography. *Mar Biol* **36:** 291-302.

Hurlbert SM (1971) The non-concept of species diversity: a critique and alternative parameters. *Ecology* **52:** 577-586

Jannasch HW (1969) Estimations of bacterial growth rates in natural waters. *J Bacteriol* **99:** 156-160

Jannasch HW and Jones GE (1959) Bacterial populations in seawater as determined by different methods of enumeration. *Limnol Oceanogr* **4:** 128-139

Kemp PF, Lee S, LaRoche J (1993) Estimating the growth rate of slowly growing marine bacteria from RNA content. *Appl Environ Microbiol* **59:** 2594-2601

Kerkhof L and Ward BB (1993) Comparison of nucleic acid hybridization and fluorometry for measurement of the relationship between RNA/DNA ratio and growth rate in a marine bacterium. *Appl Environ Microbiol* **59:** 1303-1309

Kjellberg S, Humphrey BA, Marshall KC (1983) Initial phase of starvation and activity of bacteria at surfaces. *Appl Environ Microbiol* **46:** 978-984

Kjellberg S, Humphrey BA, Marshall KC (1982) Effect of interfaces on small, starved marine bacteria. *Appl Environ Microbiol* **46:** 1166-1172

Kramer JG and Singleton FL (1993) Measurement of rRNA variations in natural communities of microorganisms on the southeastern U.S. continental shelf. *Applied Environ Microbiol* **59:** 2430-2436.

Kurath G and Morita RY (1983) Starvation-survival physiological studies of a marine Pseudomonad sp. *Appl Environ Microbiol* **45:** 1206-1211

Larsen N, Olsen GJ, Maidak BL, McCaughey MJ, Overbeek RT, Macke J, Marsh TL, Woese CR (1993) The Ribosomal Database Project. *Nucleic Acids Res* **21:** 3021-3023

Lee S and Fuhrman JA (1990) DNA hybridization to compare species compositions of natural bacterioplankton assemblages. *Appl Environ Microbiol* **56:** 739-746

Lee S, Malone C, Kemp PF (1993) Use of multiple 16S rRNA-targeted fluorescent probes to increase signal strength and measure cellular RNA from natural planktonic bacteria. *Mar Ecol Prog Ser* **101:** 193-201

Ling LL, Keohavong P, Dias C, Thilly WG (1991) Optimization of the polymerase chain reaction with regard to fidelity: modified T7, *Taq*, and vent DNA polymerases. *PCR Methods and Applications 1:* 63-69

Martin P and Macleod RA (1984) Observations on the distinction between oligotrophic and eutrophic marine bacteria. *Appl Environ Microbiol* **47:** 1017-1022.

Morita RY (1982) Starvation-survival of heterotrophs in the marine environment. *Adv Microb Ecology* **6:** 171-198

Mullins TD, Britschgi TB, Krest RL, Giovannoni SJ (1995) Genetic comparisons reveal the same unknown bacterial lineages in Atlantic and Pacific bacterioplankton communities. *Limnol Oceanogr* **40:** 148-158

Novitsky JA and Morita RY (1978) Possible strategy for the survival of marine bacteria under starvation conditions. *Mar Biol* **48:** 289-295

Ochman H and Wilson AC (1987) In *Escherischia coli and Salmonella typhimurium*, Neidhardt FC (ed) Am Soc Microbiol, Washington, DC 1649-1654

Paul JH, Cazares L, Thurmond J (1990) Amplification of the *rbcL* gene from dissolved and particulate DNA from aquatic environments. *Appl Environ Microbiol* **56:** 1963-1966

Postgate JR (1976) Death in macrobes and microbes. In *The survival of vegetative microbes*, Gray TRG and Postgate JR (eds) Cambridge University Press, Cambridge, UK 1-19.

Rehnstam AS, Backman S, Smith DC, Azam F, Hagstrom A (1993) Blooms of sequence-specific culturable bacteria in the sea. *FEMS Microbiol Ecol* **102:** 161-166

Reysenbach A-L, Giver LJ, Wickham GS, Pace NR (1992) Differential amplification of rRNA genes by polymerase chain reaction. *Appl Environ Microbiol* **58:** 3417-3418

Ribbons DW and Dawes EA (1963) Environmental and growth conditions affecting the endogenous metabolism of bacteria. *Ann NY Acad Sci* **102:** 564-586

Roszak DB and Colwell RR (1987) Survival strategies of bacteria in the natural environment. *Microbiol Rev 51:* 365-379

Saiki RK, Gelfand DH, Stoffel S, Scharf SJ, Higuchi R, Horn GT, Mullis KB, Erlich HA (1988) Primer-directed enzymatic amplification of DNA with a thermostable DNA polymerase. *Science* **239:** 487-491

Schmidt TE, DeLong EF, Pace NR (1991) Analysis of a marine picoplankton community by 16S rRNA gene cloning and sequencing. *J Bacteriol* **173:** 4371-4378

Shuldiner AR, Nirula A, Roth J (1989) Hybrid DNA artifact from PCR of closely related target sequences. *Nucleic Acids Res.* **17:** 4409

Sieburth J (1979) *Sea microbes* (Oxford University Press, New York), pp 491

Stahl DA, Flesher B, Mansfield HW, Montgomery L (1988) Use of phylogenetically-based hybridization probes for studies of ruminant microbial ecology. *Appl Environ Microbiol* **54:** 1079-1084

Stahl DA and Urbance JW (1990) The division between fast- and slow-growing species correspond to natural relationships among the mycobacteria. *J Bacteriol* **172:** 116-124

Staley JT and Konopka A (1985) Measurement of in situ activities of nonphotosynthetic microorganisms in aquatic and terrestrial habitats. *Ann Rev Microbiol* **39:** 321-346

Urbach E (1994) Evolution and Population Genetics of *Prochlorococcus marinus.* (Ph.D. thesis) Massachusetts Institute of Technology, Cambridge, MA.

Ward DM, Bateson MM, Weller R, Ruff-Roberts AL (1992) Ribosomal RNA analysis of microorganisms as they occur in nature. *Adv Microb Ecol* **12:** 219-286

Waterbury JB and Rippka R (1989) in *Bergey's Manual of Systematic Bacteriology*, Staley JT, Bryant MP, Pfenig N and Holt JG (eds) Williams and Wilkins, Baltimore. **3:** 1728-1746.

Waterbury JB, Watson SW, Guillard RRL, Brand LE (1979) Widespread occurrence of a unicellular, marine, planktonic, cyanobacterium. *Nature* **277:** 293-294

Waterbury JB, Watson SW, Valois FW, Franks DG (1987) Biological and ecological characterization of the marine unicellular cyanobacterium *Synechococcus. Can Bull Fish Aquat Sci* **214:** 159-204

Viruses - the New Players in the Game; Their Ecological Role and Could They Mediate Genetic Exchange by Transduction?

Gunnar Bratbak and Mikal Heldal
Department of Microbiology
University of Bergen
Jahnebakken 5
N-5020 Bergen
Norway

Abstract
The abundance of virus like particles in marine ecosystems range from $<10^4$ ml^{-1} to $>10^8$ ml^{-1}. Their distribution in time and space parallels that of other biological parameters such as bacterial abundance and chlorophyll. The abundance of viral particles changes on a seasonal and a diel basis, but the change in viral abundance may also be much faster, resembling synchronous "one-step" growth curves. Few studies have been carried out, but viruses appear, at least in some cases, to have a significant impact on carbon and nutrient flow in microbial food webs. Virus has also been demonstrated to exert a species-specific control of both bacteria and phytoplankton populations in natural waters. With model phage-host systems, it has been demonstrated that transduction can take place in aquatic environments. From what is known about transduction in culture systems we estimate that transduction potentially may occur at a significant rate in natural communities, but it is uncertain whether the potential is realized or not. Virus infection and lysogenization of bacteria are known to give some bacteria a competitive advantage and it is possible that similar mechanisms are important in natural waters.

Introduction
The presence of bacteriophages and other viruses in natural waters has been known for decades, but it is only during the last few years we have considered them to play an ecological role (Fuhrman, 1992; Fuhrman and Suttle, 1993; Bratbak *et al.*, 1994). The reason for the current interest in the ecological role of viruses is due to the observation that the total abundance of virus like particles in natural unpolluted waters is much higher than previously realized (Proctor *et al.*, 1988; Bergh *et al.*, 1989; Proctor and Fuhrman, 1990). This is in fact a parallel to

NATO ASI Series, Vol. G 38
Molecular Ecology of Aquatic Microbes
Edited by I. Joint

the increased interest in the ecological role of bacteria in the mid seventies when total counts of bacteria was found to be much higher than the number of colony forming units (Zimmermann and Meyer-Reil, 1974; Hobbie *et al.*, 1977).

The general properties of bacteriophages and other viruses are known from many years of research on viruses and virus-host-systems cultured in the laboratory. The viruses that are growing in natural environments on native hosts may have properties that are different from or extreme compared to laboratory strains, but we may anticipate that their basic properties are the same. We may thus make educated guesses about what viruses potentially do, and about the roles they potentially play in natural environments. However, the challenge of viral ecology is to investigate the properties that are realized by native viral communities in natural environments and to assess the quantitative significance of the various aspects of viral activity.

Some important properties of viruses

The general features that form the basis for the ecological role we attribute to viruses in natural environments include their host specificity, their mode of reproduction and their ability to transfer genetic information between cells. Virus are host specific in the sense that they have the ability to infect and reproduce in only a limited number of hosts. The hosts may belong to one species, to a few closely related species or to certain strains within a species.

As obligate parasites, virus reproduce only within infected host cells. We may distinguish between three different modes of viral reproduction. 1) In the lytic cycle the infection of the host cells immediately results in production of progeny viral particles which are released when the cells lyse. 2) The DNA of temperate viruses may reproduce along with the host cell for several generations by being inserted into the host genome. The host cells, which are called lysogenic, produce a repressor protein that inhibits the production of new virus particles. This repressor protein makes the cells immune against the virus they carry, but it does not inhibit additional viral adsorption and injection of viral nucleic acids. Production of progeny viral particles takes place only when the viral DNA for some reason resumes it's independent activity as in the lytic cycle. 3) The feature of a latent infection is that the host cell produce and release new viral particles continuously, without being killed and lysed.

A basic property of viruses is that they transfer DNA or RNA from the host cells where they are produced, to the cells they infect. In most cases this transfer nucleic acids only includes the viral genome itself. Transduction of host DNA occurs when a piece of the host genome by chance is packed into the viral capsid and transferred to a new cell.

Ecological roles of viruses

Knowing the basic properties of viruses we may assess the significance of their potential ecological roles from their abundance and activity.

ABUNDANCE

The abundance of virus-like particles (VLP) in seawater has been reported in the range from $<10^4$ ml^{-1} to $>10^8$ ml^{-1} (Bratbak *et al.*, 1994). Their distribution in time and space parallels that of many other biological parameters such as chlorophyll concentration, particulate DNA, bacterial abundance *etc* (Cochlan *et al.*, 1993; Boehme *et al.*, 1993). The highest abundances are thus generally found in coastal surface waters during the productive part of the year while the lowest are found in the open ocean, in deep waters and during the unproductive part of the year (Bratbak *et al.*, 1994). The distribution of VLP suggests that they are related to some kind of biological activity. Their association with algae and bacteria, as demonstrated by careful preparation and transmission electron microscopy (Børsheim *et al.*, 1990; Heldal and Bratbak, 1991; Bratbak *et al.*, 1992b), suggests that they are produced by these microorganisms. We can, however, not say for sure that these particles really are viruses. Moreover, there is no way to tell if these particles are infective or not.

The abundance in seawater of phages that infect specific marine host bacteria is usually reported to be between 0 and 10 ml^{-1}. Values in the order of 10^3 ml^{-1} are not exceptional (Moebus, 1992), and the highest reported are about $4x10^4$ ml^{-1} (Ahrens, 1971). These figures are based on plaque forming units (PFU). They include all phages that infect and lyse a particular host and the phages may therefore be different. However, many phages may have lost their infectivity so the total number of each phage form (infective + noninfective) may be much higher.

VBR RATIO

From the ratio between abundance of virus and bacteria (virus-to-bacteria ratio, VBR) we may infer that virus production is related to the nutrient status of the cells and to the diversity of the bacterial community. The correlation between viral and bacterial abundances, which is found when comparing data from different marine environments (Fig. 1a, Cochlan *et al.*, 1993; Boehme *et al.*, 1993), may not only be due to a simple relationship between the two parameters alone. Fig. 1b suggests that the correlation may be due to higher virus-to-bacteria ratios (VBR) in eutrophic waters as compared to more oligotrophic waters. The data on virus and bacterial abundances reported by Hara *et al.*, 1991; Weinbauer *et al.*, 1993; Cochlan *et al.*, 1993; and by Boehme *et al.*, 1993, may be interpreted similarly. Thus, bacteria may produce more free virus particles when the environmental conditions support high growth rate and production, than they do when growing under sub-optimal conditions.

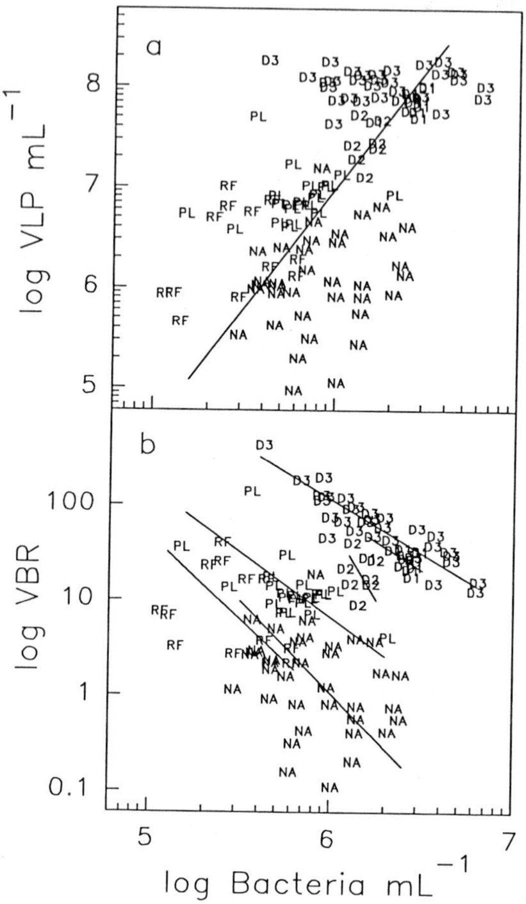

Fig. 1a): Abundance of virus-like-particles (VLP), and **1b)**: Virus-to-bacteria-ratio (VBR) as function of bacterial abundance in various marine ecosystems. Lines are functional regression. High viral and bacterial abundances are found in nutrient rich Danish coastal waters (D1, D2 and D3), while lower values are found in oligotrophic coastal waters (RF; Raunefjorden, Western Norway and PL; Plymouth, UK) and in the open ocean (NA; North Atlantic). The decrease in VBR with increasing bacterial abundance may be interpreted to indicate that high bacterial abundance in a given system involves a decrease in bacterial diversity. The positive correlation between VLP and bacteria in a) is significant (R=0.575, P<0.001). The negative correlation between VBR and bacteria is significant for PL (R=-0.600, P=0.002) and D3 (R=-0.895, P<0.001), while for the other sampling sits the correlation was insignificant (P>0.05).

Moreover, Fig. 1b suggests that there is a tendency for the VBR ratios to decrease as the bacterial abundance increases within the different ecosystems studied. Increased bacterial abundances in a given ecosystem may involve a decrease in community diversity for example during blooms of one or a few species. Thus, for a given ecosystem, a relatively low VBR ratio may indicate that the bacterial diversity is low. One possible explanation for this is as follows: The rate of absorbance of virus to its specific host is proportional to the product between virus concentration and host concentration. This implies that the total abundance of virus can be higher in a system where the host diversity is high with many different host species, as compared to a system dominated by one or a few different types of host bacteria. This explanation is based on the assumption that specific absorption is a major factor for viral loss and for regulation of viral abundance in natural systems.

ACTIVITY

Viral activity manifests itself by lysis of host cells and production of viral particles. Mature viral particles can be observed in the host cells for only a small fraction of the whole lytic cycle. Using pure cultures of marine bacteria, Proctor *et al.*, (1993) found that the fraction of the bacteria that was involved in virus production was 3.70-7.14 times the fraction that could be observed by thin sectioning to contain viral particles. Based on this and on thin sectioning of bacteria from natural marine communities they estimated that 3 - 31% of free-living marine bacteria were involved in virus production. Cells in late stages in the lytic cycle may be observed directly in the electron microscope as cells containing mature viral particles and as lysed cells surrounded by viral particles (Børsheim *et al.*, 1990; Heldal and Bratbak, 1991). From <1 to 40% of the bacterial community in natural waters has been observed to be in lysis (Bratbak *et al.*, 1994). We have also observed that the fraction of lysing cells may change on a diurnal basis from 12 to 29 % (Bratbak *et al.*, 1992b). This suggests that viral lysis to some extent may be synchronous, that the lysing population has a high turnover rate, and thus that the maximum values for the fraction of lysed and lysing cells include most cells in the lytic cycle.

The methods that have been applied to assess production of viral particles include incorporation of ^{32}P into viral DNA (Steward *et al.*, 1992), net change in viral abundance over time (Børsheim *et al.*, 1990, Bratbak *et al.*, 1990), and viral decay rate (Heldal and Bratbak 1991; Bratbak *et al.*, 1993). These methods address different aspects of virus production and turnover and they employ incubation times and sampling intervals ranging from <1h to several days. The rates that have been estimated in experiments with sampling intervals of 5h or more, are in the order of 1 d^{-1} or less. In experiments with sampling intervals between 2h and 10 min, rates in the order of 10 - 100 d^{-1} have been observed (Fig. 2, Bratbak *et al.*, 1994). Some of this variation is of course related to seasonal variations and to the particular environment investigated. However, the results also suggest that there at least are two different populations of viral particles, one with unstable particles and a very high turnover rate and one with more stable particles. The dynamics of the first population may be observed during short time incubations such as in the decay experiments (Heldal and Bratbak, 1991), as net change in viral abundance in bottle incubations (Fig. 2) and in mesocosm diel experiments (Bratbak *et al.*, 1992b; Jiang and Paul, 1994). The dynamics of the latter population is demonstrated in seasonal studies (Fig. 3, Jiang and Paul, 1994).

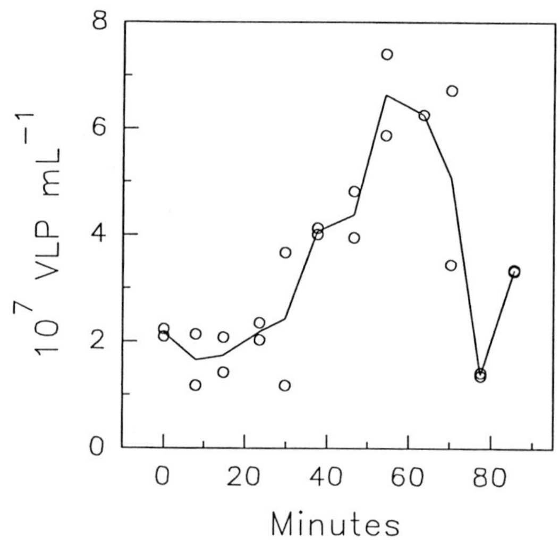

Fig. 2 Short-time variation in virus-like-particles (VLP) in 1 litre bottles in Osterfjorden, Western Norway, September 1993. Two parallel bottles were incubated *in situ* and subsampled every 7-8 min.

The factors regulating viral abundance on a time scale of days to weeks are presumably related to season, to changes in population density (blooms) and to community composition and diversity. The production may be due to lytic infection of susceptible hosts or to induction of lysogenic hosts, while the fate of the virus particles may be determined by adsorption to larger particles (Murray and Jackson, 1992; Suttle and Chen, 1992) or by disintegration (Heldal and Bratbak, 1991). Rates of change in the order of 10 - 100 d^{-1}, which may be observed for brief periods of time during short time incubations, are presumably related to synchronous mass lysis of some subpopulation of host cells. The lysis may be due to a preceding mass infection of a susceptible host population or to a

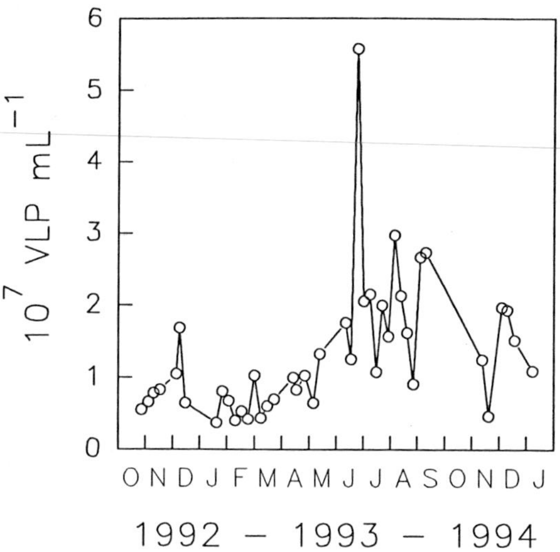

Fig. 3 Seasonal variation in virus-like particles (VLP) at station L4 located off Plymouth, UK, at 50.15N, 4.13 W.

mass induction of lysogenic hosts. The latter are the favoured hypothesis as most bacteria in natural waters probably are lysogenic. This view is motivated by the facts that most known viruses are able to lysogenize their hosts (Freifelder, 1987), that non-lysogenic strains are lysogenized when incubated together with lysogenic strains (Miller and Sayler, 1992), and that it is possible to induce virus production in a large fraction (ca 40%) of isolated marine bacteria (Jiang and Paul, 1994). The particles released disappear at a rate that is often comparable to the rate with which they appear (Fig. 2, Bratbak *et al.*, 1993; Jiang and Paul, 1994). The main process responsible for their removal is presumably disintegration. Adsorption to larger particles may in most cases be excluded as the number of target particles must exceed 2.8 10^6 ml^{-1} before the virus adsorption rate becomes 10 d^{-1}. This estimate is based on the equation

$$dP/dt = k \cdot P \cdot H$$

where P is the number of virus particles, H the number of target particles and k the adsorption rate constant 2.5 10^{-9} cm^3 min^{-1}, which imply that every collision results in adsorption (Stent, 1963).

CARBON AND NUTRIENT FLOW

The role of viruses with respect to carbon and nutrient flow in microbial food-webs is basically that, by causing cell lysis, they divert some of the organic material flowing towards higher trophic levels and return it to the bottom level as a mixture of particulate and dissolved dead organic material (Thingstad *et al.*,

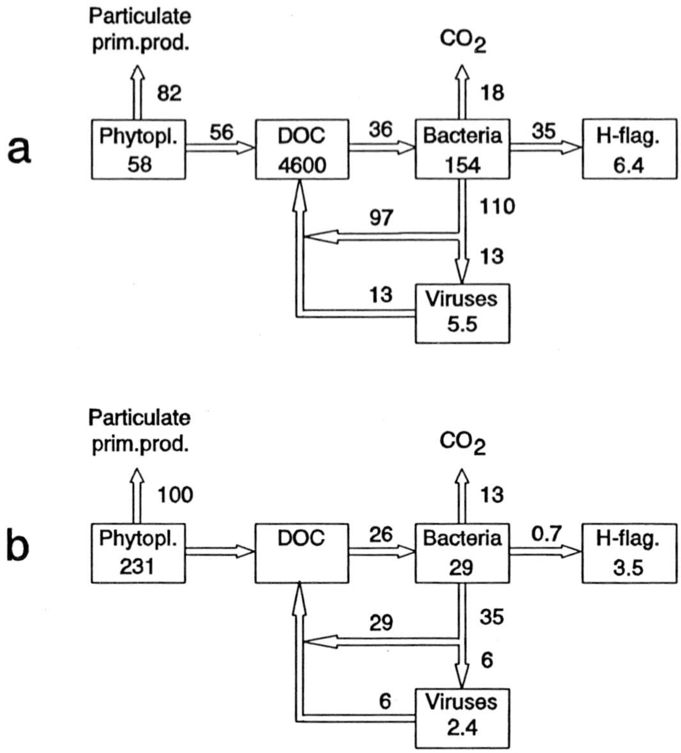

Fig. 4 Carbon flow in the viral loop in two diel mesocosm experiments in Knebel Vig, Denmark in a) 1991 and b) in 1992. The unit for flowrate (arrows) is $\mu gCl^{-1}d^{-1}$ and the unit for mass (boxes) is μgCl^{-1}. For further explanations see text.

1993). Few studies have been carried out but they support the view that viral activity may be quantitatively significance for the carbon and nutrient flow. From the fraction of the bacterial community that was found to be involved in virus production, Proctor *et al.* (1993) used a steady state model to estimate that 6 - 62% of the bacterial mortality was due to viruses. González and Suttle (1993) found that the clearance rate of viruses by natural nanoflagellate assemblages was about 4% of those for bacteria. Although grazing on viral particles thus does not seem not be an important factor for the control of viral abundance, it may be significant in the nanoflagellate nutrition. Assuming a bacterial abundance of 10^6 ml^{-1} and a viral abundance of 10^7 - 10^8 ml^{-1} González and Suttle (1993) estimated that virus may represent 0.2 - 9% of the carbon, 0.3 - 14% of the nitrogen and 0.6 - 28% of the phosphorus that the flagellates obtain from ingestion of bacteria.

In a diel mesocosm experiment Bratbak *et al.* (1993) estimated that 72 %

of the bacterial population was removed by viral lysis per day. Viral lysis of bacteria exceeded the bacterial production by a factor of about 6 and the flagellate grazing rate by a factor of about 3. (Fig. 4a). A similar diel study was conducted one year later (Fig. 4b). In this case we found that the viral abundance decreased throughout a 24 hour period (Fig. 5) and the measurements of viral decay (Heldal and Bratbak, 1991) showed no difference between the bottles where virus production was stopped by cyanide and the control bottles (Fig. 5).

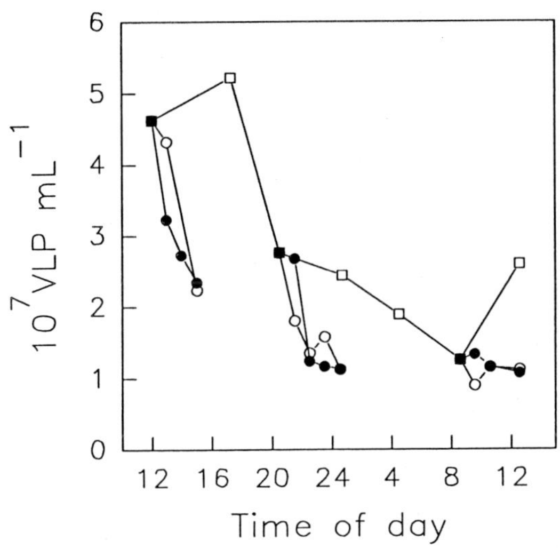

Fig. 5 Diel variation in VLP (virus-like particles) abundance during a mesocosm experiments in Knebel Vig, Denmark in 1992 (□) (see Fig. 4b). The results from three decay experiments show that the VLPs may disappear rapidly in water samples during bottle incubation both in the control bottle (O) and in the bottle where cyanide has been added to stop virus production (●). For further explanations see text.

These results might be interpreted to indicate that the viral activity was low or nil, but there is also another interpretation. If enclosing a water sample in a bottle means perturbing the ecosystem this may restrain the virus production. The control bottle may thus behave more like the bottle intentionally treated to inhibit virus production by the addition of cyanide, than as a valid control which is supposed to mimic *in situ* changes. On the other hand, it is also possible that the bottle incubation affects the stability of the viruses and causes an increased rate of decay in both the CN and the control bottle. Considering all the incubation experiments we have carried out during the last years we have indications that bottle incubations may affect the production of viruses, while there is no indication that it may affect the stability of the viruses (Børsheim *et al.*, 1990;

Heldal and Bratbak, 1991; unpublished results). It is, however, difficult to prove for each and every decay experiment. If virus production is calculated as the difference between the change observed *in situ* and the decay in the cyanide bottles (Fig. 5) we arrive at a rate of 0.3 h^{-1} (range 0.21 - 0.34 d^{-1}). Calculated as for the first diel experiment (Bratbak *et al.*, 1993) we then estimate that viral lysis of bacteria in this case removed 120% of the bacterial population per day and that it exceeded the bacterial production by a factor of about 3 (Fig. 4b).

The first diel study was conducted at the end of a 3 week-long mesocosm experiment when the plankton community was well developed and close to a successional climax. The latter study was initiated only 12 h after the mesocosm was set up and in this case the plankton community was about to reorganize after the perturbation brought about by the filling of the mesocosm bags. Some of the differences between the two diel studies may thus be due to differences in environmental conditions and in composition and succession stage of the plankton community studied. Although there are some uncertainties related to the methods used to assess microbial activities and biomasses at the various trophic levels, the above results lead us to the general conclusion that viral activity, at least in some cases, may be very important.

POPULATION CONTROL

Virus are host specific and their impact on the microbial community is related to specific host populations. Miller and Sayler (1992) have pointed out that population control exerted by virulent viruses in lytic systems is very different from that exerted by temperate viruses in lysogenic systems. Under steady state conditions, lysogenic systems do not affect the population density, while in lytic systems the host population density is reduced by 2 to 4 orders of magnitude. We argued above that most, if not all, bacteria in natural marine systems were lysogenic and thus immune against reinfection of the virus they carry. Episodic mass inductions of lysogenic bacteria and spontaneous formation of lytic mutants in lysogenic host-virus systems are thus the two mechanisms that may explain the dynamics of viral abundance in natural waters.

A major problem in viral ecology is to establish the link between a host population and a virus population in mixed natural communities. For bacteria which have no characteristic morphology there is no simple way to sort out one host population in a mixed community. This problem is less for phytoplankton and heterotrophic flagellates, which may be identified in the light microscope on the basis of morphology. Some forms may also be recognized in the electron microscope on the basis of their morphology or ultrastructure and for these it may be possible to detect the frequency of lysing cells and cells containing mature viral particles (Sieburth *et al.*, 1988; Bratbak *et al.*, 1993). Some viruses have a form or a size that make them characteristic and these may be traced in mixed viral communities (Pienaar, 1976; Bratbak *et al.*, 1992a; Bratbak *et al.*, 1993), but there is in general no simple way to distinguish between different kinds of viruses in natural communities.

Based on morphology, we have once been able to relate the decimation of one population of bacteria in a mixed community to viral lysis and increase in viral abundance (Bratbak *et al.*, 1990). The presence of viral particles in phytoplankton cells collected in natural waters suggests that viral mortality may play a significant role in the population dynamics of some species (Pienaar, 1976; Mayer and Taylor, 1979; Johnson and Sieburth, 1982; Sieburth *et al.*, 1988; Proctor and Fuhrman, 1991). Viruses have also been found in apochlorotic flagellates collected in natural waters (Nagasaki *et al.*, 1993). Viral termination of algal blooms has been demonstrated for the marine alga *Emiliania huxleyi* (Bratbak *et al.*, 1993). In this case the characteristic size of the free viruses and the presence of similar viral particles around lysing cells and inside thin sectioned cells made it possible to draw the connection between the collapse of the bloom and the concurrent increase in viral abundance.

Gene transfer

The advantage of mobile genetic properties and horizontal gene transfer is that it increases the metabolic flexibility of the population without increasing the amount of DNA that must be carried by each cell. Mobile genes encoding for a variety of different biochemical traits may be carried by different cell-lines in a population. The possibility for horizontal gene transfer ensures that these traits are available to the entire population when a change in the environmental conditions selects for it. Moreover, mobile genes open the possibility for interspecies gene transfer.

Virus infection involves transfer of DNA or RNA produced in one host cell to another cell. In most cases this will be the nucleic acids of the virus itself, but plasmids and fragments of chromosomal DNA from the host cell may by chance transferred in addition to, or instead of, the viral nucleic acid. The cell being infected may respond in various ways depending on it's physiological status and on the nature of the nucleic acids injected. 1) It may metabolize the injected nucleic acids as a useful nutrient. 2) It may express the genetic information of the nucleic acids, produce new viral particles and lyse. 3) It may insert the injected DNA into its own chromosomal DNA or retain it as an extrachromosomal element. The genetic information transferred may be expressed and entail a change in the biochemical properties of the cell. Acquisition of phenotypic traits encoded for by the phage genome is called lysogenic conversion while phage meditated transfer of bacterial DNA or plasmids from one cell to another is called transduction (Hayes, 1968; Freifelder, 1987).

It is apparently a common feature among bacteria in natural communities that, in addition to their own DNA, they carry prophages or plasmids (Kokjohn and Miller, 1992 and references therein). This means that these cells must have acquired or inherited extra DNA, and it is conceivable that this extra DNA in the long run involves a selective advantage. Lysogenization of *Escherichia coli* by phage lambda P1, P2 or Mu has for example been shown in experimental studies to give this bacterium an increased reproductive fitness (Edlin, 1977; Lin, 1977). The same may be the case for other lysogenic systems. Lysogens are immune to superinfection by phages related to the phage they carry as a prophage and lysogenic conversion may in some cases also result in increased resistance to other phages (Reanney, 1976; Kuzio and Kropinski, 1983). The selective advantage of many plasmids is obvious as they may encode for a variety of properties including antibiotic resistance, heavy metal resistance, the ability to metabolize certain organic compounds *etc* (Stanisich, 1988).

It is well known that bacteriophage gene products may be toxic. Although the toxin gene in many cases serves no obvious role in bacterial physiology and the selective advantage to the bacteria may be obscure (Bishai and Murphy, 1988), lysogenic conversion of bacteria may have significant consequences in the environment in which the bacteria live. Several human diseases are for example caused only by bacteria that are infected with a specific phage and not by the bacterium alone, *cf* diphtheria: *Corynebacterium diphtheriae*, botulism: *Clostridium botulinum*, scarlet fever: *Streptococcus pyogenes*, food poisoning: *Staphylococcus aureus*, haemorrhagic colitis: *Escherichia coli* (Bishai and Murphy, 1988). It has also been suggested that the toxicity of some cyanobacterial strains may be due to viral infection (Vance, 1977).

Using *Pseudomonas aeruginosa* and its generalized transducing phage F116 as a model system, Miller and coworkers have demonstrated that transduction may occur at significant rates under conditions simulating natural aquatic environments (Morrison *et al.*, 1978; Saye *et al.*, 1987; Saye *et al.*, 1990). Transducing phages have been isolated for a great variety of bacterial genera and the potential for transductional gene exchange is likely to be universal among bacteria (Kokjohn, 1989). Thus, there is no reason to believe that transduction between native bacteria in natural communities does not take place. The rate or frequency of transduction in natural communities is however unknown.

Transductions of bacterial DNA depends on the frequency of formation of transducing particles, on the rate of infection and on the frequency of recombination and integration into the recipients chromosome. The different processes of transduction have been described in detail for only a few phage-host systems and it is only under the assumption that the processes occurring within natural bacterial communities are similar to these, we can try to assess the rate of transduction in aquatic communities.

The frequency of formation of generalized transducing particles is depending on the number of packing site analogues on the bacterial chromosome (Weinstock, 1991). The phage DNA-packing apparatus may erroneously use these analogues instead of the specific packing sequence on the phage genome at which the DNA packing is supposed to be initiated. Transducing particles are formed at a frequency of about 10^{-3} while a particular gene may be present at a frequency of 10^{-5} to 10^{-7} (Hayes, 1968; Freifelder 1987). Specialized transducing particles are formed when a prophage is imprecisely excised from the bacterial chromosome. This occurs at a frequency of 10^{-6} to 10^{-7} (Freifelder, 1987). Generalized transducing phages with an alteration in the packing site recognition protein and dilysogenic bacteria carrying one defective specialized transducing phage and one wild-type helper phage may produce high frequency transduction lysates in which as much as 50 % of the phage particles are transducing (Freifelder, 1987; Weinstock, 1991).

The rate of infection (dP/dt) depends on the rate of virus adsorption which is proportional to the concentrations of hosts (H) and virus (P), *ie* $dP/dt = k \cdot P \cdot H$, where k is the adsorption rate constant. The value of k has been determined experimentally for T4 phages and theoretically from diffusion theory assuming that every collision results in phage adsorption to be 2.5 10^{-9} cm^3 min^{-1} (Stent, 1963). We assume that the total abundance of bacteria and of phages is 10^6 and 10^7 ml^{-1} respectively, and that the community comprises 10^2 - 10^5 different phage host systems of the same size (Bergh *et al.*, 1989; Bratbak *et al.*, 1994). From these figures we estimate that 0.036-36% of the bacterial community may experience a viral infection per day. If the frequency of transducing particles is assumed to be 10^{-3}, we estimate that 0.36 - 360 of the bacteria in a community of 10^6 ml^{-1} may be transduced per day. Roughly 2 - 5 % of these may incorporate the transduced DNA into the chromosome and become recombinants (Birge, 1981; Weinstock, 1991). The majority will be abortive transductions in which the transduced DNA fails to recombine but remains as a persistent nonreplicating DNA fragment.

An important uncertainty factor in the above assessments is that, in order to be transducing, the phage must encode for nucleases that are fragmenting the host DNA and not degrading it. For three different marine phages, Wikner *et al.*, (1993) found that most of the nucleotides for phage DNA synthesis were derived from the host cell nucleic acid (DNA and RNA) rather than from *de novo* synthesis. As an adaptation to a nutrient poor environment, degradation of the host DNA may be a common metabolic strategy for marine phages. If this is the case, the frequency of formation of transducing particles and the rate of transduction in natural aquatic communities will be much less than calculated above.

Acknowledgements
This work was supported by funding from The Research Council of Norway to the MAST-II project MEICE, contract number MAS2-CT92-0031. We wish to thank Gudni Alfredsson, Emilio Fernández, Roger Harris, Harry Havskum and Bo Riemann for sharing their water samples and data with us.

References

Ahrens R (1971) Untersuchungen zur Verbreitung von Phagen der Gattung *Agrobacterium* in der Ostsee. *Kieler Meeresforsch* **27:** 102-112

Bergh Ø, Børsheim KY, Bratbak G, Heldal M (1989) High abundance of viruses found in aquatic environments. *Nature (London)* **340:** 467-468

Birge EA (1981) *Bacterial and bacteriophage genetics.* Springer-Verlag New York

Bishai WR, Murphy JR (1988) Bacteriophage gene products that cause human disease. In *The Bacteriophages Vol 2* .Calendar R (ed) Plenum Press New York 683-724

Boehme J, Frisher ME, Jiang SC, Kellogg CA, Pichard S, Rose JB, Steinway C, Paul JH (1993) Viruses, bacterioplankton, and phytoplankton in the southeastern Gulf of Mexico: distribution and contribution to oceanic DNA pools. *Mar Ecol Prog Ser* **97:** 1-10

Børsheim Y, Bratbak G, Heldal H (1990) Enumeration and biomass estimation of planktonic bacteria and viruses by transmission electron microscopy. *Appl Environ Microbiol* **56:** 352-366

Bratbak G, Heldal M, Norland S, Thingstad TF (1990) Viruses as partners in spring bloom microbial trophodynamics. *Appl Environ Microbiol* **56:** 1400-5

Bratbak G, Haslund OH, Heldal M, Næss A, Røeggen T (1992a) Giant marine viruses? *Mar Ecol Prog Ser* **85:** 201-202

Bratbak G, Heldal M, Thingstad TF, Riemann B, Haslund OH (1992b) Incorporation of viruses into the budget of microbial C-transfer. A first approach. *Mar Ecol Prog Ser* **83:** 273-280

Bratbak G, Egge JK, Heldal M (1993) Viral mortality of the marine alga *Emiliania huxleyi* (Haptophyceae) and termination of algal blooms. *Mar Ecol Prog Ser* **93:** 39-48

Bratbak G, Heldal M, Thingstad, TF (1994) Virus and the microbial loop. Microb Ecol (in press)

Cochlan WP, Wikner J, Steward GF, Smith DC, Azam F (1993) Spatial distribution of viruses, bacteria and chlorophyll *a* in neritic, oceanic and estuarine environments, *Mar Ecol Prog Ser* **92:** 77-87

Edlin G, Lin L, Bitner R (1977) Reproductive fitness of P1, P2 and Mu lysogens *Escherichia coli. J Virol* **21:** 560-564

Freifelder G (1987) *Microbial genetics.* Jones and Bartlett Publishers Inc Boston??pp??

Fuhrman J (1992) Bacterioplankton roles in cycling of organic matter: the microbial food web. In *Primary productivity and biogeochemical cycles in the sea.* Falkowski PG, Woodhead AD (eds) Plenum Press New York 361-383

Fuhrman JA and Suttle CA (1993) Viruses in marine planktonic systems. *Oceanography* **6:** 51-63

González JM and Suttle CA (1993) Grazing by marine nanoflagellates on viruses and virus-sized particles: ingestion and digestion. *Mar Ecol Prog Ser* **94:** 1-10

Hara S, Terauchi K, Koike I (1991) Abundance of viruses in marine waters: assessment by epifluorescence and transmission electron microscopy. *Appl Environ Microbiol* **57:** 2731-2734

Hayes WRS (1968) *The genetics of bacteria and their viruses.* (2nd edn) Blackwell Scientific Publications Oxford

Heldal M and Bratbak G (1991) Production and decay of viruses in marine waters. *Mar Ecol Prog Ser* **72:** 205-212

Hobbie JE, Daley RJ, Jasper S (1977) Use of Nuclepore filters for counting bacteria by fluorescence microscopy. *Appl Environ Microbiol* **33:** 1225-1228

Jiang SC and Paul JH (1994) Seasonal and diel abundance of viruses and occurrence of lysogeny/bacteriocinogeny in the marine environment. *Mar Ecol Prog Ser* **104:** 163-172

Johnson PW and Sieburth JMcN (1982) *In situ* morphology and occurrence of eucaryotic phototrophs of bacterial size in the picoplankton of estuarine and oceanic waters. *J Phycol* **18:** 318-27

Kokjohn TA and Miller RV (1992) Gene transfer in the environment: transduction. In *Release of genetically engineered and other micro-organisms.* Fry JC, Day MJ (eds) Cambridge University Press Cambridge 54-81

Kokjohn TA (1989) Transduction: Mechanism and potential for gene transfer in the environment. In *Gene transfer in the environment.* Levy SB, Miller RV (eds) McGraw-Hill Publishing Company New York 73-97

Kuzio J and Kropinski AM (1983) O-antigen conversion in *Pseudomonas aeruginosa* strain PAO by bacteriophage D3. *J Bacteriol* **155:** 203-212

Lin L, Bitner R, Edlin G (1977) Increased reproductive fitness of *Escherichia coli* lambda lysogens. *J Virol* **21:** 564-559

Mayer JA and Taylor FJR (1979) A virus which lyses the marine nanoflagellate *Micromonas pusilla. Nature (London)* **281:** 299-301

Miller RV and Sayler GS (1992) Bacteriophage-host interactions in aquatic systems. In: *Genetic Interactions among microorganisms in the natural environment.* Wellington EMH, van Elsas JD (eds) Pergamon Press Oxford 176-193

Moebus K (1992) Further investigations on the concentration of marine bacteriophages in the water around Helgoland, with reference to the phage-host systems encountered. *Helgoländer Meeresunter* **46:** 275-292

Morrison WD, Miller RV, Sayler GS (1978) Frequency of F116 mediated transduction of *Pseudomonas aeruginosa. Appl Environ Microbiol* **36:** 724-730

Murray AG and Jackson GA (1992) Viral dynamics: a model of the effects of size, shape, motion and abundance of single-celled planktonic organisms and other particles. *Mar Ecol Prog Ser* **89:** 103-116

Nagasaki K, Ado M, Imai I, Itakura S, Ishida Y (1993) Virus-like particles found in an apochlorotic flagellate in Hiroshima Bay, Japan. *Mar Ecol Prog Ser* **96:** 307-110

Pienaar RN (1976) Virus-like particles in three species of phytoplankton from San Juan Island, Washington. *Phycologia* **15:** 185-90

Proctor LM and Fuhrman JA (1990) Viral mortality of marine bacteria and cyanobacteria *Nature (London)* **343:** 60-62

Proctor LM and Fuhrman JA (1991) Roles of viral infection in organic particle flux. *Mar Ecol Prog Ser* **69:** 133-142

Proctor LM, Fuhrman JA, Ledbetter MC (1988) Marine bacteriophages and bacterial mortality. *EOS* **69:** 1111-1112

Proctor LM, Okubo A, Fuhrman JA (1993) Calibrating estimates of phage-induced mortality in marine bacteria: ultrastructural studies of marine bacteriophage development from one-step growth experiments. *Microb Ecol* **25:** 161-182

Reanney DC (1976) Extrachromosomal elements as possible agents of adaption and development. *Bacteriol Rev* **40:** 552-590

Saye DJ, Ogunseitan O, Sayler GS, Miller RV (1987) Potential for transduction of plasmid in natural freshwater environment: effects of plasmid donor concentration and a natural microbial community on transduction in *Pseudomonas aeruginosa. Appl Environ Microbiol* **53:** 987-995

Saye DJ, Ogunseitan O, Sayler GS, Miller RV (1990) Transduction of linked chromosomal genes between *Pseudomonas aeruginosa* during incubation *in situ* in a freshwater habitat. *Appl Environ Microbiol* **56:** 140-145

Sieburth JMcN, Johnson PW, Hargraves PE (1988) Ultrastructure and ecology of *Aureococcus anophagefferens* gen. et sp. nov. (Chrysophyceae): The dominant picoplankter during a bloom in Narragansett Bay, Rhode Island, summer 1985. *J Phycol* **24:** 416-25

Stanisich VA (1988) Identification and analysis of plasmids at the genetic level. In *Methods in microbiology Vol 21 Plasmid technology* (2nd edn). Grinsted J, Bennett PM (eds) Academic Press London 11-47

Stent GS (1963) *Molecular biology of bacterial viruses.* Freeman and Company San Francisco

Steward GF Wikner J, Cochlan WP, Smith DC, Azam F (1992) Estimation of virus production in the sea: II. Field results. Mar Microb Food Webs **6:**79-90

Suttle CA and Chen F (1992) Mechanisms and rates of decay of marine viruses in seawater. *Appl Environ Microbiol* **58:**3721-3729

Thingstad TF, Heldal M, Bratbak G, Dundas I (1993) Are viruses important partners in pelagic food webs? *Trends in Ecology and Evolution.* **8:**209-213

Vance BD (1977) Prophage induction in toxic *Microcystis aeruginosa* NRC-1. *Meeting of Phycology Society of America.* p70

Weinbauer MG, Fuks D, Peduzzi (1993) Distribution of viruses and dissolved DNA along a coastal trophic gradient in the Northern Adriatic Sea. *Appl Environ Microb* **59:** 4074-4082

Weinstock GM (1991) Transduction in gram-negative bacteria. In *Modern microbial genetics.* Streips UN, Yasbin RE (eds) Wiley-Liss New York 253-270

Wikner J, Vallino JJ, Steward GF, Smith DC, Azam F (1993) Nucleic acids from the host bacterium as a major source of nucleotides for three marine bacteriophages. *FEMS Microb Ecol* **12:** 237-248

Zimmermann R and Meyer-Reil LA (1974) A new method for fluorescence staining of bacterial populations on membrane filters. *Kieler Meeresforsch* **31:** 24-27

Molecular Analysis of Plastid Evolution

Wolfgang Löffelhardt
Institute of Biochemistry and Molecular Cell Biology
and Ludwig-Boltzmann-Forschungsstelle für Biochemie
University of Vienna Biocenter
Dr. Bohr-Gasse 9
A-1030 Vienna
Austria

Introduction

The endosymbiotic origin of plastids from cyanobacterial endosymbionts is now generally accepted. Early formulations of this theory (Schimper, 1885; Mereschkowski, 1905) were induced by the mere similarity in morphology, chlorophyll pigments and photosynthetic function between the photosynthetic organelles of autotrophic eukaryotes and their ancestors and extant relatives, then termed "blue-green algae". This pioneering idea formulated in the absence of any data in comparative biochemistry became forgotten for many years and had to compete after its revival (Margulis, 1970) with hypotheses postulating an autogenous origin of the eukaryotic cell through intracellular compartmentalization and functional specialization (Cavalier-Smith, 1975; Uzzell and Spolsky, 1974). Now a wealth of conclusive biochemical data is available. These should be reviewed briefly before entering into the molecular data, were doubtless responsible for the breakthrough of the endosymbiotic theory in the 1980s.

Research during the past 30 years led to the recognition of strong parallels in the organization of the photosynthetic apparatus between chloroplasts and cyanobacteria. Both perform oxygenic photosynthesis, possess chlorophyll *a* and two photosystems, and contrast with the purple and green bacteria which are unable to perform photolysis of water in their single photosystem. The components of the thylakoid electron transport chain are almost identical, cytochrome c_{553} (fulfilling the function of plastocyanin) being absent from the chloroplasts of higher plants but present in those from green algae (Merchant and Bogorad, 1987). The three-dimensional structure of cyanobacterial photosystem I has been elucidated and can serve as a model for the counterpart from chloroplasts (Krauss *et al.*, 1993). The envelope membranes of higher plant

NATO ASI Series, Vol. G 38
Molecular Ecology of Aquatic Microbes
Edited by I. Joint

chloroplasts correspond to the inner (IEM) and outer (OEM) envelope membranes of cyanobacteria (Cavalier-Smith, 1987). OEM porins of analogous properties have been detected in plastid and cyanobacterial envelopes (Hansel *et al.*, 1994). An important clue is the lipid composition of the membranes; in the plant cell the chloroplast membranes are the only source for the cyanobacterial glycolipids monogalactosyldiacylglycerol, digalactosyldiacyl-glycerol, and sulfoquinovosyldiacylglycerol (Kates, 1990). *De novo* fatty acid biosynthesis in plants is confined to the plastids and catalyzed by a prokaryotic-type fatty acid synthetase multi-enzyme complex including acyl carrier protein as opposed to the multifunctional enzymes in the cytoplasm of animals and fungi (Slabas and Fawcett, 1992). Likewise, eukaryotic acetyl-CoA carboxylases are large biotin-binding polyproteins (>200kDa) whereas in bacteria this enzyme activity is performed by three smaller subunits. Recently both carboxylases were identified in pea leaves; the prokaryotic enzyme resides in the chloroplast whereas the eukaryotic enzyme is compartmentalized to the cytoplasm where it participates in processes other than *de novo* fatty acid biosynthesis (Konishi and Sasaki, 1994). The plastid is the exclusive site of carotenoid biosynthesis in the plant cell (Kleinig, 1989). Protein biosynthesis in plastids occurs on 70S ribosomes and is sensitive to inhibitors like chloramphenicol that interfere with prokaryotic translation only (Gray and Doolittle, 1982). The plastid is important for nitrogen and sulphur assimilation (Kindl, 1994) and is responsible for the biosynthesis of the amino acids needed for its own and the cytoplasmic translation apparatus (Last, 1993).

In fact the metabolic autonomy of the plastids is remarkable (for a review see Emes and Tobin, 1993). They supply the cell with a multitude of products synthesized within the organelle and demand only a small number of donations from the cytoplasm. One example found recently is plastoquinone which is synthesized in the Golgi apparatus of spinach leaves and obviously has to be imported into the chloroplast (Swiezewska *et al.*, 1993). The frequently used term "semiautonomous organelle" refers to the necessity for the plastids to import about 90% of their constituent polypeptides from the cytoplasm, due to the inability to harbour the respective genes on the reduced genome. Thus we encounter the paradox situation that prokaryotic pathways in the plastids are catalyzed by the products of eukaryotic (nuclear) genes. The host cell took over large parts of the endosymbiont metabolism, either acquiring new functions *eg* carotenoid biogenesis or deleting its own set of enzymes as in the case of fatty acid and amino acid biosynthesis. Gene transfer from endosymbiont to host nucleus was accompanied by deletion of a certain set of host genes. On the other hand, endosymbiont genes for enzymes of respiration and the tricarboxylate cycle were deleted, since these were contributed by the prokaryotic ancestor of mitochondria. Finally in some cases, analogous metabolic features from host and endosymbiont were retained *eg* glycolysis and the oxidative pentose phosphate pathway (Kindl, 1994).

The chloroplasts *sensu stricto*, the chlorophyll *b*-containing plastids of higher plants and green algae, share common features of morphology, starch deposition, a two-membrane envelope and comparable gene content and organization of their genomes. On the other hand algae without chlorophyll *b* constitute a very interesting group comprising two to four envelope membranes. They are distinct through different antenna systems, *eg* phycobilisomes (PBS), morphological peculiarities, storage of assimilatory starch outside of the plastid, and in general a complement of protein genes higher by about 50% than that of chloroplasts (Table 1)

Table 1: Examples for structural and pigment diversity among algal plastids

	Number of envelope membranes		
Plastid type	**Two**	**Three**	**Four**
chl a/b	Chlorophyta	Euglenophyta	
chl a/PBS	Rhodophyta Glaucophyta[a]		
chl a/c		Dinophyta	Chrysophyta Cryptophyta[b] Xanthophyta Phaeophyta

[a] Plastids (cyanelles) are surrounded by a peptidoglycan wall.
[b] In addition they contain phycobiliproteins not assembled into PBS.

The plastid genome

The presence of DNA in plastids and mitochondria is evidence for the endosymbiotic origin of these organelles which is less bolstered for peroxisomes lacking a residual genome of their own (Gray, 1992). In our case, it is important to ask if the plastid genome is more related to the genome of its postulated prokaryotic ancestor, as manifested through extant cyanobacteria, than to the nuclear genome in the same cell?

Enormous progress has been made during the past 15 years. The complete sequence of chloroplast DNAs from several higher plants (*eg* tobacco, rice, and the liverwort *Marchantia polymorpha*) is known (for a review see Sugiura, 1992). The emerging features are those of a prokaryotic gene structure and organization in a quasi condensed form of a (cyano)bacterial genome. Most plastid DNAs fall into the size range of 120 to 160 kbp corresponding to about 3-5% of the genome size of unicellular cyanobacteria. They encode all rRNAs, 30 to 32 tRNAs (Tsudzuki *et al.*, 1994) thought to be sufficient for the organelle translation apparatus and an almost invariant set of proteins, the majority of which is involved in transcription/translation and in thylakoid-bound photosynthetic processes (Table 2). The remaining genes are scattered with respect to function; there is rarely more than one gene per pathway, *eg rbcL* (large subunit, RUBISCO) representing the chloroplasts genetic contribution to the Calvin cycle. Interestingly functional *ndh* genes have been lost from the black pine chloroplast genome (Wakasugi *et al.*, 1994).

Table 2. Protein genes on the chloroplast DNAs of higher plants (according to Sugiura, 1992)

Function of the gene product	Number of genes
Ribosomal proteins	21
Photosystem I and II	17
Cytochrome b_6/f complex	4
ATP synthase	6
NADH dehydrogenase	11
RNA polymerase	4

Lack of *ndh* genes and the presence of *tufA* (elongation factor Tu, encoded by the nuclear genome in higher plants) is typical for the plastomes from green algae and euglenoids (Hallick *et al.*, 1993) that otherwise show a gene composition similar to that given in Table 2. Transcription units corresponding to known counterparts from bacterial genomes are typical for plastid genomes. In the case of the large ribosomal protein gene cluster, the three adjacent S10, *spc*-, and α-operons from *E. coli* (Lindahl and Zengel, 1986) that presumably are organized in an analogous way in cyanobacteria, have been fused after transfer of approximately 50% of the genes to the nuclear genome and after shifting of two genes to other loci on the plastid genome (Fig. 1). The order of the genes remaining in the large cluster follows exactly that observed on the *E. coli* operons with the exception of the interspersed *infA*.

E. coli	**5'-S10-*spc*-α-3'**
S10 operon	*rps10-rpl3-rpl4-rpl23-rpl2-rps19-rpl22-rps3-rpl16-rpl29-rps17*
spc operon	*rpl14-rpl24-rpl5-rps14-rps8-rpl6-rpl18-rps5-rpl30-rpl15-secY-rpl36*
α operon	*rps13-rps11-rps4-rpoA-rpl17*

Nicotiana tabacum	**Chloroplast genome**	
Large cluster	*rpl23-rpl2-rps19-rpl22-rps3-rpl16-rpl14-rps8-infA-rpl36-rps11-rpoA*	
Other loci	*rps14*	*rps4*

Fig. 1: A comparison of ribosomal protein gene organization in *E. coli* and the chloroplasts of higher plants.

In cases where the organization of the respective operons is also known for cyanobacteria and is different from that observed for *E. coli*, the chloroplasts as expected follow the pattern of their prokaryotic ancestors. Examples are the gene for the β' subunit of RNA polymerase which is split into *rpoC1* and *rpoC2* in cyanobacteria as well as in chloroplasts and the *atp* operon (Cozens and Walker, 1987) that contains all the genes in *E. coli* but is bipartite at distant loci in cyanobacteria and chloroplasts (Figure 2). In *E. coli*, F_0 contains three subunits only (gene *atpG* is missing). The genes *atpC*, *atpD*, and *atpG* for the chloroplast ATP synthase reside in the nuclear genome. Again the order is preserved.

E. coli	5'-*atpI-atpH-atpF-atpD-atpA-atpC-atpB-atpE*-3'	
Cyanobacteria	5'-*atpI-atpH-atpF-atpG-atpD-atpA-atpC*-3'	5'-*atpB-atpE*-3'
Chloroplast genome (higher plants)	5'-*atpI-atpH-atpF-atpA*-3'	5'-*atpB-atpE*-3'

Fig. 2: Organization of the ATP synthase subunit genes.

Evidence also at the protein level was obtained for a prokaryotic type core RNA polymerase (α, β, β') that recognizes plastid promoters that often come very close to the canonical "-10" and "-35" boxes known for *E. coli*. The finding that cyanobacterial promoters often appear less related to plastid promoters reflects the fact that transcriptional regulation plays an important role in the expression of cyanobacterial genes, whereas plastid genes are predominantly regulated at the post transcriptional level (Curtis, 1994). In addition a second, thus far uncharacterized form of plastid RNA polymerase is postulated, that is nucleus-encoded (Igloi and Kössel, 1992). The presence of a ribosome binding site is another prokaryotic feature of many chloroplast genes (Bonham-Smith and Bourque, 1989).

Phylogenetic analyses based on rRNA and protein sequences leave no doubt that plastids are derived from prokaryotic endosymbionts (Gray, 1992). However, there is still controversy about details of plastid evolution; a monophyletic origin would mean that cyanobacteria are the photosynthetic prokaryotes ancestral to all plastid types. A polyphyletic origin, as first formulated by Raven (1970), emphasizes the differences among the various plastid types outlined in Table 1, as the result of their origin from different photosynthetic prokaryotes; cyanobacteria would be the ancestors of rhodophyte plastids only, whereas the chlorophyll *a/b* containing prochlorophytes gave rise to chlorophyte plastids and heliobacteria-like organisms to the yellow-brown chromophyte plastids. At present the results of phylogenetic analysis do not lend much support to a polyphyletic theory (Turner *et al.*, 1989; Palenik and Haselkorn, 1991; Witt and Stackebrandt, 1988). Another point is, if there was a singular or multiple primary endosymbiotic events, *ie* if all types of plastids are derived from a common (semiautonomous) endosymbiotic precursor. Here comparison of plastid genome organization in different algal phyla might be more useful than phylogenetic analysis of single traits (Kowallik, 1992). Recently a wealth of information on sequences and organization of plastid genes became available for the glaucophyte *Cyanophora paradoxa* (Löffelhardt and Bohnert, 1994; DA Bryant, personal communication), the rhodophytes *Porphyra purpurea* (Reith and Munholland, 1993) *Porphyridium aerugineum*, *Galdieria* spec., *Antithamnion* spec. (Zetsche and Valentin, 1994), and the chromophyte *Cryptomonas* Φ (Douglas, 1992). With the exception of the *ndh* genes the full gene complement of higher plant chloroplast DNAs seems to be present plus an additional set of more than 50 genes that either are not present in higher plants or nucleus-encoded. This gene surplus is quite similar in composition among the algal species mentioned above, thus the peptidoglycan-surrounded cyanelles and especially the rhodoplasts from *Porphyra* could be considered as ancient or primitive plastids as opposed to the derived chloroplasts of higher plants and green algae (Reith and Munholland, 1993). The small subunit of RUBISCO, ferredoxin, two additional subunits of ATP synthase, and more than 12 additional ribosomal proteins are encoded on the genomes of rhodophyte and chromophyte plastids and the *Cyanophora* cyanelle. Interestingly a number of genes involved in the biosynthesis of amino acids and fatty acids is also present (Table 3). Thus in the case of algae devoid of chlorophyll *b*, at least some reactions of the prokaryotic pathways mentioned above are still performed by the products of prokaryotic genes. In addition, genes for chaperonins and components of a prokaryotic protein translocation apparatus are listed in Tablc 3.

Table 3: Unusual genes encountered on the plastid genomes of rhodophyte, glaucophyte, and chromophyte algae.

Gene	Function
apcA	Acyl carrier protein
fabH	3-Ketoacyl-ACP synthase III
glsF	Glutamate synthase
argB	N-acetylglutamate kinase
ilvB	Acetolactate synthase
trpA	Tryptophan synthetase subunit
secA	Preprotein translocase subunit
secY	Preprotein translocase subunit
dnaK	Chaperonin 70
groEL	Chaperonin 60
groES	Chaperonin 10

Through these new data it became apparent that plastid types differing in pigmentation, the number of surrounding membranes and gene complement nevertheless shared transcription units that obviously are derived from fusion of operons, *eg* the *atpI-C* operon depicted in Fig. 2, that are not adjacent in extant cyanobacteria (Fig. 3). The gene products corresponding to the respective operons are assembled into different structures within the organelle and fulfil different functions, rendering a convergent evolution very unlikely. There seems to be no selective advantage to maintain such composite operons and thus they are not universal. In plastids known to possess high recombinational activity as those from *C. reinhardtii* the transcription units shown in Fig. 3 as well as others are disrupted due to plastome rearrangements (Woessner *et al.*, 1987). Such widely distributed features of plastid genome organization can best be explained through the assumption of a common precursor for all plastid types (Kowallik, 1992; Reith and Munholland, 1993; Löffelhardt and Bohnert, 1994). In *P. purpurea* only one gene (*atpC*) is missing from the three prokaryotic clusters. The *tsf* gene specifies elongation factor Ts and is known to form an operon with *rps2* in *E. coli* (Lindahl and Zengel, 1986).

P. purpurea	5'-*rpoB-rpoC1-rpoC2-rps2-tsf-atpI-atpH-atpG-atpF-atpD-atpA*-3'
C. paradoxa	*tsf atpI*
Higher plants	*tsf* *atpG* *atpD*
P.purpurea	
C. paradoxa	*psbB*-ORF31-*psbN-psbH*
Higher plants	

Fig. 3: Gene clusters diagnostic for plastid evolution. Genes missing in some plastid types are underlined.

If a common precursor, *ie* a semiautonomous endosymbiont originating from a cyanobacterial ancestor through transfer of the majority of its genes to the host cell nuclear genome, is accepted, plastid origin necessarily would be not only monophyletic but would also be based on a singular successful primary endosymbiotic event.

Gene transfer

The original purpose of gene transfer to the nucleus might have been to prevent the endosymbiont from escaping and to establish a hereditary endosymbiosis. This complicated and lengthy process requires the conversion of prokaryotic genes into eukaryotic ones, the recruiting of suitable (*eg* light-regulated) promoters and the acquisition of targeting signals that ensure sorting of the gene product to the plastids. As long as all these goals were not completely achieved the respective gene would remain functional in the plastome. While nothing is known about the mechanism, it is clear that gene transfer has occurred between the different genomes in the plant cell and that it is a still ongoing process. Large sections of the plastid genome appear to have integrated into the mitochondrial genome in many plants (Schuster and Brennicke, 1988). While part of this "promiscuous DNA" seems to be non-functional, some mitochondrial tRNAs certainly are transcribed from transferred plastid genes (Dietrich *et al.*, 1993). Almost all spinach chloroplast DNA sequences can be traced on the nuclear genome (Timmis and Scott, 1983), quasi as a result of a thus far unsuccessful gene transfer. Gene transfer events can be used as evolutionary markers and converted into a time scale: the *tufA* gene migrated to the nucleus upon the origin of land plants. Charophycean algae considered as bridge organisms to land plants indeed show the whole spectrum of possibilities. Some species possess a nuclear *tufA* gene only, whereas in one case the non-functional remnant of the plastid gene is still present (Baldauf *et al.*, 1990).

Secondary endosymbiotic events

Secondary endosymbiosis involving an eukaryotic alga (already containing a functional semiautonomous plastid) as the photosynthetic invader of a heterotrophic host cell is invoked to explain the origin of the chromophyte plastids that are surrounded by four membranes. Results of phylogenetic analysis (Douglas, 1992) and comparison of plastid genome organization render a red alga like *Porphyra purpurea* (Reith and Munholland, 1993) the most likely candidate for the eukaryotic endosymbiont. A rhodophyte (Douglas and Turner, 1991) or a chlorophyte alga (Gibbs, 1978) are discussed as ancestral to the plastid of *Euglena* (three envelope membranes) where also a different type of (non photosynthetic) host cell is postulated. Thus, regardless of the superimposed secondary endosymbiotic event(s) that well might have been polyphyletic and multiple, the plastid interrelationships are mainly imprinted by the common singular primary event. The gene transfer also necessary upon secondary endosymbiosis most likely took place between the nuclei of the primary and the secondary host cell only. In most cases the nucleus and cytoplasm of the primary host cell have been completely lost and one or two additional membranes are the sole morphologic indication of the complex evolutionary history of these plastids (Whatley, 1993). Only in cryptomonads, the periplastidal space between the two outermost membranes and the real plastid envelope still contains remnants of the nucleo-cytoplasmic compartment of the eukaryotic endosymbiont: the nucleomorph, most likely its vestigial nucleus, starch grains, and 80S ribosomes different from those encountered in the cytoplasm (Greenwood *et al.*, 1977). The 660kbp nucleomorph DNA of *Pyrenomonas salina* is contained in three chromosomes and codes for 18S and 25S rRNA of the periplastidal ribosomes and for some proteins, *eg* a chaperonin (Hofmann *et al.*, 1994). Upon phylogenetic analysis of nuclear and nucleomorph 18S rDNA from *Cryptomonas* Φ it appeared that the former grouped with the counterpart from *Acanthamoeba castellanii* whereas the latter was closely related to 18S rDNA from red algae (Douglas *et al.*, 1991).

Protein targeting to plastids

The "conservative sorting" hypothesis (Hartl and Neupert, 1990; Smeekens *et al.*, 1990) postulates that eukaryotic cell organelles derived from prokaryotic endosymbionts: i) had to evolve an import system for cytoplasmically synthesized precursor polypeptides, presumably from pre-existing structures like phage receptors or colicin receptors, and ii) retained a prokaryotic preprotein translocase for routing within the organelle.

The plastids resulting from the singular primary endosymbiotic event, *ie* the chloroplasts of green algae and higher plants, the rhodoplasts of red algae and the cyanelles of the glaucocystophytes that are surrounded by two envelope membranes only, should then possess an analogous import apparatus recognizing similar targeting signals. The transit sequences (stroma targeting peptide, STP) of precursors destined for import into higher plant chloroplasts are overall rather hydrophilic with the exception of the amino-terminal hydrophobic part, show a positive net charge and are rich in hydroxylated amino acids (von Heijne and Nishikawa, 1990). STPs from a red alga and from *C. paradoxa* were shown to conform to these characteristics and to direct the *in vitro* import of proteins into pea chloroplasts (Apt *et al.*, 1993; Löffelhardt and Bohnert, 1994). The peptidoglycan layer present between the inner and the outer membrane of cyanelles seems to be no barrier to protein translocation, thus the semiautonomous endosymbiont ancestral to all plastids might still have contained a prokaryotic sacculus.

Nucleus-encoded lumenal proteins like plastocyanin that have to cross the two envelope membranes and the thylakoid membrane possess a bipartite targeting signal: a STP at the amino-terminus followed by a shorter hydrophobic leader sequence. The former is processed by a (eukaryotic) stromal transit peptidase, the latter by a thylakoid-bound (prokaryotic) leader peptidase. Cyanobacterial plastocyanin is synthesized as a precursor containing a leader sequence. Import into the thylakoid lumen is equivalent to export to the periplasm and is directed by analogous targeting signals. Upon endosymbiosis, such genes have obviously been transferred together with their prokaryotic targeting signal to the nuclear genome where they received in addition an eukaryotic targeting signal (Smeekens *et al.*, 1990). The second protein translocation system in plastids is thought to correspond to the bacterial Sec-dependent preprotein translocase (Wickner *et al.*, 1991). This is corroborated by the recent detection of *sec* genes on the plastomes of *C. paradoxa*, rhodophytes, and chromophytes (Table 3). Transformation of an *E. coli* mutant producing a thermosensitive SecY protein with the cyanelle *secY* gene restored growth at the restrictive temperature (Flachmann *et al.*, 1993).

Plastids originating from a secondary endosymbiotic event are expected to possess an import mechanism responsive to targeting signals different from those mentioned above. The outermost membranes, believed to be related to the endoplasmic reticulum (ER), often appear to be covered with ribosomes (Whatley, 1993), rendering a cotranslational mechanism likely, analogous to the eukaryotic secretory pathway. In *Euglena gracilis* a polyprotein precursor for RUBISCO SSU was observed (Chan *et al.*, 1990) that contained an unusual targeting signal, a N-terminal hydrophobic signal sequence followed by a STP. This is an arrangement opposite to that observed for plastocyanin and corresponds to the order and nature of the membranes that the precursor has to cross. A sole signal sequence seems to be sufficient for plastid import of a fucoxanthin chlorophyll *a/c* protein precursor from the diatom *Phaeodactylum tricornutum*. Cotranslational import across the ER followed by vesicular transport to the real plastid envelope is supposed to occur in this case (Bhaya and Grossman, 1991).

Conclusions

The data available at present on biochemistry, molecular biology, and cell biology of plastids leave no doubt about their endosymbiotic origin. Taken into account that the primary endosymbiotic event already led to the establishment of the semiautonomous photosynthetic organelle, whereas secondary endosymbioses resulted in modifications only (at the genomic and/or structural level), the statement seems justified that all plastids finally are derived from a common precursor which we might call the "protoplastid". The major part of the gene transfer from the ancestral cyanobacterium to the nucleus of the primary host cell took place until this stage was reached. Extant plastids coming close are the *Porphyra* rhodoplasts and the cyanelles of *C. paradoxa*. Gene transfer continued afterwards but at a different rate in the various lines derived from the protoplastid either directly or via a secondary endosymbiotic event; chlorophyll *b* containing plastids possess a complement of genes smaller than that found in rhodophytes, chromophytes, and cyanelles. Loss of the *ndh* genes from all algal plastid genomes occurred after the separation of the land plants which show considerable stability of the plastome. Peculiarities in the organization of some plastid genomes *eg* the vast number of introns present in *Euglena gracilis* (Hallick *et al.*, 1993) should not be overemphasized compared to the numerous parallels revealed by phylogenetic analysis and, especially, by the increasing amount of data on plastid genome organization in different algal phyla. Differences in pigment composition are now considered to be less crucial. It is now clear that chlorophyll *b* could be gained at quite different evolutionary levels: at the prokaryotic level (Palenik and Haselkorn, 1991), after the primary endosymbiotic event, and after the secondary endosymbiotic event. Alternatively, the cyanobacteria-like ancestral photosynthetic prokaryote could be imagined to possess both types of light-harvesting systems, chlorophyll *a/b* antennae as well as phycobilisomes (Bryant, 1992). In this scenario loss of one or the other function would have occurred in the different evolutionary lines. Remnants of chlorophyll *a/b* protein genes on cyanobacterial genomes and in biliprotein-containing algae or unfunctional genes for phycobilisome components in chlorophytes would lend support to this interesting hypothesis.

References

Apt K, Hoffman NE, Grossman AR (1993) The γ subunit of R-phycoerythrin and its possible mode of transport into the plastid of red algae. *J Biol Chem* **268:** 16208-16215

Baldauf SL, Manhart JR, Palmer JD (1990) Different fates of the chloroplast *tufA* gene following its transfer to the nucleus in green algae. *Proc Natl Acad Sci USA* **87:** 5317-5321

Bhaya D and Grossman AR (1991) Targeting proteins to diatom plastids involves transport through an endoplasmic reticulum. *Mol Gen Genet* **229:** 400-404

Bonham-Smith PC and Bourque DP (1989) Translation of chloroplast-encoded mRNA: potential initiation and termination signals. *Nucleic Acids Res* **17:** 2057-2080

Bryant DA (1992) Puzzles of chloroplast ancestry. *Curr Biol* **2:** 240-242

Cavalier-Smith T (1975) The origin of nuclei and of eukaryotic cells. *Nature* **256:** 463-468

Cavalier-Smith T (1987) The simultaneous symbiotic origin of mitochondria, chloroplasts, and microbodies. *Ann N Y Acad Sci* **503:** 55-71

Chan RL, Keller M, Canady Y, Weil JH, Imbault P (1990) Eight small subunits of *Euglena* ribulose-1,5-bisphosphate carboxylase/oxygenase are translated from a large mRNA as a polyprotein. *EMBO J* **9:** 333-338

Cozens AL and Walker JE (1987) The organization and sequence of the genes for ATP synthase subunits in the cyanobacterium *Synechococcus* 6301. Support for an endosymbiont origin of chloroplasts. *J Mol Biol* **194:** 359-383

Curtis S (1994) Transcription apparatus and transcriptional regulation. In: *The Molecular Biology of Cyanobacteria* (Bryant DA, ed) Kluwer Academic Publishers Dordrecht, in press

Dietrich A, Weil JH, Marechal-Drouard L (1992) Nuclear-encoded transfer RNAs in plant mitochondria. *Ann Rev Cell Biol* **8:** 115-131

Douglas SE (1992) Probable evolutionary history of cryptomonad algae. In: *Origins of Plastids* Lewin RA (ed) Chapman and Hall New York and London 265-289

Douglas SE and Turner S (1991) Molecular evidence for the origin of plastids from a cyanobacterium-like ancestor. *J Mol Evol* **33:** 267-273

Douglas SE, Murphy CA, Spencer DF, Gray MW (1991) Molecular evidence that cryptomonad algae are evolutionary chimaeras of two phylogenetically distinct unicellular eukaryotes. *Nature* **350:** 148-151

Emes MJ and Tobin AK (1993) Control of metabolism and development in higher plant plastids. In: *International Review of Cytology,* Vol **145** Jeon KW, Jarvik J, (eds) Academic Press Orlando 149-216

Flachmann R, Michalowski CB, Löffelhardt W, Bohnert HJ (1993) SecY, an integral subunit of the bacterial preprotein translocase is encoded by a plastid genome. *J Biol Chem* **268:** 7514-7519

Gibbs SP (1978) The chloroplasts of *Euglena* may have evolved from symbiotic green algae. *Can J Bot* **56:** 2883-2889

Gray MW (1992) The endosymbiont hypothesis revisited. In: *International Review of Cytology*, Vol **141** Jeon KW, Jarvik J (eds) Academic Press Orlando 233-357

Gray MW and Doolittle WF (1982) Has the endosymbiont hypothesis been proven? *Microbiol Rev* **46:** 1-42

Greenwood AD, Griffith HB, Santore UJ (1977) Chloroplasts and cell compartments in cryptophyceae. *Brit Phycol J* **13:** 119-124

Hallick RB, Hong L, Drager RG, Favreau MR, Montfort A, Orsat B, Spielmann A, Stutz E (1993) Complete sequence of *Euglena gracilis* chloroplast DNA. *Nucleic Acids Res* **21:** 3537-3544

Hansel A, Schmidt A, Tadros MH, Jürgens UJ (1994) Isolation and characterization of porin from the outer membrane of *Synechococcus* PCC 6301. *Arch Microbiol* **161:** 163-167

Hartl FU and Neupert W (1990) Protein sorting to mitochondria: evolutionary conservation of folding and assembly. *Science* **247:** 930-938

Hofmann CJB, Rensing SA, Häuber MM, Martin WF, Müller SB, Couch J, Mc Fadden GI, Igloi GL, Maier UG (1994) Smallest known eukaryotic genomes encode a protein: towards the understanding of nucleomorph functions. *Mol Gen Genet,* in press

Igloi G and Kössel H (1992) The transcriptional apparatus of chloroplasts. *Critical Reviews in Plant Sciences* **10:** 525-558

Kates M (ed) (1990) *Handbook of lipid research* Vol 6. Plenum Press New York London

Kindl H (1994) *Biochemie der Pflanzen.* Springer Heidelberg

Kleinig H (1989) The role of plastids in isoprenoid biosynthesis. *Annu Rev Plant Physiol* **40:** 39-59

Konishi T, Sasaki Y (1994) Compartmentalization of two forms of acetyl-CoA carboxylase in plants and the origin of their tolerance towards herbicides. *Proc Natl Acad Sci USA* **91:** 3598-3601

Kowallik K (1992) Origin and evolution of plastids from chlorophyll $a+c$ containing algae: Suggested ancestral relationships to red and green algal plastids. In: *Origins of Plastids* Lewin RA (ed) Chapman and Hall New York and London 223-263

Krauss N, Hinrichs W, Witt I, Fromme P, Pritzkow W, Dauter Z, Betzel C, Wilson KS, Witt HT, Saenger W (1993) Three-dimensional structure of system I of photosynthesis at 6 A resolution. *Nature* **361:** 326-331

Lindahl L and Zengel JM (1986) Ribosomal genes in *Escherichia coli. Annu Rev Genet* **202:** 186-193

Last RL (1993) The genetics of nitrogen assimilation and amino acid biosynthesis in flowering plants: progress and prospects. In: *International Review of Cytology*, Vol **143** Jeon KW and Jarvik J (eds) Academic Press Orlando 297-333

Löffelhardt W and Bohnert HJ (1994) Structure and function of the cyanelle genome. In: *International Review of Cytology*, Vol **151** Jeon KW, Jarvik J, (eds) Academic Press Orlando 29-66

Margulis L (1970) *The origin of eukaryotic cells.* Yale University Press, New Haven

Margulis L (1981) *Symbiosis in cell evolution.* WH Freeman and Co San Francisco

Merchant S, Bogorad L (1987) The Cu(II)-repressible plastidic cytochrome *c*. *J Biol Chem* **262:** 9062-9067

Mereschkowski C (1905) Über Natur und Ursprung der Chromatophoren im Pflanzenreiche. *Biol Centralbl* **25:**593-604

Palenik B and Haselkorn R (1991) Multiple evolutionary origins of prochlorophytes, the chlorophyll *b*-containing prokaryotes. *Nature* **355:** 265-267.

Raven P (1970) A multiple origin of chloroplasts and mitochondria. *Science* **169:** 641-646

Reith M and Munholland J (1993) A high-resolution gene map of the chloroplast genome of the red alga *Porphyra purpurea. Plant Cell* **5:** 465-475

Schimper AFW (1883) Über die Entwicklung der Chlorophyllkörner und Farbkörner. *Bot Z* **41:** 105-114

Schuster W and Brennicke A (1988) Interorganellar sequence transfer: plant mitochondrial DNA is nuclear, is plastid , is mitochondrial. *Plant Sci* **54:** 1-10

Slabas AR and Fawcett T (1992) The biochemistry and molecular biology of plant lipid biosynthesis. *Plant Mol Biol* **19:** 169-191

Smeekens S, Weisbeek P, Robinson C (1990) Protein transport into and within chloroplasts. *Trends Biochem Sci* **15:** 73-76

Sugiura M (1992) The chloroplast genome. *Plant Mol Biol* **19:** 149-168

Swiezewska E, Dallner G, Andersson B, Ernster L (1993) Biosynthesis of ubiquinone and plastoquinone in the endoplasmic reticulum-Golgi membranes of spinach leaves. *J Biol Chem* **268:** 1494-1499

Timmis JN and Scott NS (1983) Sequence homology between spinach nuclear and chloroplast genomes. *Nature* **305:** 65-67

Tsudzuki J, Ito S, Tsudzuki T, Wakasugi T, Sugiura M (1994) A new gene encoding $tRNA^{Pro}$(GGG) is present in the chloroplast genome of black pine: a compilation of 32 tRNA genes from black pine chloroplasts. *Curr Genet* **26:** 153-158

Turner S, Burger-Wiersma T, Giovannoni SJ, Muur LR, Pace NR (1989) The relationship of a prochlorophyte, *Prochlorothrix hollandica,* to green chloroplasts. *Nature* **337:** 380-382

Uzzell T and Spolsky C (1974) Mitochondria and plastids as endosymbionts. A revival of special creation? *Am Sci* **62:** 334-343

von Heijne G and Nishikawa K (1990) Chloroplast transit peptides. The perfect random coil? *FEBS Lett* **278:** 1-3

Whatley JM (1993) The endosymbiotic origin of chloroplasts. In: *International Review of Cytology,* Vol **144** Jeon KW, Jarvik J (eds) Academic Press Orlando 259-299

Wakasugi T, Tsudzuki J, Ito S, Nakashima K, Tsudzuki T, Sugiura M (1994) Loss of all *ndh* genes as revealed by sequencing the entire chloroplast genome of black pine, *Pinus thunbergii. Proc Natl Acad Sci USA,* **91:** 9794-9798

Wickner W, Driessen AJM, Hartl FU (1991) The enzymology of protein translocation across the *Escherichia coli* plasma membrane. *Annu Rev Biochem* **60:**101-124

Witt D and Stackebrandt E (1988) Disproving the hypothesis of a common origin of the *Ochromonas danica* chrysoplast and *Heliobacterium chlorum. Arch Microbiol* **150:** 244-248

Woessner JP, Gillham NW, Boynton JE (1987) Chloroplast genes encoding subunits of the H^+-ATPase complex of *Chlamydomonas reinhardtii* are rearranged compared to higher plants: sequence of the *atpE* gene and location of the *atpF* and *atpI* genes. *Plant Mol Biol* **8:** 151-158

Zetsche K and Valentin K (1994) Structure, coding capacity and gene sequences of the plastid genome from red algae. *Endocytobiosis Cell Res* **10:** 107-127

Can We Estimate Bacterial Growth Rates from Ribosomal RNA Content?

Paul F. Kemp
Oceanographic and Atmospheric Sciences
Brookhaven National Laboratory
Upton, NY 11973-5000
USA

Abstract

Several studies have demonstrated a strong relationship between the quantity of RNA in bacterial cells and their growth rate under laboratory conditions. It may be possible to use this relationship to provide information on the activity of natural bacterial communities, and in particular on growth rate. However, if this approach is to provide reliably interpretable information, the relationship between RNA content and growth rate must be well-understood. In particular, a requisite of such applications is that the relationship must be universal among bacteria, or alternately that the relationship can be determined and measured for specific bacterial taxa.

The RNA-growth rate relationship has not been used to evaluate bacterial growth in field studies, although rRNA content has been measured in single cells and in bulk extracts of field samples taken from coastal environments. These measurements have been treated as probable indicators of bacterial activity, but have not yet been interpreted as estimators of growth rate. The primary obstacle to such interpretations is a lack of information on biological and environmental factors that affect the RNA-growth rate relationship.

In this paper, the available data on the RNA-growth rate relationship in bacteria will be reviewed, including hypotheses regarding the regulation of RNA synthesis and degradation as a function of growth rate and environmental factors; *ie* the basic mechanisms for maintaining RNA content in proportion to growth rate. An assessment of the published laboratory and field data, the current status of this research area, and some of the remaining questions will be presented.

Introduction

Effective methods for measuring bacterial growth in environmental samples were introduced in the early 1980's (Fuhrman and Azam, 1980, 1982; Moriarty and

NATO ASI Series, Vol. G 38
Molecular Ecology of Aquatic Microbes
Edited by I. Joint

Pollard, 1981, 1982), and allowed more accurate evaluation of the role of bacteria as both producers and consumers of particulate organic carbon. Growth rates, or related measures such as cell or biomass production, remain a common quantitative tool used to evaluate the bacterial role in aquatic systems.

Two primary approaches are used to measure bacterial growth in natural samples. The more frequently used approach is based on the incorporation of radiolabelled precursors into newly formed macromolecules. It is axiomatic that cell production is normally accompanied by DNA replication, protein synthesis and synthesis of other macromolecules. In principle, macromolecular synthesis rates are suitable indirect measures of bacterial growth (especially during balanced growth), and they can be estimated by measuring the incorporation of appropriate radiolabelled precursors. The procedures used typically require short incubation times, and in water samples require relatively little sample disturbance.

Radiotracer methodologies are known to have numerous problems, some of which are specific to particular procedures, and considerable effort has gone toward refining and testing these methods. A brief catalogue of some of the problems is: the requirement for conversion factors to estimate growth rates from radioisotope incorporation, and the known uncertainties and variability of those factors (Coveney and Wetzel, 1988); the unknown effects of community composition on conversion factors; the laborious nature of experiments required to obtain conversion factors; apparent non-specific labelling due to artefacts of the DNA extraction and purification process (Burnison and Nuttley, 1990; Servais *et al.*, 1987); real non-specific labelling due to the use of precursors as energy sources rather than as macromolecule precursors under conditions of nutrient limitation (Hollibaugh, 1988; Iriberri *et al.*, 1990; Carman *et al.*, 1988a; Servais *et al.*, 1987; Fallon and Newell, 1986); poor detection of anaerobic and possibly of oligotrophic bacterial growth (Moriarty, 1986); high background counts and variability of uptake in heterogenous sediment matrices (Kemp, 1990); and the requirement for an incubation period after sampling, with concomitant sample manipulations which may result in disturbance artefacts (LaRock *et al.*, 1988), especially in sediment (Findlay *et al.*, 1985; Carman *et al.*, 1988a,b).

Currently, the principle alternative to radioisotopic methods is incubation of bacteria after either removal of grazers, or suppression of grazing activity with metabolic poisons. The use of selective metabolic inhibitors to suppress grazing, with a few exceptions (*eg* Newell *et al.*, 1983) has not been particularly successful (Kemp, 1990; Tremaine and Mills, 1987; Sherr *et al.*, 1986), because inhibitors are commonly found to be partially non-specific or only partially effective when used in natural samples. Long incubations are required to detect an increase in bacterial abundance in the absence of grazing, whether grazing is reduced by selective filtration, dilution with filtered seawater (*eg* Kirchman *et al.*, 1982; Wright and Coffin, 1984) or by metabolic inhibitors. Although "bottle effects" may be minor during incubations of bacteria in sufficiently large containers (Lee and Fuhrman, 1991), long incubations severely limit the rate at which data can be acquired, and certainly are difficult to interpret as representing instantaneous, in-situ growth rates.

Despite these problems, extensive reviews of bacterial production data, largely based on radioisotope procedures (*eg* Ducklow and Carlson, 1992), clearly show that these methods yield data that are interpretable and appear to make ecological sense, the benchmark test of any method. One can conclude that the methods are either not as prone to error as a catalogue of problems might suggest, or that the magnitude of error introduced by these problems does not overshadow the real patterns in data obtained by conventional methods. What can estimates of growth rate based on RNA content contribute to the existing information on bacterial growth in aquatic environments?

The best conceivable application of conventional methods nonetheless has considerable limitations. Foremost among these is the inability to consider the growth and activity of individual components (*eg* species) of the total microbial community, especially for bacteria. The existing methods measure total community turnover rates only. Estimated growth rates are dominated by growth of the most active cells, but in most studies no information is obtained which would allow some assessment of the proportion of cells which are active. Using these community-level methods, it is effectively impossible to examine such fundamental ecological problems as species diversity and succession, or the relative contributions of various specific taxa to carbon remineralization. Finally, one may reasonably argue that when methods are known to have potentially serious errors, it is worthwhile to complement them with independently derived data.

The strong correlation between bacterial RNA content and growth rate was first observed more than 40 years ago, and has been demonstrated repeatedly to occur in various bacterial species used as laboratory model systems (particularly *Escherichia coli* and *Salmonella typhimurium*). A partial compilation of the published data on the RNA-growth rate relationship is shown in Fig. 1 Considering the variety of methods used to determine RNA content, the number of bacterial isolates represented, the diversity in culture conditions, and the broad range of growth rates observed, it is remarkable that the correlation of growth rate with RNA content is so strong (overall $r^2 = 0.85$).

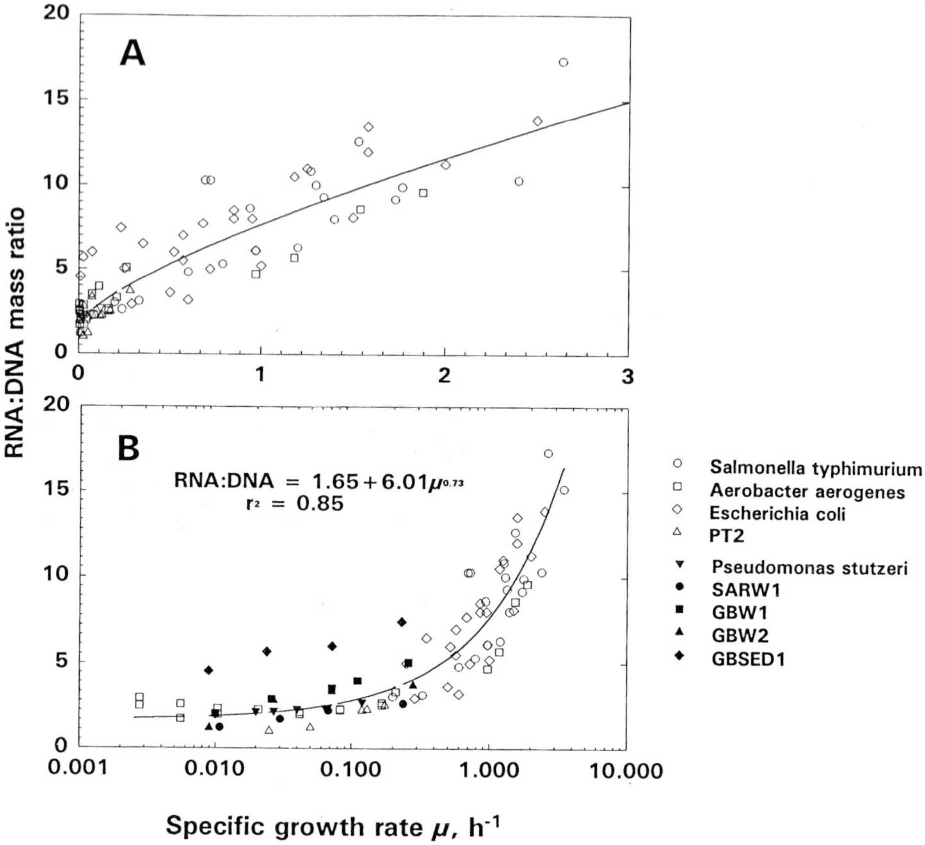

Fig. 1. Partial compilation of published data on the RNA:DNA ratio versus specific growth rate. Data for marine species in filled symbols, freshwater or enteric in open symbols. Original data sources are: *Escherichia coli*, Rossett *et al.* (1966), Churchward *et al.* (1982), Neidhardt (1987); *Salmonella typhimurium*, Rossett *et al.* (1966), Schaechter *et al.* (1958), Kjeldgaard and Kurland (1963); *Aerobacter aerogenes*, Rossett *et al.* (1966); strain PT2 Poulsen *et al.* (1993); *Pseudomonas stutzeri*, Kerkhof and Ward (1993); strains SARW1, GBW1, GBW2, and GBSED1, Kemp *et al.* (1993). A) shows specific growth rate on a linear scale; B) shows growth rate on a log scale.

The strength of this correlation suggests that the underlying causal relationship may be relatively universal, at least among bacterial taxa similar to those represented (Kerkhof and Ward, 1993). If so, then it may be possible to use measurements of RNA content in mixed aquatic bacterial communities in a relatively simple and straightforward manner. [Note that because most of the RNA in cells is ribosomal RNA (rRNA), total cellular RNA content is practically synonymous with rRNA content. In this paper, the term rRNA will be used only when the more specific term is justified.]

The following discussion will consider: the molecular and physiological mechanisms involved in regulating RNA synthesis and content, and the environmental factors that may influence this regulation in nature; whether the RNA-growth rate relationship is universal, or at least whether there is a universal tendency for a strong correlation between RNA and growth rate; in broad terms, the methods that have been used to measure RNA content, and their advantages and limitations; and field applications of RNA measurements as indicators of bacterial activity.

Regulation of rRNA synthesis and content

Considerable effort has gone into understanding the physiological, biochemical and molecular responses of model bacteria to growth-limiting conditions, including those responses that determine the rate of synthesis and the concentration of cellular rRNA. This body of work has repeatedly demonstrated that rRNA synthesis and degradation rates are adjusted by complex, independent feedback mechanisms, such that the resulting concentration of ribosomes is neither in excess nor insufficient to support the current requirements for protein synthesis in the cell (Nomura *et al.*, 1984; Lindahl and Zengel, 1986). Teleologically speaking, cells are designed to contain approximately as many ribosomes as are needed to maintain the growth rate that the environmental conditions will allow. The following is a brief synopsis of the feedback mechanisms that are hypothesized to maintain ribosomal concentration in proportion to growth rate.

Nutrient limitation of several bacterial species (*eg E. coli, Paracoccus denitrificans, Bacillus polymixa, B. licheniformis, Clostridium beijerinckii*) results in characteristic patterns of growth which reflect the degree of nutrient stress. The effects of nutrient availability can be divided into non-limited, chronic starvation, and starvation conditions, with a surprising consistency among these model bacteria in the growth rates which are typically associated with transitions between nutrient limitation states (Chesbro, 1988; Chesbro *et al.*, 1990).

These transitions can be characterized by changes in the synthesis rate and abundance of regulatory nucleotides, especially guanosine 5'-diphosphate 3'-diphosphate (ppGpp, or guanosine tetraphosphate). In the Bacteria, ppGpp influences, among other processes, translational error correction during protein synthesis (Gallant and Foley, 1979), the initiation of rRNA synthesis (Glass *et al.*, 1986) and the protein chain elongation rate (Rojas *et al.*, 1984). The combined effect is a reduction in cellular ribosome concentration and in protein synthesis as growth rate declines.

The cellular concentration of ppGpp is strongly and negatively correlated with growth rate and positively correlated with the proportion of RNA synthesis devoted to rRNA (Ryals *et al.*, 1982a). In non-limiting conditions, ppGpp is at a basal level and growth is exponential. During the transition to chronic starvation condition (typically at doubling times near 15-20 h, Chesbro *et al.*, 1979), ppGpp concentration increases, rRNA synthesis decreases and growth slows. With increasingly severe chronic starvation, the need to increase translational error correction, in particular, occupies an increasingly large fraction of cellular energy costs. As a consequence, there is a dramatic reduction in biomass synthesis in favour of redirecting energy flow to increasing maintenance requirements. At doubling times near 50-60 h, ppGpp levels increase substantially, rRNA synthesis is strongly reduced and growth rate decreases correspondingly (Van Verseveldt *et al.*, 1984; Arbige and Chesbro, 1982). Finally, under starvation conditions ppGpp levels remain somewhat elevated, and growth is near-zero. Similar changes in ppGpp concentration are observed when cells are shifted into starvation by metabolic inhibitors (Ryals *et al.*, 1982a), rather than entering starvation due to long-term exhaustion of resources.

The onset of severely reduced growth resulting from the shift of energy resources to maintenance functions has been termed the stringent response, originally for the specific case of the response to amino acid deficiency (Fiil *et al.*, 1977). During and following the stringent response, specific proteins are induced which appear to increase the long-term survival of starved cells (Nystrom *et al.*, 1990; Albertson *et al.*, 1990). However, the correlation between ppGpp levels and growth rate occurs not only during the onset of starvation, but as a general phenomenon over a very broad range of growth rates (Fig. 2 of Ryals *et al.*, 1982a).

The cellular abundance of ppGpp is controlled at both the production and degradation levels. ppGpp synthesis involves at least two routes, one of which is controlled by a ribosome-bound pyrophosphate transferase that requires mRNA and uncharged tRNA (*ie* tRNA not carrying an amino acid) (Goldman and Jakubowski, 1990). ppGpp degradation also is dependent on a ribosome-bound enzyme (guanosine 3',5'-bis[diphosphate] 3'-pyrophosphohydrolase, or ppGppase) that is inhibited by uncharged tRNA (Fiil *et al.*, 1977; Richter, 1980) and by glucose exhaustion (DeBoer *et al.*, 1975; Tetu *et al.*, 1980). Note that it is the relative proportions of uncharged tRNA and ribosomes, rather than the absolute concentrations, that trigger the synthesis or degradation of ppGpp. This dependency on uncharged tRNA couples both synthesis and degradation of ppGpp to cellular amino acid abundance, and therefore to the supply of energy, carbon, nitrogen and sulphur (Arbige and Chesbro, 1982). Because the pyrophosphate transferase and hydrolase are both ribosome-bound, the synthesis and degradation of ppGpp are mediated in part by the abundance of ribosomes.

ppGpp appears to control rRNA synthesis at the transcriptional level by influencing the binding of RNA polymerase to RNA gene promoters. Ryals *et al.* (1982a) presented a model which explains the differential effects of ppGpp on stable RNA synthesis and mRNA synthesis. In their model, ppGpp-controlled conformational changes in RNA polymerase preferentially shift its binding from stable RNA (tRNA and rRNA) promotors to mRNA promotors. This model is supported by similar changes in RNA synthesis which occur in mutations that affect the sensitivity of RNA polymerase to ppGpp (Little *et al.*, 1983), or that produce conformational changes in RNA polymerase.

Jensen and Pedersen (1990) present an alternative model for the differential effects of ppGpp on stringently controlled and other genes. They hypothesize that promotors of stringently controlled genes have lower affinity for RNA polymerase than promotors of non-stringently controlled genes. As growth rate slows and the concentration of RNA polymerase decreases, stringently controlled gene promotors lose RNA polymerase molecules to competing promotors with higher affinity for RNA polymerase. rRNA synthesis decreases as a consequence, and the concentration of ribosomes declines until the pool of charged tRNA is sufficient to engage most ribosomes in synthesis. This triggers decreased ppGpp synthesis and increased ppGpp degradation. Steady state growth in their model is a compromise between the abundance of ribosomes, and the availability of substrate per ribosome at a given growth rate.

Severe reductions in stable RNA content and growth also have been observed for starved bacteria in the absence of ppGpp accumulation, implying that there are independent mechanisms which operate in response to starvation (Acosta and Lueking, 1987; Gaal and Gourse, 1990). In eukaryotic organisms such as the yeast *Saccharomyces cerevisiae*, ribosomal concentration is correlated with growth rate and there are strong responses to starvation that parallel the stringent response, but these occur in the absence of ppGpp (Warner, 1989). The basic requirement for maintaining a balance between ribosomal concentration and growth rate is met, but not necessarily through the ppGpp mechanism.

Ribosomal protein synthesis is also regulated by feedback mechanisms. Ribosomal protein synthesis is repressed when ribosomal proteins are present in excess relative to rRNA (Jinks-Robertson and Nomura, 1987), because free ribosomal proteins (not assembled into ribosomes) bind to the promotors of their own genes and thereby competitively inhibit transcription. Ribosomal proteins also appear to act to protect rRNA against degradation, once assembled into ribosomes. For example, when rRNA is overproduced in a conditional gene-expression system (Gourse *et al.*, 1985), the turnover of newly synthesized rRNA is much higher than normal. Thus, mechanisms for ribosomal proteins and ribosomal RNA interact to prevent either component of intact ribosomes from being synthesized in excess.

An additional feedback regulatory mechanism involves the FIS (factor for inversion stimulation, Nilsson *et al.*, 1992) protein, the product of the *fis* gene. The FIS protein activates the transcription of both rRNA and tRNA operons when cells are exposed to improved nutritrional conditions and shift into higher growth rates. It appears to be more important at higher growth rates (Nilsson *et al.*, 1992), although as yet its overall importance is not well understood.

Growth rate regulation in marine bacteria

These growth-regulating mechanisms have been examined in several model bacteria and may also be general to many aquatic bacteria. However, few studies of the causes and mechanisms of bacterial growth regulation have been conducted for aquatic systems. Studies of the biochemical consequences of substrate limitation in aquatic bacteria have focused on the "starvation-survival" phenomenon in marine bacteria (Morita, 1982, 1988; Amy *et al.*, 1983; Baker *et al.*, 1983; Marden *et al.*, 1985). In response to extremely low nutrient availability, marine bacteria enter a condition in which growth rates are effectively zero and biochemical adaptations occur that maximize survival rather than growth. This phenomenon is equivalent to starvation as described above for cultured model bacteria. The starvation-survival phenomenon is known in marine systems primarily from deep-ocean and extremely oligotrophic environments.

Nystrom *et al.* (1990) examined the response of a marine *Vibrio* strain to complete starvation for single and multiple nutrients. Metabolic changes consistent with a stringent response mechanism (*ie* increased ppGpp levels and concomitant decreased RNA synthesis) occurred under amino acid starvation. This provided the first direct evidence that the stringent response occurs in marine bacteria, and that ppGpp is implicated in the response. Nystrom *et al.* found that a total of 66 proteins were starvation-inducible, and that each starvation condition tested (*ie* glucose, amino acids, ammonium, or phosphate deficiency) resulted in expression of an almost completely unique subset of these proteins. Synthesis of additional proteins were induced by multiple-nutrient starvation. If the expression of starvation-survival proteins was inhibited, continued survival was greatly decreased. Similar responses to starvation have been observed for *E. coli* and *S. typhimurium* (Schultz and Matin, 1991; Groat and Matin, 1986; Foster and Spector, 1986; Groat *et al.*, 1986). A few of these starvation-induced proteins have been isolated and identified or sequenced (Groat and Matin, 1986; Jenkins *et al.*, 1988).

In summary, the regulatory mechanisms controlling bacterial growth under nutrient limitation are reasonably well understood for model bacteria, and comparable mechanisms appear to operate in at least some marine bacteria.

While this is encouraging, it is important to recognize that we have no detailed information on the average physiological status of bacteria in natural aquatic systems, let alone their exact metabolic responses to different growth conditions. One cannot assume that stringent or starvation-survival responses are necessarily common, or even occur in bacteria in aquatic systems. However, as noted in the following section, there is considerable indirect evidence to suggest that these or alternative regulatory mechanisms do operate in most aquatic bacteria, and indeed in organisms other than bacteria, and regulate rRNA concentration in direct proportion to growth rate.

How universal is the RNA-growth rate relationship?

The data in Fig. 1 are a subset of all published data, where the data were or could be expressed in common units (specifically, as an RNA:DNA ratio). Nine members of the Bacteria are represented; three are enteric species, one is a sulphate-reducing anaerobe, and five are isolates from marine environments. Only one is known to be a member of the α-Proteobacteria; the others are known or assumed to be γ-Proteobacteria. In other words, these data, while voluminous, are severely limited in terms of the phylogenetic diversity represented. What information is available to suggest that a similar relationship between RNA and growth rate will exist for bacteria other than the small subset that have been examined?

RNA:DNA ratios have been used as a measure of growth rate for a number of other organisms. Sutcliffe proposed as early as 1965 that ribosomal concentration could be used to predict the growth rate of marine invertebrates (Sutcliffe, 1965). Dortch *et al.* (1983) cited a number of examples of the use of RNA:DNA ratios to measure growth in several large invertebrates (see references cited), and found that the RNA:DNA ratio was linearly related to growth rate for marine phytoplankton. Buckley (1984) demonstrated that the RNA:DNA ratio was an excellent predictor of growth in larval fish. DiRuggiero *et al.* (1993) found that rRNA content increased with increasing growth rate of a hyperthermophilic Archaean. rRNA content is also proportional to growth rate in the yeast *Saccharomycetes cerevisiae* (Waldron and Lacroute, 1975; Kief and Warner, 1983; Waldron, 1977; Warner, 1989). In short, there are abundant data indicating that an RNA-growth rate relationship is a common, perhaps universal property of living organisms.

The form of that relationship may vary when taxa are compared in detail. Kemp *et al.* (1993) found that there were statistically significant differences among the RNA-growth rate equations determined for four marine bacterial isolates in chemostat culture. These isolates are included in the data shown in Fig. 1, and suggest that at least some of the variance in Fig. 1 is due to real differences among bacterial taxa. However, Kemp *et al.* attributed at least some of the differences among isolates to differences in their characteristic size; when cell size was taken into account, the statistical differences among isolates were greatly reduced. RNA content varies with cell size for cells growing at the same rates. Larger cells on the average contain a greater biomass and require more ribosomes to reproduce that larger biomass than do smaller cells growing at the same rate (Kemp *et al.*, 1993). Thus, measurement methods that compensate for cell size (Poulsen *et al.*, 1993) reduce the apparent variability of the RNA-growth rate relationship.

Environmental factors

Based on the data in Fig. 1, it appears that a single, mathematically simple equation (a simple linear model fits the data in Fig. 1 nearly as well) might be used to describe the relationship that cells maintain between rRNA concentration and growth rate. Clearly, there is substantial variation also evident in Fig. 1. Some of that variation may be due to variations in characteristic cell size, as discussed above. At least some of that variation is due to environmental factors. For field applications, these factors might be accounted for or eliminated. What environmental variables are likely to occur and may influence the RNA-growth rate relationship? Obvious candidates are variations in food resources, and in physical conditions such as temperature, salinity, pH, and other water properties.

Bacterial growth in aquatic environments is usually assumed to be limited by the abundance of labile dissolved organic matter (DOM), which provides the link between primary and bacterial production (Azam *et al.*, 1983; and many others). This assumption is supported by tight coupling between the release and bacterial uptake of amino acids (Fuhrman, 1987), and various correlative and inferential data. For example, low concentrations of free amino acids are often attributed to high bacterial uptake and utilization (*eg* Jorgensen *et al.*, 1983; Mopper and Lindroth, 1982). Diel and longer cycles in bacterial activity are often attributed to concurrent cycles in phytoplankton production and release of dissolved organic matter (Painting *et al.*, 1989; Fuhrman *et al.*, 1985; Turley and Lochte, 1986, Lancelot and Billen, 1984; Herndl and Malacic, 1987). For the most part, the DOM used by bacteria is uncharacterized, and little is known regarding what specific DOM components are in limiting supply.

It is fortunate, then, that the cause of nutrient limitation (*eg* carbon *vs* nitrogen starvation, single amino acid deficiency *vs* multiple deficiencies) does not appear to influence the form of the rRNA-growth rate relationship. Indeed, the most common method for manipulating bacterial growth rates in the studies represented in Fig. 1 was to alter food quality as well as quantity, for example by altering the carbon and energy source provided in a nutrient medium. One way of phrasing it is that the effects of nutrient limitation on bacterial growth rate appear to be interpreted through their effects on the activity, synthesis, and concentration of ribosomes. Although much more work needs to be done to test this hypothesis for aquatic bacteria, the data for cultured bacteria supports it. What other environmental variables influence the rRNA-growth rate relationship and are likely to be important in aquatic systems?

The specific growth rate μ of bacteria can be described by a theoretical relationship which incorporates the synthesis rates of RNA, RNA polymerase and protein (rewritten from eqns. in Bremer, 1975; Ryals *et al.*, 1982b):

$$\mu = k_0 \sqrt{((\varepsilon_r \alpha_p \Phi_s c_s \beta_p k_1)/(k_2 k_3 k_4))} \qquad \textbf{(1)}$$

where ε_r = rate of protein production per ribosome; α_p = proportion of synthesized protein which is RNA polymerase; Φ_s = fraction of RNA polymerase synthesizing ribosomal RNA (rRNA); c_s = stable RNA chain elongation rate per RNA polymerase; and β_p = active fraction of RNA polymerase. The constants k_0-k_4 are: k_0 = dimensional constant; k_1 = ratio of rRNA to total stable RNA; k_2 = ratio of length of complete stable RNA transcripts (including spacer regions) to mature stable RNA; k_3 = rRNA nucleotides per ribosome; and k_4 = amino acids per RNA polymerase molecule.

Nutrient limitation has its primary effect in reducing rRNA synthesis rates (c_s), at least partly mediated by the effects of charged and uncharged amino acid concentrations on ppGpp concentration, as discussed above. Reduced rRNA synthesis rates result in a reduction in cellular rRNA content and consequently in protein synthesis. In addition, nutrient limitation can also suppress other growth-limiting processes such as peptide chain elongation rate (ε_r), because ribosomes in cells starved for a specific amino acid tend to "pause" in translation when that amino acid is required in the growing peptide chain.

Temperature has a strong effect on both the rRNA and protein chain elongation rates (c_s and ε_r). In *E. coli*, ribosomal and RNA polymerase activity both have typical, and rather similar, Q_{10} responses to changes in temperature. For a given rRNA concentration, higher temperatures result in higher rRNA and protein synthesis rates, and consequently higher growth rates. Ryals *et al.* (1982b) noted that the effect of temperature on growth rate appeared quite simple, but suggest that this apparent simplicity is actually the result of interactions among more complex effects. As a practical matter, however, temperature can be incorporated into an rRNA-growth rate equation as a simple linear variable, orthogonal to rRNA content. Buckley (1984) was able to predict larval fish growth using an empirical, predictive equation for larval fish, where growth rate was a linear function of both temperature and rRNA content. The equation explained most of the variance in weight-specific growth rates of several species in aquaculture. Similarly, Juinio *et al.* (1992) found that temperature effects must be taken into account to use RNA:DNA ratios as a measure of the nutritional status of postlarval lobsters; the RNA:DNA ratio was inversely correlated with temperature and positively correlated with food ration.

In relatively short-term culture conditions, rRNA content per cell is invariant with changes in temperature for *E. coli* (Ryals *et al.*, 1982b) and *S. typhimurium* (Schaechter *et al.*, 1958). Under field conditions, however, the average rRNA content of cells appears to decline with increasing temperature (Lee *et al.* 1993) over seasonal time scales, suggesting that fewer ribosomes are needed at higher temperatures to maintain the same average growth rate. Alternatively, community composition may change over seasonal time scales (Lee and Fuhrman, 1991), and the seasonal changes in average rRNA concentration observed by Lee *et al.* may relate more to changes in species than to changes in temperature. At present, it seems likely that it will be necessary to incorporate temperature effects into predictive rRNA-growth rate equations for natural aquatic communities; however, seasonal changes in response to temperature may prove to complicate matters.

There is now limited evidence that rRNA content varies with diel periodicity in some organisms, including bacteria. Kramer and Singleton (1993) found strong diel periodicity in *Synechococcus* (phototrophic cyanobacterium) populations in coastal marine waters, such that ribosome concentrations were highest at noon and lowest at night. It was not known whether this diel change was in proportion to growth rate, but the existence of diel patterns points out that measurements of RNA content must be interpreted in the context of periodic changes in the environment.

It also points out that ribosomal concentration can alter over relatively short time scales. We have found that rRNA of marine bacterial isolates increased to a new equilibrium level after growth rate upshifts, over times less than or equal to one generation time (Kemp *et al.*, 1993); *ie* within hours for faster growing cells. For cells undergoing growth rate downshifts, rRNA content of marine isolates decreased to the new level equally rapidly (Kemp, unpublished data). The rapidity with which ribosomal concentration is adjusted means that rRNA concentrations at any given time should reflect the current (or at worst very recently past) growth rate.

Kramer and Singleton (1992), however, found that the rate at which two marine *Vibrio* species reduced ribosome concentration depended on the species and the cause of the nutrient downshift. *V. alginolyticus* shifted from nutrient-replete to nutrient-depleted medium lost rRNA slowly, in comparison to cells allowed to exhaust the nutrients in the original medium. In *V. furnissii*, however, cells entering starvation by either route lost rRNA rapidly. Thus, there is evidence to suggest that under certain conditions, some bacteria may retain excess rRNA (not necessarily as functional ribosomes) over extended periods. This is of some concern, because it implies that the rRNA content of these bacteria does not necessarily reflect the current growth rate, but rather may be a relict of earlier events. More work is certainly needed to test whether the phenomenon of ribosome retention is common or atypical. To date, we have not observed it during extensive work with four marine γ-Proteobacteria isolates (Kemp, unpublished data). Work is underway to look for this phenomenon in a larger suite of isolates.

Oxygen clearly also has a strong influence on cell metabolism. Dramatic changes in metabolic pathways occur when organisms switch between aerobic and anaerobic metabolism, and there is relatively little information available to judge whether these changes affects the regulation of rRNA synthesis and the RNA-growth rate relationship. Georgellis *et al.* (1993) found that there was a large increase in RNA stability when *E. coli* cells were grown anaerobically, relative to cells grown aerobically. However, there was no difference in rRNA concentration in aerobically and anaerobically grown *E. coli* at comparable growth rates, because rRNA synthesis was reduced in anaerobic cells, thus compensating for the increased half-life of the RNA. This suggests that the cells use compensatory changes in synthesis and degradation rates to maintain the same proportionality between ribosome concentration and growth rate as they switch between aerobic and anaerobic metabolism (Georgellis *et al.*, 1993). This is particularly encouraging because at least some conventional methods for measuring bacterial growth rate are believed to be unreliable in anaerobic conditions (*eg* Moriarty, 1986). RNA-based methods may not be as problematic as conventional methods when used in anaerobic environments.

In summary, the single environmental factor known to be important under at least some conditions is temperature. Diel changes in environmental conditions, especially light, may be important to photosynthetic bacteria. Food quality does not appear to alter the form of the RNA-growth rate relationship. Cell size is important and must be taken into account. There are little or no data to judge whether other environmental variables such as oxygen, salinity and pH will present problems in field applications.

Methods used

Measurements of RNA content can be acquired by several alternate approaches. In broad terms, these approaches can be divided into biochemical measurements, and nucleic acid hybridization methods. The various specific methods differ in the information returned, their sensitivity and detection limits, and their ease of use.

Biochemical methods utilize colorimetric reactions, or alternatively, nucleic acid-specific fluorochromes. Much of the published data on rRNA content and growth of bacteria is based on measurements of bulk RNA in nucleic acid extracts, originally through colorimetric assays, and more recently using fluorochromes such as ethidium bromide. These data are typically normalized to cell numbers, in recognition that it is the RNA content per cell mass that is proportional to growth rate. In many cases, cell counts are not available, or are difficult to relate to the quantity of RNA extracted because the efficiency of extraction is not measured or is not well known (*ie* it is not certain how many cells are actually represented in a quantity of extracted nucleic acids). The most common solution is to normalize RNA to DNA in the same nucleic acid extract. This treats DNA as a surrogate for cell abundance, assuming that the DNA content of bacterial cells is much less variable than their RNA content.

On the positive side, RNA:DNA ratios have the advantage of correcting for variations in the efficiency of nucleic acid extraction, because much of the variation in extraction efficiency appears to occur at the level of co-purification of RNA and DNA from intact cells, rather than later in the extraction when RNA and DNA are separated (Kemp *et al.*, 1993). The disadvantage is that to the extent that DNA content also varies with growth rate, the RNA:DNA ratio approach will actually result in a loss of dynamic range of the variable used to predict growth rate. That is, the slope of a line describing the change in RNA:DNA ratio per unit change in growth rate is proportionally less steep than the slope of a line describing the change in RNA per unit change in growth rate.

More recently, hybridization methods are finding increasing use because they provide a greater capacity to target specific taxa and to measure rRNA content in individual cells. These methods are based on hybridization of nucleic acid probes to target sites on the rRNA molecules. The probes carry a radioisotopic or fluorescent label, or alternatively carry a target for a secondary detection probe.

Recent reviews detail the molecular basis for hybridization methods (*eg* Ward *et al.*, 1992; Wolcott, 1992). In brief, these methods capitalize on the evolutionarily conservative nature of ribosomal RNA, and the resulting ability to select target locations on the rRNA molecule whose nucleotide sequences are diagnostic for a specific taxon. Segments of the rRNA molecules (5S, 16S, and 23S rRNA are the three rRNA molecules incorporated into bacterial ribosomes) vary in the degree to which their nucleotide sequences are evolutionarily conserved. Nucleotide sequences in conservative segments are common to relatively broad phyla, whereas nucleotide sequences in non-conservative regions may be unique even at the species or subspecies level. Consequently, probe sequences can be selected which are characteristic of specific phylogenetic groups, ranging from universal probes, to kingdom-level (Bacteria, Archaea, or Eukarya), to extremely specific probes. DeLong *et al.* (1989), for example, used rRNA probes to distinguish between two closely related bacterial species. Gobel *et al.* (1987) developed rRNA probes capable of discriminating between congeneric bacteria. Salama *et al.* (1991) describe the development of rRNA probes to distinguish *Lactococcus lactis* sub. *cremoris* from its conspecifics. Specificity is manipulated both by target site selection, and by altering the conditions of the hybridization.

Hybridization methods can be applied to bulk extracts of nucleic acids, *eg* by hybridizing DNA probes against RNA extracted from natural samples. Like biochemical methods discussed earlier, these methods yield total RNA per sample, which is then normalized to either cell numbers or to a surrogate measure, such as total DNA abundance, or the abundance of rDNA genes (Kerkhof and Ward, 1993). The important difference from data that is obtained using biochemical measurements is that hybridization techniques incorporate taxonomic specificity. Ideally, these methods can be used, for example, to measure the RNA:DNA ratio of a specific bacterium in a mixed community (*eg* Kerkhof and Ward, 1993).

An alternative form of hybridization methods uses the same principle of targeting sites on the rRNA molecule with a labelled DNA probe, but this is done in intact cells rather than extracted nucleic acids. Most commonly, the probe carries a fluorochrome and hybridization is detected using microfluorometry. These "*in situ* hybridizations" are now used commonly to detect and enumerate targeted bacteria in natural and laboratory samples (*eg* Amann *et al.*, 1990; Manz *et al.*, 1993; McSweeney *et al.*, 1993; Kemp, 1994; Wagner *et al.*, 1993). It is a natural extension to that work to measure the fluorescence signal obtained after hybridization, to learn to interpret that fluorescence intensity as a measure of rRNA content, and to explore the meaning of that rRNA content as an indicator of growth rate.

The advantage to this approach is that the rRNA content of individual cells is measured with the same taxonomic specificity as in bulk nucleic acid hybridization methods. Ideally, such data can be interpreted as indicators of the growth rate of individual cells, and it should be possible to determine the distribution of growth rate among cells in bacterial populations. Such cell-specific data have been difficult to acquire with conventional methods (*eg* using microautoradiography, Tabor and Neihof, 1982). Indeed, it is not yet known how useful rRNA-frequency distributions may be as measures of the ecological status and responses of aquatic bacteria.

A principle disadvantage to the in-situ hybridization approach is that the fluorescence signal is low for relatively inactive cells containing few ribosomes, and therefore the method tends to be biased in favour of active cells. Solutions for this problem are to enhance the fluorescence signal, increase the sensitivity of the detector (*eg* cooled-CCD camera systems rather than optical photometers), or simply accept that the method is biased. The bias, it should be noted, exists as well for virtually any growth rate method that pools entire populations of cells, including RNA methods and macromolecular labelling methods (*eg* thymidine and leucine labelling) based on bulk nucleic acid extracts.

Enhancement methods are problematic. Multiple independent probes act to increase fluorescence additively and have been used successfully in field samples (Lee *et al.*, 1993; Lee and Kemp, 1994). For a given target group, it is probably feasible to design a few independent probes of similar specificity and similar requirements for hybridization (a function of the base sequence of probes, their length, and the number of mismatches expected from the target sequence), but this produces only a limited enhancement of 2-5x. Other methods that have been explored include: labelling with multiple fluors; attachment of long polythymidine "tails" detected with polyadenine secondary probes; and the use of biotinylated probes detected with an fluorochrome-labelled avidin secondary probe. The former two methods have been found to result in variable signal strength, depending on cell permeability, and undesirable non-specific probe binding. The biotin-avidin method, however, has worked well for labelling rRNA in marine protists (Lim *et al.*, 1993), and is a promising approach for aquatic bacteria as well.

Improving instrumentation is an effective method for increasing sensitivity and improving detection limits. Cooled-CCD cameras are much more sensitive than optical photometers, in our experience; indeed, typical cooled-CCD cameras are considerably more sensitive to low fluorescence than the human eye. Using these detectors, it is common to find that cells are seen in captured images that were not visible to the eye. With an optical microphotometer, we were able to detect as little as 0.3 femtograms of RNA per cell (Lee and Kemp, 1994). The cooled-CCD camera we now employ is substantially more sensitive, although other factors are involved in setting the absolute detection limit, including background and non-specific fluorescence, and software limitations in analyzing captured images.

Field applications: examples

There are, as yet, relatively few examples of the use of rRNA measurements to study natural aquatic bacterial communities. The rRNA-growth rate relationship has not been used to evaluate bacterial growth in field studies. Instead, RNA measurements have been treated as probable indicators of bacterial activity rather than as estimators of growth rate. This level of interpretation is appropriate. The rRNA-growth rate relationship is not yet sufficiently well understood to predict the growth rate of complex natural communities, based solely on the rRNA-growth relationship determined for model organisms. However, a more complete interpretation waits only on additional information. The most urgently needed are data on the RNA-growth rate relationship for bacterial taxa actually present in natural populations, and comparisons with other measures of growth rate.

Published studies of variations in bacterial rRNA content under field conditions are so far limited to exploratory work. Kramer and Singleton (1993) measured variations in rRNA content of heterotrophic bacterial communities and of photosynthetic *Synechococcus* in coastal marine waters. The approach used was probing bulk-extracted rRNA with either Bacteria- or Cyanobacteria-specific 16S rRNA oligonucleotide probes. In this demonstration study, they were able to detect substantial diel variations in rRNA content of *Synechococcus*. They also found that heterotrophic bacteria were stimulated, in terms of increased rRNA content, by additions of glucose but not by additions of amino acids. Their work demonstrated the potential for rRNA measurements to reveal patterns in bacterial activity, including changes in activity induced by changes in the nutrient field.

In our laboratory, we have begun an extensive series of measurements of rRNA content in field populations of marine bacteria, using in-situ hybridization and analyses of rRNA content in individual cells by microfluorometry. Lee *et al.* (1993) and Lee and Kemp (1994) reported on the effectiveness of multiple independent probes to detect the rRNA content of individual cells in nearshore bacterial communities. Data were presented showing that there are pronounced changes in mean rRNA content in these communities over seasonal time scales. To some extent (but not entirely), these changes are consistent with the effects expected from seasonal changes in mean water temperature.

In more recent work, Kemp (1994) presented a preliminary data set showing a section plot of bacterial rRNA content across the continental shelf. Variations in rRNA were compared to changes in temperature, salinity, nutrients, chlorophyll, and other water properties. rRNA content was also compared to measurements of thymidine incorporation and leucine incorporation. In general, rRNA content showed very strong spatial pattern that was to a considerable extent statistically explained by variations in other measured variables. For example, there was a large increase in the mean rRNA content of bacteria in locations associated with the outflow of the Chesapeake Bay estuary, marked by high temperature and low salinity. A second site of high rRNA content coincided with an area of high chlorophyll concentration at mid-shelf near the bottom.

Thymidine and leucine incorporation showed equally strong spatial pattern, and in general terms were elevated in the same water masses. However, the three different measures of bacterial activity were weakly intercorrelated, indicating that each represented some different aspect of the relationship of bacterial activity to the environment, such that detailed point-by-point comparisons revealed differences in spatial distributions. This points out the value of having multiple independent measures of activity, each of which provides some unique glimpse into the relationship of bacteria to their environment.

We have found that, on the average, approximately 65% of bacterial cells in coastal marine communities contain sufficient rRNA to be detected and measured using an in-situ hybridization approach and multiple independent probes. The remainder contain less than ca. 0.3 fg rRNA per cell, and we assume that these represent the least active cells. Most are small and coccoidal (Lee and Kemp, 1994).

Conclusions

Clearly, there is a large body of laboratory data that supports the hypothesis that rRNA content is strongly and consistently related to growth rate in bacteria, and there is ample evidence to suggest that the existence of such a relationship is a universal property. However, there are as yet insufficient data to assume that the exact form of an equation describing the relationship is also universal among bacteria. Existing data indicate that such an equation must explicitly or implicitly recognize at least rRNA content, temperature, and cell size. Other environmental factors may prove to be important as well. The available data suggest that a simple (*eg* linear) proportionality of growth rate to rRNA content (expressed as rRNA per cell biomass or other measure of cell size) and temperature is sufficient. With these measurements, it is now possible to use RNA content measurements to estimate growth rate, for specific cultures in laboratory conditions.

Early field work with *in-situ* hybridization and bulk hybridization approaches have demonstrated that measurements of rRNA content are useful indicators of bacterial responses to environmental conditions. However, the current understanding of the rRNA-growth rate relationship is still too primitive to justify more than a very cautious interpretation of rRNA content as a predictor of growth rate for field populations. The current status of the method might be described as having moved from laboratory demonstration to field validation. All data now available suggest that the approach will prove robust, and a useful complement to conventional methods for measuring bacterial growth rate.

Acknowledgement

This research was performed under the auspices of the United States Department of Energy under contract no. DE-AC02-76CH00016. By acceptance of this article, the publisher and/or recipient acknowledges the US. Government's right to retain a nonexclusive, royalty-free license in and to any copyright covering this paper.

References

Acosta R and Lueking DR (1987) Stringency in the absence of ppGpp accumulation in *Rhodobacter sphaeroides*. *J Bacteriol* **169:** 908-912

Albertson NH, Nystrom T, Kjelleberg S (1990) Starvation-induced modulations in binding protein-dependent glucose transport by the marine *Vibrio* sp. S14. *FEMS Microb Lett* **70:** 205-210

Amann RI, Binder BJ, Olson RJ, Chisholm SW, Deveraux R, Stahl DA (1990) Combination of 16S rRNA-targeted oligonucleotide probes with flow cytometry for analyzing mixed microbial populations. *Appl Environ Microb* **56:** 1919-1925

Amy PS, Pauling C, Morita RY (1983) Starvation-survival processes of a marine vibrio. *Appl Environ Microbiol* **45:** 1041-1048

Arbige M and Chesbro WR (1982) *rel A* and related loci are growth rate determinants for *Escherichia coli* in a recycling fermentor. *J Gen Microbiol* **128:** 693-703.

Azam F, Fenchel T, Field JG, Gray J, Meyer-Reil L-A, Thingstad F. (1983) The ecological role of water-column microbes in the sea. *Mar Ecol Prog Ser* **10:** 257-263.

Baker RM, Singleton C, Hood MA (1983) Effects of nutrient deprivation on *Vibrio cholerae*. *Appl Environ Microb* **46:** 930-940

Bremer H (1975) Parameters affecting the rate of synthesis of ribosomes and RNA polymerase in bacteria. *J Theor Biol* **53:** 115-124

Buckley LJ (1984) RNA-DNA ratio: an index of larval fish growth in the sea. *Mar Biol* **80:** 291-298.

Burnison BK and Nuttley DJ (1990) Purification of DNA for bacterial productivity estimates. *Appl Environ Microb* **56:** 362-365

Carman KR, Dobbs FC, Guckert JB (1988a) Consequences of thymidine catabolism for estimates of bacterial production: An example from a coastal marine sediment. *Limnol Oceanogr* **33:** 1595-1606

Carman KR, Dobbs FC, Guckert JB (1988b) Non-specific labelling, catabolism, and recycling of [methyl]^{3}H- and [methyl]^{14}C-thymidine in a marine sediment. *Limnol Oceanogr* **33:** 1595-1606

Chesbro W (1988) The domains of slow bacterial growth. *Can J Microbiol* **34:**427-435.

Chesbro WR, Evans T, Eifert R (1979) Very slow growth of *Escherichia coli. J Bacteriol* **139:** 625-636.

Chesbro W, Arbige M, Eifert R (1990) When nutrient limitation places bacteria in the domains of slow growth: metabolic, morphologic and cell cycle behavior. *FEMS Microb Ecol* **74:** 103-120

Churchward G, Bremer H, Young R (1982) Transcription in bacteria at different DNA concentrations. *J Bacteriol* **150:** 572-581

Coveney MF, Wetzel RG (1988) Experimental evaluation of conversion factors for the [3H]thymidine incorporation assay of bacterial secondary productivity. *Appl Environ Microbiol* **54:** 2018-2026.

DeBoer H, Bakker A, Weyer W, Gruber M (1975) The role of energy generating processes in the degradation of guanosine tetraphosphate, ppGpp, in *Escherichia coli*. *Biochim Biophys Acta* **432:** 361-368

DeLong EF, Wickham GS, Pace NR (1989) Phylogenetic stains: ribosomal RNA-based probes for the identification of single cells. *Science* **243:** 1360-1363

DiRuggiero J, Achenbach LA, Brown SH, Kelly RM, Robb FT (1993) Regulation of ribosomal RNA transcription by growth rate in the hyperthermophilic Archaeon, *Pyrococcus furiosus*. *FEMS Microb Let* **111:** 159-164

Dortch Q, Roberts TL, Clayton JR Jr, Ahmad SI (1983) RNA/DNA ratios and DNA concentrations as indicators of growth rate and biomass in planktonic marine organisms. *Mar Ecol Prog Ser* **13:** 61-71

Ducklow HW and Carlson CA (1992) Oceanic bacterial production. *Adv Microbial Ecol* **12:** 113-181

Fallon RD and Newell SY (1986) Thymidine incorporation by the microbial community of standing-dead *Spartina alterniflora*. *Appl Environ Microb* **52:** 1206-1208

Fiil N, Willumsen B, Friesen J, von Meyenberg K (1977) Interaction of alleles of the rel A, rel C, and spoT genes in *Escherichia coli*: analyses of the interconversion of GTP, ppGpp, and pppGpp. *Mol Gen Genet* **150:** 87-101

Findlay RH, Pollard PC, Moriarty DJW, White DC (1985) Quantitative determination of microbial activity and community nutritional status in estuarine sediments: evidence for a disturbance artefact. *Can J Microb* **31:** 493-498

Foster JW and Spector MP (1986) Phosphate starvation regulon of *Salmonella typhimurium*. *J Bacteriol* **166:** 666-669

Fuhrman J (1987) Close coupling between release and uptake of dissolved free amino acids in seawater studied by an isotope dilution approach. *Mar Ecol Prog Ser* **37:** 45-52

Fuhrman JA and Azam F (1980) Bacterioplankton secondary production estimates for coastal waters of British Columbia, Antarctica, and California. *Appl Environ Microbiol* **39:** 1085-1095

Fuhrman JA, and Azam F (1982) Thymidine incorporation as a measure of heterotrophic bacterioplankton production in marine surface waters: evaluation and field results. *Mar Biol* **66:** 109

Fuhrman J, Eppley RW, Hagstrom A, Azam F (1985) Diel variations in bacterioplankton, phytoplankton, and related parameters in the Southern California Bight. *Mar Ecol Prog Ser* **27:** 9-20

Gaal T and Ghourse RL (1990) Guanosine 3'-diphosphate-5'-diphosphate is not required for growth rate-dependent control of rRNA synthesis in *Escherichia coli*. *Proc Nat Acad Sci* **87:** 5533-5537

Gallant J and Foley D (1979) Stringent control of translational accuracy. In *Regulation of macromolecular synthesis by low molecular weight mediators*, Koch G, Richter D (eds) Academic Press, Toronto 5-23

Georgellis D, Barlow T, Arvidson S, von Gabain A (1993) Retarded RNA turnover in *Escherichia coli*: a means of maintaining gene expression during anaerobiosis. *Mol Microbiol* **9:** 375-381

Glass RE, Jones ST, Ishihama A (1986) Genetic studies on the beta subunit of

Escherichia coli RNA polymerase. VII. RNA polymerase as a target for guanosine tetraphosphate. *Mol Gen Genet* **203:** 265-268

Gobel UB, Geiser A, Stanbridge EJ (1987) Oligonucleotide probes complementary to variable regions of ribosomal RNA discriminate between *Mycoplasma* species. *J Gen Microb* **133:** 1969-1974

Goldman E and Jakubowski H (1990) Uncharged tRNA, protein synthesis, and the bacterial stringent response. *Mol Microb* **4:** 2035-2040

Gourse RL, Takebe Y, Sharrock RA, Nomura M (1985) Feedback regulation of rRNA and tRNA synthesis and accumulation of free ribosomes after conditional expression of rRNA genes. *Proc Nat Acad Sci* **82:** 1069-1073

Groat RG and Matin A (1986) Synthesis of unique polypeptides at the onset of starvation in *Escherichia coli. J Indust Microbiol* **1:** 69-73

Groat RG, Schultz JE, Zychlinsky E, Bockman A, Matin A (1986) Starvation proteins in *Escherichia coli*: kinetics of synthesis and role in starvation survival. *J Bacteriol* **168:** 486-493

Herndl GJ and Malacic V (1987) Impact of the pycnocline layer on bacterioplankton: diel and spatial variations in microbial parameters in the stratified water column of the Gulf of Trieste (Northern Adriatic Sea). *Mar Ecol Prog Ser* **38:** 295-303

Hollibaugh JT (1988) Limitations of the [^{3}H]thymidine method for estimating bacterial productivity due to thymidine metabolism, *Mar Ecol Prog Ser* **43:** 19-30

Iriberri J, Unanue M, Ayo B, Barcina I, Egea L (1990) Bacterial production and growth rate estimation from [^{3}H]thymidine incorporation for attached and free-living bacteria in aquatic systems. *Appl Environ Microb* **56:** 483-487

Jenkins DE, Schultz JE, Matin A (1988) Starvation-induced cross protection against heat or H_2O_2 challenge in *E. coli. J Bacteriol* **172:** 3910-3914

Jensen KF and Pedersen S (1990) Metabolic growth rate control in *Escherichia coli* may be a consequence of subsaturation of the macromolecular biosynthetic apparatus with substrates and catalytic components. *Microb Rev* **54:** 89-100

Jinks-Robertson S and Nomura M (1987) Ribosomes and tRNA. In *Escherichia coli and Salmonella typhimurium: cellular and molecular biology*. Neidhardt FC, Ingraham JL, Low KB, Magasanik B, Schaechter M, Umbarger HE (eds.) Washington DC, American Society for Microbiology, 1358-1385.

Jorgensen, NOG, Sondergaard M, Hansen HJ, Bosselmann S, Riemann B (1983) Diel variation in concentration, assimilation and respiration of dissolved free amino acids in relation to planktonic primary and secondary production in two eutrophic lakes. *Hydrobiol* **107:** 107-122

Juinio MAR, Cobb JS, Bengtson D, Johnson M (1992) Changes in nucleic acids over the molt cycle in relation to food availability and temperature in *Homarus americanus* postlarvae. *Mar Biol* **114:** 1-10

Kemp PF (1990) The fate of benthic bacterial production. *Rev Aquatic Sci* **2:**109-124

Kemp PF, Lee S, LaRoche J (1993) Estimating the growth rate of slow-growing marine bacteria from RNA content. *Appl Environ Microb* **59:** 2594-2601

Kemp PF (1994) A philosophy of methods development: the assimilation of new methods and information into aquatic microbial ecology. *Microb Ecol* **28:** 159-162

Kerkhof L and Ward BB (1993) Comparison of nucleic acid hybridization and fluorometry for measurement of the relationship between RNA/DNA ratio and growth rate in a marine bacterium. *Appl Environ Microb* **59:** 1303-1309

Kirchman D, Ducklow HW, Mitchell R (1982) Estimates of bacterial production and growth rates based on changes in biomass and uptake rates. *Appl Environ Microb* **44:** 1296-1307

Kjeldgaard NO, and Kurland CG (1963) The distribution of soluble and ribosomal RNA as a function of growth rate. *J Mol Biol* **6:** 341-348

Kramer JG and Singleton FL (1992) Variations in rRNA content of marine Vibrio spp. during starvation-survival and recovery. *Appl Environ Microb* **58:** 201-207

Kramer JG and Singleton FL (1993) Measurement of rRNA variations in natural communities of microorganisms on the southeastern U.S. continental shelf. *Appl Environ Microb* **59:** 2430-2436

Lancelot C and Billen G (1984) Activity of heterotrophic bacteria and the coupling to primary production during the spring phytoplankton bloom in the Southern Bight of the North Sea. *Limnol Oceanogr* **29:** 721-730

LaRock PA, Schwartz JR, Hofer KG (1988) Pulse labelling: a method for measuring microbial growth rates in the ocean. *J Microbiol Meth* **8:** 281-297

Lee S and Fuhrman JA (1991) Species composition shift of confined bacterioplankton studies at the level of community DNA. *Mar Ecol Prog Ser* **101:** 193-201

Lee S, Malone C, Kemp PF (1993) Use of multiple 16S rRNA-targeted fluorescent probes to increase signal strength and measure cellular RNA from natural planktonic bacteria. *Mar Ecol Prog Ser* **101:** 193-201

Lee S and Kemp PF (1994) Single-cell RNA content of natural marine planktonic bacteria measured by hybridization with multiple 16S rRNA-targeted fluorescent probes. *Limnol Oceanogr* **39:** 869-879

Lim EL, Amaral LA, Caron DA, DeLong EF (1993) Application of rRNA-based probes for observing marine nanoplanktonic protists. *Appl Environ Microbiol* **59:** 1647-1655

Lindahl L and Zengel JM (1986) Ribosomal genes in *Escherichia coli. Ann Rev Genetics* **20:** 297-236.

Little R, Ryals J, Bremer H (1983) rpoB mutation in *Escherichia coli* alters control of ribosome sysnthesis by guanosine tetraphosphate. *J Bacteriol* **154:** 787-792

Manz W, Szewzyk U, Ericsson P, Amann R, Schleifer K-H, Stenstrom T-A (1993) *In situ* identification of bacteria in drinking water and adjoining biofilms by hybridization with 16S and 23S rRNA-directed fluorescent oligonucleotide probes. *Appl Environ Microb* **59:** 2293-2298

Marden P, Tunlid A, Malmcrona-Friberg K, Odham G, Kjelleberg S (1985) Physiological and morphological changes during short term starvation of marine bacterial isolates. *Arch Microbiol* **149:** 326-332

McSweeney CS, Mackie RI, Odenyo AA, Stahl DA (1993) Development of an oligonucleotide probe targeting 16S rRNA and its application for detection and quantitation of the ruminal bacterium *Synergistes jonesii* in a mixed-population chemostat. *Appl Environ Microbiol* **59:** 1607-1612

Mopper K and Lindroth P (1982) Diel and depth variations in dissolved free amino acids and ammonium in the Baltic Sea determined by shipboard HPLC analysis. *Limnol Oceanogr* **27:** 336-347

Moriarty DJW (1986) Measurement of microbial growth rates in aquatic systems using rates of nucleic acid synthesis, *Adv Microb Ecol* **9:** 245-292

Moriarty DJW and Pollard PC (1981) DNA synthesis as a measure of bacterial productivity in seagrass sediments. *Mar Ecol Prog Ser* **5:** 151-156

Moriarty DJW and Pollard PC (1982) Diel variation of bacterial productivity in seagrass Zostera capricorni) beds measured by rate of thymidine incorporation into DNA. *Mar Biol* **72:** 165-173

Morita RY (1982) Starvation-survival of heterotrophs in the marine environment. *Advances in Microbial Ecology* **6:** 171-198

Morita RY (1988) Bioavailability of energy and its relationship to growth and starvation survival in nature. *Can J Microbiol* **34:** 436-441

Neidhardt F (1987) Chemical composition of *Escherichia coli*, In. *Escherichia coli and Salmonella typhimurium: cellular and molecular biology,* Neidhardt FC, Ingraham JL, Low KB, Magasanik B, Schaechter M, Umbarger HE (eds) Am. Soc. Microbiol., Washington, DC 3-6

Newell SY, Sherr BF, Sherr EB, Fallon RD (1983) bacterial response to presence of eukaryote inhibitors in water from a coastal marine environment. *Mar Environ Res* **10:** 147-157

Nilsson L, Verbeek H, Hoffman U, Haupt M, Bosch L (1992) Inactivation of the *fis* gene leads to reduced growth rate. *FEMS Microb Let* **99:** 85-88

Nomura M, Ghourse R, Baughman G (1984) Regulation of the synthesis of ribosomes and ribosomal components. *Ann Rev Biochemistry* **53:** 75-117

Nystrom T, Flardh K, Kjelleberg S (1990) Responses to multiple-nutrient starvation in marine *Vibrio* sp. strain CCUG 15956. *J Bacteriol* **172:** 7085-7097

Painting SJ, Lucas MI, Muir DG (1989) Fluctuations in heterotrophic bacterial community structure, activity and production in response to development and decay of phytoplankton in a microcosm. *Mar Ecol Prog Ser* **53:** 129-141

Poulsen LK, Ballard G, Stahl DA (1993) Use of rRNA fluorescence *in situ* hybridization for measuring the activity of single cells in young and established biofilms. *Appl Environ Microbiol* **59:** 1354-1360

Richter D (1980) Uncharged tRNA inhibits guanosine 3'5'-bis(diphosphate) 3'-pyrophosphohydrolase (ppGppase), the spoT gene product from *Escherichia coli. Mol Gen Genet* **178:** 325-327

Rojas AM, Ehrenberg M, Andersson SGE, Kurland CG (1984) ppGpp inhibition of elongation factors Tu, G and Ts during polypeptide synthesis. *Mol Gen Genet* **197:** 36-45

Rossett R, Julien J, Monier R (1966) Ribonucleic acid composition of bacteria as a function of growth rate. *J Mol Biol* **18:** 308-320

Ryals J, Little R, Bremer H (1982a) Control of rRNA and tRNA syntheses in *Escherichia coli* by guanosine tetraphosphate. *J Bact* **151:** 1261-1268

Ryals J, Little R, Bremer H (1982b) Temperature dependence of RNA synthesis parameters in *Escherichia coli. J Bacteriol* **151:** 879-887

Salama M, Sandine W, Giovannoni S (1991) Development and application of oligonucleotide probes for identification of *Lactococcus lactis* subsp. *cremoris*. *Appl Environ Microb* **57:** 1313-1318

Schaechter E, Maaloe O, Kjeldgaard NO (1958) Dependence on medium and temperature of cell size and chemical composition during balanced growth of *Salmonella typhimurium*. *J Gen Microbiol* **19:** 592-606

Schultz JE and Matin A (1991) Molecular and functional characterization of a carbon starvation gene of *Escherischia coli*. *J Mol Biol* **218:** 129-140

Servais P, Martinez J, Billen G, Vives-Rego J (1987) Determining [^{3}H]]thymidine incorporation into bacterioplankton DNA: improvement of the method by DNAse treatment. *Appl Environ Microbiol* **53:** 1977-1979

Sherr BF, Sherr EB, Andrew TL, Fallon RD, Newell SY (1986) Trophic interactions between heterotrophic protozoa and bacterioplankton in estuarine water analyzed with selective metabolic inhibitors. *Mar Ecol Prog Ser* **32:** 169-179

Sutcliffe WH Jr (1965) Growth estimates from ribonucleic acid content in small organisms. *Limnol Oceanogr* **10S:** R253-R258

Tabor PS and Neihof RA (1982) Improved microautoradiographic method to determine individual microorganisms active in substrate uptake in natural waters. *Appl Environ Microb* **44:** 945-953

Tetu C, Dassa E, Bouquet P (1980) The energy dependent degradation of guanosine 5'-diphosphate 3'-diphosphate in *Escherichia coli*. *J Bacteriol* **141:** 1148-1156

Tremaine SC and Mills AL (1987) Inadequacy of the eucaryote inhibitor cycloheximide in studies of protozoan grazing on bacteria at the freshwater-sediment interface. *Appl Environ Microb* **53:** 1969-1972

Turley CM and Lochte K (1986) Diel changes in the specific growth rate and mean cell volume of natural bacterial communities in two different water masses in the Irish Sea. *Microb Ecol* **12:** 271-282

Van Verseveldt HW, Chesbro WR, Braster M, Stouthamer AH (1984) Eubacteria have 3 growth modes keyed to nutrient flow. *Arch Microbiol* **137:** 176-184

Wagner M, Amann R, Lemmer H, Scleifer K-H (1993) Probing activated sludge with oligonucleotides specific for Proteobacteria: inadequacy of culture-dependent methods for describing microbial community structure. *Appl Environ Microbiol* **59:** 1520-1525

Waldron C (1977) Synthesis of ribosomal and transfer ribonucleic acid in yeast during a nutritional shiftup. *J Gen Microb* **98:** 215-221

Waldron C and Lacroute F (1975) Effect of growth rate on the amounts of ribosomal and transfer ribonucleic acids in yeast. *J Bacteriol* **122**: 855

Ward DM, Bateson MM, Weller R, Ruff-Roberts AL (1992) Ribosomal RNA analysis of microorganisms as they occur in nature *Adv Microb Ecol* **12:** 219-286

Warner JR (1989) Synthesis of ribosomes in S*accharomyces cerevisiae*. *Microbiol Reviews* **53:** 256-273

Wolcott MJ (1992) Advances in nucleic acid-based detection methods. *Clinical Microbiology Rev* **5:** 370-386

Wright RT and Coffin RB (1984) Measuring microzooplankton grazing of planktonic marine bacteria by its impact on bacterial production. *Microb Ecol* **10:** 137-149

The Cell Cycle of Phytoplankton: Coupling Cell Growth to Population Growth

Daniel Vaulot
CNRS et Université Paris VI,
Station Biologique BP74, 29682 Roscoff Cx
France

Introduction

The cell cycle is the succession of steps that are completed between the birth of a cell and its division into two daughter cells. It plays a central role in cell biology since it encompasses key events such as DNA replication, chromosome segregation and cell division. In less than ten years, there has been a tremendous increase in the knowledge of the fundamental biological mechanisms that drive the cell cycle, in particular at the molecular level. The cell cycle engine appears almost perfectly conserved among all eukaryotes from yeast to man. In contrast, cell cycle mechanisms appear more diverse among bacteria although some of the basic controls are probably universal.

The cell cycle is of particular relevance to aquatic ecology since cell division translates cell growth into population growth. In response to light, cells take up carbon, and accumulate protein and other key metabolites. As they grow bigger, cells ultimately divide, which results in population increase, expressed as the cell division rate or population growth rate μ (day^{-1}). Cell metabolism is usually assessed in aquatic systems by monitoring carbon uptake using the classical ^{14}C method. In some respects, carbon uptake measurements are difficult to interpret. First, they do not distinguish between net and gross production. Second, carbon uptake is obtained as mgC per unit time and is difficult to convert into a true rate (with dimension T^{-1}) since the denominator (*ie* the photosynthetic fraction of the particulate carbon) is not easy to estimate. In contrast, the population growth rate is the net result of all metabolic processes and reflects the actual speed at which a given population is turning over. The measurement of phytoplankton growth rates has challenged aquatic ecologists for a long time (Furnas 1990). Cell cycle-based methods offer an elegant way to circumvent this road block.

I will first briefly review recent advances in fundamental cell cycle research, drawing largely from the excellent book recently written by Murray and Hunt (1993) to which I refer the reader who wants to gain a deeper understanding of

NATO ASI Series, Vol. G 38
Molecular Ecology of Aquatic Microbes
Edited by I. Joint

the cell cycle. I will then focus on phytoplankton. I will detail cell cycle control by external factors and how we can use this knowledge to interrogate the status of natural populations. Finally I will examine how the natural synchrony of aquatic populations to the diel cycle can be used to estimate their growth rate.

The cell cycle of eukaryotes

The succession of events that make up the cell cycle contain a few steps that are easily recognizable by classical microscopical observation (Fig. 1A), in particular mitosis (segregation of chromosome), karyokinesis (nuclear division) and cytokinesis (cell division). In 1951, Howard and Pelc demonstrated that DNA synthesis was not continuous in eukaryotes but restricted to a specific phase they called S, for synthesis (Fig. 1A).

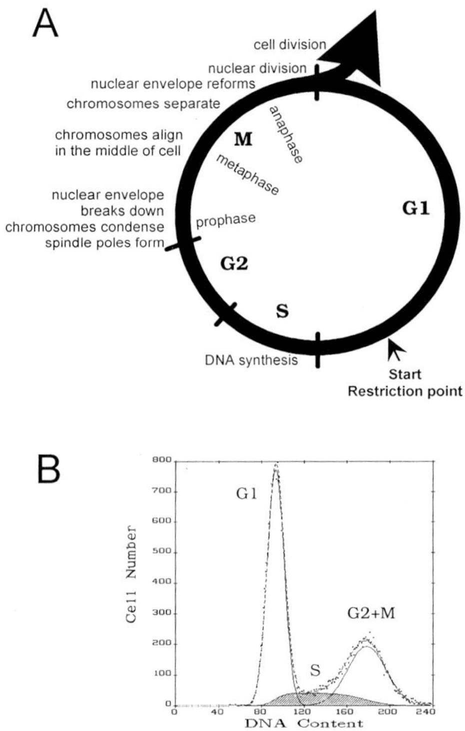

Fig. 1 (A) Some of the key events of the eukaryotic cell cycle (see text for details). (B) Use of flow cytometry to determine the cell cycle distribution of a population. X axis corresponds to DNA content and y axis to the number of cells with a given DNA content. The two peaks correspond to cells in G_1 and G_2+M, with the cells in between belonging to S. Software analysis (*eg* MCYCLE from Phoenix Flow Systems) permits computation of the percentage of cells in each phase.

This phase is preceded by a first gap (G_1) and separated from mitosis (M) by a second gap (G_2). Gametic cells of diploid organisms undergo a differentiated cycle called meiosis in which a round of DNA replication is followed by two rounds of cell division yielding haploid cells.

In order to study the cell cycle, it is critical to determine the localization of cells in the different cell cycle phases. The method of choice is analysis by flow cytometry of cells stained with a DNA fluorochrome such as propidium iodide, DAPI or Hoechst 33342 (Darzynkiewicz and Crissman 1990). The distribution obtained can be deconvoluted into the different populations G_1, S and G_2 (Fig. 1B). Other cell characteristics such as RNA, protein, or the level of a given antigen can be determined simultaneously to the cell cycle stage.

THE UNIVERSAL ENGINE

Our current understanding of the cell cycle originates from the merging of two completely different approaches which developed over the past 25 years. In amphibians and marine invertebrates, eggs are large, very well synchronized and therefore easy to study. Masui and Markert (1971) showed that a frog egg, which is stimulated by progesterone and induced to enter meiosis I, contains a substance they called maturation promoting factor (MPF). MPF induces meiosis I when injected into an unstimulated egg. MPF activity was shown to oscillate regularly in cells with a peak at mitosis and a trough during interphase. In eggs, the oscillation was independent of spindle assembly and of DNA synthesis, demonstrating that it was a key upstream event. Later Evans *et al.* (1983) discovered a protein named cyclin that was continuously synthesized but periodically destroyed during the cell cycle. This led to the hypothesis that high cyclin levels activate MPF. Since then, different types of cyclin have been found, some specific of M (cyclin B), some of S (cyclin A) and some of G_1 (cyclins C to F).

During the same period, research on both the budding yeast (baker's yeast, *Saccharomyces cerevisiae*) and the fission yeast (*Schizosaccharomyces pombe*) produced a number of temperature-sensitive mutants (*ie* with an aberrant phenotype at 37°C, but not at 25°C) that exhibited defects in the cell cycle. For example in fission yeast, two mutants $cdc2^{ts}$ and $cdc25^{ts}$ failed to divide at the restrictive temperature. In contrast, another mutants $wee1^{ts}$ divided at a smaller size than the wild type. Intriguingly the double mutant $cdc25^{ts}wee1^{ts}$ divided normally. Careful analysis demonstrated that while the cdc2 gene was required for mitosis, wee1 and cdc25 respectively inhibited and stimulated its activity, explaining why double mutants of the later genes recovered the wild phenotype. Cloning revealed that cdc2 encoded a 34 kDa protein kinase (often called p34^{cdc2}). Genes analogous to cdc2 were later found in many organisms ranging from the budding yeast (CDC28) to *Homo sapiens*.

The real breakthrough in cell cycle research came with the purification of MPF (Lohka *et al.* 1988) from frog eggs. This established that MPF is made of two units, one analogous to p34^{cdc2}, previously identified in yeast, the other to cyclin B, identified in sea urchin, and demonstrated the highly conserved nature of cell cycle components. This allowed the reconstruction of the basic engine of the cell cycle (Fig. 2). In brief, MPF is only active when p34^{cdc2} is phosphorylated on one of two phosphorylation sites (threonine 161). In this state, it induces key mitotic events such as spindle assembly. Just after mitosis, cyclin B is very rapidly degraded, leading to the destruction of the MPF complex; at the same time, p34^{cdc2} is dephosphorylated. During interphase, cyclin accumulation leads to the building up of new MPF. However, since the two sites of p34^{cdc2} are dephosphorylated, several kinases (wee1) and phosphatases (cdc25) are involved in restoring MPF back to its activated state. Still many details of the cell cycle engine remain obscure: in particular other kinases may be required for its successful functioning.

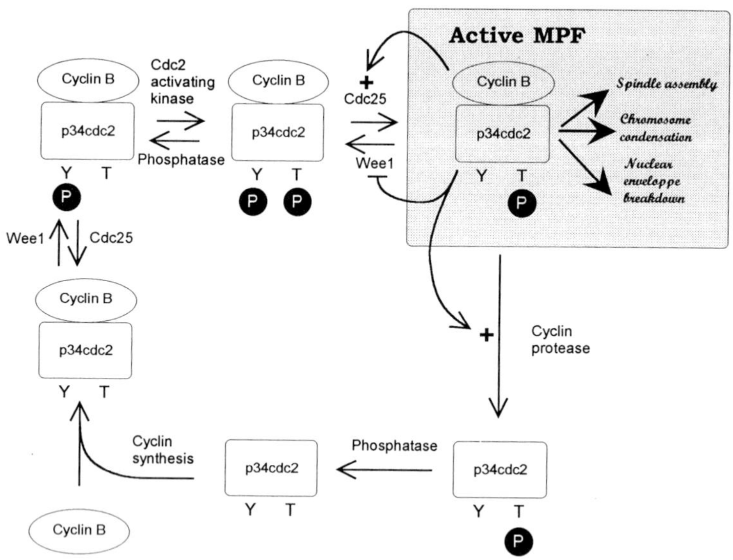

Fig. 2 The basic engine of the cell cycle (see text for details, redrawn from Murray and Hunt 1993).

CONTROLS AND CHECKPOINTS

Both internal and external controls are required for the cell cycle to work correctly. On the one hand, it is crucial for cell integrity that events such as DNA

synthesis and mitosis be coordinated: for example DNA synthesis must not occur twice in a row, without an intervening round of mitosis. On the other hand, the cell cycle needs to respond to the external environment so that it can slow down or even completely stop under adverse conditions, such as the absence of key nutrients or growth factors, in order to prevent irreversible damages.
The need to prevent re-replication of DNA is a good example of internal control. For this purpose, cells use what is called a licensing factor which is bound to chromosomes and is needed for DNA replication. This molecule is destroyed during replication and can only enter the nucleus when the nuclear membrane is either permeable or absent, a condition that only occurs during mitosis. Therefore in the absence of mitosis, DNA cannot replicate again. Moreover in yeast, the licensing factor appears to be inactivated as long as G_1 is not completed; therefore although the licensing factor is bound to chromosomes, cells cannot replicate DNA at the very beginning of the cell cycle.

External factors do not exert their influence on the whole cell cycle but only on certain of its parts. For example, in both budding and fission yeasts, the G_1 phase contains a critical point called Start (Fig. 1), beyond which cells are committed to replicate their DNA and divide. In fission yeast, cells can pass Start only when they reach a critical size; in mammalian cells, the same type of control exists and is called restriction point. Once cells are located beyond the restriction point, they are committed to complete the current cycle even if the necessary growth factors are absent. In contrast, G_1 cells, that have not yet reached the restriction point and are deprived of a growth factor, enter an inactive state called G_0. The details of the pathway that leads cells to pass through the restriction point when stimulated by the presence of optimal conditions are still to be fully worked out. In many cases, protein kinase cascades appear to be involved. A protein kinase becomes active once phosphorylated; in turn, it phosphorylates the next kinase in the cascade. Such cascades probably result in rapid amplification of the initial triggering signal.

The cell cycle of prokaryotes

The main difference between prokaryotic and eukaryotic cells is the absence of distinct cellular compartments. In particular, the chromatin is not enclosed inside the nucleus but is loosely arranged within the cytoplasm, forming what is called the nucleoid (Robinow and Kellenberger 1994); moreover there is no structure such as the cytoskeleton to move cell components within the cell. However, with respect to the cell cycle, bacteria still require some coordination between the DNA replication cycle and the division cycle to ensure that all dividing cells contain at least one copy of the genome. A full discussion of the prokaryotic cell cycle can be found in a recent book by Cooper (1991).

In contrast to eukaryotes, prokaryotes may display a cell division cycle that is shorter than the DNA replication cycle (Helmstetter and Cooper 1968). In *E. coli* at slow growth rates, DNA replication only occupies a fraction of the cell cycle and phases analogous to G_1, S and G_2 can be delineated (Fig. 3). At high growth rates however chromosome replication takes longer than the division cycle and spans more than one cell generation. For this to be possible, new replication forks are initiated on chromosomes that have not yet finished the previous round of replication (Fig. 3). The end of one round of replication triggers chromosome separation and cell division.

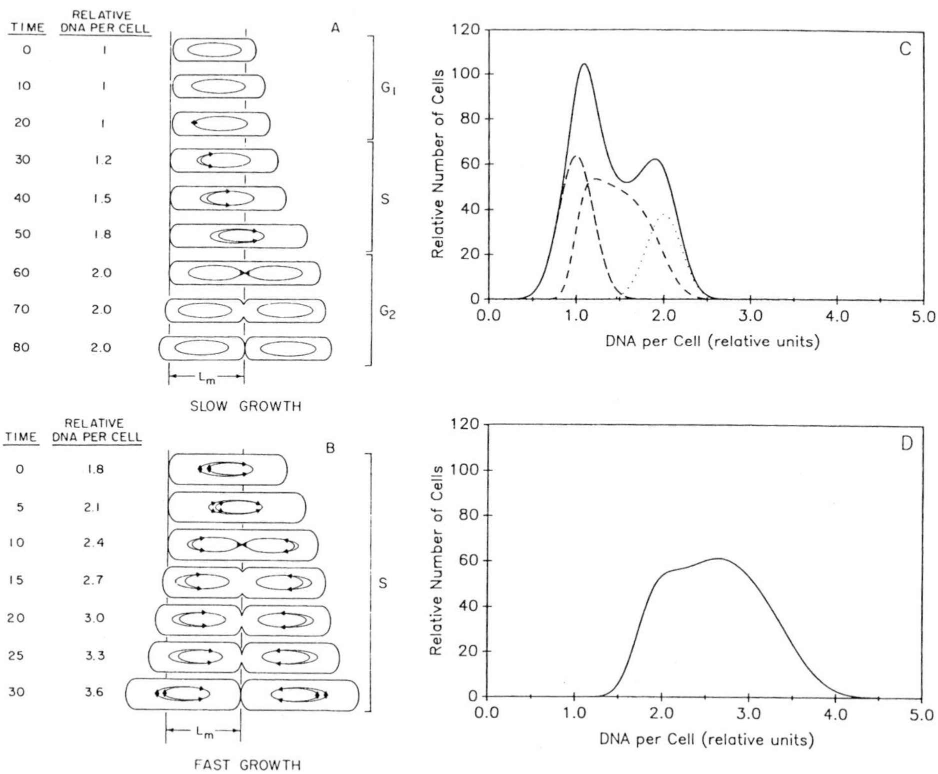

Fig. 3 The prokaryotic cell cycle. In slow-growing populations (A), DNA replication is shorter than the division cycle yielding flow cytometric DNA distributions (C) analogous to that of eukaryotes. In fast-growing populations (B), DNA replication is longer than the division cycle. New-born cells have more than one chromosome and DNA replications is continuous, yielding flow cytometric DNA distributions (D) without well-defined peaks (reprinted with permission from Armbrust *et al.* 1989).

Despite these differences, some controls appear similar between prokaryotes and eukaryotes. For example, new replication forks can be initiated only when the ratio of the cell mass to the number of replication forks reaches a given threshold (Bremer and Churchward 1994), a reminder of Start control in yeast. At the genetic level, some of the key events such as the initiation of DNA replication or cell division are well understood (see review by Donachie 1993). However, the underlying oscillators (potentially analogous to MPF in eukaryotes) are unknown. If such oscillators exist, it is not clear whether they are phylogenetically conserved across all prokaryotes. For example, some of the mechanisms observed in *E. coli* model (such as the synchronous initiation of replication forks, Fig. 3) clearly do not apply in other eubacteria, such as the cyanobacterium *Synechococcus* PCC 6301 (Binder and Chisholm 1990).

The cell cycle of phytoplankton

Cell cycle basic mechanisms and regulations have only been investigated in detail in a few unicellular photosynthetic organisms such as *Chlamydomonas*, *Chlorella* or *Euglena* (*eg* Donnan *et al.* 1985). However these cellular models, with their very special form of division (in particular the potential to form multiple daughter cells: 2, 4, 8, 16) may not be relevant to species found in natural phytoplankton communities, especially in oceanic waters. Recently the use of flow cytometry has considerably advanced our knowledge of phytoplankton cell cycle both in culture (Olson *et al.* 1983) and in nature (Boucher *et al.* 1991).

Within eukaryotic phytoplankton, all organisms investigated to date (see Table 1) display a typical eukaryotic cell cycle with clearly defined G_1, S and G_2 phases. In the past, there had been some speculations that dinoflagellates, which have permanently condensed chromosomes, continuously synthesize DNA throughout the cell cycle (Karentz 1983). Recent evidence, especially from studies using single cell DNA measurements, do not support this hypothesis and indicate a clearly defined S phase (*eg* Bhaud *et al.* 1991). Interestingly, of all phytoplankton species studied to date, only the dinoflagellate *Gyrodinium uncatenum* displays a cell cycle with a very long G_2 phase (Cetta and Anderson 1990).

In photosynthetic prokaryotes, the few cases studied indicate in general the presence of phases analogous to G_1, S and G_2, suggesting DNA replication is shorter than the division cycle (Fig. 3). However, several *Synechococcus* strains, in particular a freshwater one PCC 6301, display an intriguing behaviour. Cells may have more than 2 copies of the genome even at relatively low light-limited growth rates. Moreover copy numbers of completed chromosomes are equal to 1, 2, 3, 4, 5 *etc* and do not correspond to powers of two, *ie* 1, 2, 4, 8 *etc*, as predicted by the *E. coli* model (Binder and Chisholm 1990, submitted). Such behaviour has also been observed in *Prochlorococcus*, where up to 5 copies of the genome may occur in the same cell (Vaulot, unpublished data).

Table 1: Cell cycle dependence on external factors in phytoplankton. Data obtained mainly from starvation experiments. When limitation has a different effect than starvation, it is mentioned (*eg* for silica in diatoms).

Factor	**Group**	**Species**	**G_1**	**S**	**G_2**	**References**
Temperature	Diatoms	*Thalassiosira weissflogii*	+	+	+	Olson *et al.* 1986
	Prymnesiophytes	*Pleurochrysis carterae*	+	+	+	"
	Dinoflagellates	*Amphidinium carterae*	+	+	+	"
Light	Diatoms	*Thalassiosira weissflogii*	+		+	Olson *et al.* 1986; Vaulot *et al.* 1986
		Cylindrotheca fusiformis	+		+	Brzezinski *et al.* 1990
		Thalassiosira pseudononana	+		+	"
		Minutocellus polymorphus	+		+	"
		Chaetoceros simplex	+			"
		Chaetoceros muellerii	+		+	"
		Phaeodactylum tricornutum	+			"
	Chlorophytes	*Chlamydomonas reinhardtii*	+			Spudich &. Sager 1980
	Prymnesiophytes	*Pleurochrysis carterae*	+			Olson *et al.* 1986; Vaulot *et al.* 1986
	Dinoflagellates	*Amphidinium carteri*	+			Olson &. Chisholm 1986; Olson *et al.* 1986
	Euglenophytes	*Euglena gracilis*	+			Yee &. Bartholomew 1988
	Cyanobacteria	*Synechococcus* WH8101	+		+	Armbrust *et al.* 1989
		Synechococcus WH8103	+		+	Binder &. Chisholm submitted
		Synechococcus WH7805	+		+	"
		Synechococcus PCC6301	+		+[1]	"
		Synechococcus WH7803	+		+[1]	"
	Prochlorophytes	*Prochlorococcus* strain MED	+	+[2]	+[2]	Vaulot, unpublished data
Nitrogen	Diatoms	*Thalassiosira weissflogii*	+			Olson *et al.* 1986; Vaulot *et al.* 1987
	Prymnesiophytes	*Pleurochrysis carterae*	+			"
	Dinoflagellates	*Amphidinium carteri*	+			Olson *et al.* 1986
	Prochlorophytes	*Prochlorococcus* strain MED	+			Parpais 1994
Phosphorus	Prochlorophytes	*Prochlorococcus* strain MED	+	+[3]	+	Parpais 1994
Silica	Diatoms	*Thalassiosira weissflogii*	+		+[4]	Vaulot *et al.* 1987; Brzezinski *et al.* 1990
		Cylindrotheca fusiformis	+		+	Brzezinski *et al.* 1990
		Thalassiosira pseudononana	+		+	"
		Minutocellus polymorphus	+		+	"
		Chaetoceros simplex	+		+[5]	"
		Chaetoceros muellerii	+		+	"
		Phaeodactylum tricornutum[6]				"
Vitamin B12	Euglenophytes	*Euglena gracilis*		+	?	Lefort-Tran *et al.* 1987

[1]Cells may have more than 2 copies of the chromosome [2] Light limitation does not affect S and G_2. Darkness induces cells to arrest also in S and G_2 [3] Arrest in S is irreversible [4] Only Si starvation, not Si limitation, has effect on G_2 [5] Only Si limitation, not Si starvation, has effect on G_2 [6] Si has no effect on the cell cycle

The interpretation of these observations is that initiation of new replication origins is not synchronous and that the DNA replication cycle and the division cycle are loosely coupled in these genera (Binder and Chisholm submitted).

Regulation of phytoplankton cell cycle by environmental factors

CULTURE STUDIES

In marine phytoplankton, several studies have tried to assess how phytoplankton cell cycle responds to its environment. Is the concept of restriction point (see above) applicable to phytoplankton and, if yes, how do the different stimuli map onto the cell cycle? Two methods can be used to address these questions.

One can **starve** cells for a given factor (*eg* maintain cells in darkness) and locate the portion of the cell cycle where cells accumulate. The reversibility of the arrest can be verified by restoring the missing factor (*eg* put cells back into light). In order to map more precisely the location of the restriction point within the cell cycle, selective inhibitors may be used: for example hydroxyurea that prevents DNA synthesis and blocks S phase cells, allowing distinction between G_1 and G_2 restriction points (*eg* Vaulot *et al.* 1986). Alternatively, one may **limit** cells by decreasing the level of the factor under study without removing it totally (*eg* put cells under low light conditions) and determine the phases which are preferentially lengthened.

Although starvation and limitation experiments yield in general similar answers, it is not always the case. For example Brzezinski *et al.* (1990) found that the diatom *Thalassiosira weissflogii* arrests in G_2 after silica starvation, but does not show any increase in G_2 during silica limitation.

Temperature appears to have the most uniform action of all factors investigated (Table 1). When temperature decreases, all phases are lengthened in equal proportions. This suggests that the effect of lowering the temperature on the cell cycle is a uniform slowing down of biochemical reactions. Although it has not been studied, temperature stress may have a more specific effect, such as the induction of heat shock proteins (Alexandrov 1994).

The effect of **light** on the cell cycle of *Chlamydomonas* was investigated in detail by Spudich and Sager (1980). They found a restriction point located in mid-G_1. They also demonstrated that DCMU, an inhibitor of photosynthetic electron transport, had the same effect than darkness, pleading for a direct coupling between reductant pools and the cell cycle. The existence of a light-dependent restriction point was also demonstrated in the diatom *Thalassiosira pseudonana* (Heath and Spencer 1985) and in the green alga *Chlamydomonas eugametos* (Zachleder and Van den Ende 1992). Recent studies using flow cytometry have demonstrated that light limitation and starvation induce phytoplankton cell arrest in G_1 (Table 1). Two groups of organisms, diatoms and cyanobacteria, present a second control point in G_2. In diatoms, this late control appears to be linked to the light dependence of silica uptake during frustule formation (Vaulot *et al.* 1986; Brzezinski *et al.* 1990). In *Synechococcus*, darkness blocks cell division and prevents the initiation of new replication forks (Mann and Carr 1974; Binder and Chisholm 1990). In *Prochlorococcus*, a genus phylogenetically close to *Synechococcus*, light limitation only affects G_1 but darkness blocks cells in all cell cycle phases including S (Vaulot, unpublished data).

One key question with respect to the effect of light on the cell cycle concerns circadian rhythms. It is well known that phytoplankton division is synchronized to the light:dark cycle, both in cultures and in the natural environment (Chisholm 1981). Many other cellular processes such as photosynthesis or gene expression are also synchronized. Does this synchrony result from the direct entrainment of the cell cycle by light or is the light signal relayed through a cellular clock? Strong evidence for a circadian clock exists in some eukaryotes such as *Euglena* (Edmunds and Adams 1981), but also in some prokaryotes, such as *Synechococcus* (Kondo *et al.* 1993). However, the synchrony observed in most phytoplankton species can be explained without resorting to an endogeneous clock (Vaulot and Chisholm 1987).

In all species investigated to date, **nitrogen** supply is only required in G_1 (Table 1) which suggests either that cells do not require nitrogen once DNA synthesis has started (most likely because protein synthesis is not required during the S phase, Ronning and Lindmo 1983), or that they have a control mechanism that prevents them from entering S, when they are nitrogen-depleted. In striking contrast, the only study dealing with **phosphorus** starvation (done on the prokaryote *Prochlorococcus*, Parpais 1994) revealed that, not only it induced cells to arrest in all cell cycle phases, but also that cells arrested in S could not resume their cycle once resupplied with phosphorus, while those arrested in G_1 and G_2 could. This behaviour may be due to the absence of strong checkpoints in prokaryotes, but comparable work needs to be done in eukaryotes. **Silica** is a requirement specific to diatoms and appears needed both in G_1 and G_2 for most, but not all species (Table 1). The G_1 control is apparently due to a silica requirement for DNA synthesis (Darley and Volcani 1969), while the G_2 control is linked to frustule formation (Volcani 1981).

Natural populations

Only very recently, have natural populations been interrogated with respect to their cell cycle status. Vaulot and Partensky (1992) found that, in the Mediterranean Sea in winter, some surface *Prochlorococcus* populations had very few cells in $S+G_2$. These were found at stations where ambient nitrate concentration was below $0.3\mu M$. Incubating samples with the addition of $1\mu M$

nitrogen (as nitrate or ammonia) induced a large fraction of cells to enter $S+G_2$ after 24 hrs. The cell cycle response decreased with depth, indicating that light also plays a key role in the cell cycle control of these natural populations. In contrast, we observed that, in summer in the Mediterranean Sea, the cell cycle of surface *Synechococcus* populations did not respond to nitrogen addition. It did respond to phosphorus addition even at very low levels of the order of 20nM (Vaulot *et al.*, unpublished). These data suggest an alternation between nitrogen and phosphorus limitation in this region depending on the time of the year.

Using the cell cycle to estimate the growth rate of natural populations

Phytoplankton populations are naturally entrained to the alternating light:dark cycle. In particular, the percentage of dividing cells fluctuates during the day with a maximum occurring in general at night and a minimum during the day (Chisholm 1981). If cells that are dividing or close to division (*eg* with two nuclei) can be discriminated, then one has a handle on estimating the fraction of the population that will divide every day. Gough as early as 1905 used this method to estimate that the dinoflagellate *Ceratium fusus* divided on average once every two days in the English Channel. As will be seen later, the so-called mitotic index method is not as simple as it seems and suffers severe limitations. Recent refinements using cell cycle measurements should make it more useful and easily applicable.

MODELLING OF CELL POPULATIONS

The mitotic index method relies on the formalism of age-structured populations (Nisbet and Gurney 1982). A population with cell concentration $N(t)$, is best expressed in terms of its age distribution $N(t,a)$ such that $N(t,a).da$ is the number of cells between ages a and $a+da$ at time t:

$$\int_0^\infty N(t,a).da = N(t)$$

When the sole source of mortality is cell division, "conservation of cells" can be expressed as:

$$\frac{\partial N}{\partial t} + \frac{\partial N}{\partial a} + \mu(t,a).N = 0 \qquad \textbf{(1)}$$

with the assumption that cells divide into two new-born cells of age 0:

$$N(t,0) = 2\int_0^\infty N(t,a).\mu(t,a).da = 2\mu(t) \qquad \textbf{(2)}$$

$\mu(t,a)$ is the age-dependent division rate and $\mu(t)$, the population division rate such that:

$$\frac{1}{N}\frac{dN}{dt} = \mu(t)$$

If the dynamics of the population do not depend on age but on another cellular variable such as cell size or the concentration of a given protein, it is easy to express equations 1 and 2 as a function of the driving cellular variable (see Vaulot and Chisholm 1987)

As an example, one can solve easily equations 1 and 2 if conditions are constant (*eg* continuous light) and if one assumes that all cells divide at a given age tg, also called the generation time (in fact, this hypothesis is never strictly realized, since there is always some cell-to-cell variability, Vaulot and Chisholm 1987). Then $\mu(t,a) = \delta(a-tg)$, δ being the Dirac function. Using the normalized age density such that:

$$n(t,a) = \frac{N(t,a)}{N(t)}$$

Then simple calculus gives the equation:

$$\frac{\partial n}{\partial t} + \frac{\partial n}{\partial a} = -\mu(t).n(t) \qquad \textbf{(3)}$$

with boundary condition:

$$\boldsymbol{n(0,t) = 2n(t, tg) = 2\mu(t)} \qquad \textbf{(4)}$$

Since the hypothesis was made of constant conditions, then $n(t,a)$ and $\mu(t)$ do not depend on t and the solution to equations 3 and 4 is:

$$\boldsymbol{n(a) = 2\mu \exp(-\mu a)} \qquad \textbf{(5)}$$
$$\boldsymbol{\mu = \ln(2)/tg}$$

Equation 5 can be used to estimate the percentage of cells in each cell cycle phase as a function of the stage duration under steady state conditions (Slater *et al.* 1977). For example, if the duration of G_2 + M is t_{G2+M} and the fraction of cells in G_2 + M is f_{G2+M} then:

$$f_{G2+M} = \int_{t_g - t_{G2+M}}^{t_g} n(a) \cdot da = \exp(\mu . t_{G2+M}) - 1$$

Under non-steady state conditions (*ie* when $n(t,a)$ is a function of both a and t), such as light:dark cycles, equation 1 and its boundary condition, equation 2, are in general not solvable in closed analytical form. Two alternatives can be used to investigate solutions:

1. Monte-Carlo simulations. In this case, each individual cell is followed as it responds to variations in external conditions (*eg* in the dark, cells progress in their cell cycle only if they are in S or in G_2 +M but not if they are in G_1). The population behaviour is reconstructed by summing the behaviour of a statistically significant number of cells.
2. Numerical solution. It is possible, for example, to solve equations 1 and 2 along characteristics (Vaulot and Chisholm 1987; Chang and Carpenter 1990).

THE MITOTIC INDEX METHOD

This method has been formalized by McDuff and Chisholm (1982) and Vaulot (1992). The idea is to transform the curve of the frequency of cells observed in an easy-to-recognize terminal event of the cell cycle into the division rate curve (Fig. 4). Mathematically, this can be expressed as deducing $\mu(t)$ from

$f(t) = \int_{t_g - t_d}^{t_g} n(t,a).da$, where t_d is the duration of a terminal phase of the cell cycle, such as the phase between mitosis and division. Two conditions are necessary for the method to apply. First the phase chosen must be terminated by cell division.

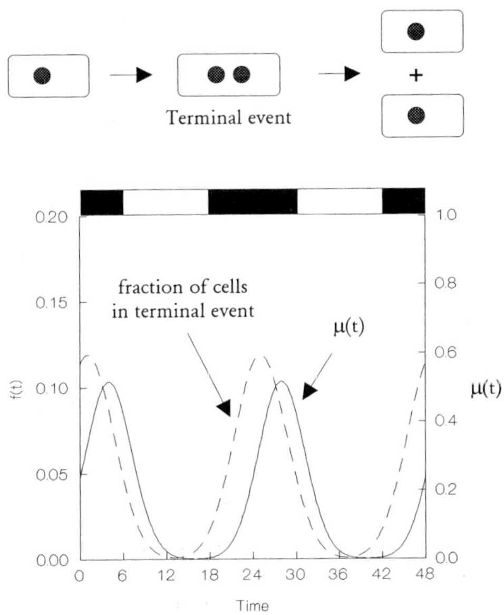

Fig. 4 The principle of the mitotic index method. The fraction of cells in a well-recognizable terminal phase of the cell cycle (*eg* post-mitotic cells with 2 nuclei) is counted during a diel cycle [$f(t)$, dashed curve]. From $f(t)$, the average value of the division rate [$\mu(t)$, continuous curve] is estimated (see text).

Second, its duration must be constant for a given set of conditions. For example if cells are under light:dark conditions, light-born cells must not spend 2 hours between mitosis and division and dark-born cells 8 hours.

Under the condition that t_d is invariant, use of equations 1 and 2 yields (McDuff and Chisholm 1982):

$$\ln[1+f(t)] = \int_{t}^{t+t_d} \mu(\tau)\mathrm{d}\tau \qquad (6)$$

If $f(t)$ is periodic, as observed in nature, one can find an approximate equation :

$$\mu_{\mathrm{p}} \approx \frac{1}{m t_d} \sum_{i=1}^{m} \ln(1+f_i) \qquad \textbf{(7)}$$

where m is the number of samples collected at fixed intervals during a given entraining period, t_p (usually 24 h), and μ_{p} the average growth over the entraining period.

Vaulot (1992) established further that:

$$\mu_{\mathrm{p}} > \frac{1}{t_p} \ln\left(\frac{1+f_{\max}}{1+f_{\min}}\right) = \mu_{\mathbf{fmin\ fmax}} \qquad \textbf{(8)}$$

where $f_{\min}$ and $f_{\max}$ are respectively the maximum and the minimum of the fraction of cells $f(t)$ during a daily cycle. In most case, $f_{\min} = 0$, and then equation 8 reduces to:

$$\mu_{\mathrm{p}} > \frac{1}{t_p} \ln(1+f_{\max}) = \mu_{\mathbf{fmax}} \qquad \textbf{(9)}$$

μ_{fmax} was used by early investigator as an estimate of μ_{p} (Swift *et al.* 1976; Weiler and Chisholm 1976). However, it may severely underestimate the actual division rate. This is easily explained by the fact that only considering the maximum value of $f(t)$ may miss cells that have already left the terminal phase before this maximum or will enter it afterwards. Underestimation increases both as population asynchrony increases and as the duration of the terminal phase chosen is shorter (Vaulot 1992).

In contrast, equation 7 is clearly more accurate and has been used in more recent studies (*eg* Braunwarth and Sommer 1985). The major problem associated with equation 7 lies in the estimation of t_d, which has to be extrapolated from culture work (*eg* Campbell and Carpenter 1986). However, first, the cell cycle of natural populations may be regulated by factors not tested in culture and that induce unexpected change in the duration of t_d. Second, natural populations may be genetically different from cultured populations and exhibit different t_d duration. To make matters worse, in some cases such as the dinoflagellate *Gyrodinium aureolum*, the duration of the post-mitotic phase is not constant but linearly related to the generation time: in such case equation 7 is useless (Videau and Partensky 1990).

A derivative of the mitotic index method, called the frequency of dividing cells method (FDC, Hagström *et al.* 1979), has been used to estimate the division rate of natural assemblages of bacteria. It consists in establishing, usually under controlled conditions, a relation between the fraction of bacterial cells in division (*ie* those exhibiting a septum) and cell division rate. This empirical relationship (that should have the form of equation 7, with f_i constant if cell division is asynchronous) is then used for field populations. As pointed out by Newell *et al.* (1986), the FDC method has very severe flaws. In particular, equation 7 is only valid for a genetically homogeneous population in which all cells behave similarly. This is obviously not the case for natural bacterial populations (*eg* Giovannoni *et al.* 1990).

THE CELL CYCLE METHOD

In order to circumvent the problems of the mitotic index method, and in particular the need to make assumptions on t_d, Carpenter and Chang (1988) came up with the elegant idea of using cell cycle data to compute directly t_d from field data. Instead of one terminal event (*eg* mitosis), they considered two cell cycle events namely S and G_2+M. Using a simple population model, they demonstrated that under periodic conditions (*eg* light:dark), t_{S+G2+M} is equal to twice the difference between $t_{G2+M\max}$ and $t_{S\max}$, the times at which f_S and f_{G2+M} are respectively maximum:

$$t_{S+G2+M} = 2\,(t_{G2+M\max} - t_{S\max})$$

Moreover, t_S and t_{G2+M} can be computed using the following equation, where i and m have the same meaning than in Eq. 7:

$$\frac{t_S}{t_{G2+M}} = \frac{\sum_{i=1}^{m}(1+f_S+f_{G2+M})}{\sum_{i=1}^{m}(1+f_{G2+M})} - 1$$

One can then estimate μ_p from equation 7 using either G_2 +M or S + G_2 + M as the terminal phase (*eg* replacing t_d by t_{G2+M} and f by f_{G2+M}). Moreover, equation 9 can be used to obtain a lower bound to the division rate.

In the original work of Carpenter's group, $f_S(t)$ and $f_{G2+M}(t)$ were determined by microscopic image analysis of cell nuclei stained with the DNA fluorochrome DAPI. The cell cycle method has been validated both using population models (Chang and Carpenter 1990; Chang and Dam 1993) and dinoflagellate cultures (Chang and Carpenter 1988; Antia *et al.* 1990; Chang and Dam 1993). The following remarks apply: 1. The use of a longer terminal event (*ie* S + G_2 + M instead of G_2 + M) gives more accurate estimates. 2. A 2h sampling interval provides a good compromise between a reasonable number of samples and good accuracy. 3. Grazing does not influence the estimate unless it is strongly cell cycle stage specific (*eg* if grazers only prey on the larger cells that are close to division).

Initial field applications using image analysis were realized for large species (diatoms, dinoflagellates) both in coastal (Chang and Carpenter 1991) and oligotrophic waters (Chang and Carpenter 1994). The estimated rates were in good agreement with those obtained from changes in cell number in caged cultures, but much larger than those obtained by the mitotic index method, especially in the case of oligotrophic waters. The failure of the latter method is probably due to the fact that mitosis is too short for these dinoflagellates (Vaulot 1992; Chang and Carpenter 1994).

Recently, flow cytometry has been used to obtain time series of cell cycle distributions of the marine photosynthetic prokaryotes *Prochlorococcus* and *Synechococcus* in several oligotrophic areas including the Pacific (Vaulot *et al.*; Campbell *et al.* unpublished data), Atlantic (Partensky *et al.* unpublished data) and Indian Oceans (Veldhuis *et al.* unpublished data) and the Mediterranean Sea (Vaulot *et al.* unpublished data). Advantages of flow cytometry over image analysis include increased accuracy of cell cycle stage determination, larger number of cells analyzed, and higher speed. These data revealed that these prokaryotes are highly synchronized to the daily light cycle with division rates of the order of 1 doubling per day (0.69 day^{-1}). Such tight synchrony of some of the key photosynthetic producers in oligotrophic waters may echo throughout the whole food web, as suggested by the synchronous division of tintinnides observed off Hawaii by Heinbokel (1987).

Widespread application of this technique should allow oceanographers to obtain highly accurate estimates of division rates of dominant phytoplanktonic populations as well as to better understand how the cell cycle is regulated by the daily light cycle in nature.

Perspectives

As should be now evident from reading this chapter, still very little is known on the mechanisms and regulations of phytoplankton cell cycle. However, cell cycle analysis of both cultured and natural populations is very informative about their environmental status and their growth rate. Therefore basic studies are needed to interpret these data.

In this respect, the use of molecular techniques should be very useful, but almost everything has to be done. For example, virtually nothing is known on the basic cell cycle engine in unicellular algae (see however John *et al.* 1989, 1991; Rodriguez *et al.* 1993; Lin *et al.* 1994). Another area of research is the characterization of genes that are activated or de-activated at specific points in the cell cycle (*eg* Gagné and Guertin 1992), as well on how environmental signals are transduced to the cell cycle. Finally, a fascinating topic is the relationship between the cell cycle and the sexual cycle (Armbrust *et al.* 1990).

Acknowledgements

I wish to thank D. Marie for performing flow cytometry analyses, F. Partensky, L. Campbell, M. Veldhuis, R.J. Olson and S.W. Chisholm for discussions and communication of unpublished data. This work was supported in part by the European Community MAST II project MEDIPELAGOS (MAS2-CT93-0063)

References

Alexandrov VY (1994) Functionnal aspects of cell response to heat shock. In: *International Review of Cytology*, **148:** . Academic Press, Inc, San Diego

Antia AN, Carpenter EJ, Chang J (1990) Species-specific phytoplankton growth rates via diel DNA synthesis cycles. III. Accuracy of growth rate measurement in the dinoflagellate *Prorocentrum minutum*. *Mar Ecol Prog Ser* **63:** 273-279

Armbrust EV, Bowen JD, Olson RJ, Chisholm SW (1989) Effect of light on the cell cycle of a marine *Synechococcus* strain. *Appl Environ Microbiol* **55:** 425-432

Armbrust EV, Chisholm SW, Olson RJ (1990) Role of light and the cell cycle in the induction of spermatogenesis in a centric diatom. *J Phycol* **26:** 470-478

Bhaud Y, Salmon J-M, Soyer-Gobillard M-O (1991) The complex cell cycle of the dinoflagellate protoctist *Crypthecodinium cohnii* as studied *in vivo* and by cytofluorimetry. *J Cell Sci* **100:** 675-682

Binder BJ and Chisholm SW (1990) Relationship between DNA cycle and growth rate in *Synechococcus* sp. strain PCC 6301. *J Bacteriol* **172:** 2313-2319

Binder BJ and Chisholm SW (submitted) Cell cycle regulation in marine *Synechococcus* sp. strains. *Appl Environ Microbiol*

Boucher N, Vaulot D, Partensky F (1991) Flow cytometric determination of phytoplankton DNA in cultures and oceanic populations. *Mar Ecol Prog Ser* **71:** 75-84

Braunwarth C and Sommer U (1985) Analyses of the *in situ* growth rates of Cryptophyceae by use of the mitotic index technique. *Limnol Oceanogr* **30:** 893-897

Bremer H and Churchward G (1994) Control of cyclic chromosome replication in *Escherichia coli*. *Microbiol Rev* **55:** 459-475

Brzezinski MA, Olson RJ, Chisholm SW (1990) Silicon availability and cell cycle progression in marine diatoms. *Mar Ecol Prog Ser* **67:** 83-96

Campbell L and Carpenter EJ (1986) Diel patterns of cell division in marine *Synechococcus* spp. (Cyanobacteria): use of the frequency of dividing cells technique to measure growth rate. *Mar Ecol Prog Ser* **32:** 139-148

Carpenter EJ, Chang J (1988) Species-specific phytoplankton growth rates via diel DNA synthesis cycles. Concept of the method. *Mar Ecol Prog Ser* **43:** 105-111

Cetta CM and Anderson DM (1990) Cell cycle studies of the dinoflagellates *Gonyaulax polyedra* Stein and *Gyrodinium uncatenum* Hulburt during asexual reproduction. *J exp mar Biol Ecol* **135:** 69-84

Chang J and Carpenter EJ (1988) Species-specific phytoplankton growth rates via diel DNA synthesis cycles. II. DNA quantification and model verification in the dinoflagellate *Heterocapsa triquetra*. *Mar Ecol Prog Ser* **44:** 287-296

Chang J and Carpenter EJ (1990) Species-specific phytoplankton growth-gates via diel DNA-synthesis cycles IV. Evaluation of the magnitude of error with computer-simulated cell-populations. *Mar Ecol Prog Ser* **65:** 293-304

Chang J and Carpenter EJ (1991) Species-specific phytoplankton growth rates via diel DNA synthesis cycles. V. Application to natural populations in Long Island Sound. *Mar Ecol Prog Ser* **78:** 115-122

Chang J and Carpenter EJ (1994) Active growth of the oceanic dinoflagellate *Ceratium teres* in the Caribbean and Sargasso Seas estimated by cell cycle analysis. *J Phycol* **30:** 375-381

Chang J and Dam HG (1993) The influence of grazing on the estimation of phytoplankton growth rate via cell cycle analysis: modeling and experimental evidences. *Limnol Oceanogr* **38:** 202-212

Chisholm SW (1981) Temporal patterns of cell division in unicellular algae. In: *Physiological Bases of Phytoplankton Ecology* (Platt T ed) *Canadian Bulletin of Fisheries and Aquatic Sciences*, **210:** 150-181.

Cooper S (1991) *Bacterial growth and division.* Academic Press, Inc., San Diego

Darley WM and Volcani BE (1969) Role of silicon in diatom metabolism. A silicon requirement for deoxyribonucleic acid synthesis in the diatom *Cylindrotheca fusiformis* Reimann and Lewin. *Exp Cell Res* **58:** 334-342

Darzynkiewicz and Crissman HA (eds) (1990) *Methods in cell biology*, 33: Flow Cytometry. Academic Press, inc., San Diego

Donachie WD (1993) The cell cycle of *E coli*. *Ann Rev Microbiol* **47:** 199-230

Donnan L, Carvill EP, Gilliland TJ, John PCL (1985) The cell cycles of *Chlamydomonas* and *Chlorella*. *New Phytol* **99:** 1-40

Edmunds LN and Adams KJ (1981) Clocked cell cycle clocks. *Science* **211:** 1002-1013

Evans T, Rosenthal ET, Youngblom J, Distel D, Hunt T (1983) Cyclin: a protein specified by maternal mRNA in sea urchin eggs that is destroyed at each cleavage division. *Cell* **33:** 389-396

Furnas MJ (1990) *In situ* growth rates of marine phytoplankton: approaches to measurement, community, species growth rates. *J Plankt Res* **12:** 1117-1151

Gagné G and Guertin M (1992) The early genetic response of light in the green unicellular alga *Chlamydomonas eugametos* grown under light/dark cycles involves genes that represent direct responses to light and photosynthesis. *Plant Mol Biol* **18:** 429-445

Giovannoni SJ, Britschgi TB, Moyer CL, Field KG (1990) Genetic diversity in Sargasso Sea bacterioplankton. *Nature, Lond* **345:** 60-63

Gough LH (1905) Report on the plankton of the English Channel in 1903. *Rep N Sea Fish Invest Comm (S Area)* **1:** 325-377

Hagström A, Larsson U, Hörstedt P, Normark S (1979) Frequency of dividing cells, a new approach to the determination of bacterial growth rates in aquatic environments. *Appl Environ Microbiol* **37:** 805-812

Heath MR and Spencer CP (1985) A model of the cell cycle and cell division phasing in a marine diatom. *J Gen Microbiol* **131:** 411-425

Heinbokel JF (1987) Diel periodicities and rates of reproduction in natural populations of tintinnines in the oligotrophic waters off Hawaii, September 1982. *Mar Microb Food Webs* **2:** 1-14

Helmstetter CE and Cooper S (1968) DNA synthesis during the division cycle of rapidly growing *Escherichia coli* B/r. *J Mol Biol* **31:** 507-518

Howard A, Pelc SR (1951) Nuclear incorporation of ^{32}P as demonstrated by autoradiographs. *Exp Cell Res* **2:** 178-187

John PCL, Sek FJ, Lee MG (1989) A homolog of the cell cycle control protein $p34^{cdc2}$ participates in the division cycle of *Chlamydomonas*, and a similar protein is detectable in higher plants and remote taxa. *Plant Cell* **1:** 1185-1193

John PCL, Sek FJ, Hayles J (1991) Association of the plant $p34^{cdc2}$-like protein with $p13^{suc1}$: implications for control of cell division cycles in plants. *Protoplasma* **161:** 70-74

Karentz D (1983) Patterns of DNA synthesis and cell division in marine dinoflagellates. *J Protozool* **30:** 581-588

Kondo T, Strayer CA, Kulkarni RD, Taylor W, Ishiura M, Golden SS, Johnson CA (1993) Circadian rythms in prokaryotes: luciferase as a reporter circadian gene expression in cyanobacteria. *Proc Natl Acad Sci USA* **90:** 5672-5676

Lefort-Tran M, Bre MH, Pouphile M, Manigault P (1987) DNA flow cytometry of control *Euglena* and cell cycle blockade of vitamin B_{12}-starved cells. *Cytometry* **8:** 46-54

Lin S, Chang J, Carpenter EJ (1994) Detection of proliferating cell nuclear antigen analog in four species of marine phytoplankton. *J Phycol* **30:** 449-456

Lohka MJ, Hayes MK, Maller JL (1988) Purification of maturation promoting factor, an intracellular regulator of early mitotic events. *Proc Natl Acad Sci USA* **85:** 3009-3013

Mann NH and Carr NG (1974) Control of macromolecular composition and cell division in the bluegreen alga *Anacystis nidulans*. *J Gen Microbiol* **83:** 399-405

Masui Y and Markert CL (1971) Cytoplasmic control of nuclear behavior during meiotic maturation of frog oocytes. *J Exp Zool* **177:** 129-145

McDuff RE and Chisholm SW (1982) The calculation of in situ growth rates of phytoplankton populations from fractions of cells undergoing mitosis: A clarification. *Limnol Oceanogr* **27:** 783-788

Murray A and Hunt T (1993) *The cell cycle: an introduction.* W.H. Freeman and company, New York

Newell SY, Fallon RD, Tabor PS (1986) Direct microscopy of natural assemblages. In: Bacteria in nature, vol 2 (Pointdexter JS, Leadbetter ER eds) Plenum, NY

Nisbet RM and Gurney WSC (1982) *Modelling of Fluctuating Populations* Wiley, NY

Olson RJ and Chisholm SW (1986) Effect of light and nitrogen limitation on the cell cycle of the dinoflagellate *Amphidinium carteri*. *J Plankton Res* **8:** 785-793

Olson RJ, Frankel SL, Chisholm SW, Shapiro HM (1983) An inexpensive flow cytometer for the analysis of fluorescence signals in phytoplankton: chlorophyll and DNA distributions. *J exp mar Biol Ecol* **68:** 129-144

Olson RJ, Vaulot D, Chisholm SW (1986) Effect of environmental stresses on the cell cycle of two marine phytoplankton species. *Plant Physiol* **80:** 918-925

Parpais J (1994) Controle du cycle cellulaire de *Prochlorococcus* par le phosphore. DEA Dissertation, Université Pierre et Marie Curie, Paris.

Robinow C and Kellenberger E (1994) The bacterial nucleoid revisited. *Microbiol Rev* **58:** 211-232

Rodriguez M, Cho JW, Sauer HW, Rizzo PJ (1993) Evidence for the presence of a *cdc*2-like protein kinase in the dinoflagellate *Crypthecodinium cohnii*. *J Euk Microbiol* **40:** 91-96

Ronning O and Lindmo T (1983) Progress through G_1 and S in relation to net protein accumulation in human NHIK 3025 cells. *Exp Cell Res* **144:** 171-179

Slater ML, Sharrow SO, Gart JJ (1977) Cell cycle of *Saccharomyces cerevisiae* in populations growing at different rates. *Proc Natl Acad Sci USA* **74:** 3850-3854

Spudich J and Sager R (1980) Regulation of *Chlamydomonas* cell cycle by light and dark. *J Cell Biol* **85:** 136-145

Swift E, Stuart M, Meunier V (1976) The *in situ* growth rates of some deep-living oceanic dinoflagellates: *Pyrocystis fusiformis* and *Pyrocystis noctiluca*. *Limnol Oceanogr* **21:** 418-426

Vaulot D (1992) Estimate of phytoplankton division rates by the mitotic index method: the *fmax* approach revisited. *Limnol Oceanogr* **37:** 644-649

Vaulot D and Chisholm SW (1987) A simple model of the growth of phytoplankton populations in light:dark cycles. *J Plankton Res* **9:** 345-366

Vaulot D and Partensky F (1992) Cell cycle distributions of prochlorophytes in the North Western Mediterranean Sea. *Deep-Sea Res* **39:** 727-742

Vaulot D, Olson RJ, Chisholm SW (1986) Light and dark control of the cell cycle in two phytoplankton species. *Exp Cell Res* **167:** 38-52

Vaulot D, Olson RJ, Merkel SM, Chisholm SW (1987) Cell cycle response to nutrient starvation in two marine phytoplankton species. *Mar Biol* **95:** 625-630

Videau C and Partensky F (1990) Variability in the growth characteristics of *Gymnodinium* cf. *nagasakiense* (Dinophyceae) and its consequences on the determination of *in situ* growth rates. *J exp mar Biol Ecol* **142:** 169-182

Volcani BE (1981) Cell wall formation in diatoms: morphogenesis and biochemestry. In: *Silicon and siliceous structures in biological systems.* Simpson TL, Volcani BE (eds) Springer-Verlag, New York.

Weiler CS and Chisholm SW (1976) Phased cell division in natural populations of marine dinoflagellates from ship-board cultures. *J exp mar Biol Ecol* **25:** 239-247

Yee MC and Bartholomew JC (1988) Light regulation of the cell cycle in *Euglena gracilis bacillaris*. *Cytometry* **9:** 387-393

Zachleder V and Van den Ende H (1992) Cell cycle events in the green alga *Chlamydomonas eugametos* and their control by environmental factors. *J Cell Sci* **102:** 469-474

Response of Photosynthetic Microorganisms to Changing Ambient Concentration of CO_2

Michal Ronen-Tarazi, Rakefet Schwarz, Anne Bouevitch, Judy Lieman-Hurwitz, Jonathan Erez and Aaron Kaplan
The Hebrew University of Jerusalem,
91904 Jerusalem,
Israel

Abstract
In this manuscript we briefly describe three aspects of our recent studies on the response of cyanobacteria to changes in ambient level of carbon dioxide. We identified several genes responding and used one of them to initiate study on the regulation of gene expression by CO_2. We also developed a method for the isolation of new mutants which demand high-CO_2 for growth and for the identification of the genes involved. These cells are being used for the elucidation of the biotic and environmental factors affecting the fractionation of stable carbon isotopes, a parameter widely used to assess the level of CO_2 during geological periods.

Introduction

The rising concentration of CO_2 in the atmosphere, consequent on the man-made alteration of the global carbon cycle by increasing utilization of fossil fuel and deforestation, may lead to serious climatic changes. It is therefore one of the most important topics in current environmental research. Photosynthetic assimilation of CO_2 by aquatic organisms followed by precipitation of organic and inorganic carbon in the ocean floor is an important mechanism of CO_2 removal from the atmosphere. Nevertheless, very little information is as yet available on the physiological and genetical processes involved in the response of aquatic organisms to changing ambient CO_2.

Photosynthetic microorganisms are capable of adapting to a wide range of ambient CO_2 concentrations. When the cells are transferred from high to low concentration of CO_2, they undergo a light-dependent syndrome of changes at various cellular levels leading to adaptation to the new ambient CO_2 concentration. As a consequence, the apparent photosynthetic affinity for external

NATO ASI Series, Vol. G 38
Molecular Ecology of Aquatic Microbes
Edited by I. Joint

inorganic carbon (Ci) is 10 to 20-fold higher in low-CO_2 grown cells compared with cells exposed to a high level of CO_2. An important aspect of the adaptation process, which has been intensively studied, is the induction of an inorganic carbon-concentrating mechanism (CCM). This mechanism enables the cells to actively accumulate Ci internally. The internal Ci pool may reach values higher than 50mM and accumulation ratios (*ie* the internal/external Ci concentrations) higher than 1000 have been reported. The activity of the CCM is of particular significance in the lower range of external CO_2 concentrations. It enables these organisms to overcome the very large difference between the concentration of CO_2 in their bulk medium and the $K_{½}$ of their carboxylating enzyme, ribulose 1,5-bisphosphate carboxylase/oxygenase (RUBISCO). The elevated internal CO_2 concentration at the site of RUBISCO raises the efficiency of photosynthetic utilization of Ci and inhibits photorespiration. The improved availability of CO_2, consequent on the activity of the CCM, may also results in reduced photoinhibitory damage to cells exposed to high light intensity.

In this manuscript we shall briefly present some of our recent studies on the means by which cyanobacteria respond to changes in the ambient concentration of CO_2. We shall focus on the regulation of gene expression by CO_2, novel CO_2 dependent mutants obtained with the aid of an inactivation library; and fractionation of stable carbon isotopes, as affected by the growth conditions. The reader is referred to Kaplan *et al.* 1991; Badger and Price, 1992; Ogawa, 1993; Kaplan *et al.* 1994; Goericke *et al.* 1994 for recent reviews and literature citations.

CO_2-dependent gene expression

As indicated above, very little information is available on the mechanisms whereby gene expression in cyanobacteria is regulated by the concentration of CO_2. Several experimental approaches were, nevertheless, taken in an attempt to identify CO_2-responding genes. CO_2-dependent promoters were identified in *Synechococcus* sp. strain PCC 7942 with the aid of *lacZ* as a reporter gene (Scanlan *et al.* 1990) but the nature and role of the relevant genes were not clarified.

The expression of several genes in *Synechococcus* sp. strain PCC 7942 is strongly affected by the concentration of CO_2 in the environment (Fig. 1). Transcripts originating from *purK* (encoding subunit II of phosphoribosyl aminoimidazole carboxylase involved in purine biosynthesis, Schwarz *et al.* 1992) and *cmpA* (encoding a cytoplasmic-membrane-located 42 kDa polypeptide which accumulate following treatment with low CO_2, (Omata *et al.* 1990) or high light intensity -Reddy *et al.* 1989) were only observed when the cells were exposed to low CO_2. Higher level of transcripts following transition from high to low CO_2 has been noted in the cases of *ndhB* (encoding subunit II of NADH dehydrogenase, Marco *et al.* 1993), *rbc* (encoding the large and small subunits of RUBISCO) and the operon *ccmN-ccmO* (involved in the correct organization of the carboxysomes and in the regulation of gene expression under low CO_2, see Kaplan *et al.* 1994).

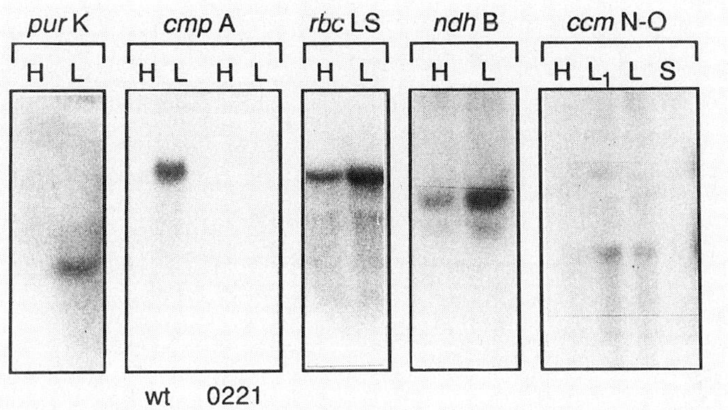

Fig. 1. Northern analyses of RNA isolated from high (H) CO_2 grown *Synechococcus* sp. strain PCC 7942 (WT) and probed with fragments bearing parts of *purK*, *cmpA*, *rbcLS*, *ndhB* and the operon *ccmN-O*. In the cases of *purK*, *rbcLS* and *ndhB*, the RNA was also isolated from cells exposed to low (L) CO_2 for 60-90 min. In the case of the operon *ccmN-O* the cells were exposed to 60 min. (L1) or 12 hours of low CO_2 (L) or non-aerated standing culture for 12 hours (S). Hybridizations with the probe originating from *cmpA* were also carried out with RNA isolated from mutant 0221 exposed to low CO_2 for 12 hours.

We selected *cmpA* as a model system to study the mechanism of regulation of gene expression by ambient CO_2 for three reasons: i) Transcripts originating from *cmpA* were undetectable in high-CO_2-grown cells but accumulated to a very high level following short exposure to low-CO_2 (see Fig. 1): ii) A mutant, O221, which does not accumulate the 42 kDa polypeptide (the gene product of *cmpA*, Schwarz *et al.* 1988) is available; the lesion in this mutant was mapped as a substitution of one nucleotide in *ccmN* located upstream of *rbc* (Friedberg *et al.* 1989), *ie* in a different genomic region than *cmpA*: iii) Inactivation of *cmpA* did not result in a selectable phenotype (Omata *et al.* 1990) and, therefore, it is possible to modify it. Modification of other CO_2-dependent genes such as *purK* resulted in a mutant which demanded high CO_2 for growth. Under these conditions it is difficult to test the response to low concentrations of CO_2. The genomic region upstream of *cmpA* is being analyzed by fusing different segments from it to a promoter-less *cat* (encoding chloramphenicol-resistance, Cm^r) used as a reporter gene and to a cartridge conferring kanamycin-resistance (Km^r) as a transformation selection. The constructs were inserted in a genomic site downstream of *rbc* in *Synechococcus* sp. PCC 7942 (see Fig. 2). Insertion of the Km^r cartridge in this site did not result in a selectable phenotype, other than Km^r.

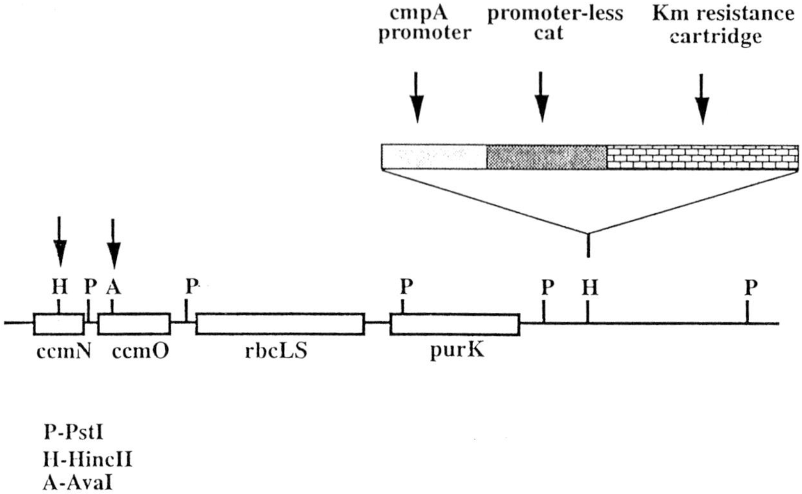

Fig. 2. The construct used for the promoter analyses and the genomic site where they were inserted in *Synechococcus* sp. PCC 7942. The *Hinc*II and *Ava*I sites where the cartridge was inserted in mutants ccmN/spc and ccmO/spc are indicated.

When the 700 or 380 bp (designated L and M, respectively, in Fig. 3 and 4) upstream from the coding region of *cmpA* were fused to *cat* the cells were able to grow in the presence of chloramphenicol only under low CO_2 (mixture of 1:1 air and CO_2-free air), not under high (5% CO_2 in air) CO_2. On the other hand, when the 150 bp (designated S in Fig. 3 and 4) upstream of the 5' initiating codon of *cmpA* were fused to *cat*, the cells were able to grow in the presence of chloramphenicol under both low and high CO_2 (Fig. 3). The simplest interpretation of these data was that the region between 380 and 150 bp upstream of the coding region of *cmpA* contains elements which suppress transcription under high CO_2. Removal of these elements enabled growth in the presence of chloramphenicol under these conditions. In order to draw a firm conclusion, however, it is essential to determine the level of the gene product of the reporter gene.

A commercially available ELISA kit was used to determine the level of chloramphenicol acetyl transferase (CAT). The results (Fig. 4) clearly show that the level of CAT is strongly affected by the length of the *cmpA* promoter segment fused to *cat* and by the ambient concentration of CO_2.

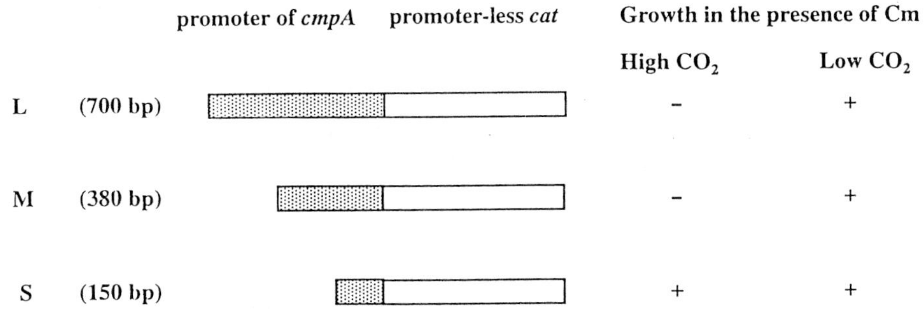

Fig. 3. The effect of the length of the promoter region of *cmpA* inserted in *Synechococcus* sp. PCC 7942 (see Fig. 2) on their ability to grow in the presence of chloramphenicol under high or low CO_2.

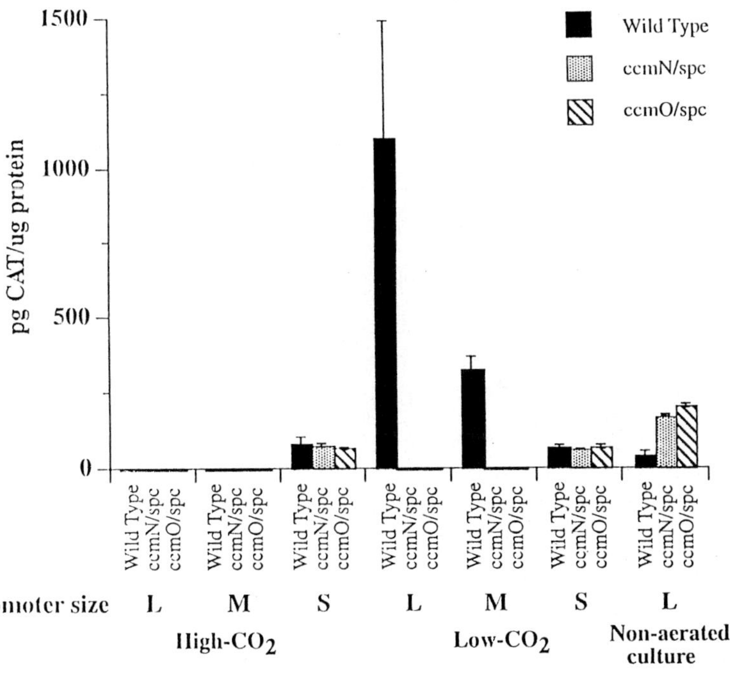

Fig. 4. The effect of the length of the *cmpA* promoter fused to the promoter-less *cat* and the growth conditions on the level of chloramphenicol acetyl transferase (CAT). High-CO_2-grown *Synechococcus* sp. PCC 7942 and the mutants ccmN/spc and ccmO/spc thereof, all bearing the constructs shown in Fig. 2 were exposed to low CO_2 or to non-aerated conditions for 12 hours.

Under low CO_2, the highest level of CAT was obtained with the longest promoter (L) and the level of CAT gradually decreased in the cells possessing shorter promoter regions. This effect of promoter length on the expression of CAT was not detected in the growth experiments (Fig. 3). In agreement with these experiments, we did not detect a significant level of CAT when the cells were exposed to a high level of CO_2, apart from the case where *cat* was fused to the shortest promoter. The data also indicated that a low level of CAT (such as present in cells bearing the short promoter, S) is sufficient to enable growth in the presence of chloramphenicol and demonstrated the need to quantify the level of CAT.

Mutants where either *ccmN* or *ccmO*, located upstream of *rbc* (see Fig. 2), were inactivated by the insertion of a cartridge conferring spectinomycin-resistance (designated ccmN/spc and ccmO/spc in Fig. 4) were also used in these experiments. These mutants demand high CO_2 for growth like their counterparts, mutants O221 (Friedberg *et al.* 1989) and N1 (Marco *et al.* 1994) where these genes were modified, respectively. The constructs bearing the different fragments from the *cmpA* promoter (L, M, and S) fused to *cat* and to the Km^r cartridge were inserted in these mutants in the same *Hinc*II site used in the case of the wild type (see Fig. 2). When exposed to high CO_2, the expression of *cat* in these mutants was similar to that of the wild type bearing the same constructs. On the other hand, when exposed to low CO_2 for 12 hours the mutants bearing L and M did not express CAT whereas those bearing S expressed it like the wild type bearing the same construct. Exposure of the cells bearing the L promoter to "non-aerated conditions", a treatment under which *Synechococcus* UTEX 625 exhibits typical physiological characters (see McKay *et al.*, 1992; Espie and Kandasamy. 1992), resulted in a severe inhibition of CAT expression in the wild type but significant stimulation of CAT expression in the mutants. Northern analyses (not shown) demonstrated that these differences between the wild type and the mutants were expressed at the level of transcription.

These experiments indicated the presence of positive and negative regulatory elements in the promoter region of *cmpA* and the significance of the gene product of *ccmN-ccmO* in the regulation of gene expression by CO_2. It is important to note that these genes form an operon (see Fig. 1). It is not surprising, therefore, that the mutants where they were inactivated gave identical results.

Isolation of mutants with the aid of an inactivation library

Chemical mutagenesis and transposon inactivations in the genomic region of *rbc* of *Synechococcus* sp. PCC 7942 resulted in high-CO_2-requiring mutants and enabled the identification of some of the genes involved in their ability to grow under low CO_2 (Kaplan *et al.* 1994). However, other relevant genes such as those directly involved in Ci uptake were not identified.

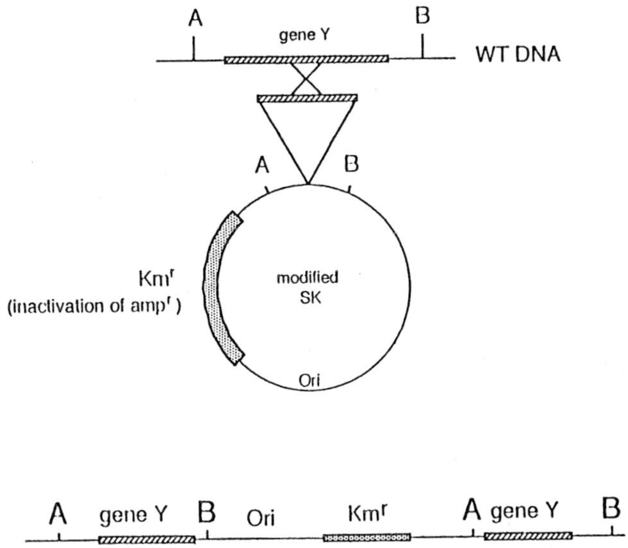

Fig. 5. Schematic presentation of the inactivation library used to isolate mutants of *Synechococcus* sp. PCC 7942 where gene Y (involved in the ability to grow under low Ci) was inactivated, and the organization of the genomic DNA in the mutant (see text for detailed explanation).

We are using an "inactivation library" in order to isolate new mutants and identify the relevant genes (see the scheme in Fig. 5, and Dolganov and Grossman 1993). The library was constructed by complete digestion of the genomic DNA of *Synechococcus* sp. PCC 7942 with *Taq*I (which cleaves the DNA very frequently). The fragments were ligated within the *Acc*I site of a modified Bluescript SK plasmid where we inactivated the gene conferring ampicillin resistance by the insertion of a gene conferring Km^r). Since the genomic fragments ligated in the vector are short and hence, in most cases do not cover the entire length of the gene, single crossover recombination events between the genomic DNA and the homologous fragment in the vector is likely to inactivate the respective gene.

Several high-CO_2-requiring mutants (including IL-3, IL-7 and IL-11, presented here) were isolated as follows: After the transfection with the inactivation library, the cultures of *Synechococcus* were grown under high CO_2 in the presence of kanamycin. The Km^r cells obtained were transferred to low CO_2 for 5 hours of adaptation followed by ampicillin treatment. Cells capable of adapting and dividing under low CO_2 were killed by the ampicillin treatment whereas mutants which were unable to divide under low CO_2 survived, and were isolated following the removal of ampicillin. Basically, this procedure may be applied for the isolation of various mutants and identification of the relevant genes, if the ampicillin treatment, *ie* cessation of growth of the mutant under the selection conditions, is applicable.

Physiological characterization of the high-CO_2-requiring mutants is in progress. When grown in the presence of high CO_2, some of the mutants exhibited normal ability to take up CO_2. When exposed to low CO_2, they failed to adapt to the new conditions by, for instance, raising the uptake of CO_2, and even more so of bicarbonate, as does the wild type (Table 1).

Table 1. Uptake of inorganic carbon by *Synechococcus* sp. PCC 7942 and mutants obtained with the aid of the inactivation library (see Fig. 5). The cells were exposed to low Ci for 10 hours were provided with ^{14}Ci as carbon dioxide or bicarbonate for 10 sec. Uptake is expressed in nmole Ci μg $Chl^{-1}h^{-1}$.

	CO_2		HCO_3^-	
	High	Low	High	Low
WT	31.6	53.9	30.9	53.9
IL-3	16.8	38.9	30.6	53.9
IL7	18.5	30.5	53.9	53.9
IL-11	20.3	37.4	53.9	53.9

A major advantage of the method whereby these mutants were isolated is that the relevant genes are "tagged" with the vector and Km^r sequences. Fig. 5 shows the organization of the relevant genomic region in the mutants. In order to isolate and analyze the genomic region flanking the inserted plasmid in the different mutants, *ie* the inactivated gene(s) in the respective cases, DNA from the mutants was isolated and cleaved with enzymes which digest the vector at either end (marked A and B in Fig. 5).

Self-ligation followed by transformation to competent cells of *E. coli* and selection on kanamycin resulted in the isolation of plasmids bearing the relevant genomic fragments and hence in direct cloning of the relevant genes. Sequence analysis of the inserts is in progress. Northern analyses (Fig. 6) demonstrated that the respective genes in these mutants are transcribed and that the level of transcript in some of them depends on the ambient concentration of CO_2.

Fractionation of stable carbon isotopes in *Synechococcus*

This fractionation (measured as $\delta^{13}C$ difference between the Ci and organic matter) is being used as a natural tracer to study past and present changes in the global carbon cycling (see Jasper and Hayes 1990; Raven 1991; Freeman and Hayes 1992; Guy *et al.* 1993 and citations therein). The fractionation results

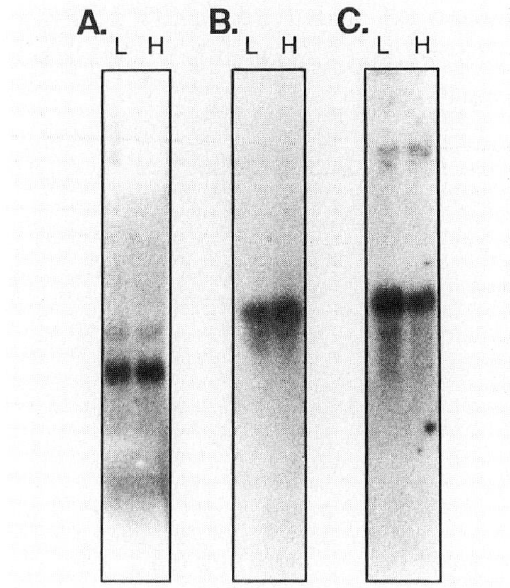

Fig. 6. Northern analyses of RNA isolated from high-Ci-grown (H) *Synechococcus* sp. PCC 7942 and cells exposed to low Ci for 60 min. (L), probed with fragments bearing parts of the inserts used to raise mutants IL-3 (A), IL-7 (B) and IL-11 (C).

mainly from the activity of RUBISCO which discriminates against $^{13}CO_2$ (see Guy *et al.* 1993) leading to a lower content of ^{13}C in the photosynthetic products than in the bulk CO_2. Changes in the composition of atmospheric CO_2 are recorded in fossil organic matter. However, the application of carbon isotopes as a tool to assess changes in the global carbon cycle is hampered by our poor understanding of the biological processes involved in carbon isotope fractionation. There are noticeable variations in the $\delta^{13}C$ values observed in different environmental conditions and habitats as well as between organisms representing different taxonomic groups. Studies by Rau *et al.* (1989) indicated a strong correlation between the concentration of dissolved CO_2 and the extent of fractionation. The mechanisms leading to these variations, however, are poorly understood (see Goericke *et al.* 1994).

Our working hypothesis is that the $\delta^{13}C$ values observed in the organic matter are strongly affected by the flux balance of Ci associated with the activity of the CCM in photosynthetic microorganisms. Environmental and biotic factors which differentially affect the ratio between CO_2 and HCO_3^- uptake or between the unidirectional Ci influx and photosynthetic utilization of Ci (and hence efflux) will have a considerable effect on the observed fractionation. This hypothesis is being examined by measuring the activity of the CCM (in wild type and mutants impaired in various aspects of the CCM) and the carbon isotope fractionation as affected by the growth conditions.

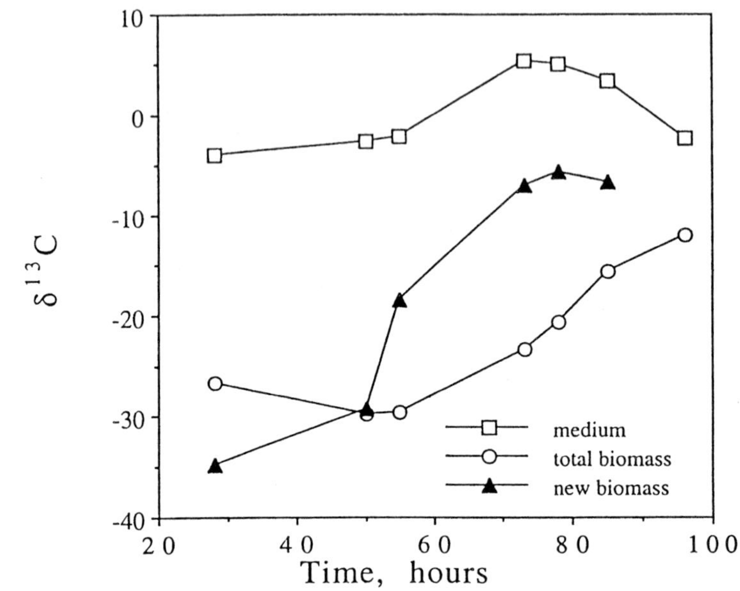

Fig. 7. The $\delta^{13}C$ in the organic matter and in the medium of high-Ci-grown *Synechococcus* sp. strain PCC 7942 cells aerated with air time zero. The cells were first placed in medium BG11, pH8.0 and 5mM Ci. Samples were withdrawn and the amount of organic carbon accumulating, the alkalinity of the medium (not shown) and the fractionation of the organic matter and the medium were determined. The fractionation of the new organic carbon was calculated from the changes with time of the amount of organic carbon and its isotopic composition.

Fig. 7 presents one example where we followed the $\delta^{13}C$ in the organic carbon and in the medium as affected by time in a batch culture of *Synechococcus* sp. PCC 7942. The calculated $\delta^{13}C$ value of the newly formed organic carbon increased dramatically following the adaptation of the cells to low CO_2 and with the rising cell density in the culture. Simultaneous measurements of the CCM activity and the flux balance of Ci will provide important information on the effect of the environmental conditions, including the concentration of CO_2, on the $\delta^{13}C$.

Acknowledgements
This research was supported by grants from the USA-Israel Binational Science Foundation (BSF), Jerusalem, Israel and the New Energy and Industrial Technology Development Organization (NEDO), Japan.

References

Badger MR and Price GD (1992) The CO_2 concentrating mechanism of cyanobacteria and green algae. *Physiol Plant* **84:** 606-615

Dolganov N and Grossman AR (1993) Insertional inactivation of genes to isolate mutants of *Synechococcus* sp. strain PCC 7942: Isolation of filamentous strains. *J Bacteriol* **175:** 7644-7651

Espie GS and Kandasamy RA (1992) Na^+-independent HCO_3^- transport and accumulation in the cyanobacterium *Synechococcus* UTEX 625. *Plant Physiol* **98:** 560-568

Freeman KH and Hayes JM (1992) Fractionation of carbon isotopes by phytoplankton and estimates of ancient CO_2 levels. *Global Biogeochem Cycles* **6:** 185-198

Friedberg D, Kaplan A, Ariel R, Kessel M, Seijffers J (1989) The 5' flanking region of the gene encoding the large subunit of ribulose-1,5-bisphosphate carboxylase/oxygenase is crucial for growth of the cyanobacterium *Synechococcus* PCC7942 in air level of CO_2. *J Bacteriol* **171:** 6069-6076

Goericke R, Montoya JP, Fry B (1994) Physiology of isotopic fractionation in algae and cyanobacteria. In: *Stable isotopes in ecology and environmental science* Lajtha K, Michener RH (eds.). Blackwell Scientific Pub. Oxford, pp 187-221

Guy RD, Fogel ML, Berry JA (1993) Photosynthetic fractionation of the stable isotopes of oxygen and carbon. *Plant Physiol* **101:** 37-47

Jasper JP and Hayes JM (1990) A carbon isotope record of CO_2 levels during the late quaternary. *Nature* **347:** 462-464

Kaplan A, Schwarz R, Lieman-Hurwitz J, Reinhold L (1991) Physiological and molecular aspects of the inorganic carbon concentrating mechanism in cyanobacteria. *Plant Physiol* **97:** 851-855

Kaplan A, Schwarz R, Lieman-Hurwitz J, Reinhold L (1994) Physiological and molecular studies on the response of cyanobacteria to changes in the ambient inorganic carbon concentration. In: *The Molecular Biology of the cyanobacteria*, Bryant D (ed.) Kluwer Academic Pub., Dordrecht, The Netherlands (in press).

Marco M, Ohad N, Schwarz R, Lieman-Hurwitz J, Gabay C, Kaplan A (1993) High CO_2 concentration alleviates the block in photosynthetic electron transport in an *ndhB*-inactivated mutant of *Synechococcus* sp. PCC 7942. *Plant Physiol* **101:** 1047-1053

Marco E, Martinez I, Ronen-Tarazi M, Orus MI, Kaplan A (1994) Inactivation of *ccmO* in *Synechococcus* sp. strain PCC7942 results in a mutant requiring high level of CO_2. *App Environ Microbiol* **60:** 1018-1020

McKay RML, Gibbs SP, Espie GS (1992) Effect of dissolved inorganic carbon on the expression of carboxysomes, localization of RUBISCO and the mode of inorganic carbon transport in cells of the cyanobacterium *Synechococcus* UTEX 625. *Arch Microbiol* **159:** 21-29

Ogawa T (1993) Molecular analysis of the CO_2-concentrating mechanism in cyanobacteria. In: *Photosynthetic Responses to the Environment*, Yamamoto H and, Smith C (eds.) American Society of Plant Physiologist Series pp 113-125

Omata T, Carlson TJ, Ogawa T, Pierce J (1990) Sequencing and modification of the gene encoding the 42-kilodalton protein in the cytoplasmic membrane of *Synechococcus* PCC 7942. *Plant Physiol* **93:** 305-311

Rau GH, Takahashi T, Des Marais DJ (1989) latitudinal variations in plankton $\delta^{13}C$: Implications for CO_2 and productivity in past oceans. *Nature* **341:** 516-518

Raven JA (1991) Implications of inorganic C utilization: ecology, evolution and geochemistry. *Can J Bot* **69:** 908-924

Reddy KJ, Masamoto K, Sherman DM, Sherman LA (1989) DNA sequence and regulation of the gene (*cpbA*) encoding the 42-kilodalton cytoplasmic membrane carotenoprotein of the cyanobacterium sp. strain PCC 7942. *J Bacteriol* **171:** 3486-3493

Scanlan DJ, Bloye SA, Mann NH, Hodgson DA, Carr NG (1990) Construction of *lacZ* promoter probe vectors for use in *Synechococcus*: Application to the identification of CO_2- regulated promoters. *Gene* **90:** 43-49

Schwarz R, Friedberg D, Reinhold L, Kaplan A (1988) Is there a role for the 42kDa polypeptide in inorganic carbon uptake by cyanobacteria?. *Plant Physiol* **88:** 284-288

Schwarz R, Lieman-Hurwitz J, Hassidim M, Kaplan A (1992) Phenotypic Complementation of High-CO_2-Requiring Mutants of the Cyanobacterium *Synechococcus* sp. strain PCC 7942 by Inosine 5'-Monophosphate. *Plant Physiol* **100:** 1987-1993

Nitrogen Fixation in the Sea: Why Only *Trichodesmium*?

Jonathan P. Zehr
Biology Department
Rensselaer Polytechnic Institute
Troy, New York 12180
USA

Abstract
The relative importance of different nutrients in limiting primary production in the sea continues to be the subject of debate, but it is clear that several nutrients are often in short supply in many regions of the oceans. Nitrogen fixation capabilities should provide an ecological advantage to microorganisms in the oceanic environment, regardless of the primary nutrient limiting productivity. *Trichodesmium* is a filamentous nonheterocystous nitrogen-fixing cyanobacterium which is a conspicuous component of tropical and subtropical oceans, and appears to play a major role in carbon and nitrogen-fixation in regions where it is found. It is not intuitively obvious which characteristics of *Trichodesmium* confer an ecological advantage such that it is the predominant organism to exploit nitrogen fixation as a mechanism to obtain nitrogen in nitrogen-deficient oligotrophic oceans. Furthermore, there still remains the question of whether *Trichodesmium* is truly one of only a few species to fix nitrogen, or whether other nitrogen fixing organisms exist in the open ocean, but have yet to be cultivated. Molecular approaches have provided a way to examine both of these issues: what is the distribution of nitrogen fixing microorganisms in the marine environment, and what are the molecular and biochemical features that determine the ecological success of *Trichodesmium* in the open ocean environment?

Introduction
Early studies recognized that the open ocean environment is a nutrient-poor environment, and that the productivity of the oceans relies upon fluxes and cycling of the elements composing biological material (Redfield 1958). In particular, the concentrations of the major nutrients nitrogen (N) and phosphorus (P) are in short supply (Redfield *et al.* 1963, Thomas 1970), as are some micronutrients such as iron (Fe) (Martin *et al.* 1991). The issues of how these

NATO ASI Series, Vol. G 38
Molecular Ecology of Aquatic Microbes
Edited by I. Joint

nutrients regulate oceanic primary production are central to understanding the role of marine carbon fixation in relation to global carbon dynamics, yet we still have an incomplete understanding of how nutrients limit growth in the sea (Cullen *et al.* 1991). Consideration of N as a primary limiting nutrient raises a perplexing problem. In contrast to the biogeochemical cycle of phosphorus, the N cycle includes a gaseous phase (N_2) which is a product of denitrification, and is consumed or fixed into biologically available ammonium through nitrogen (N_2) fixation. N_2 is by far the most abundant form of N on earth, and is present in concentrations at least two orders of magnitude greater than any other form in the oceans (Sharp 1983). Therefore, the Earth's atmosphere provides a large potential reservoir of N for productivity in the marine environment. Biological N_2 fixation can be performed by diverse groups of prokaryotic organisms, representatives of which are numerous in the marine environment (Carpenter 1983, Capone 1983, Paerl 1990). These facts have provided the basis for arguments that N should not be the factor which ultimately limits productivity in the ocean, since N limiting conditions should select for N_2 fixing organisms which would make up for deficits in fixed N, and drive the system to limitation by another nutrient (Smith 1984, Redfield 1958). The issue of whether or not N_2 fixation occurs becomes critical in discussions based on the concept of "new" and "regenerated" N (Dugdale and Goering 1967), since N_2 fixation is one potential source of "new" N. In general, measurements of N_2 fixation rates, and isolation of N_2 fixing organisms indicates that N_2 fixation is not quantitatively important, but this poses a contradiction to the argument that N_2 fixation should ameliorate N limitation in the sea. This paradox is made even more confusing by the fact that *Trichodesmium*, a cyanobacterium common to tropical and subtropical waters, is present in relatively high abundance and fixes N_2 (Carpenter 1983, Carpenter and Romans 1991). So why are there not more N_2 fixing microorganisms in the sea?

Trichodesmium is a filamentous nonheterocystous cyanobacterium which is a visually and functionally prominent feature of tropical and subtropical oceans (Carpenter 1983). It is a curious organism from a number of perspectives, and often has been described as an enigma or a paradox, largely because of its ability to fix N_2 aerobically, while photosynthesizing, without an obvious mechanism for separating the O_2-sensitive N_2 fixation reactions from the O_2 evolved through photosynthesis. *Trichodesmium* is abundant and important in the primary productivity of the environments in which it occurs (Carpenter and Romans 1991), and seemingly is the only, or at least the most abundant, free-living N_2 fixing organism in tropical and subtropical oceans. So beyond the basic biological question of simultaneous N_2 fixation and oxygenic photosynthesis lies an equally perplexing and nagging ecological question: What makes *Trichodesmium* the dominant N_2 fixing organism in subtropical and tropical seas?

This broad ecological question involves an assumption and a hypothesis. First it is assumed that indeed *Trichodesmium* is the only free-living N_2 fixing organism present and that we have not missed other potentially important N_2 fixing species because of our inability to culture isolates or measure N_2 fixation associated with other organisms. Second, it is hypothesized that *Trichodesmium*, as a N_2 fixing organism, has special adaptations or features that allow it to exploit the low N environment of the oligotrophic oceans, whereas other less-suitably adapted N_2 fixing species fail.

Rapid technological advances in the biological sciences within the past two decades have provided the basis for an integrated understanding of the way organisms function and respond to their environment. Molecular characterization of genetic material and biochemical, biophysical and physiological characterization of organism structure and function make it possible to link genetic potential with phenotypic expression. In this chapter, the molecular biology and biochemistry of *Trichodesmium* will be discussed with respect to the factors that limit N_2 fixation in the sea in order to evaluate our progress in addressing the issue: Why only *Trichodesmium*?

The nitrogen fixation appatatus from an ecological perspective

The N_2 fixation apparatus consists of a set of proteins including the structural components of the enzyme nitrogenase, proteins involved in processing or modifying the structural components, electron carriers and regulatory proteins. The reaction catalyzed by nitrogenase requires large amounts of ATP and reducing power since the N_2 molecule is very unreactive (Postgate 1987). For this reason the N_2 fixation process is not a general feature of microbial metabolism, but is restricted to certain taxa who tightly regulate nitrogenase activity in order not to waste energy when N_2 fixing conditions do not exist. On the other hand, N_2 fixation genes are found in representatives of diverse prokaryotic taxa, from cyanobacteria to archaeobacteria, including representatives of the α, β, γ, and δ subdivisions of the proteobacteria (Young 1992).

The dispersion of N_2 fixation genes throughout the prokaryotic kingdom suggests that the genetic potential for N_2 fixation would be expected to occur in virtually any habitat (Paerl 1990). The factors that limit N_2 fixation in ecosystems are often discussed in the context of how environmental parameters affect N_2 fixation rates (Howarth *et al.* 1988). However, the factors that limit the genetic potential and expression of N_2 fixation are different for different types of organisms. Organic carbon substrate availability or the presence of reduced microzones and turbulence may be the major factors limiting the distribution and activity of heterotrophic diazotrophs in the sea, whereas light, inorganic nutrients other than nitrogen and possibly turbulence may limit the distribution of diazotrophic cyanobacteria.

Expression of nitrogenase activity is also dependent upon the availability of molybdenum and/or vanadium/iron for the core of the nitrogenase enzyme proteins (Postgate 1987). Mo concentrations are relatively high in the sea, but it has been proposed that the uptake of Mo could be reduced because of competitive uptake of sulphate (Howarth and Cole 1985). Nitrogenase is extremely sensitive to inactivation by O_2 (Gallon 1992), which is a particularly important and highly variable parameter in the environment. Finally, N_2 fixation can be regulated by the availability of other nutrients, as N_2 fixation provides no ecological advantage if cell growth is severely constrained by the availability of other nutrients.

The purpose of this chapter is to address the question of why *Trichodesmium* appears to be one of only a few N_2 fixing organisms in the open ocean. It is an attempt to bring the *Trichodesmium* studies full circle: from ecology to physiology, biochemistry, molecular biology back to ecology. As has been the case for many organisms, research has taken an increasingly reductionist course, borne out of the ecological questions, but rarely completing the quest by applying the cellular information at the ecological level (Walsby 1994). It is likely that this chapter will provide more questions than answers, but hopefully, they will be questions that help to focus future research on *Trichodesmium* molecular biology, biochemistry, and ecology.

The broad topic of "Why Only *Trichodesmium*" will be addressed by focusing on the following questions: Is *Trichodesmium* one of only a few N_2 fixing organisms? Who is *Trichodesmium*? Is *Trichodesmium* truly a free-living organism? What are the characteristics of the N_2 fixation apparatus in *Trichodesmium*? Does *Trichodesmium* possess unusual features that are involved in its ecological success in the open ocean environment?

1. IS TRICHODESMIUM THE ONLY FREE-LIVING NITROGEN FIXING ORGANISM IN OLIGOTROPHIC OCEANS? IS IT THE ONLY FREE-LIVING NITROGEN FIXING CYANOBACTERIUM?

The question of why there are not more N_2 fixing organisms in the sea, whereas N_2 fixation is a common feature of terrestrial and freshwater ecosystems, has long perplexed the ecologist (Vitousek and Howarth 1991). N_2 fixing organisms are found in many soil environments (Sprent and Sprent 1990), and N_2 fixing cyanobacterial blooms are often characteristic of N limited lakes (Horne and Goldman 1972, Paerl 1990, Howarth *et al.* 1988). This difference in the presence of N_2 fixing populations implies some feature of the marine environment that selects against or constrains the growth and/or activity of N_2 fixing species

(Howarth *et al.* 1988). Salinity can be a factor in restricting the distribution of some diazotrophic species (Fernandes *et al.* 1993), but does not explain the dearth of N_2 fixing organisms, since there are a few species found in open ocean environments and numerous species in marine benthic mats and sediments. *Trichodesmium* appears to be the dominant N_2 fixing organism in the open ocean, but less abundant N_2 fixing organisms are found in certain regions and at certain times. Other open ocean diazotrophs include the heterocystous cyanobacterial symbionts found in certain species of the diatoms *Rhizosolenia* (Venrick 1974) and *Hemiaulus* (Villareal 1994). Heterotrophic N_2 fixing bacteria have also been reported to be associated with *Rhizosolenia* mats in the Pacific (Martínez *et al.* 1983). *Dichothrix*, a heterocystous cyanobacterium, is often found associated with the macroalgae *Sargassum*, in the Sargasso Sea (Carpenter 1972). Blooms of *Nodularia* and *Aphanizomenon* occur in the Baltic Sea (Niemi 1979), and there is a rich flora of cyanobacterial and heterotrophic N_2 fixing organisms in benthic mats (Potts and Whitton 1977, Paerl *et al.* 1981), coral reefs (Wiebe *et al.* 1975), and sediments (Capone 1988). Only a relatively limited number of planktonic diazotrophs have been detected or characterized from the open ocean. However, prokaryotic organisms are abundant in the open ocean, including heterotrophic bacteria (10^5-$10^6 ml^{-1}$), unicellular cyanobacteria (10^3-$10^4 ml^{-1}$; Waterbury *et al.* 1986) and prochlorophytes (10^4-$10^5 ml^{-1}$; Olson *et al.* 1990). It is not clear whether any of these prokaryotic organisms have the potential to fix N_2, but do not express nitrogenase, or that the marine environment has selected against the genetic potential for N_2 fixation.

Marine N_2 fixation potential has been addressed by direct measurement of N_2 fixation activity (Dugdale *et al.* 1964, Mague *et al.* 1974, Saino and Hattori 1978), or by enriching or culturing isolates on N-free media (Wynn-Williams and Rhodes 1974, Guerinot and Colwell 1985, Kawai and Sugahara 1971). The success of culturing techniques is limited by the ability to culture organisms from the environment, and we know that many times the dominant organisms in natural populations are not represented in culture collections (Giovannoni *et al.* 1990). The measurement of N_2 fixation activity, using the acetylene reduction technique or using ^{15}N as a tracer, is more direct and provides estimates of *in situ* rates. In the dilute open ocean, the measurement of N_2 fixation is more difficult, and N_2 fixation may be overlooked unless sufficient cell concentration is achieved to increase the sensitivity of the assay for measuring N_2 fixation rates by organisms that are not extremely abundant. These organisms could still be important in N flux, for example, N_2 fixation by *Trichodesmium* has been estimated to be twice that of nitrate flux across the thermocline (Carpenter and Romans 1991), and yet abundances are only 1000 colonies m^{-3}, which is equivalent to about 24 cells ml^{-1}. With prokaryotic cell abundances of 10^5-$10^6 ml^{-1}$, only 0.005% of the prokaryotic cell population could fix an equivalent amount of N_2 as *Trichodesmium*. Typically, ten colonies of *Trichodesmium* are used for the acetylene reduction assay, which is equivalent to concentrating 10 litres of water into a 10ml assay vial. Although this is relatively easy to do with *Trichodesmium*, as the colonies can be concentrated with a plankton net, visualized with the unaided eye and handled with a pipette, it is not so easy to concentrate other microbes for the acetylene reduction assay. Even if appropriate cell concentrations can be achieved, incubation conditions may affect the measurement of N_2 fixation, including "bottle effects" or temporal variations in N_2 fixation activity such as diel cycles. Finally, measuring rates provides no information on whether the genetic potential for N_2 fixation is present, but not expressed, or what types of organisms can or are involved in N_2 fixation. For this reason, we still cannot be certain whether N_2 fixing species, other than *Trichodesmium*, are abundant in the open ocean. Molecular approaches offer the ability to differentiate between the presence of genetic potential (the presence of nitrogenase genes) and the expression of nitrogenase activity (nitrogenase mRNA, protein and activity).

Prokaryotic populations in the open ocean have been characterized by phylogenetic analysis of 16S rRNA sequences (Giovannoni *et al.* 1990, Schmidt *et al.* 1991, DeLong *et al.* 1993). The picoplankton are composed primarily of α and γ proteobacteria, cyanobacteria and prochlorophytes. There is little evidence of N_2 fixation in prochlorophytes (Lewin and Cheng 1989). However, numerous genera of α and γ proteobacteria and cyanobacteria can fix N_2 (Young 1992), although by no means do all genera fix N_2. Heterotrophic (Guerinot and Colwell 1985) and unicellular cyanobacterial (Duerr 1981, Waterbury *et al.* 1988) N_2 fixing strains have been isolated from the marine environment, and relative abundances of prokaryotic organisms have been derived from 16S rRNA characterization of picoplankton DNA (Giovannoni *et al.* 1990, Schmidt *et al.* 1991, Britschgi and Giovannoni 1991). Although it is difficult to assign any biochemical function to organisms identified by 16S rRNA sequencing, it is impossible to assign N_2 fixation to any species identified by 16S rRNA because of the seemingly random dispersion of nitrogenase genes among prokaryotic taxa.

Coupling the measurement of activity to the detection of the genes encoding the N_2 fixation apparatus can provide information on the distribution and abundance of N_2 fixing organisms, as well as the rates of N_2 fixation. In order to do this, a probe is needed that will detect N_2 fixation genes in diverse types of organisms. The probe technique must be sensitive enough that cells can be detected at very low concentrations. The polymerase chain reaction (PCR) provides a method for rapid detection and characterization of N_2 fixation genes. By careful design of the primers, providing for variations in codon utilization and

GC content, the *nifH* gene, which encodes the Fe protein of nitrogenase, can be amplified from diverse prokaryotic organisms (Zehr and McReynolds 1989, Kirshtein *et al.* 1991). Amplified fragments can be cloned, and characterized by hybridization to probes by DNA sequencing (Ben-Porath and Zehr 1994).

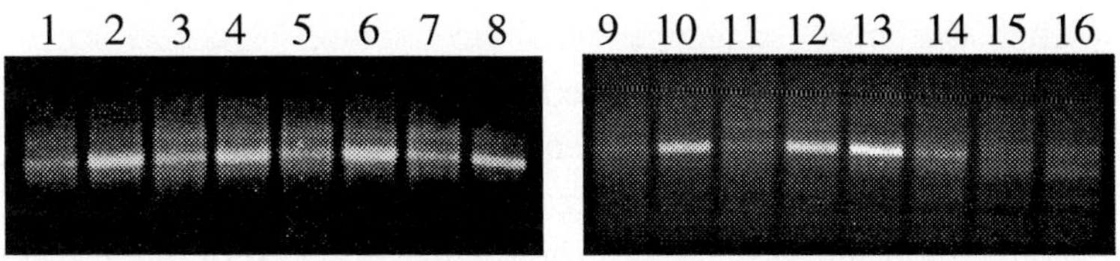

Fig.1. Amplification of 359 bp fragment of *nifH* from picoplankton from the Atlantic Ocean (approximately 15°N, 50°W). Internal controls were included in duplicates of each sample (lanes 2,4,6,8,10 and 12). Lanes 13-16 are 10^5, 10^3, $10^2$and 0 copies, respectively. 1, 2: 170m; 3,4: 140m; 5,6: 100m; 7,8: 70m; 9,10: 40m; 11,12: surface. Unamended picoplankton samples contain approximately 1000 copies.

DNA extracted from DNA water column samples can be challenged for the presence of the *nifH* gene. The extent of inhibition of the sample matrix is tested by including a positive control consisting of a plasmid containing a previously amplified *nifH* gene (Fig. 1). Amplification of *nifH* has been obtained from samples from the Atlantic and the Sargasso Sea east of the Bahaman Islands (Fig. 1). Typically, 10ng of DNA is used in the PCR reaction. An estimate of the number of diazotrophic organisms can be calculated by quantifying the amount of PCR product and the number of cells challenged in the PCR (Fig. 1). A slight positive signal is obtained throughout the mixed layer (Fig. 1). The most conservative estimate of the number of cells containing nitrogenase genes is obtained by quantifying the amount of DNA used in the reaction divided by an average amount of DNA per prokaryotic cell (5fg $cell^{-1}$; Simon and Azam 1989). Diazotrophic organisms appear to be present in up to .05% of the prokaryotic genomes in surface and deep water samples, which is equivalent to about 2,500,000 cells l^{-1}. These results set an upper limit on the genetic potential for N_2 fixation by free-living picoplankton, but also indicate that the abundance of N_2 fixing organisms could be on the order of magnitude of *Trichodesmium*. The issue will not be resolved until the *nifH* amplification products are cloned and sequenced. Furthermore, the number of *nifH* genes per litre varies in space and time, and seasonal studies are needed to assess the changes in abundance of diazotrophic populations over time.

In contrast to the free-living picoplankton, N_2 fixing organisms may be more abundant in particles such as marine snow. Particles may contain microaerophilic zones where N_2 fixation could occur (Alldredge and Silver 1988, Paerl and Prufert 1987). In a detrital aggregate composed of *Rhizosolenia* (non-symbiont containing species) and *Phaeocystis* from the Santa Barbara Channel (DNA generously provided by Ed DeLong), N_2 fixation genes were detected, cloned, and sequenced. Analysis of the sequence of the amplified *nifH* gene showed that the *nifH* sequence was most closely related to those of α proteobacteria (Fig. 2). This aggregate had previously been characterized by 16S rRNA sequence analysis, but α proteobacterial 16S rRNA sequences were not found (DeLong *et al.* 1993). Therefore, it is likely that the N_2 fixing α proteobacteria represent a small fraction of the prokaryotic populations in these aggregates.

2. WHO IS *TRICHODESMIUM*?

Trichodesmium is a cyanobacterium of the order *Oscillatoriales*, which includes all filamentous cyanobacteria which undergo binary fission in a single plane and produce only vegetative cells (Castenholz 1989). It is closely related to the members of the genus *Oscillatoria*, the primary difference being that it is a marine planktonic form, whereas most planktonic *Oscillatoria* occur in freshwater. The *Oscillatoria* and *Trichodesmium* genera are distinguished from the *Lyngbya-Plectonema-Phormidium* group by the almost invisible sheath (Castenholz 1989). Phylogenetic analysis of 16S rRNA shows that species given the genus name *Oscillatoria* do not form a tight cluster (Giovannoni *et al.* 1988, Wilmotte *et al.* 1994). Phylogenetic analysis of the 16S rRNA sequence of *Trichodesmium* sp. NIBB 1067 places *Trichodesmium* close to an *Oscillatoria* species (strain PCC 7515) and a *Lyngbya* strain (PCC 7419) (Wilmotte *et al.* 1994).

Phylogeny based on the DNA sequence of a fragment of *nifH* shows that the heterocystous strains form a relatively closely related cluster, whereas the nonheterocystous species do not form a cluster, and are more distantly related to each other than are the heterocystous species (Fig. 3; Ben-Porath and Zehr 1994). It is interesting to note that *Trichodesmium* and *Gloeothece*, both which can fix N_2 aerobically, are distantly related to the other nonheterocystous and heterocystous strains (Fig.3). It will be interesting to see, when more DNA sequence is available, it there are characteristics of the Fe protein primary sequence that are related to the capability for aerobic N_2 fixation.

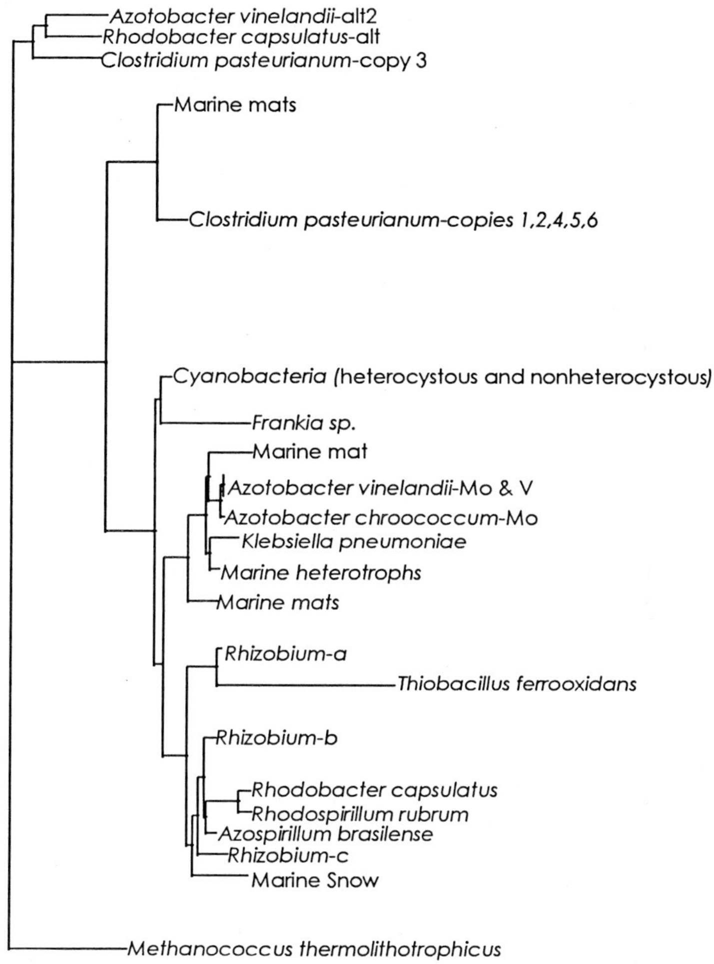

Fig. 2. Phylogenetic analysis of *nifH* deduced amino acid sequences (from 325 bp internal DNA fragment) from cultivated organisms and *nifH* sequences amplified from marine snow and marine mats. Tree created by Neighbor-Joining algorithm using the PHYLIP 3.5 package (Felsenstein 1993). Cyanobacterial species branches are collapsed to show position of group on tree. Note group of alternative nitrogenases in distinct cluster. *Rhizobium*-a, -b and -c refer to clusters of *Rhizobium* strain *nifH* sequences. Marine snow sequence groups with α proteobacteria.

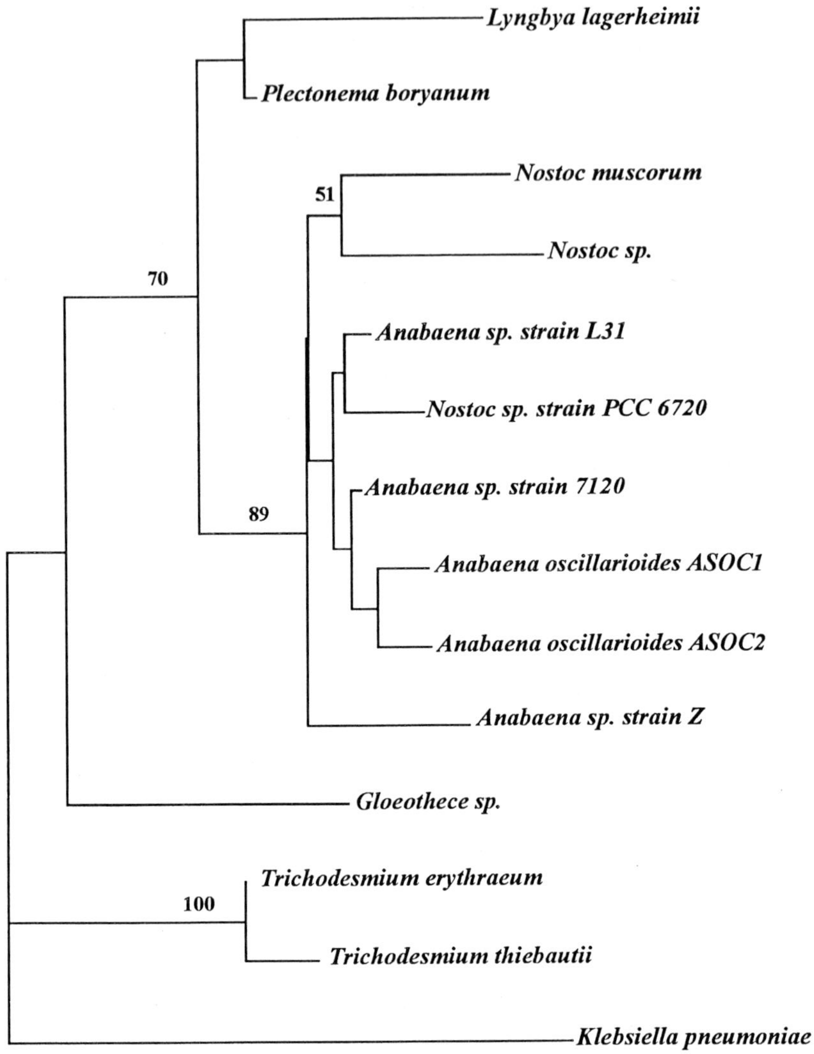

Fig. 3. Phylogenetic analysis of *nifH* deduced amino acid sequences (from 325 bp internal DNA fragment) from cyanobacteria. Data was boot- strapped and tree was created by Neighbor-Joining algorithm using the PHYLIP 3.5 package (Felsenstein 1993). Bootstrap values greater than 50% are shown.

The most appropriate question is: Who are *Trichodesmium*? There have been a number of species of *Trichodesmium* described, although the most commonly reported species are *Trichodesmium thiebautii* and *T. erythraeum*. The species designations are based on cell dimensions, cell shapes and colony morphology, although all of these characteristics vary in nature such that it is not always possible to assign species names to natural populations or cultivated strains (Siddiqui *et al.* 1992). The DNA sequences of *nifH* from some of these strains were compared to determine the relationships of different forms in nature to a cultivated strain (*Trichodesmium* sp. NIBB 1067). It was shown that there were at least two distinctive genotypes based on slight variations in *nifH* sequence (*T. thiebautii* and *T. erythraeum*), and that the cultivated strain was most closely related to *T. erythraeum* (Ben-Porath *et al.* 1993).

Recently it has been shown that the ITS region of the rRNA genes may also be useful to assess genotypic variability in natural populations (Wilmotte *et al.* 1994, In press). It will be interesting to see if genotypic differences based on *nifH* are supported by analysis of the ITS region. It is important to assess the genotypic variability of natural populations since there are substantial differences in the physiology (Carpenter *et al.* 1993, Siddiqui *et al.* 1992), toxicity (Hawser *et al.* 1992), and molecular biology (see Section 5) among different strains of *Trichodesmium*.

3. IS *TRICHODESMIUM* A FREE-LIVING ORGANISM?

Microscopic examination of *Trichodesmium* colonies demonstrate a high abundance of other prokaryotic organisms. These organisms range from uncharacterized filamentous cells and eubacteria, to other species of filamentous nonheterocystous cyanobacteria (Siddiqui *et al.* 1992). The relationship among these organisms and the role they play in the dynamics of the *Trichodesmium* colony are unclear, and only a few organisms have been cultured from *Trichodesmium* colonies (Paerl *et al.* 1989). It is likely that there are exchanges of carbon and nitrogen between these organisms (Bryceson and Fay 1981, Capone *et al.* submitted), and there may exist symbiotic relationships as exist between freshwater cyanobacteria and associated bacteria (Paerl *et al.* 1990). The associations between *Trichodesmium* and species of associated bacteria may be extremely important in the success of *Trichodesmium* in the sea (Paerl 1990, Sellner 1992), and yet we know very little of the nature of these bacterial species.

Using fluorescently labelled DNA probes for 16S rRNA, major groups of bacteria associated with *Trichodesmium* colonies were identified (Fig. 4). Total numbers of cells per *Trichodesmium* filament ranged from 76 to 297. This is approximately 1-2 bacterial cells per *Trichodesmium* cell (assuming an average of 120 cells per filament; Carpenter and Romans 1991). Approximately 40 filamentous bacterial cells (not autofluorescent) were associated with each trichome. The filamentous cells did not hybridize to any of the probes used (high or low GC firmicutes, enterics, vibrio, flavobacteria, α proteobacteria, β proteobacteria). Flavobacteria were the most abundant of the bacteria that could be identified (Fig. 4). Flavobacteria are chemoorganotrophic, and some species require amino acids. *Trichodesmium* excretes amino acids (Carpenter *et al.* 1992, Capone *et al.* submitted), and DON and ammonium concentrations are higher in *Trichodesmium* bloom patches than outside patches (Karl *et al.* 1992). *Flavobacteria* are common in particles (DeLong *et al.* 1993) as well as free-living, and it will be interesting to determine whether the *Flavobacteria* associated with *Trichodesmium* are different species than the free-living forms, indicating that they are adapted to the unique environment of the *Trichodesmium* colony. Enterics are also abundant in the *Trichodesmium* colony. Some species of this group can fix N_2, such as *Klebsiella*, and a N_2 fixing *Klebsiella*-like strain has been isolated from *Trichodesmium* assemblages (Paerl *et al.* 1989). No cells hybridized to the α proteobacteria 16S rRNA probe, which is unexpected since α proteobacteria were the most abundant in phytodetrital aggregates in the Santa Barbara Channel (DeLong *et al.* 1993). Only a few β proteobacteria were enumerated, and β proteobacteria were not detected in marine snow aggregates (DeLong *et al.* 1993). It should be interesting to determine what the β proteobacterial species are and what role they play in *Trichodesmium* aggregates.

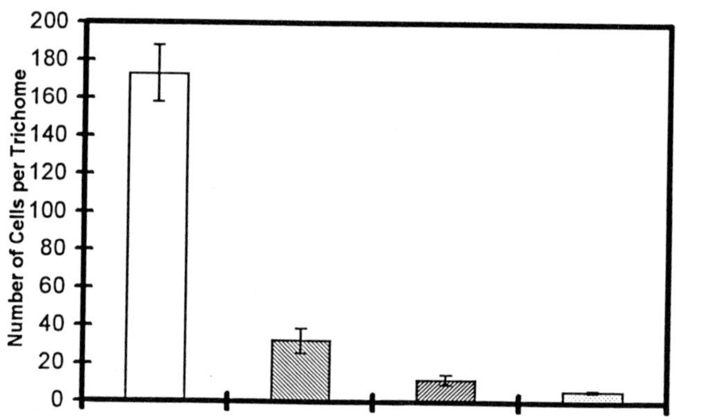

Fig. 4. Identification of bacteria associated with *Trichodesmium* using 16S rRNA directed fluorescent DNA probes.

The presence of heterotrophic organisms in *Trichodesmium* bundles raises the intriguing question of the possibility of symbiotic or consortial interactions. Mutually beneficial associations between *Trichodesmium* and heterotrophs could explain why *Trichodesmium* is so difficult to cultivate. At least some of the bacteria appear to play a role in decomposition, being attached primarily to senescent filaments (unpublished data). *Trichodesmium* can rarely be maintained during incubations lasting more than 24 hours. During incubation it has been observed that colonies become overgrown with bacteria and protists, resulting in lysis and bleaching of the *Trichodesmium* filaments. Apparently, intact colonies can inhibit the growth of these organisms, and incubation conditions affect the biochemical or physiological mechanisms that prevent bacterial growth. Interestingly, the addition of 10μM EDTA or Fe plus EDTA extends the viability of filaments during incubation and bacterial growth is substantially reduced. This is a particularly interesting observation as it suggests that the physiological state of *Trichodesmium* is related to the extent of bacterial colonization. It is unlikely that this concentration of EDTA inhibits the growth of bacteria directly, and the EDTA clearly affects the physiology of *Trichodesmium*, since it substantially changes the diel cycle of N_2 fixation (Zehr and Capone, unpublished data).

Bacteria associated with *Trichodesmium* probably play multiple roles in the *Trichodesmium* colony, from mutually beneficial interactions to decomposition and parasitism (Sellner 1992). Characterization of 16S rRNA sequences in *Trichodesmium* aggregates and the use of 16S rRNA directed DNA probes will help to elucidate the complex interactions between *Trichodesmium* and its associated microflora. Identification of associated bacteria in healthy and senescing colonies should make it possible to distinguish between bacteria that play a mutualistic or consortial role, from those that are opportunists or are involved in the decomposition of *Trichodesmium* filaments. Consortial interactions may be a key parameter in the success of *Trichodesmium* in the sea.

4. Is the nitrogen fixation apparatus of *Trichodesmium* particularly well-suited for the open ocean environment?

N_2 fixation is catalyzed by the enzyme nitrogenase. The N_2 fixation apparatus, including the enzyme nitrogenase, is a tightly controlled set of structural and regulatory proteins. Some features of the apparatus are highly conserved among organisms, but there are differences particularly in the way that organisms regulate synthesis and activity of the N_2 fixation proteins. The enzyme nitrogenase is composed of two different multi-subunit proteins, nitrogenase and nitrogenase reductase. These proteins, and the genes that encode these proteins (*nifH*, *D* and *K*), have been highly conserved through evolution (Postgate 1987). Yet N_2 fixing organisms demonstrate a high diversity of physiological strategies and tolerances to environmental conditions. The characteristics of the nitrogenase proteins, and the regulatory mechanisms involved, can be important in determining the ability of N_2 fixing organisms to exist and perform N_2 fixation in different environments.

The N_2 fixation operon in *Trichodesmium* contains the genes for the "conventional" nitrogenase, *ie* the genes for the Mo containing nitrogenase protein. There are no indications of multiple *nif* operons, based on the number of bands seen in Southern blots using the *Anabaena nifH*, *D*, and *K* genes as probes (Zehr *et al.* 1991a). The genes for *nifH*, *D*, and *K* in *Trichodesmium* are contiguous, as is the case for heterocysts in heterocystous cyanobacteria and for nonheterocystous cyanobacterial genes (Zehr *et al.* 1991a). In heterocystous species, the *nif* operon undergoes two gene rearrangements, one of which excises a transposon within *nifD* (Haselkorn *et al.* 1990) Although this rearrangement does not regulate transcription of the *nif* genes, it is part of the sequence of events that leads to expression of nitrogenase in specialized cells. Earlier studies indicated that *Trichodesmium* may restrict N_2 fixation to cells in the internal regions of the colony, analogous to the specialized activities of the heterocyst in heterocystous species (Carpenter and Price 1976). The restriction map of the *nif* operon was identical when *Trichodesmium* was grown with fixed N or under N_2 fixing conditions, and there were not multiple bands as would be the case if two types of gene arrangements existed in the filaments (Zehr *et al.* 1991a). If *Trichodesmium* does fix N_2 in specific cells or filaments, it does so by a different sequence of events than used by heterocystous cyanobacteria.

Comparison of the sequence of a fragment of *nifH* to other cyanobacterial sequences shows that *Trichodesmium* and *Gloeothece nifH* sequences are not closely related to other cyanobacterial *nifH* sequences and both genera are capable of aerobic N_2 fixation (Fig. 3). This leaves open the possibility that the *Trichodesmium nifH* gene sequence, and therefore, the gene product is significantly different than the *nifH* genes and gene products of other nonheterocystous filamentous cyanobacteria. Differences in the structure of the Fe protein could have important biochemical or physiological implications for N_2 fixation in the aerobic environment of the surface waters of the ocean. Comparison of the biochemical features of *Trichodesmium* nitrogenase to nitrogenase of other organisms must await the results of *in vitro* characterization of *Trichodesmium* nitrogenase, but these studies have been hampered by the difficulty of obtaining active protein preparations from natural populations and slow-growing cultures (P. Ludden, personal communication).

It has long been recognized that *Trichodesmium* fixes N_2 only during the day (Saino and Hattori 1978). This is unusual for nonheterocystous cyanobacteria, who usually fix N_2 microaerophilically or during the night (Khamees *et al.* 1987, Griffiths *et al.* 1987), although recently a benthic *Lyngbya* strain has been described which also fixes N aerobically in the light (Jones 1990). It is possible that the cycle of N_2 fixation is due to a circadian rhythm (Roenneberg and Carpenter 1993), as is believed to be the case for *Synechococcus* RF-1 (Grobbelaar *et al.* 1986). Roenneberg and Carpenter (1993) found that the compensation point followed a diel pattern that was maintained under constant dark or light conditions. This pattern was not demonstrated in other respiratory or photosynthetic parameters (Roenneberg and Carpenter 1993), and attempts to maintain the N_2 fixation pattern under constant illumination failed (Ohki *et al.* 1992). It is clear that light is necessary for the daily cycle in N_2 fixation (Ohki *et al.* 1992), but light could serve only to set the clock (see Sweeney and Borgese 1988) and is necessary since N_2 fixation in *Trichodesmium* seems to be directly dependent upon photosynthesis. A tantalizing observation is that the level of *nif* mRNA transcripts increases in early morning, hours before dawn (Wyman *et al.* submitted).

It is counterintuitive that *Trichodesmium* fixes N_2 during the day, when photosynthetic O_2 evolution is greatest (Fay 1992, Gallon and Stal 1992). The obvious advantage is that N_2 fixation can be coupled to the ATP and reductant generated by photosynthesis, and yet there must be a physiological cost in protecting nitrogenase from O_2 inactivation. High respiratory rates may be important in ameliorating the effects of O_2 evolved during photosynthesis, but may prevent the build-up of carbon reserves that would allow the fixation of N_2 during the dark phase. The timing of N_2 fixation to the light phase, which allows direct coupling of N_2 fixation to photosynthesis, may be one of the important features of *Trichodesmium* for N_2 fixation in the open ocean.

The diel cycle of N_2 fixation in natural populations of *Trichodesmium* is correlated with synthesis of the Fe and Mo proteins of nitrogenase, as well as the change in apparent molecular mass of the Fe protein of nitrogenase (Capone *et al.* 1990, Zehr *et al.* 1993). The Fe protein changes to the higher molecular mass form (inactive) during the night (Zehr *et al.* 1993, Ohki *et al.* 1992). The amount of N_2 fixed during the day appears to be a factor in determining the timing of switch-off of activity (and change of the Fe protein to high molecular mass form), and light is necessary for initiating synthesis and return of the Fe protein to the active, low molecular mass form (Ohki *et al.* 1992, Zehr *et al.* 1993). The change of the apparent molecular mass of the Fe protein appears to be a key factor in the timing of N_2 fixation to the light phase, although the underlying mechanism for the change in the apparent molecular mass of the Fe protein is not yet known.

In *Rhodospirillum rubrum*, two forms of the Fe protein have been shown to be the result of an ADP-ribosylation of one of the Fe protein subunits, and nitrogenase activity ceases when half of the subunits are ribosylated (Pope *et al.* 1985, Kanemoto and Ludden 1984). This post-translational modification in *R. rubrum* is reversible and the modification and demodification are enzymatically mediated (Roberts *et al.* 1990). The Fe protein is modified in response to elevated nitrogen and O_2 concentrations and appears to be a regulatory mechanism. Cyanobacteria, including *Trichodesmium and Oscillatoria*, also have two protein bands that are cross-reactive with Fe protein antisera (Reich and Böger 1989, Tabita *et al.* 1988, Du and Gallon 1993, Ohki *et al.* 1991, Stal and Bergman

1990, Zehr *et al.* 1993). The mechanism of regulation seems at first to be similar to that in *R. rubrum*, in that the apparent high molecular mass form is present when nitrogenase is not active, and the lower molecular mass form is present when nitrogenase is active (Zehr *et al.* 1993, Ohki *et al.* 1991). Recent studies suggest that the modification of the Fe protein in cyanobacteria occurs by a different mechanism than the ADP-ribosylation that has been described in *R. rubrum* (Durner *et al.* 1994). Attempts to label the modified band with ^{32}P in *Anabaena variabilis* were unsuccessful (Durner *et al.* 1994), as were attempts with *Trichodesmium* (J. Zehr, unpub. data). Furthermore, the Fe protein from *A. variabilis* could not be modified or demodified with activating/deactivating enzymes (DRAT and DRAG) cloned from *R. rubrum* (Durner *et al.* 1994).

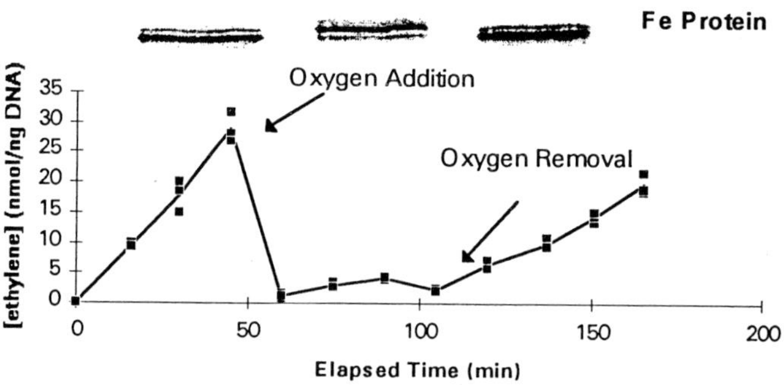

Fig. 5. Recovery of nitrogen fixation in *Trichodesmium* in response to incubation with oxygen. Protein synthesis was inhibited with chloramphenicol. Incubation under 40% oxygen results in cessation of activity and change in apparent molecular mass of Fe protein to higher molecular mass form. Incubation under nitrogen following eposure to oxygen results in recovery of activity and return of Fe protein to lower molecular mass form.

The Fe protein modification occurs in response to elevated O_2 concentrations (Fig. 5) as well as being correlated with the diel change in activity (Zehr *et al.* 1993, Ohki *et al.* 1992). The modification may play a role in the protection of the Fe protein from inactivation, as the protein is modified when the O_2 concentration is elevated, and returns to the unmodified state when O_2 concentrations are lowered (Fig. 5). The modification and demodification occurs even in the presence of chloramphenicol (Fig 5), showing that the two forms are not different proteins. The return of the Fe protein to the unmodified state following short-term incubation under elevated O_2, is correlated with the recovery of N_2 fixation (Fig. 5). *Trichodesmium* colonies are entrained in surface waters, and particularly during blooms, experience high irradiances near the surface (Lewis *et al.* 1988). Although the modification/demodification of the Fe protein does not explain simultaneous photosynthesis and N_2 fixation, it is likely to play a role in the protection of nitrogenase from inactivation by O_2 under certain conditions (Zehr *et al.* 1993). Furthermore, the switch-on and switch-off of nitrogenase activity in *Trichodesmium* occurs more quickly than does the modification, suggesting that there is a second mechanism involved in regulation (Ohki *et al.* 1992). Recently, a similar response has been described in *Rhodobacter capsulatus* (Pierrard *et al.* 1993).

The mechanism for simultaneous N_2 fixation and photosynthesis in *Trichodesmium* has yet to be elucidated. Nitrogenase could be continuously synthesized to replace inactivated nitrogenase (Gallon 1992), but inhibition of protein synthesis with chloramphenicol during the day does not inhibit N_2 fixation rates (Capone *et al.* 1990). Therefore, the turnover of nitrogenase during the day appears to be slow and does not explain simultaneous N_2 fixation and photosynthesis. High respiration rates (Kana 1993, Carpenter *et al.* 1990) are probably important in minimizing the O_2 problem. High rates of pseudocyclic electron flow around PSI, which reduces O_2, have been observed and may contribute to reducing inactivation of nitrogenase by O_2 evolved through photosynthesis (Kana 1993). In *Gloeothece* spp. PCC 6909, nitrogenase activity is supported by respiration, even in the light (Maryan *et al.* 1986). Under high light, intense nitrogenase activity was stimulated by DCMU (Maryan *et al.* 1986), however, in *Trichodesmium* DCMU inhibits nitrogenase activity (Ohki and Fujita 1988). Although it is usually assumed that interaction between O_2 and nitrogenase leads to inactivation, it has been shown that nitrogenase can itself reduce O_2 (Thorneley and Ashby 1989). This mechanism, called "autoprotection", was suggested to be of potential importance for aerobic N_2 fixation in genetically engineered plants (Thorneley and Ashby 1989), but is dependent upon a 4-fold molar excess of reduced Fe protein over O_2 (Thorneley and Ashby 1989).

Early studies suggested the intriguing possibility that N_2 fixation was restricted to the interior of *Trichodesmium* colonies analogous to the restriction of N_2 fixation to heterocysts in heterocystous species (Carpenter and Price 1976, Bryceson and Fay 1981). As discussed above, it is unlikely that this occurs at the level of cellular differentiation as in heterocystous species, but nitrogenase could possibly be expressed at different levels in different cells. Immunolocalization of the Fe protein of nitrogenase showed no consistent pattern in labelling along the trichomes, indicating that the protein is expressed along the entire length of the filament (Paerl *et al.* 1989, Bergman and Carpenter 1991). Even when the nitrogenase protein is expressed, it could be regulated by a post-translational mechanism that is involved in the diel cycle of activity (Ohki *et al.* 1992, Zehr *et al.* 1993). The protein from the natural populations is consistently in the entirely active form during the day (Zehr *et al.* 1993), which eliminates the possibility that there are variations in activation state among the cells in the colony. Pigmentation and DCMU enhanced fluorescence were evenly distributed along trichomes, which indicates that there is not differential expression of PSII along trichomes (Carpenter *et al.* 1990). Measurements of O_2 concentrations within and outside *Trichodesmium* aggregates showed that O_2 levels could be reduced in the centre of bundles, even when the bundles were illuminated at saturating intensities (Paerl and Bebout 1988). The bundles may help to limit photosynthetic rates by self-shading, but probably a variety of intracellular biochemical mechanisms are involved in protecting nitrogenase from O_2 inactivation in *Trichodesmium* (Carpenter *et al.* 1990).

There appears to be differences in extent of immunolabeling of nitrogenase among filaments in the same colony, which has been suggested to reflect partitioning of N_2 fixation activity among filaments (Bergman and Carpenter 1991). This phenomenon does not necessarily provide an explanation for simultaneous nitrogen fixation and photosynthesis, since the filaments which label differentially may also differ in physiological state and photosynthetic rates. The amount of nitrogenase is correlated with the amount of cytochrome oxidase based on dual immunolabeling experiments, which has been interpreted to mean that respiration is involved in protection of nitrogenase from O_2 inactivation (Bergman *et al.* 1993). One interpretation of these results is that trichomes cycle through N_2 fixing and photosynthetic phases. If trichomes cycled through a N_2 fixing stage (where they presumably would reduce or repress photosynthesis), it would be consistent with other observed features of *Trichodesmium*. Respiration rates are very high in *Trichodesmium*, with a compensation point of about 300µEinstein $m^{-2}s^{-1}$ (Kana 1993). If cultures and natural populations contain a mixture of filaments going through N_2 fixation and photosynthetic phases, the measured respiration rates would reflect all cells, whereas photosynthesis rates would reflect a fraction of the cells. The result would be that net O_2 exchange for the entire colony (or culture) would appear to have a high compensation point. Measured growth rates of natural populations are low (Carpenter *et al.* 1987), and *Trichodesmium* grows slowly in culture as well. This could reflect the necessity for having different stages of growth (N_2 fixing and photosynthesizing) with overall growth rate reduced. This would be similar to the link between N_2 fixation and photosynthesis in the cell cycle of a unicellular cyanobacterium (Mitsui *et al.* 1986).

5. DOES *TRICHODESMIUM* HAVE SPECIALIZATIONS FOR THE OPEN OCEAN HABITAT THAT ARE UNUSUAL FOR NITROGEN FIXING SPECIES?

Morphologically, *Trichodesmium* is typical of filamentous nonheterocystous cyanobacteria. *Trichodesmium* has several characteristics that may be particularly relevant to its planktonic existence. First, *Trichodesmium* forms colonies. The colonies may play an important role in allowing self-shading to reduce photoinhibition (Lewis *et al.* 1988), modification of sinking or flotation rates to maintain position in the water column (Walsby 1992), facilitating the creation of reduced O_2 zones (Paerl and Bebout 1988), or providing a habitat for associated bacteria that may play a symbiotic role or consume O_2 (Paerl 1990). Second, *Trichodesmium* synthesizes gas vacuoles that are substantially stronger than gas vacuoles in freshwater planktonic cyanobacteria (Walsby 1992). *Trichodesmium* contains a toxin, that restricts grazing by invertebrate herbivores (Hawser *et al.* 1992). The potential significance of the gas vacuoles, colony formation and associated bacteria have been reviewed (Paerl 1990, Carpenter 1983, Walsby 1992) and will not be discussed here. However, another unusual biochemical feature of *Trichodesmium* has recently been described that may be important in the ecology of *Trichodesmium*.

The DNA of many prokaryotic and eukaryotic organisms contains methylated nucleotide bases. Methylation may occur as part of restriction-modification systems believed to be involved in protection from infection by viruses (Arber and Linn 1969). Cyanobacterial species often contain methylated cytosine and adenine residues (Lambert and Carr 1984). In *Trichodesmium*, it was found that about 14% of the adenines were modified, and HPLC analysis showed that the modified base was not modified at the usual N^6 position (Zehr *et al.* 1991b). It was suggested that this rather extensive methylation could be a factor in the ecological success of *Trichodesmium* (Zehr *et al.* 1991b). Viruses have recently been suggested to be important in phytoplankton dynamics in the sea (Suttle *et al.* 1991, Proctor and Fuhrman 1990). A large number of cyanophages that infect *Synechococcus* have been isolated from the marine environment, many of which have broad host ranges (Waterbury and Valois 1993). It was shown that the viruses had little effect on *Synechococcus* population dynamics because the populations were largely resistant to infection and the concentration of sensitive cells relative to phage concentration was too low (Waterbury and Valois 1993). *Trichodesmium*, with its high concentration of cells localized in colonies, might provide a good host for cyanophages. The extensive modification of adenine may be part of the mechanism that inhibits infection of *Trichodesmium* by cyanophage. Methylation is usually only a part of the protection mechanism against virus infection. The companion restriction endonuclease degrades viral DNA, and the methylase protects the host DNA from digestion by the endonuclease. In the case of *Trichodesmium*, the methylase must have a very degenerate recognition site in order to methylate 14% of the adenines (Zehr *et al.* 1991b), and it seems unlikely that a restriction enzyme would have such a degenerate recognition site. Attempts to assay restriction enzyme activity in *Trichodesmium* thus far have been unsuccessful (J. Zehr, unpub. data). However, phage-encoded restriction enzymes have been reported (Xia *et al.* 1987), and analysis of DNA methylation patterns in marine cyanophage have suggested that such systems may be present (Wilson *et al.* 1993). The extensive methylation in *Trichodesmium* would provide a good defence against viral-encoded restriction enzymes.

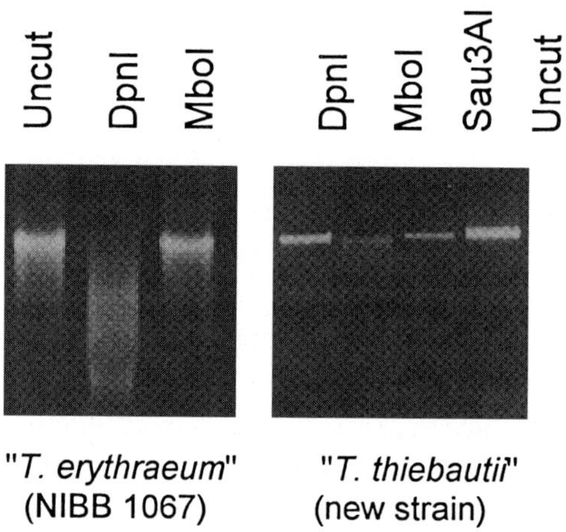

Fig. 6. Methylation patterns in two *Trichodesmium* strains. DNA of *Trichodesmium* sp. NIBB 1067 is highly methylated by modification by adenine, which is demonstrated by the degree of digestion with DpnI. In contrast, a new strain of *Trichodesmium thiebautii* is resistant to digestion with DpnI and Sau3AI, which suggests methylation of cytosine, but not as extensive adenine methylation.

Recently, a new isolate of *Trichodesmium* has been obtained from the Caribbean Sea (Prufert-Bebout *et al.* 1993), which morphologically seems to be more closely related to *T. thiebautii*, than to *T. erythraeum*. The unusual modification was initially determined with DNA from *Trichodesmium* sp. NIBB 1067, which appears to be most closely related to *T. erythraeum* (Ben-Porath *et al.* 1993). Surprisingly, the DNA of this culture is methylated differently to that of *Trichodesmium* sp. NIBB 1067 (Fig. 6). The pattern of methylation of DNA from natural populations of *T. thiebautii* seems to be more similar to the isolate from the Caribbean Sea. The fact that the DNA is not digested by the restriction enzyme Sau3AI, but is digested by either DpnI or MboI, indicates that cytosine methylation is more extensive in this new isolate than in *Trichodesmium* sp. NIBB 1067. *T. thiebautii* is much more abundant than *T. erythraeum* in the Caribbean and Sargasso Seas. It could very well be that the extent and nature of DNA methylation, and subsequent resistance to phage infection could play a role in determining the outcome of competition between these species.

The modification of DNA seems likely to be an important characteristic of *Trichodesmium* ecology, but the methylation of DNA is not necessarily only related to phage resistance. Modification of DNA may confer stability to high UV irradiances in surface waters, or may be important in DNA repair mechanisms. The role of DNA modification in *Trichodesmium* ecology may be clarified by experimental work, once the methylating enzyme is cloned and can be used in studies of the regulation of its expression.

Limitation of nitrogen fixation in the sea

The factors proposed to control N_2 fixation in the sea include DOC, O_2, phosphorus or N/P ratios, and Mo uptake (Howarth *et al.* 1988). These factors probably all play a role in regulating the expression of activity by *in situ* populations at different times and places. Results indicate that the genetic potential for N_2 fixation, at least by the free-living picoplankton in the sea, is limited (Fig. 1). The answer to the question of why the genetic potential is so restricted involves long time scales and large distances. Why are there so few N_2 fixation genes in the marine environment and what enables *Trichodesmium* to maintain its position as a dominant N_2 fixing organism in the sea? The answer probably lies in the large scale of the open ocean environment, its relatively homogenous and stable physical and chemical characteristics and the restricted conditions necessary for N_2 fixation. In order for N_2 fixation activity to be expressed, a "seed" population must exist, which can grow and express nitrogenase genes under N_2 fixing conditions (Fig.7).

Seed populations can be introduced from adjacent habitats, but the open ocean is separated from potential seed populations in sediments, or terrestrial or freshwater ecosystems by large distances. N_2 fixing seed populations could only be maintained in the open ocean by association with organisms or particles with a long residence time, or by long-term starvation/survival mechanisms. Microhabitats such as zooplankton guts could provide a refuge for seed populations, but this possibility has yet to be investigated. The potential for *Trichodesmium* to maintain seed populations may provide an explanation for its ecological success.

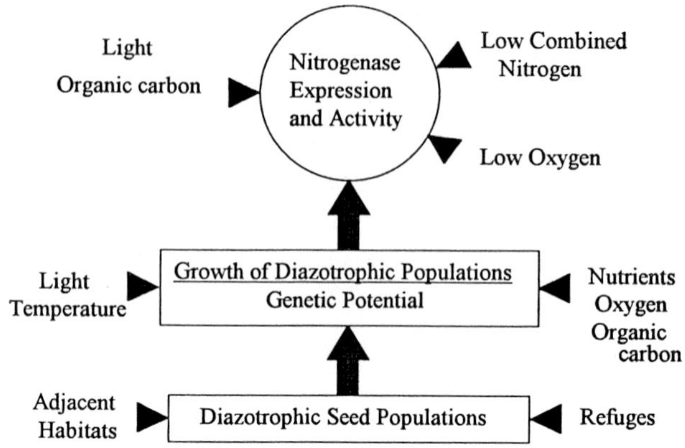

Fig. 7. Model for regulation of nitrogen fixation potential and expression in the environment. Nitrogen fixing seed populations must be introduced from adjacent habitats or be present in refuges within the habitat, in order for nitrogen fixing populations to develop when nitrogen fixing conditions exist. Nitrogen fixation is then regulated at the transcriptional and translational levels by environmental conditions.

Trichodesmium grows slowly and can remain viable for great lengths of time. *Trichodesmium* colonies have been observed as far as 59°N in the Atlantic Ocean (Carpenter 1983b), well out of its normal range. The gas vacuoles of *Trichodesmium* ensure that *Trichodesmium* remains in the water column and can be carried great distances, rather than settling out of the water column (Carpenter *et al.* 1987) following collapse of blooms. The buoyancy of cyanobacteria is often affected by light and nutritional status (Walsby 1994), but *Trichodesmium* appears to remain buoyant even when nutrient stressed. In nearshore and freshwater environments, cyanobacteria can settle to the bottom and provide seed populations for blooms in subsequent years or when conditions change (Huber 1984, Reynolds *et al.* 1981). If *Trichodesmium* lost buoyancy when conditions were suboptimal for growth and N_2 fixation, it would take a long time for populations to develop when N_2 fixing conditions returned. Very little is known about the regulatory mechanisms controlling gas vacuole synthesis in cyanobacteria (Walsby 1994), but it may be that the regulatory mechanisms in *Trichodesmium* are different than in many cyanobacterial species which form blooms in lakes or shallow waters. It seems highly probable that the ability to maintain viability and buoyancy, coupled to its unique biochemical and physiological characteristics, are critical factors in the ecological success of *Trichodesmium* as a N_2 fixing organism in the sea.

Acknowledgements
I greatly acknowledge the excellent work of M. Mellon, S. Braun, and B. Dominic in my laboratory. M. Williams, S. Danielsen, and S. Nierzwicki-Bauer made the 16S rRNA probe work possible. D. Capone, K. Ohki, E. Carpenter, and H. Paerl provided many discussions and much encouragement over the years. E. DeLong generously provided marine snow DNA. L. Prufert deserves many thanks for successfully cultivating *Trichodesmium thiebautii* and making the strain available. M. A. Sessions was invaluable in manuscript preparation. This work was supported by NSF Grants OCE 9101399 and OCE 9202106.

References

Alldredge AL and Silver MW (1988) Characteristics, dynamics, and significance of marine snow. *Prog Oceanogr* **20:** 41-82

Arber W and Linn S (1969) DNA modification and restriction. *Annu Rev Biochem* **38:** 467-500

Ben-Porath J and Zehr JP (1994) Detection and characterization of cyanobacterial *nifH* genes. *Appl Environ Microbiol* **60,** 880-887

Ben-Porath Carpenter EJ, Zehr ZP (1993) Genotypic relationships in *Trichodesmium* (cyanophyceae) based on *nifH* sequence comparisons. *J Phycol* **29:** 806-810

Bergman B and Carpenter EJ (1991) Nitrogenase confined to randomly distributed trichomes in the marine cyanobacterium *Trichodesmium thiebautii. J Phycol* **27:** 158-165

Bergman B, Siddiqui PJA, Carpenter EJ and Peschek GA (1993) Cytochrome oxidase: subcellular distribution and relationship to nitrogenase expression in the nonheterocystous marine cyanobacterium *Trichodesmium thiebautii. Appl Environ Microbiol* **59:** 3239-3244

Britschgi TB and Giovannoni SJ (1991) Phylogenetic analysis of a natural marine bacterioplankton population by rRNA gene cloning and sequencing. *Appl Environ Microbiol* **57:** 1707-1713

Bryceson I and Fay P (1981) Nitrogen fixation in *Oscillatoria (Trichodesmium) erythraea* in relation to bundle formation and trichome differentiation. *Mar Biol* **61:** 159-166

Capone DG, Ferrier MD, Carpenter EJ (1995). Amino acid cycling in colonies of the planktonic marine cyanobacterium, *Trichodesmium thiebautii*. Submitted.

Capone DG (1988) Benthic nitrogen fixation. In *Nitrogen Cycling in Coastal Marine Environments* TH Blackburn and J Sørensen (eds) John Wiley & Sons Ltd 85-123

Capone DC (1983) Benthic nitrogen fixation. In *Nitrogen in the Marine Environment* Carpenter EJ and Capone DG (eds) Academic Press, Inc., New York 105-138

Capone DG, O'Neil JM, Zehr J, Carpenter EJ (1990) Basis for diel variation in nitrogenase activity in the marine planktonic cyanobacterium *Trichodesmium thiebautii. Appl Environ Microbiol* **56:** 3532-3536

Carpenter EJ (1983a) Physiology and ecology of marine planktonic *Oscillatoria (Trichodesmium). Mar Biol Lett* **4:** 69-85

Carpenter EJ (1983b) Nitrogen fixation by marine *Oscillatoria* (*Trichodesmium*) in the world's oceans. In *Nitrogen in the Marine Environment* Carpenter EJ and Capone DG (eds) Academic Press, Inc., New York 65-104

Carpenter EJ (1972) Nitrogen fixation by a blue-green epiphyte on pelagic *Sargassum*. *Science* **178:** 1207-1208

Carpenter EJ and Price CC (1976) Marine *Oscillatoria (Trichodesmium)*: Explanation for aerobic nitrogen fixation without heterocysts. *Science* **191:** 1278-1280

Carpenter EJ and Romans K (1991) Major role of the cyanobacterium *Trichodesmium* in nutrient cycling in the North Atlantic Ocean. *Science* **254:** 1356-1358

Carpenter EJ, Scranton MI, Novelli PC, Michaels A (1987) Validity of N_2 fixation rate measurements in marine *Oscillatoria (Trichodesmium)*. *J Plankton Res* **9:** 1047-1056

Carpenter EJ, Bergman B, Dawson R, Siddiqui PJA, Söderbäck E, Capone DG (1992) Glutamine synthetase and nitrogen cycling in colonies of the marine diazotrophic cyanobacteria *Trichodesmium* spp. *Appl Environ Microbiol* **58:** 3122-3129

Carpenter EJ, O'Neil JM, Dawson R, Capone DG, Siddiqui PJA, Roenneberg T, Bergman B (1993) The tropical diazotrophic phytoplankter *Trichodesmium*: biological characteristics of two common species. *Mar Ecol Prog Ser* **95:** 295-304.

Carpenter EJ, Chang J, Cottrell M, Schubauer J, Paerl HW, Bebout BM, Capone DG (1990) Re-evaluation of nitrogenase oxygen-protective mechanisms in the planktonic marine cyanobacterium *Trichodesmium*. *Mar Ecol Prog Ser* **65:** 151-158

Castenholz RW (1989) Subsection III. Order *Oscillatoriales*; subsection IV. Order *Nostocales*; subsection V. Order *Stigonematales*. In *Bergey's Manual of Systematic Bacteriology* pp 1771-91, 1794-99. Williams and Wilkins, Baltimore, MD

Cullen J, Yang X, MacIntyre HL (1991) Nutrient limitation and marine photosynthesis. In *Primary Productivity and Biogeochemical Cycles in the Sea* Falkowski PG and Woodhead AD (eds.) Plenum Press, New York 69-68

DeLong EF, Franks DG, Alldredge AL (1993) Phylogenetic diversity of aggregate-attached vs. free-living marine bacterial assemblages. *Limnol Oceanogr* **38:** 924-934

Du C and Gallon JR (1993) Modification of the Fe protein of the nitrogenase of *Gloeothece* (Nägeli) sp. ATCC 27152 during growth under alternating light and darkness. *New Phytol* **125:** 121-129

Duerr EO (1981) Aerobic nitrogen fixation by two unicellular marine cyanobacteria (*Synechococcus* spp.). Ph.D. Thesis, University of Miami, FL.

Dugdale RC and Goering JJ (1967) Uptake of new and regenerated forms of nitrogen in primary productivity. *Limnol Oceangr* **12:** 196-206

Dugdale RC, Goering JJ, Ryther JH (1964) High nitrogen fixation rates in the Sargasso Sea and the Arabian Sea. *Limnol Oceanogr* **9:** 507-510

Durner J, Böhm I, Hilz H, Böger P (1994) Posttranslational modification of nitrogenase: differences between the purple bacterium *Rhodospirillum rubrum* and the cyanobacterium *Anabaena variabilis*. *Eur J Biochem* **220:** 125-130

Fay P (1992) Oxygen relations of nitrogen fixation in cyanobacteria. *Microbiol Revs* **56:** 340-373

Felsenstein J (1993) PHYLIP-Phylogeny Inference Package (Version 3.5c). Distributed by the author. Department of Genetics, University of Washington, Seattle, WA

Fernandes TA, Iyer V, Apte SK (1993) Differential responses of nitrogen-fixing cyanobacteria to salinity and osmotic stresses. *Appl Environ Microbiol* **59,** 899-904

Gallon JR (1992) Reconciling the incompatible: N_2 fixation and O_2. *New Phytol* **122:** 571-609

Gallon JR and Stal LJ (1992) N_2 fixation in non-heterocystous cyanobacteria; an overview. In *Marine Pelagic Cyanobacteria:* Trichodesmium *and other Diazotrophs.* Carpenter EJ, CaponeDG, Rueter JG (eds) Kluwer Academic Publishers, The Netherlands 115-140.

Giovannoni SJ, Britschgi TB, Moyer CL, Field KG (1990) Genetic diversity in Sargasso Sea bacterioplankton. *Nature (London)* **345:** 60-63

Giovannoni SJ, Turner S, Olsen GJ, Barns S, Lane DJ, Pace NR (1988) Evolutionary relationships among cyanobacteria and green chloroplasts. *J Bacteriol* **170:** 3584-3592

Griffiths MSH, Gallon JR, Chaplin AE (1987) The diurnal pattern of dinitrogen fixation by cyanobacteria *in situ. New Phytol* **107:** 649-657

Grobbelaar N, Huang TC, Lin HY, Chow TJ (1986) Dinitrogen-fixing endogenous rhythm in *Synechococcus* RF-1. *FEMS Microbiol Lett* **37:** 173-177

Guerinot ML and Colwell RR (1985) Enumeration, isolation, and characterization of N_2-fixing bacteria from seawater. *Appl Environ Microbiol* **50:** 350-355

Haselkorn R, Basche M, Böhme B, Borthakur D, Borthakur PB, Buikema WJ, Mulligan ME, Norris D (1990) Nitrogen fixation in filamentous cyanobacteria. In *Nitrogen Fixation: Achievements and Objectives* Gresshoff PM, Roth LE, Stacey G, NewtonWE (eds.) Chapman and Hall, New York 497-504

Hawser SP, O'Neil JM, Roman MR, Codd GA (1992) Toxicity of blooms of the cyanobacterium *Trichodesmium* to zooplankton. *J Appl Phycol* **4:** 79-86.

Horne AJ and Goldman CR (1972) Nitrogen fixation in Clear Lake, California; 1. Seasonal variation and the role of heterocysts. *Limnol Oceanogr* **17:** 678-692

Howarth RW and Cole JJ (1985) Molybdenum availability, nitrogen limitation, and phytoplankton growth in natural waters. *Science* **229:** 653-655

Howarth RW, Marino R, Cole JJ (1988) Nitrogen fixation in freshwater, estuarine, and marine ecosystems. 2. Biogeochemical controls. *Limnol Oceanogr* **33:** 688-701

Huber AL (1984) *Nodularia (cyanobacteriaceae)* akinetes in the sediments of the Peel-Harvey estuary, Western Australia: Potential inoculum source for *Nodularia* blooms. *Appl Environ Microbiol* **47:** 234-238

Jones K (1990) Aerobic nitrogen fixation by *Lyngbya* sp., a marine tropical cyanobacterium. *Br Phycol J* **25:** 287-289

Kana TM (1993) Rapid oxygen cycling in *Trichodesmium thiebautii. Limnol Oceanogr* **38:** 18-24

Kanemoto RH and Ludden PW (1984) Effect of ammonia, darkness, and phenazine methosulfate on whole-cell nitrogenase activity and Fe protein modification in *Rhodospirillum rubrum*. *J Bacteriol* **158:** 713-720

Karl DM, Letelier R, Hebel DV, Bir DF, Winn CD (1992) *Trichodesmium* blooms and new nitrogen in the North Pacific Gyre. In *Marine Pelagic Cyanobacteria:* Trichodesmium *and other Diazotrophs* Carpenter EJ, Capone DG, Rueter JG (eds.) . Kluwer Academic Publishers, The Netherlands 219-238

Kawai A and Sugahara I (1971) Microbiological studies on nitrogen fixation in aquatic environments - III. On the nitrogen fixing bacteria in offshore regions. *Bull Japanese Soc Scientific Fisheries* **37:** 981-985

Khamees HS, Gallon JR, Chaplin AE (1987) The pattern of acetylene reduction by cyanobacteria grown under alternating light and darkness. *Br Phycol J* **22:** 55-60

Kirshtein JD, Paerl HW, Zehr J (1991) Amplification, cloning, and sequencing of a *nifH* segment from aquatic microorganisms and natural communities. *Appl Environ Microbiol* **57:** 2645-2650

Lambert GR and Carr NG (1984) Resistance of DNA from filamentous and unicellular cyanobacteria to restriction endonuclease cleavage. *Biochim Biophys Acta* **781:** 45-55

Lewin RA and Cheng L (Eds.) (1989) "*Prochloron* A Microbial Enigma", Chapman and Hall, New York.

Lewis MR, Ulloa O, Platt T (1988) Photosynthetic action, absorption, and quantum yield spectra for a natural population of *Oscillatoria* in the North Atlantic. *Limnol Oceanogr* **33:** 92-98

Mague TH, Weare MM, Holm-Hansen O (1974) Nitrogen fixation in the north Pacific Ocean. *Mar Biol* **24:** 109-119

Martin JH, Gordon RM, Fitzwater SE (1991) The case for iron. *Limnol Oceanogr* **36:** 1793-1802

Martínez LA, Silver MW, King JM, Alldredge AL (1983) Nitrogen fixation by floating diatom mats: a source of new nitrogen to oligotrophic ocean waters. *Science* **221:** 152-154

Maryan PS, Eady RR, Chaplin AE Gallon JR (1986) Nitrogen fixation by *Gloeothece* sp. PCC 6909: Respiration and not photosynthesis supports nitrogenase activity in the light. *J Gen Microbiol* **132:** 789-796

Mitsui A, Kumazawa S, Takahashi A, Ikemoto H, Cao S, Arai T (1986) Strategy by which nitrogen-fixing unicellular cyanobacteria grow photo-autotrophically. *Nature (London)* **323:** 720-722

Niemi A (1979) Blue-green algal blooms and N : P ratio in the Baltic Sea. *Acta Bot Fenn* **110:** 57-61

Ohki K and Fujita Y (1988) Aerobic nitrogenase activity measured as acetylene reduction in the marine non-heterocystous cyanobacterium *Trichodesmium* spp. grown under artificial conditions. *Mar Biol* **98:** 111-114.

Ohki K, Zehr JP, Fujita Y (1992) Regulation of nitrogenase activity in relation to the light-dark regime in the filamentous non-heterocystous cyanobacterium *Trichodesmium* sp. NIBB 1067. *J Gen Microbiol* **138:** 2679-2685

Ohki K, Zehr JP Falkowski PG, Fujita Y (1991) Regulation of nitrogen-fixation by different nitrogen sources in the marine non-heterocystous cyanobacterium *Trichodesmium* sp. NIBB1067. *Arch Microbiol* **156:** 335-337

Olson RJ, Chisholm SW, Zettler ER, Altabet M, Dusenberry J (1990) Spatial and temporal distributions of prochlorophyte picoplankton in the North Atlantic Ocean. *Deep Sea Res* **37:** 1033-1051

Paerl HW (1990) Physiological ecology and regulation of N_2 fixation in natural waters. In *Advances in Microbial Ecology* Marshall KC (ed.) Plenum Publishing Corp 305-344

Paerl HW and Bebout BM (1988) Direct measurement of O_2-depleted microzones in marine *Oscillatoria*: Relation to N_2 fixation. *Science* **241:** 442-445

Paerl HW and Prufert LE (1987) Oxygen-poor microzones as potential sites of microbial N_2 fixation in nitrogen-depleted aerobic marine waters. *Appl Environ Microbiol* **53:** 1078-1087

Paerl HW, Rudek J, Mallin MA (1990) Stimulation of phytoplankton production in coastal waters by natural rainfall inputs: nutritional and trophic implications. *Mar Biol* **107:** 247-254

Paerl HW, Priscu JC, Brawner DL (1989) Immunochemical localization of nitrogenase in marine *Trichodesmium* aggregates: relationship to N_2 fixation potential. *Appl Environ Microbiol* **55:** 2965-2975

Paerl HW, Webb KL, Baker J, Wiebe WJ (1981) Nitrogen fixation in waters. In *Nitrogen Fixation, Vol. 1, Ecolog.y* Broughton WJ (ed) pp 193-240. Claredon Press, Oxford

Pierrard J, Ludden PW, Roberts GP (1993) Posttranslational regulation of nitrogenase in *Rhodobacter capsulatus*: Existence of two independent regulatory effects of ammonium. *J Bacteriol* **175:** 1358-1366

Pope MR, Murrell SA, Ludden PW (1985) Convalent modification of the iron protein of nitrogenase from *Rhodospirillum rubrum* by adenosine diphoshoribosylation of a specific arginine residue. *Proc Natl Acad Sci USA* **82:** 3173-3177

Postgate JR (ed) (1987) *Nitrogen Fixation* 2nd Edition, Edward Arnold (Publishers) Ltd., London, England

Potts M and Whitton BA (1977) Nitrogen fixation by blue-green algal communities in the intertidal zone of the lagoon of Aldabra Atoll. *Oecologia* **27:** 275-283

Proctor LM and Fuhrman JA (1990) Viral mortality of marine bacteria and cyanobacteria. *Nature* **343:** 60-62

Prufert-Bebout L, Paerl HW, Lassen C (1993) Growth, nitrogen fixation, and spectral attenuation in cultivated *Trichodesmium* species. *Appl Environ Microbiol* **59:** 1367-1375

Redfield AC (1958) The biological control of chemical factors in the environment. *Am Sci* **46:** 205-222

Redfield AC, Ketchum BH, Richards FA (1963) The influence of organisms on the composition of sea-water. In *The Sea* Hill MN (ed.) Interscience Publishers, John Wiley and Sons, New York 26-77

Reich S and Böger P (1989) Regulation of nitrogenase activity in *Anabaena variabilis* by modification of the Fe protein. *FEMS Microbiol Lett* **58:** 81-86

Reynolds CS, Oliver RL Walsby AE (1987) Cyanobacterial dominance: the role of buoyancy regulation in dynamic lake environments. *NZ J Mar Freshwater Res* **21:** 379-390

Roberts GP, Ludden PW, Burris RH, Fitzmaurice WP, Fu H-A, Nielsen G, Liang J-H, Lehman L, Woehle D, Lies D, Wirt H, Montogmery S, Davis R, Bao Y (1990) The genetics and biochemistry of the reversible ADP-ribosylation systems of *Rhodospirillum rubrum* and *Azospirillum lipoferum.* In *Nitrogen Fixation: Achievements and Objectives.* Gresshoff PM, Roth LE, Stacey G, Newton WE (eds.) Chapman and Hall, New York 475-482

Roenneberg T and Carpenter EJ (1993) Daily rhythm of O_2-evolution in the cyanobacterium *Trichodesmium thiebautii* under natural and constant conditions. *Mar Biol* **117:** 693-697

Saino T and Hattori A (1978) Diel variation in nitrogen fixation by a marine blue-green alga, *Trichodesmium thiebautii. Deep-Sea Res* **25:** 1259-1263

Sellner KG (1992) Trophodynamics of marine cyanobacteria blooms. In *Marine Pelagic Cyanobacteria:* Trichodesmium *and other Diazotrophs.* Carpenter EJ, Capone DG, Rueter JG (eds) Kluwer Academic Publishers, The Netherlands 75-94.

Schmidt TM, DeLong EF, Pace NR (1991) Analysis of a marine picoplankton community by 16S rRNA gene cloning and sequencing. *J Bateriol* **173:** 4371-4378

Sharp JH (1983) The distributions of inorganic nitrogen and dissolved and particulate organic nitrogen in the sea. In *Nitrogen in the Marine Environment* Carpenter EJ, and Capone DG (eds) Academic Press, Inc., New York 1-36.

Siddiqui PJA, Carpenter EJ, Bergman B (1992) *Trichodesmium*: Ultrastructure and protein localization. In *Marine Pelagic Cyanobacteria: Trichodesmium and other Diazotrophs* Carpenter EJ, Capone DG, Rueter JG (eds) Kluwer Academic Publishers, The Netherlands 9-28

Simon M and Azam F (1989) Protein content and protein synthesis rates of planktonic marine bacteria. *Mar Ecol Prog Ser* **51:** 201-213

Smith SV (1984) Phosphorus versus nitrogen limitation in the marine environment. *Limnol Oceanogr* **29:** 1149-1160

Sprent JI and Sprent P (eds.) (1990) *Nitrogen Fixing Organisms, Pure and Applied Aspects* Chapman and Hall, New York

Stal LJ and Bergman B (1990) Immunological characterization of nitrogenase in the filamentous non-heterocystous cyanobacterium *Oscillatoria limosa. Planta* **182:** 287-291

Suttle CA, Chan AM, Cottrell MT (1991) Infection of phytoplankton by viruses and reduction of primary productivity. *Nature (London)* **347:** 467-469

Sweeney BM and Borgese MB (1988) A circadian rhythm in cell division in a prokaryote, the cyanobacterium *Synechococcus* WH7803. *J Phycol* **25:** 183-186

Tabita FR, Van Baalen C, Smith RL, Kumar D (1988) *In vivo* protection and recovery of nitrogenase from oxygen in heterocystous cyanobacteria. In *Nitrogen Fixation: Hundred Years After.* Bothe H, deBruijn FJ, Newton WE (eds.) Gustav Fischer, Stuttgart, New York 189-194

Thomas WH (1970) On nitrogen deficiency in tropical Pacific Oceanic phytoplankton: photosynthetic parameters in poor and rich water. *Limnol Oceangr* **15:** 380-385

Thorneley RNF and Ashby GA (1989) Oxidation of nitrogenase iron protein by dioxygen without inactivation could contribute to high respiration rates of *Azotobacter* species and facilitate nitrogen fixation in other aerobic environments. *Biochem J* **261:** 181-187

Venrick EL (1974) The distribution and significance of *Richelia intracellularis* Schmidt in the North Pacific Central Gyre. *Limnol Oceanogr* **19:** 437-445

Villareal TA (1994) Widespread occurrence of the *Hemiaulus* cyanobacterial symbiosis in the Southwest North Atlantic Ocean. *Bull Mar Sci* **54:** 1-7

Vitousek PM and Howarth RW (1991) Nitrogen limitation on land and in the sea: How can it occur? *Biogeochem* **13:** 87-115

Walsby AE (1994) Gas vesicles. *Microbiol Rev* **58:** 94-144

Walsby AE (1992) The gas vesicles and buoyancy of *Trichodesmium*. In *Marine Pelagic Cyanobacteria:* Trichodesmium *and other Diazotrophs.* Carpenter EJ, Capone DG, Rueter JG (eds) Kluwer Academic Publishers, Netherlands 141-162

Waterbury JB and Valois FW (1993) Resistance to co-occurring phages enables marine *Synechococcus* communities to coexist with cyanophages abundant in seawater. *Appl Environ Microbiol* **59:** 3393-3399

Waterbury JB, Watson SW, Valois FW (1988) Temporal separation of photosynthesis and dinitrogen fixation in the marine unicellular cyanobacterium *Erythrosphaera marin. EOS* **69:**1089

Waterbury JB, Watson SW, Valois FW, Franks DG (1986) Biological and ecological characterization of the marine unicellular cyanobacterium Synechococcus. In *Photosynthetic Picoplankton* Platt T and Li WKW (eds) *Can Bull Fisheries Aquat Sci **214:*** 71-120

Wiebe WJ, Johannes RE, Webb KL (1975) Nitrogen fixation in a coral reef community. *Science* **188:** 257-259

Wilmotte A, Neefs J-M, De Wachter R (1994) Evolutionary affiliation of the marine nitrogen-fixing cyanobacterium *Trichodesmium* sp. strain NIBB 1067, derived by 16S ribosomal RNA sequence analysis. *Microbiol.* **140:** 2159-2164

Wilson WH, Joint IR, Carr NG, Mann NH (1993) Isolation and molecular characterization of five marine cyanophages propagated on *Synechococcus* sp. Strain WH7803. *Appl Environ Microbiol* **59:** 3736-3743

Wyman M, Zehr JP, Capone, DG (1994) Temporal variability in nitrogenase gene expression in a natural population of the marine cyanobacterium *Trichodesmium thiebautii.* Submitted to *Appl Environ Microbiol*

Wynn-Williams DD and Rhodes ME (1974) Nitrogen fixation in seawater. *J Appl Bact* **37:** 203-216

Xia Y, Burbank DE, Uher L, Rabussay D, Van Etten JL (1987) IL-3A virus infection of a *Chlorella*-like green alga induces a DNA restriction endonuclease with novel sequence specificity. *Nucleic Acids Res* **15:** 6075-6090

Young JPW (1992) Phylogenetic classification of nitrogen-fixing organisms. In *Biological Nitrogen Fixation.* Stacey G, Burris RH, Evans HJ (eds) Chapman and Hall, Inc., New York 43-86

Zehr JP and McReynolds LA (1989) Use of degenerate oligonucleotides for amplification of the nifH gene from the marine cyanobacterium *Trichodesmium thiebautii*. *Appl Environ Microbiol* **55:** 2522-2526

Zehr JP, Ohki K, Fujita Y (1991a) Arrangement of nitrogenase structural genes in an aerobic filamentous nonheterocystous cyanobacterium. *J Bacteriol* **173:** 7055-7058.

Zehr JP, Ohki K,. Fujita Y, Landry D (1991b) Unique modification of adenine in genomic DNA of the marine cyanobacterium *Trichodesmium* sp. strain NIBB 1067. *J Bacteriol* **173:** 7059-7062

Zehr JP, Wyman M, Miller V, Duguay L, Capone DG (1993) Modification of the Fe protein of nitrogenase in natural populations of *Trichodesmium thiebautii*. *Appl Environ Microbiol* **59:** 669-676

Molecular Ecology of Marine Methanotrophs

J Colin Murrell and Andrew J Holmes
Department of Biological Sciences
University of Warwick
Coventry
CV4 7AL
UK

General Introduction to Methanotrophs

Methane oxidising bacteria (methanotrophs) are a unique group of methylotrophs (organisms that can utilise one-carbon compounds for growth) that grow using methane as their sole source of carbon and energy. They are generally regarded as obligate organisms and although many methanotrophs can also grow on methanol, they do not utilise multi-carbon compounds. The first report of a methanotroph was by Söhngen who isolated the pink methanotroph *Bacillus methanicus* (Söhngen, 1906) which was subsequently renamed *Pseudomonas methanica* and then *Methylomonas methanica*. This was re-isolated many years later by Dworkin and Foster (1956) but there were only a few sporadic reports on methanotrophs until the pioneering work of Whittenbury and colleagues who isolated and characterised over 100 different strains of methane-oxidising bacteria from soils, sediments and freshwater (Whittenbury *et al.*, 1970). These were all strictly aerobic, gram negative bacteria that grew on a simple mineral salts medium and methane. Based on their cell morphology, the resting stages formed, the intracytoplasmic membranes they contained and a few physiological characteristics, these bacteria were divided into five proposed genera, *Methylococcus*, *Methylobacter*, *Methylomonas*, *Methylocystis* and *Methylosinus*. Considering the relatively few diagnostic tests that could be undertaken at that stage, this classification scheme has served "methanophiles" (those who isolate and work with methane-oxidisers!) remarkably well over the last 25 years and has formed the framework for classification of subsequent isolates.

These five genera fall into two major groups. Type I methanotrophs *Methylomonas* and *Methylobacter*, related to bacteria in the γ-subdivision of the Proteobacteria, contain bundles of intracytoplasmic membranes and assimilate formaldehyde (an intermediate in the methane oxidation pathway) by the ribulose monophosphate (RuMP) pathway, whereas Type II methanotrophs *Methylocystis*

NATO ASI Series, Vol. G 38
Molecular Ecology of Aquatic Microbes
Edited by I. Joint

and *Methylosinus* that fall into the α-subdivision of the Proteobacteria have intracytoplasmic membranes arranged around the periphery of the cell and assimilate carbon into biomass at the level of formaldehyde using the serine pathway. A third grouping, Type X, has been proposed for *Methylococcus capsulatus* (Bath) since it has characteristics of both groups (Table 1).

Table 1. Characteristics of obligate methane oxidising bacteria

	Methylomonas	*Methylobacter*	*Methylococcus*	*Methylosinus*	*Methylocystis*
No. of validated species	3	6	2	2	2
Phylogenetic Position	γ-Proteo-bacteria	γ-Proteo-bacteria	γ-Proteo-bacteria	α-Proteo-bacteria	α-Proteo-bacteria
Membrane type	Type I	Type I	Type I	Type II	Type II
Pathway	RuMP	RuMP	RuMP/RuBP	Serine	Serine
Enzyme type	pMMO	pMMO	sMMO/ pMMO	sMMO/ pMMO	sMMO/ pMMO
Major Fatty Acid	16:1, 14:0	16:1	16:1, 16:0	18:1	18:1
Mol % G + C	50-54	50-54	62.5	62-63	62-63
Lysis in 2% SDS	+	+	+	-	-
N_2 fixation	V	-	+	+	+
Carotenoids	+	-	-	-	-
Growth in liquid	Pellicle	Dispersed	Dispersed	Dispersed	Dispersed

The taxonomy of methanotrophs has been extensively reviewed (Green 1992, Green 1993, Hanson *et al.*, 1993). Most recently, through extensive numerical taxonomy, DNA-DNA hybridisation and phospholipid fatty acid composition studies on environmental isolates and culture collection methanotrophs, the five genera of methanotrophs have been validated (Bowman *et al.*, 1993). This has resulted in the transfer of some *Methylomonas* strains, including marine isolates into the genus *Methylobacter*. 16S rRNA sequence analysis of many methanotrophs, largely through the efforts of Hanson, Bowman and colleagues, has also strengthened the taxonomy of these bacteria (see later).

There has been considerable interest in methanotrophs over the last few years since it is now recognised that these organisms, which appear to be ubiquitous in the natural environment, are a major sink for atmospheric methane (reviewed in Reeburgh, 1993; King, 1992, 1993; Cicerone and Oremland, 1988; Topp and Hanson, 1991). These organisms have also attracted attention due to their potential for the production of bulk chemicals, single cell protein and use in biotransformations (*eg* see Lidstrom and Stirling, 1990; Leak, 1992). More recently, their ability to degrade the ground water pollutant trichloroethylene and other halogenated hydrocarbons has also been examined in detail (Oldenhuis and Janssen, 1993; Alvarez-Cohen, 1993).

Physiology and Biochemistry of Methanotrophs

The biochemistry of methylotrophs has been described in detail (Anthony, 1982; Anthony, 1986). Methanol, produced from the oxidation of methane, is further oxidised to formaldehyde by the periplasmic pyrroloquinoline-quinone (PQQ) - linked enzyme methanol dehydrogenase. This enzyme is found in all gram negative methane and methanol utilisers. The central oxidation intermediate formaldehyde can be further oxidised to formate and then CO_2 (which is not usually fixed by ribulose bisphosphate carboxylase) by formaldehyde and formate dehydrogenases to yield reducing power for biosynthesis and the initial oxidation of methane. Formaldehyde is also assimilated into cell carbon by either the RuMP or serine pathway. Some methanotrophs have the ability to fix N_2 under reduced oxygen tensions (Murrell and Dalton, 1983). Type I methanotrophs appear to lack a complete TCA cycle. Indeed the absence of several enzymes of central metabolic pathways has been suggested as a reason for their inability to grow on multi-carbon compounds.

The first step in methane oxidation is carried out by the enzyme methane monooxygenase (MMO). The most well-characterised methane monooxygenases are the soluble (cytoplasmic) MMO from the Type II *Methylosinus trichosporium* OB3b and the Type X methanotroph *Methylococcus capsulatus* (Bath) (reviewed in Dalton, 1992; Dalton *et al.*, 1993; Froland *et al.*, 1993). Both methanotrophs synthesise sMMO with similar characteristics during copper-deficient growth. sMMO consists of 3 components. Protein A, the hydroxylase component has three subunits of 60, 45 and 20 kDa arranged in a $\alpha_2\ \beta_2\ \gamma_2$ configuration. The site of the monooxygenation reaction is believed to be a binuclear iron-oxo centre residing in Protein A α subunits. Protein B, a polypeptide of 16 kDa, acts as an effector of electron transfer in the catalytic mechanism, mediating the flow of electrons from Protein C to Protein A. Protein C, the reductase component of the sMMO complex, contains 1 mol each of FAD and a (2Fe-2S) cluster and acts as a moderator between hydride transfer from NADH and electron transfer to the hydroxylase component.

The expression of sMMO appears to be regulated by the amount of copper available for growth of *M. trichosporium* and *M. capsulatus* (Bath). At even relatively low copper-to-biomass ratios (greater than 1μmol l^{-1} per gram dry weight), the synthesis of this enzyme complex is repressed and a completely distinct membrane-bound or particulate methane monooxygenase (pMMO) is expressed. Not all methanotrophs appear to contain an sMMO complex under copper-limited growth. The only well-characterised sMMO to date have been from *M. trichosporium*, *M. capsulatus* (Bath), *Methylosinus sporium* 5 (Pilkington and Dalton, 1991), *Methylocystis* strain M (Nakajima *et al*., 1992) and *Methylomonas methanica* 68-1 (Koh *et al*., 1993). Many other methanotrophs only grow poorly in copper-depleted medium but in copper-sufficient growth conditions, all methane oxidisers appear to grow using pMMO (Dalton, 1992). Studies on this enzyme have been hampered by its lability in cell-free extracts and it has proved difficult to purify in active form from membranes (Smith and Dalton, 1989; Akent'eva and Gvozdev, 1990). Recently, biochemical and biophysical studies with *M. capsulatus* (Bath) have revealed that pMMO has an obligate requirement for copper and from SDS-polyacrylamide gel electrophoresis of membrane fractions of this organism, three major polypeptides of 46, 35 and 26 kDa have been implicated. The high correlation between copper level in membranes and pMMO activity, together with EPR studies, suggest the presence of a trinuclear copper centre which may constitute the active site of the pMMO (Chan *et al*., 1993; Nguyen *et al*., 1994). Since pMMO appears to be found in all methanotrophs and may be a unique enzyme, considerable efforts are now being directed towards the purification of this protein and also the cloning of the genes encoding pMMO since their DNA sequences will be invaluable for future molecular ecology studies (see later). Why some methanotrophs should possess both enzyme systems is unclear. The sMMO has a very wide substrate range and it is able to co-oxidise a wide range of alkanes, alkenes and aromatic compounds such as naphthalene. The pMMO has a rather narrow substrate range and does not oxidise larger organic compounds. There is no apparent advantage in having the additional sMMO system other than the capacity to grow in copper-deficient medium. The ecological significance of this is not known but is currently under investigation in the Authors' laboratory.

Molecular Biology and Genetics of Methanotrophs

The genetics and molecular biology of the methane oxidising bacteria has been reviewed recently (Murrell, 1992; Murrell, 1994; Barta and Hanson, 1993) and therefore will only be summarised here. The relatively slow growth of methanotrophs and their restricted mode of growth has hampered genetic studies on methanotrophs. Conventional mutagenesis techniques have not been particularly successful but more recently the use of the 'suicide substrate' dichloromethane (DCM) has allowed the selection of methane oxidation minus mutants of *Methylosinus* and *Methylomonas*. This is based on the ability of MMO to co-oxidise DCM to carbon monoxide which becomes toxic to the cell. MMO^- mutants, therefore, only grow on methanol and are DCM resistant (Reviewed in Murrell, 1994). sMMO constitutive mutants of *M. trichosporium* which fail to express pMMO, have been isolated by this method (Phelps *et al.*, 1992). Subsequent characterisation of these mutants has indicated that this mutant phenotype arises from defects in copper uptake and metabolism rather than from changes in sMMO expression or enzyme stability (Fitch *et al.*, 1993).

Gene transfer systems for methanotrophs are not particularly well-developed since conventional transformation procedures do not appear to work with these organisms. Conjugation methods are however available for transfer of broad host-range plasmids of the IncP and IncQ groups to several methanotrophs at reasonable frequencies. This has allowed the construction of sMMO-minus mutants by marker-exchange mutagenesis (Murrell, 1993). Most methanotrophs also contain plasmids, the functions of which are unknown. Bacteriophages have also been isolated that are specific for a number of methanotrophs (reviewed in Murrell, 1992).

The genes encoding methanol dehydrogenase from the methanol-utilisers *Methylobacterium extorquens* and *Paracoccus denitrificans* have been extensively studied in the laboratories of Hanson, Lidstrom, Harms and Goodwin. One of these genes, *moxF*, encoding the large subunit structural gene of methanol dehydrogenase (the second enzyme in the methane oxidation pathway) has been used as a heterologous probe to identify and clone the corresponding gene in methanotrophs (Stephens *et al.*, 1988). The *moxF* gene appears to be highly conserved in all gram negative methylotrophs and is therefore a good indicator of the presence of methylotrophs in the natural environment (see later).

The purification of sMMO from *M. capsulatus* (Bath) by Dalton and colleagues has resulted in the use of N-terminal amino acid sequence information of sMMO polypeptides to design oligonucleotide probes which has enabled us to clone the genes encoding sMMO from this organism. DNA sequencing has shown that the genes encoding the α, β and γ subunits of Protein A (*mmoX*, *Y* and *Z*), Protein B (*mmoB*) and Protein C (*mmoC*) are all linked on the chromosome.

This sMMO gene cluster has been used as a probe for sMMO genes in a number of representative strains of methanotrophs and uncharacterised isolates. sMMO homologs were only detected in *Methylosinus* and *Methylococcus* strains and were absent from the *Methylocystis*, *Methylobacter* and *Methylomonas* strains tested (Stainthorpe *et al.*, 1990). Subsequently we also cloned and DNA sequenced the entire sMMO gene cluster from *Methylosinus trichosporium* OB3b. The genes are also clustered in the same order as in *Methylococcus* and there is a high degree of similarity between deduced polypeptide sequences of both sMMOs. Screening of these DNA and derived protein sequences indicate that with the exception of 1) a region of the Protein A α subunit, which has identity with the Glu-X-X-His sequence motif that is known to coordinate the two irons in the R2 protein of ribonucleotide reductase (the putative active site of the sMMO hydroxylase; Dalton *et al.*, 1993) and; 2) the N-terminal 90 amino acids of Protein C, which has significant homologies with ferredoxins from plants and bacteria, these genes and the proteins they encode are unique and may be used as indicators for the presence of methanotrophs in the natural environment. Indeed, sequence information from analysis of these sMMO gene clusters has been used to design polymerase chain reaction (PCR) primers that have allowed the cloning of sMMO genes from several other methanotrophs (Murrell *et al.*, unpublished; also see later). The organisation, regulation of expression of these genes and promoter analysis has been discussed in detail (Murrell, 1992; 1993; 1994).

Molecular Ecology of Methanotrophs

It is now generally accepted that methanotrophs form the major biological sink for methane in the biosphere and methane oxidising bacteria have been isolated from a wide variety of soils, sediments, freshwater and saline environments.

Aerobic methane oxidation activity is generally at its maximum close to the oxic and anoxic interface of freshwater environments (Rudd and Taylor, 1980). This is generally the zone where methane and oxygen concentrations are optimum for methanotrophs. However, a number of other environmental variables, such as nitrogen source availability, pH and light should also been considered. Microbes that carry out the oxidation of methane in the absence of oxygen have yet to be isolated although there is some evidence for their existence (see later).

There have been many excellent reviews on the cycling of methane in aquatic environments and the role that methanotrophs may play in this cycle (*eg* see Rudd and Taylor, 1980; Hanson, 1980; Cicerone and Oremland, 1988; Kiene, 1991; King, 1992; King, 1993; Reeburgh, 1993; Hanson and Wattenburg, 1991; Hanson *et al.*, 1993) and therefore we will confine our attentions to the molecular ecological techniques that have already been used to identify methanotrophs in these environments and focus on the use of these techniques to the study of marine methanotrophs.

Traditionally, the methods that have involved the identification and quantification of methanotrophs in the environment have included variations on the classical enrichment techniques of Whittenbury and others, (Whittenbury *et al.*, 1970). The shortcomings of culture-based techniques for microbial ecology are well-documented (see Ward *et al.*, 1992) and for methanotrophs include the problem of the slow growth of some methanotrophs or scavenging of non-methanotrophs on even "purified agar" plates which may cause further errors with these techniques. Estimates of the biomass and activities of methanotrophs can be made by measuring the evolution of $^{14}CO_2$ and incorporation of $^{14}CH_4$ added to soil or water samples but this will not allow analysis of community structure. Phospholipid analysis is another way of assessing the types of methanotrophs in natural environmental samples or enrichment cultures, but this technique is relatively expensive, requires sophisticated equipment, and the data base for lipopolysaccharide hydroxy fatty acid and the phospholipid fatty acid profile data base for methylotrophs is not comprehensive. The detection and identification of methanotrophs in environmental samples has also been attempted using antibodies, which potentially is an attractive idea. However, antisera (to whole cells) must be prepared against culturable methanotrophs which must then react with all methanotrophs of interest and yet not detect non-methanotrophs. Rigorous controls with many culture collection isolates would still not guarantee, however, that all methanotrophs in a natural sample could be detected. Reed and Dugan (1978) were able to detect *Methylosinus trichosporium* and *Methylomonas methanica* in Cleveland Harbor samples by indirect fluorescent-antibody-membrane filter staining techniques. However, controls with seeded samples indicated variable recovery rates. Similar methods have been used by Gal'chenko *et al.* (1988) to look at species composition of aerobic methanotrophs in the Black Sea. Antibodies against key enzymes of methane oxidation such as methanol dehydrogenase or soluble methane monooxygenase could also be of use in detection methods (reviewed in Hanson and Wattenberg, 1991). More recently Sieburth and colleagues (Sieburth *et al.*, 1993b) have used antisera raised against the marine methanotroph *Methylobacter pelagicus*, methanol dehydrogenase and the methylamine utiliser *Methylophaga marina* in indirect immunofluorescence procedures to analyse uncharacterised pure cultures of marine C_1 utilisers and heterotrophs and compare them serologically with culture collection isolates. This extensive study showed that of the 32 methanotrophic enrichments set up using Sargasso sea water, over 70% yielded methane-oxidisers. Nearly all of the 27 isolates obtained in pure culture from these samples were identical to *Methylobacter pelagicus*. In contrast, Chesapeake Bay enrichments, yielded 54 estuarine methane oxidisers, of which only 13% appeared to be identical to *M. pelagicus* with 87% (46 enrichment cultures) displaying little or no antigenic cross-reactivity with anti-sera raised against *M. pelagicus*, suggesting that these were potentially novel marine methanotrophs and highlighting the difficulty in developing antisera capable of detecting novel strains.

Anti-MDH antiserum was not particularly effective in these experiments since it displayed cross-reactivity with only 23% of the methanotroph cultures (and 15% of methylamine-oxidising cultures). This was unexpected since it was assumed that all these methylotrophs would have the typical PQQ-linked gram negative methanol dehydrogenase. Lack of cell permeability to antibody or different types of MDH (*eg* gram positive alcohol dehydrogenase - like enzymes) are possible reasons for this.

The limitations of the techniques described above revolve around the need for cultivation of the organisms. The database of cultured organisms relative to uncultured is so small that one can't be confident of detecting novel strains which may be present in the environment. Researchers, therefore have started to develop molecular biological methods for the analysis of microbes in the natural environment which allow detection and identification of uncultivated organisms (Ward *et al.*, 1992). Application of these molecular biology methods to methanotrophs has been helped greatly by the sequencing of a number of 16S rRNA genes of both methanotrophs and methylotrophs (Tsuji *et al.*, 1990; Bratina *et al.*, 1992; Bowman, 1992) and the cloning of several methanotroph-specific genes (Murrell, 1992).

Type I methanotrophs such as *Methylomonas* and *Methylobacter* and Type X methanotrophs *eg Methylococcus capsulatus* appear to cluster in the γ-subdivision of the purple bacteria whereas Type II methanotrophs like *Methylosinus* and *Methylocystis* are found in the α-2 subdivision of the class Proteobacteria (reviewed in Hanson *et al.*, 1993; Boulygina *et al.*, 1993). Hanson and colleagues used these sequence data to design oligonucleotide probes of broad specificity for methylotrophs in natural samples. One probe, designated 9α is complementary to target sequences in 16S rRNA of serine pathway methylotrophs (which would include the Type II methanotrophs) and appears to be specific for these organisms. Another probe, 10γ is specific for methylotrophs, including Type I methanotrophs that utilise the RuMP pathway of formaldehyde assimilation. When used with two different fluors, these two probes can differentiate between mixed cultures of Type I and Type II methanotrophs in *in situ* hybridisations (Tsien *et al.*, 1990; Hanson and Wattenberg, 1991; Hanson *et al.*, 1993).

Further 16S rRNA sequence analysis of 36 methylotrophic bacteria and 10 reference organisms from subdivisions within the Proteobacteria has allowed

Hanson and colleagues to construct phylogenetic trees which show that methanotrophs are separated into coherent clusters that share common physiological characteristics. Sequence data were then used to design five new oligonucleotide probes which, when labelled with digoxigenin-11-dd UTP, could be used to detect the different groups of methanotrophs present in soils samples that had been enriched in a methane-air atmosphere for one month. In extensive control experiments, the probes hybridised to RNA extracted from representative bacteria of each of the five clusters. Furthermore, isolation and purification of RNA directly from these soils, and Northern slot blot techniques showed that these probes could be used to detect 16S rRNA sequences complementary to probes for both serine and RuMP pathway methanotrophs in a soil sample when 5μg of total RNA extracted from soil was used. The non-radioactive detection system also performed reliably with detection limits of around 10ng of RNA from pure cultures (Brusseau *et al.*, 1994). This was the first report of specific detection of the activity of uncultivated methanotrophs *in situ*.

Restriction Fragment Length Polymorphism (RFLP) in conjugation with hybridisation to *mox*F sequences (the methanol dehydrogenase structural gene) has been used to distinguish methanotrophs (Tsuji *et al.*, 1989). Such techniques have also been used to detect trichloroethylene-(TCE) degrading methanotrophs in environmental samples and bioreactors using a fragment of the sMMO gene cluster from *M. trichosporium* OB3b (Tsien and Hanson, 1992). This sMMO gene probe has also been used to confirm the presence of soluble methane monooxygenase in a Type I methanotroph *Methylomonas methanica* 68-1 (Koh *et al.*, 1993). Analysis of a TCE-and chloroform-degrading bacterial consortium using a combination of 16S rRNA oligo probing and sMMO gene probing, has also allowed the molecular characterisation of a methanotroph from a bioreactor (Alvarez-Cohen *et al.*, 1992).

We have utilised the apparently highly conserved nature of *moxF* and sMMO genes in methanotrophs to detect these organisms in environmental samples. Sets of PCR primers have been designed for each of the five sMMO structural genes (McDonald *et al.*, 1994) and *moxF* (Murrell *et al.*, 1992). PCR amplification of each of these genes is now possible from cultured methanotrophs known to contain sMMO. In addition, when total bacterial DNA from a variety of freshwater, marine, soil and peat samples was used as template with these sMMO-specific primers, a PCR product of the correct size was obtained. When probed with the appropriate sMMO gene from *Methylococcus* or *Methylosinus*, a high degree of homology was observed. Subsequent DNA sequencing of a number of these PCR products has proved that methanotroph-specific sMMO-encoding DNA has been amplified from environmental DNA samples and that these genes are very similar to, but not identical to known sMMO gene sequences, illustrating the diversity of methanotrophs in these environments (McDonald *et al.*, 1994).

Methane in the Marine Environment

It is now generally accepted that the oceans are supersaturated with respect to atmospheric methane and that this phenomenon appears to occur in a wide variety of geographical locations, making it a world wide phenomenon (Conrad and Seiler, 1988). In addition to methane it has been suggested that at least 15 C_1 compounds can be found in the marine environment making it a potential habitat for many different methylotrophic organisms (Kiene, 1993). In turn, methylotrophs could therefore make a significant contribution to oceanic carbon cycling. Cicerone and Oremland (1988) have estimated that the contribution of the oceans is only around 1-4% (5 to 20Tg per year) of the total global atmospheric flux of methane and if atmospheric CH_4 concentrations continue to rise the oceans may eventually become a sink for methane. Maximum methane concentrations are generally observed considerably below the air/sea interface and it has been suggested that this is due to *in situ* production of CH_4 rather than lateral transport of CH_4 (Scranton and Brewer, 1977). However biological methane production by methanogens is generally regarded as a strictly anaerobic process. One view ("The ocean methane paradox"; Kiene, 1991) is that strictly anaerobic methanogens might be producing this methane in anoxic microniches of suspended particulates that accumulate in oxic zones and more recently evidence for this has emerged.

Sieburth (1993) and colleagues in studies of C_1 bacteria in the water column of Chesapeake bay have shown that 0_2- and H_2S- tolerant obligately anaerobic methylotrophic methanogens can occur in microniches that are developed by consortia including aerobic methylamine utilisers. Detailed analysis of these methanogenic bacterial consortia (MBC) has allowed the 'top down' working hypothesis whereby bacterial syntrophies can explain the presence of the different groups of bacteria found in methanogenic enrichments carried out under three different regimes and the existence of reduced microniches due to bacterial activity (Sieburth, 1991; Sieburth, 1993; Sieburth *et al.*, 1993a). These consortia have been analysed by immunochemical methods which suggest the presence of new taxa of methanogens, and it would be interesting to apply 16S rRNA community analysis methodology to these consortia.

In a recent study in the North Pacific Ocean, Karl and Tilbrook (1994) have demonstrated that CH_4 is associated with sinking particles which then exchanges with the water column as particles sink. The release of this methane, which is presumably dissolved in interstitial fluids of particulate biogenic materials, provides a mechanism for the active transport of CH_4 in the water column. These particle-to-seawater fluxes could account in part for characteristic methane supersaturations in the upper water column.

Estuarine inputs of CH_4 to seawater may also be important in the marine methane cycle, however, it is not known if this CH_4 is mainly oxidised by microbes or distributed by physical processes which would make the turnover of this CH_4 in sea water relatively slow. Estimates of the concentrations of CH_4 and aerobic CH_4 oxidation rates in a variety of aerobic sea water environments have been made (Reviewed in Kiene, 1991). Generally turnover rates are low which suggests that CH_4 concentrations may be too low to sustain a large population of methanotrophs. Other areas of considerable interest are those that are methane rich for example deep-sea hydrothermal plumes of the Juan de Fuca Ridge where recent unusually high methane oxidation rates have been reported (De Angelis *et al.*, 1993) and methane seeps of "bubbling reefs" such as those of the Kattegatt (Jensen *et al.*, 1992).

Methane oxidation rates and methane fluxes in deep anoxic waters and the surface of the Cariaco basin have been examined. Methane oxidation activity occurred at all depths with low activities at the surface, a small increase at the oxic/anoxic interface and highest rates in the deepest anaerobic part of the water column (Ward *et al.*, 1987). The low turnover rates of methane in surface waters suggest that perhaps methanotrophs are not particularly active in this environment. However, we and others (see later) have readily isolated methanotrophs from such environments and therefore their relative importance in methane cycling remains to be determined. In a recent report, Scranton *et al.* (1993) investigating the control of methane fluxes in an anoxic basin of the Pettaquamscutt concluded that methane oxidation in aerobic or anaerobic regions of the water column was of minor importance in removal of methane from the system, and that methane distributions and fluxes could rapidly fluctuate in response to episodic events ranging from elution of bubbles to intrusion of water into the deeper layers of the basin. Anaerobic methane oxidation has also been a subject of much debate in the past but there is now clearly good biogeochemical and biological evidence that it occurs. This has also been reviewed previously (*eg* see Reeburgh, 1993; King, 1992; Kiene, 1991; Hanson and Wattenberg, 1991) and to date, to the Authors' knowledge, no pure cultures or consortia of anaerobic methane oxidisers have been described in the literature. This area remains an extremely fruitful one for future research.

Marine methanotrophs

The first report of the isolation of methanotrophs from the marine environment was by Hutton and ZoBell (1949). Using aged sea water or mineral salts medium they reported the enrichment of methanotrophs from a large number of different coastal mud surface samples. Pure cultures were obtained but no detailed characterisations were undertaken. Type I and Type II marine methanotrophs were reported by Heyer *et al.* (1984) but no characteristics of these organisms were given. The first detailed report of the isolation and characterisation of marine methanotrophs was by Sieburth *et al.* (1987) who described *Methylomonas pelagica* from the upper mixing layer of the Sargasso Sea. 23 different enrichments yielded the same non-pigmented obligate Type I methanotroph which required NaCl for growth. Pellicular growth was often encountered in enrichments and the addition of cycloheximide in enrichments to inhibit grazing of methanotrophs by nanoflagellates and ciliates was recommended for such enrichments. It grew only on methane or methanol and was inhibited by sunlight. Electron microscopy showed that it had a single polar flagellum and Type I intracytoplasmic membranes. % G + C content for all three isolates examined in detail ranged from 48.8 to 49.1. Key enzymes for the serine pathway, RuMP and RuBISCO were assayed for. Results indicated that formaldehyde was fixed via the RuMP pathway and that RuBISCO was absent. Ammonia and nitrate were used as nitrogen sources but there was no N_2 fixation by these strains. Antiserum was raised against *M. pelagicus* and this did not cross react with terrestrial methanotrophs (also see later). The type strain of *Methylomonas pelagica* AA-23 (NCIMB 2265) has now been examined in more taxonomic detail and has been validated and reclassified as *Methylobacter pelagicus* (Bowman *et al.*, 1993).

Lidstrom (1988) has also described the enrichment and isolation of four new marine methanotrophs from a sewage outfall sediment near Los Angeles. They required NaCl for growth and exhibit characteristics typical of Type I methanotrophs. All grew well on methane on plates and in liquid culture with doubling times of 3-4 hours. At least one strain has been trained to grow well on methanol and is being used extensively for molecular genetics of C_1 metabolism (Lidstrom, pers. comm.). All strains grew at 37°C, were motile, gave brown or creamy colonies and had a DNA % G + C ratio of 54-56. They also, unusually, contained key enzyme activities of both the RuMP and serine pathways of formaldehyde assimilation, but no RuBISCO was detected. These marine strains did not fix N_2. The type strain ACM 4717, originally referred to as A4 has now been validated as *Methylobacter marinus*. The methanol adapted strain is now referred to as strain A45 (Bowman *et al.*, 1993).

We have also isolated marine methanotrophs from sea water samples taken from Plymouth Sound at depths of 1-20m. Batch, fermenter and continuous enrichment cultures yielded methanotrophs and occasionally pellicular growth was noted. No significant problems occurred with grazing by marine eukaryotes and so cycloheximide was not employed in these enrichments. However, many enrichment cultures, clearly containing methanotrophs did not plate well even on purified agars with mineral salts and methane. This was finally overcome by the use of "sloppy agar" (0.4% w/v Noble agar plates) and as a result two marine obligate methanotrophs designated *Methylomonas* sp. IR1 and *Methylomonas* sp. DR1 were isolated. These grew well on methane or methanol at 30°C (μ max 0.2-0.23 hr^{-1}) in liquid medium with a pH optimum of 7.6. They were gram negative motile rods (1.4 x 0.8μm), contained only pMMO and assimilated formaldehyde via the RuMP pathway. DNA composition was 48-49% G+C. These marine strains were unusual in that they did not utilise ammonia as a nitrogen source. They were not N_2-fixers and employed the glutamate dehydrogenase pathway for ammonia assimilation, when the ammonia was produce intracellularly from nitrate or amino acids (Lees *et al.*, 1991). The inability of these marine strains to grow on normal agar plates (the agar was presumed to be toxic) should be noted since many marine strains could be overlooked in conventional enrichment and isolation procedures. The 16S rRNA sequence of strains IR1 and DR1 (Hanson, pers. comm.) suggest that these marine methanotrophs are phylogenetically close to *Methylobacter pelagicus*.

Another exciting area of marine methanotroph research is the recent discovery of symbiotic relationships between methanotrophic bacteria and marine invertebrates in deep sea hydrothermal vents and cold seeps (reviewed by Cavanaugh, 1993; Fisher, 1990). Several lines of evidence point to these symbionts being methanotrophs. Transmission electron microscopy has revealed the presence of gram negative bacteria with Type I (or Type X) bundles of intracytoplasmic membranes. Although MMO activity has not been detected in mytilid tissue (MMO is a labile enzyme), MDH and hexulose phosphate synthase, a key enzyme of the RuMP pathway have been measured, indicating the presence of Type I methylotrophs. ^{14}C-labelled methane incorporation and oxidation has been demonstrated. Carbon isotope composition ($\delta^{13}C$) measurements also indicate that locally produced methane is a major source of carbon (*via* bacterial methane oxidation) for these animals. A recent study of the methanotrophic symbiont in the gill of a mytilid collected 600m down on the continental slope off the coast of Louisiana was carried out. PCR amplification was used to obtain bacterial 16S rRNA gene sequences from the gill tissue of three mytilid species. Subsequent analysis of 16S rRNA sequences indicated that the symbiont of the Louisiana mussel represents a new taxon closely related but not identical to known free-living cultivable Type I methanotrophs (Distel and Cavanaugh, 1994). To date, to the Authors' knowledge, no symbionts of these mussels have been cultured in the laboratory, but it will be interesting to learn if molecular ecological techniques will reveal the presence of related, free-living methanotrophs in these environments, which may shed light on the evolution of these symbioses.

As described earlier, the strong evidence for the presence of methanotrophs in many marine environments has led to numerous attempts to isolate these organisms. Presently the only two validly described strains in pure culture are *Methylobacter pelagicus* (syn. *Methylomonas pelagica*, Sieburth *et al.*, 1987) and *Methylobacter marinus* (Lidstrom, 1987). There have also been reports of enrichment of methanotrophs in methanogenic Bacterial Consortia (MBC) (Sieburth, 1993) but there are no reports of the isolation of these organisms.

Application of molecular ecology to marine methanotrophs

As part of a study on the role of bacteria in methane cycling in the oceans we required a sensitive and reliable technique for detection of all marine methanotrophs. Previous workers had relied on the distinctive membranes found in methanotrophs or broadly reactive antisera prepared against cultivated strains for identification of methanotrophs in marine samples (Sieburth *et al.*; 1993a, Sieburth and Donaghay, 1993; DeAngelis *et al.*, 1991). However ammonia-oxidising bacteria possess similar membrane structures and the poor cultivability of marine methanotrophs makes good antisera difficult to produce. We have therefore explored the potential of nucleic acid techniques for the identification of bacteria (see Ward, 1992 for review) and applied them to the study of marine methanotrophs.

As detailed previously, methanotrophs fall into two phylogenetic clusters, within the α- and γ-Proteobacteria, based on 16S rRNA sequence analysis (Bowman, 1992; Tsuji *et al.*, 1990; Bratina *et al.*, 1992). The 16S rRNA sequences provide the largest database for design of methanotroph-specific probes. Additionally two unique enzyme systems for the oxidation of methane are found in at least some representatives of both these groups. Genes encoding these enzymes may provide additional probes. The best studied of these, the sMMO, has a restricted distribution and has been cloned and sequenced from two species (Murrell, 1983). The pMMO appears to be universally and exclusively present in methanotrophs but has proved difficult to work with and no sequence data are presently available for this enzyme complex. Together these 3 sets of genes then provide the strongest possibilities for detection and discrimination of methanotrophs using nucleic acid based approaches.

Our initial aim in this project was to broaden the diversity of marine methanotrophs in culture or at least obtain DNA sequence information from new strains to assess the validity of probes used for detection of methanotrophs in natural samples. As each of these 3 potential genetic targets has both strengths and weaknesses, we have attempted to develop an integrated approach employing all 3. Our strategy (Fig 1) revolved around two key assumptions: 1) we could enrich for methanotrophs from sea water - even if we could not isolate them, and; 2) methanotrophs belong to 2 distinct, phylogenetically defined clusters (5 genera). Methanotroph 16S rDNA sequences retrieved from enrichment cultures can then be distinguished by their branching position. Whilst it is unlikely that these two assumptions will hold true for all methanotrophic bacteria likely to be encountered (viz. anaerobic methane oxidisers) we felt that, based on present knowledge, they would allow us to dramatically increase the range of methanotrophs known from marine samples and recover the majority of methanotrophs related to the presently known forms.

ENRICHMENT STRATEGIES

Fig. 1 shows the approaches taken in this study. Enrichment cultures were established by supplying sea water with various mineral salts combinations based on the ANMS formulation of Whittenbury *et al.* (1970) and a methane:air (50:50) head space (Holmes *et al.*, 1994). From one sample the enrichment was established in a 20l fermenter to allow regular sampling with minimal disturbance of the system. Other enrichments were established in 250ml sealed flasks and sampled only at the end of the incubation period. Cells were harvested from these enrichments on Sterivex 0.22µm filters and DNA extracted by the method of Sommerville *et al.* (1989). We had previously established this technique to be appropriate for pure cultures of methanotrophs in the University of Warwick culture collection (Holmes and Murrell, unpublished).

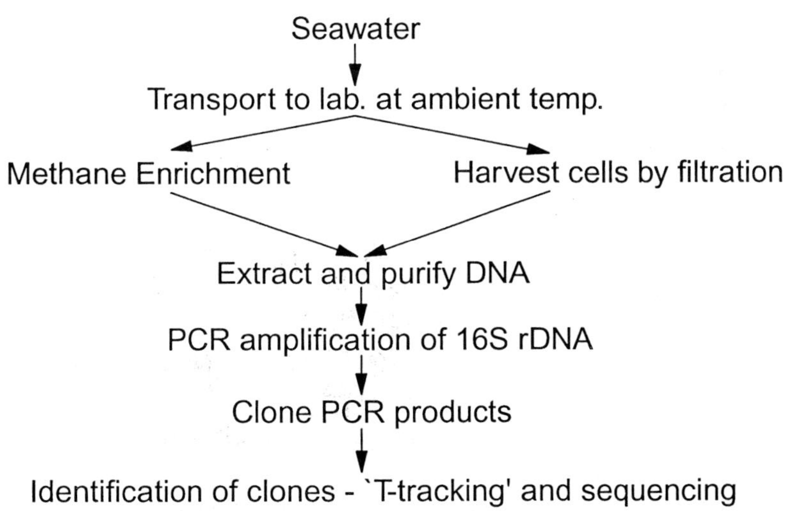

Fig 1. Experimental strategies

The bacteria-specific primers f27 and r1492 (Lane, 1991) targeting conserved regions of the 16S rDNA were used to amplify this gene from DNA samples as has been described in detail by other workers (De Long *et al.*, 1993; Britschgi and Giovannoni, 1991). PCR products were cloned via the T/A cloning kit (Invitrogen) to create 16S rRNA gene libraries representative of different stages of the culture (fermenter enrichment) and of the mature enrichments (all cultures). Clones which showed greater representation in the gene library after enrichment were considered to represent potential methanotrophs.

Many workers have expressed concern with the potential introduction of bias in environmental gene libraries by PCR (Ward *et al.*, 1992; Reysenbach *et al.*, 1992). Such biases could affect the comparisons of gene libraries in the strategy used here. To test the reproducibility of PCR, amplicons from the original sea water sample and the 1 month fermenter culture were digested with *Sau*3A and separated on a 1% agarose gel. Distinct and reproducible banding patterns were found for these two samples This reproducibility of the PCR gives us confidence that the difference between these two samples reflects a shift in bacterial population rather than an artefact introduced during amplification. However, within each sample the frequency of occurrence of a particular sequence type may not reflect exactly its original representation in the initial DNA sample due to potentially biased amplification in PCR (Reysenbach *et al.*, 1992).

The two gene libraries constructed from these samples were labelled Plym (original Plymouth sea water sample) and Month (1 month fermenter sample). Approximately 100 clones from each library were "mini-prepped" and the plasmids screened by 'T'-tracking (Schmidt *et al.*, 1991). Clones were initially classified according to their 'T' track signature and sorted into clusters. The results of this are summarised in Fig. 2.

250 bases of sequence information were read from the primer r1492. It was hoped that the relatively short stretch of sequence information would minimise the problem of chimaeric clones (which may not be recognised by single-track sequencing) giving a false impression of variability (Kopczynski *et al.*, 1994). The variable region of bases 1110-1160 was included in this analysis allowing discrimination between even closely related strains.

There was a dramatic difference in composition between the two gene libraries, confirming the population shift predicted from the RFLP-PCR experiment. One sequence type in particular, Type 20, was not represented in the Plym library but provided 22% of the clones in the month library. This dramatic increase in proportional representation is unlikely to be solely due to bias introduced in PCR. Three other sequence types also had multiple clones in the month library but were absent from the Plym library, although the numbers of these clones were not as significant. Only one sequence type, Plym 5, was found in high proportion in both libraries. Approximately half of all sequence types retrieved had only a single representative in one of the two libraries (especially for Plym library). It would be necessary to sequence many more clones or probe the original samples with specific oligonucleotides to confirm that organisms represented by sequence types 21 and 22 had indeed been enriched.

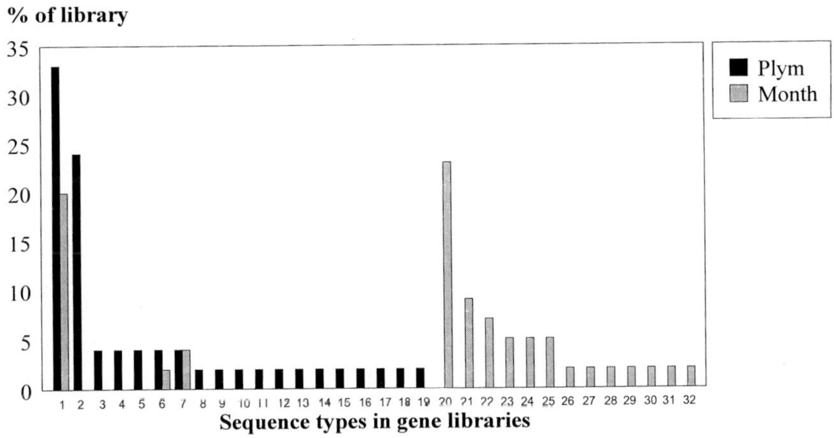

Fig. 2 Population shift in the Plym and Month gene libraries seen after enrichment with methane.

While methanotrophs could be expected to be one of the dominant cell types in 'mature' enrichments, as discussed above the frequency of occurrence in the gene library may not exactly correspond to the frequency in the original DNA samples. To attempt to confirm a methanotrophic physiology for retrieved clones, further experiments were performed as described below.

ESTABLISHING METHANOTROPHIC PHYSIOLOGY

Fermenter culture

Representative clones of each of the sequence types 1, 2, 20, 21 and 22 were selected for further analysis and fully sequenced using described 16S rRNA primers (Lane, 1991) and the Sequenase (USB) sequencing kit. Phylogenetic trees constructed from these data showed the Type 20 clones to be γ-Proteobacteria, but not to branch with the methanotrophic representatives of this subclass (Holmes *et al.*, 1994). Subcultures of this enrichment in methane/mineral salts media failed to show significant bacterial growth. Primers specific for the *moxF* gene of methanol dehydrogenase did give an amplification product from the

Month DNA however this enzyme is found in all methylotrophs and is not exclusive to methanotrophs. This gene was also amplified from most sea water samples and all successful enrichment cultures (Holmes and Murrell, unpublished). We were unable to confirm the presence of sMMO from any of the enrichments in this study. It seems likely that these organisms are not methanotrophic, although some methylotrophs are present, and that while we detected a population shift after enrichment on methane, it was not due to growth on this substrate.

Small scale enrichments.

100ml cultures in 250 ml flasks were also established with various sea water samples. From Bristol Channel samples, replicate cultures supplied with AMS, NMS or ANMS (Whittenbury *et al.*, 1970) and methane were set up. The NMS culture showed orange pigmented flocs of growth after 1 month incubation with constant shaking. Gene libraries of 16S rDNA were constructed as described earlier and 24 clones screened by T-tracking. 6 were identical and showed high similarity to Type I methanotroph sequences. One of these Pen 2 and another distinct sequence type Pen 10 were fully sequenced and phylogenetic trees constructed using programs from the PHYLIP v3.4 package (Felsenstein, 1988). Pen 2 branched within the radiation of Type I methanotroph (genera *Methylomonas*, *Methylococcus* and *Methylobacter*) in all analyses (Fig. 3). This sequence showed no evidence of a chimaeric nature by secondary structure consideration and tree construction with partial sequences. The high proportion of this sequence type in the library was confirmed by colony blots with a *Methylomonas* group specific probe. Positive clones comprised 49% of the library. To the best of our knowledge the Type I methanotrophs are a phylogenetically coherent group with no non-methanotrophic relatives reported. The phylogenetic position of Pen 2 within this cluster and its prevalence in the gene library then provide strong circumstantial evidence of a methanotrophic physiology for the source organism of this clone. Other evidence is the pigmentation of the enrichment culture and the formation of a pellicle in standing culture. These are both attributes of *Methylomonas* spp., the closest phylogenetic relatives of the PEN 2 clone, which are not known from other described species of methanotrophs (see Table 1).

Fluorescently labelled oligonucleotide probes specific for the PEN2/*Methylomonas* group were synthesised and used for *in situ* hybridisations to identify the morphology of this organism. Results with the enrichment cultures were significantly improved upon the inclusion of formamide in the hybridisation buffer (Manz *et al.*, 1993) whilst this had no significant effect in earlier experiments using pure cultures of methanotrophs. Fig. 4 shows phase contrast and fluorescence micrographs of a subculture from this enrichment. The large (4 x 2μm) rods are the only morphotype to hybridise to the group-specific probes Mm 650 and Mm 850, whilst all morphotypes hybridised to the kingdom-specific probe Eub 338 (Amann *et al*, 1990). This cell morphology is similar to typical representatives of the genus *Methylomonas* (Bowman *et al*., 1993).

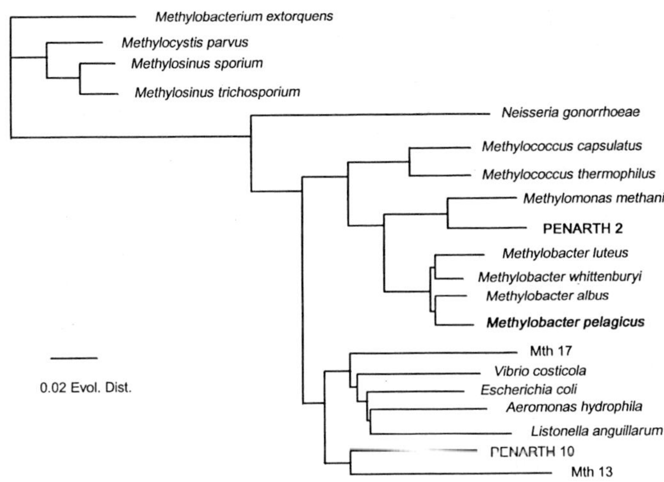

Fig 3. Phylogenetic tree showing the position of the clone Pen2 within the Type I methanotrophs and distinct from *Methylobacter pelagicus*. The tree was constructed using the programs DNADIST and FITCH from PHYLIP v3.4. Representative of sequence types 20 (Mth 17) and type 21 (Mth 13) from the Plymouth libraries show no relationship to methanotrophs

The intensity of staining of cells with rRNA-targeted probes is proportional to the number of ribosomes within the cell which can limit their effectiveness for environmental detection (Ward *et al*., 1992). Intracellular levels of rRNA are also thought to reflect the metabolic activity of a cell, especially in culture (Kemp *et al*., 1993; Lee *et al*., 1993). We have exploited this to confirm the physiology of our putative methanotroph. Cultures were grown in 3 different media; AMS/methane, 0.05% yeast extract/AMS/methane and 0.05% yeast extract/AMS. Cells from these enrichment cultures were them hybridised to our suite of probes. In the methane only cultures the Pen 2 cells are the only ones to show bright fluorescence with the Eub338 probe. In yeast extract plus methane cultures both non-Pen 2 and Pen 2 cell types show a strong signal. In yeast extract only cultures the PEN2 cells fluoresce very weakly and only the non-Pen 2 type show a strong signal. The signal of Pen 2 cells from this latter culture is increased again by supply of methane for 24 hours. These results considerably strengthen the case for the Pen 2 cells having an obligately methanotrophic physiology - as predicted from their phylogenetic position.

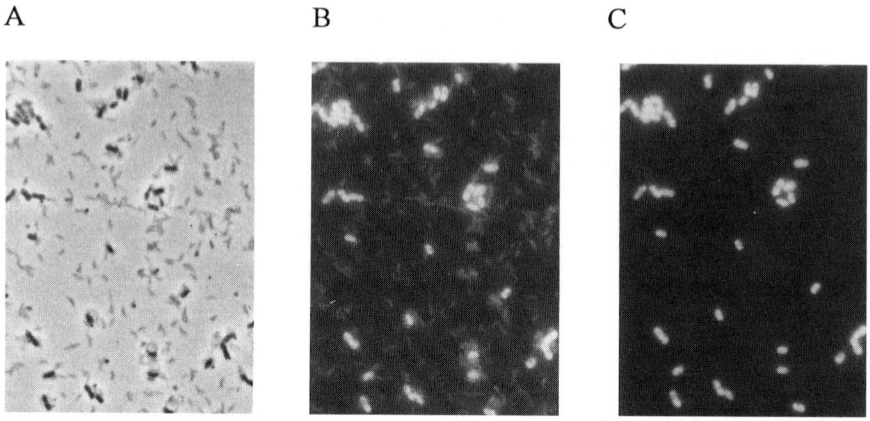

Fig 4. Penarth enrichment grown in 0.01% yeast extract, ammonium mineral salts, methane, 3 days, 30°C. Cells were hybridised to a mixture of Eub338 (Fluorescein), Mm 650 (Rhodamine) and Mm 850 (Rhodamine) in a buffer containing 20% formamide at 37°C for 2 hours. The cells were viewed on a Zeiss Axioskop 20 microscope under a) Phase contrast, b) epifluorescence with filter set 09 (fluorescein) and c) epifluorescence with filter set 15 (Rhodamine). The large rod-shaped cells stain specifically with Mm 650/Mm 850. Cell clumps typical of these cultures are visible. Sequences of the Pen 2 specific probes are Mm 650 CCTCTACTCAACTCTCTAGT and Mm 850 TACGTTAGCTCCACCACTAA.

Conclusions

The conclusive proof of an organism's physiological mode requires its isolation in pure culture. For many organisms, including some marine methanotrophs, this has not proved possible with present cultural techniques. We have obtained very strong evidence, in the absence of isolation, that the organisms represented by Clone Pen 2 have an obligately methanotrophic physiology. This evidence includes morphology, phylogenetic position and cellular ribosome content in the present and absence of methane. Our success here gives us confidence in the efficacy of the probes developed against culturable, terrestrial methanotrophs for detection of the non-culturable marine methanotrophs.

In many respects the Penarth enrichment cultures are analogous to the Methanogenic Bacterial Consortia (MBC) reported by Sieburth (1993). In the early stages of growth, our cultures showed very few 'free-swimming' Pen 2 cells.

Rather these were found in aggregated clumps of cells which always contained at least two other morphotypes, a hyphal organism and a small rod-shaped cell. Only in mature (>3day) cultures were significant numbers of free Pen 2 cells present. In plate cultures the number of colony-forming units corresponded more closely to the number of aggregated cell clumps than the number of free-swimming cells in the inoculum. Colonies which formed were always composed of 3 or more morphotypes. As no growth occurred in the absence of methane, there are apparently no methanogens involved in the cell associations seen here in contrast to the observation of Sieburth (1993). This probably reflects the different enrichment conditions used. Our results suggest that some marine methanotrophs require growth factors, in addition to methane, supplied by more than one organism. The MBC proposed by Sieburth *et al.* (1993a) could well contain such organisms. These syntrophic aggregates may be naturally occurring in marine snow. Open ocean macroaggregates have been reported to evolve methane (Karl and Tilbrook, 1994), have both aerobic and anaerobic portions, (Shanks and Reader, 1993) and contain complex bacterial communities distinct from the pelagic population (De Long *et al.*, 1993).

As discussed above, and in detail in Sieburth (1993), methanotrophs are potentially a significant component of the oceanic carbon cycle and the major sink of methane in the oceans. We plan to further refine our existing suite of probes though similar studies in a wider range of marine environment including methane seeps, organic rich marine sediment and stratified sea. The extant probes may be used in exploration of macroaggregates, for the presence of methanotrophs, to confirm the hypothesis presented here.

References

Akent'eva NF and Gvozdev RI (1988) Purification and physiochemical properties of methane monooxygenase from membrane structures of *Methylococcus capsulatus. Biokhymija* **53:** 91-96

Alvarez-Cohen L. (1993) Application of methanotrophic oxidations for the bioremediation of chlorinated organics. In: *Microbial Growth on* C_1 *Compounds* Murrell JC and Kelly DP (eds) Intercept Press, Andover, UK 337-350

Alvarez-Cohen L, McCarty PL, Boulygina E., Brusseau G, Hanson RS (1992) Cometabolic biotransformation of trichloroethylene and chloroform by a bacterial consortium grown with methane *Appl Environ Microbiol* **58**: 1886-1893

Amann RI, Binder B, Chisholm SW, Olsen R, Devereaux R, Stahl DA (1990) Combination of 16S rRNA-targetted oligonucleotide probes with flow cytometry for analysing mixed microbial populations. *Appl Environ Microbiol* **56:** 1919-1925

Anthony C. (1982) *The Biochemistry of Methylotrophs*. Academic Press, London.

Anthony C. (1986) Bacterial oxidation of methane and methanol. *Adv Micro Physiol* **27:**113-210

Barta TM and Hanson RS (1993) Genetics of methane and methanol oxidation in gram negative methylotrophic bacteria. *Anton van Leeuw 64*: 109-120

Boulygina ES, Chumakov KM, Netrusov AI (1993) Systematics of gram-negative methylotrophic bacteria based on 5S rRNA sequences In: *Microbial Growth on C_1 Compounds* Murrell JC and Kelly DP (eds) Intercept Press, Andover, UK 275-284

Bowman JP, Sly I, Nicholas PD, Hayward AC (1993) Revised taxonomy of the methanotrophs: description of *Methylobacter* gen. nov., emendation of *Methylococcus*, validation of *Methylosinus* species and a proposal that the family *Methylococcaceae* includes only group I methanotrophs. *Int J Syst Bacteriol* **43:** 735-753.

Bowman JP (1992) *The systematics of methane-utilizing bacteria.* PhD Thesis University of Queensland, Brisbane, Australia

Bratina BJ, Brusseau GA, Hanson RS (1992) Use of 16S rRNA analysis to investigate phylogeny of methylotrophic bacteria. *Int J System Bacteriol* **42:** 645-648

Britschgi TB and Giovannoni SJ (1991) Phylogenetic analysis of a natural marine bacteriaplankton population by rRNA gene cloning and sequencing. *Appl. Environ. Microbiol.* **57:** 1707-1713

Brusseau GA, Bulygina ES, Hanson RS (1994) Phylogenetic analysis and development of probes for differentiating methylotrophic bacteria. *Appl Env Microbiol* **60:** 626-636

Cavanaugh CM (1993) Methanotroph-invertebrate symbioses in the marine environment: ultrastructure, biochemical and molecular studies. In: *Microbial Growth on C_1 Compounds* Murrell JC and Kelly KP (eds) Intercept, Andover, UK 315-328

Chan SI, Nguyen H-HT, Shiemke AK, Lidstrom ME (1993) Biochemical and biophysical studies towards characterization of the membrane-associated methane monooxygenase. In: *Microbial Growth on C_1 Compounds* Murrell JC and Kelly KP (eds) Intercept Press, Andover, UK 93-107

Cicerone RJ and Oremland RS (1988) Biogeochemical aspects of atmospheric methane. *Global Biogeochem. Cycles* **2:** 299-327

Conrad R and Seiler W (1988) Influence of the surface microlayer on the flux of non conservative trace gases (CO, H_2, CH_4, N_2O) across the ocean-atmosphere interface. *J Atmosph Chem* **6:** 83-94

Dalton H, Wilkins PC, Jiang Y (1993) Structure and mechanism of action of the hydroxylase of soluble methane monooxygenase. In: *Microbial Growth on C_1 Compounds* Murrell JC and Kelly KP (eds) Intercept Press, Andover, UK 65-80

Dalton H (1992) Methane oxidation by methanotrophs, physiological and mechanistic implications. In: *Methane and Methanol Utilizers* Murrell JC and Dalton H, (eds) Plenum, New York 85-114

De Angelis MA, Lulley MD, Olsen EJ, Baross JA (1993) Methane oxidation in deep-sea hydrothermal plumes of the Endeavour segment of the Juan de Fuca Ridge. *Deep Sea Res* **40:** 1169-1186

DeAngelis MA, Reysenbach A, Baross JA (1991) Surfaces of hydrothermal vent invertebrates: Sites of elevated microbial CH_4 oxidation activity. *Limnol Oceanogr* **36:** 570-577

DeLong EF, Franks DG, Aldredge AL (1993) Phylogenetic diversity of aggregate-attached vs. free-living marine bacterial assemblages. *Limnol Oceanogr* **38:** 924-934

Distel DL and Cavanaugh CM (1994) Independent phylogenetic regions of methanotrophic and chemoautotrophic bacterial endosymbioses in marine bivalves. *J Bacteriol* **176:** 1932-1938

Dworkin M and Foster JW (1956) Studies on *Pseudomonas methanica* (Söhngen) nov. comb. *J Bacteriol* **72:** 646-659

Felsenstein J (1988) Phylogenies from molecular sequences: Inference and reliability. *Ann Rev Genet* **22:** 521-565

Fisher CR (1990) Chemoautotrophic and methanotrophic symbioses in marine invertebrates. *Rev Aquat Sci* **2:** 399-436

Fitch MW, Graham DW, Arnold RG, Agarwal SK, Phelps P, Speitel Jr GE, Georgiou G (1993) Phenotypic characterization of copper-resistant mutants of *Methylosinus trichosporium* OB3b *Appl Env Microbiol* **59:** 2771-2776

Froland WA, Andersson KK, Lee S-K, Liu Y, Lipscomb JD (1993) The catalytic cycle of methane monooxygenases and the novel roles played by protein component complexes during turnover. In: *Microbial Growth on C_1 Compounds* Murrell JC and Kelly KP (eds) Intercept Press, Andover, UK 81-92

Gal'chenko VF, Abramochkina FN, Bezrukova LV, Sokolova EN, Ivanov MW (1988) Species composition of aerobic methanotrophic microflora in the Black Sea. *Microbiologica* **57:**305-311

Green PN (1992) Taxonomy of methylotrophic bacteria. In: *Methane and Methanol Utilizers* Murrell JC and Dalton H (eds) Plenum, New York 23-84

Green PN (1993) Overview of the current state of methylotroph taxonomy. In: *Microbial Growth on C_1 Compounds* Murrell JC and Kelly KP (eds) Intercept Press, Andover, UK 253-265

Hanson RS and Wattenberg EV (1991) Ecology of methylotrophic bacteria, In: *Biology of Methylotrophs* Goldberg I and Rokem JS (eds.) Butterworth-Heinemann, USA 325-348

Hanson RS (1980) Ecology and diversity of methylotrophic organisms. *Adv Appl Microbiol* **26:** 3-39

Hanson RS, Bratina BJ, Brusseau GA (1993) Phylogeny and ecology of methylotrophic bacteria. In: *Microbial Growth on C_1 Compounds* Murrell JC and Kelly KP (eds) Intercept Press, Andover, UK 285-302

Hanson RS and Wattenberg EV (1991) Ecology of methylotrophic bacteria. In: *Biology of methylotrophs* Goldberg I. and Rokem JS (eds) Butterworth-Heinemann, Boston, USA 325-348

Heyer J, Malashenko UU, Berger U, Budkova E (1984) Verbreiting methanotropher Bakterien *Zeitschr für allg Mikrobiol* **24:** 725-744

Holmes AJ, Owens NJP, Murrell JC (1995). 16S rRNA-based methods for monitoring enrichment cultures of marine methane-oxidising bacteria (In preparation).

Hutton WE and ZoBell CE (1949) The occurrence and characteristics of methane-oxidising bacteria in marine sediments. *J Bacteriol* **58:** 463-473

Jensen P, Aagaard I, Burk RA, Dando PR, Jorgensen NO, Kuipers A, Laier T, O'Hara SCM, Schmaljohann R (1992) "Bubbling reefs" in the Kattegat: submarine landscapes of carbonate-cemented rocks support a diverse ecosystem at methane seeps. *Mar Ecol Prog Ser* **83:** 103-112

Karl DM and Tilbrook BD (1994) Production and transport of methane in oceanic particulate organic matter. *Nature* **368:** 732-734

Kemp PF, Lee S, LaRoche J (1993) Estimating the growth rate of slowly growing marine bacteria from RNA content. *Appl Environ Microbiol* **59:** 2594-2601

Kiene RP (1991) Production and consumption of methane in aquatic systems, In: *Microbial Production and Consumption of Greenhouse Gases: Methane,*

Nitrogen Oxides and Halomethanes Rogers. JE and Whitman WB, (eds) American Society for Microbiology, Washington DC 111-146

Kiene RP (1993) Microbial sources and sinks for methylated sulphur compounds in the marine environment. In: *Microbial Growth on C_1 Compounds* Murrell JC and Kelly KP (eds) Intercept, Andover, UK 15-33

King GM (1992) Ecological aspects of methane oxidation, a key determinant of global methane dynamics. *Adv Microb Ecol* **12:** 431-468

King GM (1993) Ecophysiological characteristics of obligate methanotrophic bacteria and methane oxidation *in situ.* In: *Microbial Growth on C_1 Compounds* Murrell JC and Kelly KP (eds) Intercept Press, Andover, UK 303-313

Koh S-C, Bowman JP, Sayler GS (1993) Soluble methane monooxygenase production and trichloroethylene degradation by a type 1 methanotroph *Methylomonas methanica* 68-1. *Appl Env Microbiol* **59:** 960-967

Kopczynski ED, Bateson MM, Ward DM (1994) Recognition of chimeric small-subunit ribosomal DNAs composed of genes from uncultivated microorganisms. *Appl Environ Microbiol* **60:** 746-748

Lane DJ (1991) 16S/23S rRNA sequencing. In *Nucleic acid techniques in bacterial systematics.* Stackebrandt E and Goodfellow M (eds) Wiley, Chichester 115-175

Leak DJ (1992) Biotechnological and applied aspects of methane and methanol utilizers. In: *Methane and Methanol Utilizers* Murrell JC and Dalton H (eds) Plenum Press, New York 245-279

Lee S, Malcone C, Kemp PF (1993) Use of multiple 16S rRNA-targetted fluorescent probes to increase signal strength and measure cellular RNA from natural planktonic bacteria. Mar Ecol Prog Ser **101:** 193-201

Lees V, Owens NJP, Murrell JC (1991) Nitrogen metabolism in marine methanotrophs. *Arch Microbiol* **157:** 60-65

Lidstrom ME (1988) Isolation and characterisation of marine methanotrophs. *Anton van Leeuw* **54:** 189-199

Lidstrom ME and Stirling DI (1990) Methylotrophs: genetics and commercial applications. *Ann Rev Microbiol* **44:** 27-58

Manz W, Amann R, Ludwig W, Wagner M, Schleifer K-H (1992) Phylogenetic oligodeoxynucleotide probes for the major subclasses of Proteobacteria: problems and solutions. *System Appl Microbiol* **15:** 593-600

McDonald IR, Kenna EM, Murrell JC (1994) Detection of methanotrophic bacteria in environmental samples by using the polymerase chain reaction. *Appl Env Microbiol* (In press)

Murrell JC and Dalton H (1983) Nitrogen fixation in obligate methanotrophs. J Gen Microbiol **129:** 3481-3486

Murrell JC (1994) Molecular genetics of methane oxidation. *Biodegradation.* (In Press)

Murrell JC (1993) Molecular biology of methane oxidation In: *Microbial Growth on C_1 Compounds* Murrell JC and Kelly KP (eds) Intercept, Andover, UK 109-120

Murrell JC, McGowan V, Cardy DLN (1992) Detection of methylotrophic bacteria in natural samples by molecular probing techniques. *Chemosphere* **26:** 1-11

Murrell JC (1992) Genetics and molecular biology of methanotrophs. *FEMS Microbiol Rev* **88:** 233-248

Nakajima T, Uchiyama H, Yagi O, Nakahara T (1992) Purification and properties of a soluble methane monooxygenase from *Methylocystis* sp. M. *Biosci Biotech Biochem* **56:** 736-740

Nguyen HT, Shiemke AK, Jacobs SJ, Hales BJ, Lidstrom ME, Chan SI (1994) The nature of the copper ions in the membranes containing the particulate methane monooxygenase from *Methylococcus capsulatus* (Bath). *J Biol Chem* **269:** 14995-15005

Oldenhuis R and Janssen DB (1993) Degradation of trichlorethylene by methanotrophic bacteria. In: *Microbial Growth on C_1 Compounds* Murrell JC and Kelly KP (eds) Intercept Press, Andover, UK 121-133

Phelps PA, Agarmal SK, Speital GE, Georgiou G (1992) *Methylosinus trichosporium* OB3b mutants having constitutive expression of soluble methane monooxygenase in the presence of high levels of copper. *Appl Env Microbiol* **58:** 3701-3708

Pilkington SJ and Dalton H (1991) Purification and characterization of the soluble methane monooxygenase from *Methylosinus sporium* 5 demonstrates the highly conserved nature of this enzyme in methanotrophs. *FEMS Microb Lett* **78:** 103-108

Reeburgh WS (1993) The role of methylotrophy in the global methane budget. *Microbial Growth on C_1 Compounds* Murrell JC and Kelly KP (eds)Intercept Press, Andover, UK 1-14

Reed WM and Dugan PR (1978) Distribution of *Methylomonas methanica* and *Methylosinus trichosporium* in Cleveland Harbor as determined by an indirect fluorescent antibody-membrane filter technique. *Appl Env Microbiol* **35:** 422-430

Reysenbach A-L, Giver LJ, Wickham GS, Pace NR (1992) Differential amplification of rRNA genes by Polymerase Chain Reaction. *Appl Environ Microbiol* **58:** 3417-3418

Rudd JWM and Taylor CD (1980) Methane cycling in aquatic environments. *Adv Aquat Microbiol* **2:** 77-150

Schmidt TM, DeLong EF, Pace NR (1991) Analysis of a marine picoplankton community by 16S rRNA gene cloning and sequencing. *J Bacteriol* **173:** 4371-4378

Scranton MI and Brewer PG (1977) Occurrence of methane in the near-surface waters of the western sub tropical North Atlantic *Deep-Sea Res* **24:** 127-138

Shanks AL and Reader ML (1993) Reducing microzones and sulfide production in marine snow. *Mar Ecol Prog Ser* **96:** 43-47

Sieburth J McN (1993) C_1 bacteria in the water column of Chesapeake bay, USA. I. Distribution of sub-populations of O_2-tolerant, obligately anaerobic, methylotrophic methanogens that occur in microniches reduced by their bacterial consorts. *Mar Ecol Prog Ser* **85:** 67-80

Sieburth JMcN and Donaghay PL (1993a) Planktonic methane production and oxidation within the algal maximum of the pycnocline: seasonal fine-scale observations in an anoxic estuarine basin. *Mar Ecol Prog Ser* **100:** 3-15

Sieburth JMcN, Johnson PW, Eberhardt MA, Sieracki ME, Lidstrom M, Laux D (1987) The first methane-oxidising bacterium from the upper mixing layer of the deep ocean: *Methylomonas pelagica* sp. nov. *Current Microbiol* **14:** 285-293

Sieburth JMcN, Johnson PW, Macario AJL, Conway de Macario E (1993a) C_1 bacteria in the water column of Chesapeake Bay, USA. II. The dominant O_2- and H_2S-tolerant methylotrophic methanogens, co-enriched with their oxidative and sulphate reducing bacterial consorts, are all new communotypes and probably include new taxa. *Mar Ecol Prog Ser* **95:** 81-89

Sieburth JMcN, Johnson PW, Church VM, Laux DC (1993b) C_1 bacteria in the water column of Chesapeake Bay, USA. III. Immunologic relationships in the type species of marine monomethylamine- and methane-oxidising bacteria to wild estuarine and oceanic cultures. *Mar Ecol Prog Ser* **95:** 91-102

Sieburth JMcN (1991). Methane and hydrogen sulphide in the pycnocline: a result of tight coupling of photosynthetic and "benthic" processes in stratified waters. In: *Microbial Production and Consumption of Greenhouse Gases: Methane, Nitrogen oxides and Halomethane* Rogers JE and Whitman WB, (eds) American Society for Microbiology Press, Washington DC, USA 147-174

Smith DDS and Dalton H (1989) Solubilization of methane monooxygenase from *Methylococcus capsulatus* (Bath). *Eur J Biochem* **182:** 667-671

Söhngen NL (1906) Üeber Bakerien, Welche methan als kohlenstoffnahrung und energiquelle gebrauchen. *Z Bakteriol Parasiterk (Abt.11)* **15:** 513-517

Sommerville CC, Knight IT, Strawbe WL, Colwell RR (1989) Simples, rapid methods for direct isolation of nucleic acids from aquatic environments. *Appl Environ Microbiol* **55:** 548-554

Stainthorpe AC, Salmond GPC, Dalton H, Murrell JC (1990) Screening of obligate methanotrophs for soluble methane monooxygenase genes. *FEMS Microbiol Lett* **72:** 345-348

Stephens RL, Haygood MG, Lidstrom ME (1988) Identification of putative methanol dehydrogenase (*mox*F) structural genes in methylotrophs and cloning of *mox*F genes from *Methylococcus capsulatus* (Bath) and *Methylomonas albus* BG8. *J Bacteriol* **170:** 2063-2069

Topp E and Hanson RS (1991) Metabolism of radiatively important trace gases by methane-oxidizing bacteria. In: *Microbial production and consumption of greenhouse gases: methane, nitrogen oxides and halomethanes* Rogers JE and Whitman WB (eds) American Society for Microbiology, Washington DC 71-90

Tsien HC, Bratina BJ, Tsuji K, Hanson RS (1990) Use of oligonucleotide signature probes for identification of physiological groups of methylotrophic bacteria. *Appl Environ Microbiol* **56:** 2858-2865

Tsien HC and Hanson RS (1992) A soluble methane monooxygenase component B gene probe for the identification of methanotrophs that rapidly oxidize trichloroethylene. *Appl Environ Microbiol* **58:** 953-960

Tsuji K, Tsien HC, Bratina B, Bastien C, Zhang Y, Machlin S, Hanson RS (1989) *Genetic and biochemical studies of methylotrophic bacteria.* IGT Press, Chicago, USA

Tsuji K, Tsien HC, Hanson RS, De Palma SR, Scholtz R, La Roche S (1990) 16S ribosomal RNA sequences analysis for determination of phylogenetic relationship among methylotrophs. *J Gen Microbiol* **136:** 1-10

Ward BB, Kilpatrick KA, Nordli PC, Scranton MI (1987) Methane oxidation and methane fluxes in the ocean surface layer and deep anoxic waters. *Nature* **327:** 226-229

Ward DM, Bateson MM, Weller R, Ruff-Roberts AL (1992) Ribosomal RNA analysis of microorganisms as they occur in nature. *Adv Microb Ecol* **12:** 209-286

Whittenbury R, Phillips KC, Wilkinson JF (1970) Enrichment, isolation and some properties of methane-utilizing bacteria. *J Gen Microbiol* **61:** 205-218

Microbial Cultures and Natural Populations

Noel G Carr
Department of Biological Sciences
University of Warwick
Coventry CV4 7AL
United Kingdom

For some years there have been descriptions, particularly from oligotrophic environments, of microorganisms that evade culture. Initially these were based upon the very low proportion of colonies obtained relative to the number of microscopically observed bacteria in a sample. Recently there has been several papers that, using molecular biological methods, describe a much greater phylogenetic diversity of bacteria in the open ocean than could have been forecast from traditional means of bacterial culture (see Giovannoni, 1995). Without doubt this is partly a result of our less than perfect skills in isolation but the very low recovery rate reported, often 1 in 10^3 or 10^4 leads one to think that other factors may be involved. When bacteria are brought into culture and maintained in the laboratory their growth rate and maximum culture density often increase over periods of months or even years. They 'adapt' to laboratory culture. The roots of the idea that the organisms that microbiologists study in the laboratory are often but not always only representative, rather than a cross section, of those from their natural environment is not new and goes back to what has been termed the golden age of bacteriology around the end of the nineteenth century. The very use of the term 'elective culture' and the application of the technique by Winogradsky and Beijerinck to isolate pure cultures of bacteria involved in what we now call biogeochemical cycles, implies the selection from a diverse population of the individual species which were thought to be responsible for particular chemical transformations of environmental significance. These strategies in turn had built upon the recognition of the role of specific bacteria in different mammalian and plant diseases. Such reservations regarding isolated cultures of phytoplankton were probably less important because the microscopic comparison of laboratory and natural populations was more convincing, although it is recognised that important physiological differences may be obscured by morphological similarity.

NATO ASI Series, Vol. G 38
Molecular Ecology of Aquatic Microbes
Edited by I. Joint

The isolation of a vast, and seemingly unlimited, array of axenic bacterial cultures followed during the course of this century and in many cases there was an understanding at the physiological level of the features that led to their success, as pathogens or as agents in the natural environment. There was a parallel emphasis on division rate and the ways by which this could be maximised; this was clearly important to both laboratory research using microorganisms and to their increasing industrial applications. Microorganisms became central to the study of biochemistry and were absolutely crucial in the development of molecular biology. There were however some downsides to this success story in the form of the emphasis placed on aspects of research which attracted, with some honourable exceptions, the attention of microbiologists. These included: an unhealthy confidence in the concept of a bacterial species and a resulting taxonomy; a lack of interest in the organisation of microbial communities; an exclusive enthusiasm for rapid growth kinetics; the belief that the understanding of a 'type' organism would hold for all/most others and finally a disinclination to study organisms that did not exhibit these characteristics.

What this short essay attempts to explore is the way in which the conventional description of microbial growth can be reconciled with our view of natural populations and what we may learn from the latter that could cause us to question our understanding of the former. To some extent asking such a question is pushing at an open door; many of the assumptions of traditional microbiology outlined above have and continue to be challenged by the developments in microbial ecology and these challenges are increasingly fuelled by the application of molecular biology.

The species concept as applied to bacteria has been increasingly difficult to defend for most organisms partially as a result of knowledge about lateral gene transfer *via* plasmids. There are certainly some bacteria, *Escherichia coli* for example, that can confidently be considered as species on the basis of genomic DNA exchange and the essential sequence similarity of most of their DNA. There is also of course perfectly reasonable use of the term at the identification level (see Logan 1994) and in quantifying bacterial diversity in an ecological sense (O'Donnell *et al.* 1994). But unease arises in surveying the thousands of named species in the literature; it is difficult to consider seriously the idea of a species that is divorced from relatedness. The emergence of a true phylogeny of bacteria based on the gene sequence for 16S RNA (Woese 1987) must be one of the great triumphs of modern biology and it demands reconsideration of relatedness and diversity. The construction of a phylogenetic tree for bacteria, and the relationship to that of the eukaryotes, allows rational thinking about the microbial evolution (see Doolittle and Brown 1994). A practical consequence of the establishment of a molecular phylogeny of bacteria has been the overturn of several dearly held beliefs regarding bacterial relatedness. Perhaps the most striking of these has been the collapse of the genus *Pseudomonas* as a coherent entity. It is now evident that bacteria previously classified and studied together as a group, should be seen as distributed throughout the proteo-bacterial subclass with a rather low degree of relatedness - other than actually being bacteria. Nothing better illustrates pitfalls of grouping bacteria together on the basis of a limited number of phenotypic features, no matter how interesting those are thought to be. We can now think about bacteria in evolutionary terms and this will, or should, in due course inform laboratory as well as ecological work. The well known phrase of Dobzhansky that "nothing in biology makes sense, except in the light of evolution" applies equally to bacteria as it does to higher organisms. It would be reasonable to hope that the image of the unrooted tree of linking bacterial and eukaryotic phylogeny (Woese 1987) will in due course be adopted as a contemporary icon, as has happened with that of the double helix and the satellite image of the earth.

The role of communities of microorganisms acting together to carry out a particular physiological function or geochemical change is well established in environmental microbiology. The bacterial mats sometimes found on the surface of salt marshes are an example in which multi-organism assemblages have an organised structure which facilitates their interactions. In a simple description of such structures, cyanobacteria form the outermost layer underneath which lie sulphide-utilising photosynthetic bacteria, the sulphide being produced by heterotrophic sulphate-reducing organisms which form an inner, black region. The assemblage of different bacterial types allows the partitioning of light and the interchange of metabolic products. The interaction between a limited number of observed and isolated bacterial strains in cyanobacterial mats has already been shown to be an over-simplification by the application of molecular analysis based upon 16s rRNA diversity (Weller *et al.* 1992). Two types of bacteria acting together to carry metabolic conversions without physical contact has long been recognised as an important feature in microbial ecology. The inter-species hydrogen transfer between a heterotroph and a methanogen leading to methane production being one example. The lack of apparent physical contact makes the analysis of such obligatory associations difficult. Elective continuous culture in a chemostat has been successful in the repeated isolation of a collection of six bacterial strains that were necessary for the mineralisation of the xenobiotic herbicide 2-chloroproprionamide (Slater and Bull 1982). These workers used the term 'consortia' to describe the obligatory community of six bacterial strains, without apparent physical contact, which emerged in a continuous-flow enrichment of a soil sample in a chemostat with the herbicide as growth-limiting carbon and energy source. Co-metabolism was involved in the initial attack on the herbicide and two members of the consortia had no function in the core catabolic events and may exist on excreted metabolites or the products of cell lysis.

We should consider whether or not some of the unculturable organisms in the ocean may be components of consortia which are involved in the elemental cycles which are known to occur there. The possible role of marine snow as a matrix for consortia of bacteria is an obvious example and already is an area of investigation. There have been over the last few years increasing indications that upper oceans may be the site of methane production as well as an a zone of methane oxidation (Oremland 1979). Since all the known methanogenic bacteria are extremely strict anaerobes and the upper oceans are highly oxygenated, some form of separation of the two processes should exist. Methanogenic activity has been observed in particulate fractions of sea water (Bianchi *et al.*, 1992) and recently a report of the flux of methane from sinking particles indicates that this source would be sufficient to account for the methane observed in the sea (Karl and Tilbrook 1994). Whether the sizes of marine snow particles, and the consumption of oxygen that is caused by heterotrophic bacteria, are such as to allow the development of an oxygen gradient that would produce a sufficiently anoxic environment at the centre of such particles for known methanogens to function, requires to be examined. It may be that the putative methanogenic bacteria involved will be more tolerant of higher levels of oxygen than expected from the other methane forming bacteria so far studied; if this should prove to be the case it will be a significant contribution to microbiology.

If consortia of bacteria, without the support of structured physical contact and not driven by the input of a xenobiotic molecule, are a general phenomenon in natural environments they could clearly contribute to the problem of isolating individual species. There are several ways in which this type of physiological community could be assembled. The use of interlocking electron acceptors and substrates involving more than one organism may be a case in point; ammonia formation from nitrate during anaerobic respiration providing an acceptable form of nitrogen or photosynthate from a primary producer acting as a substrate for a heterotroph being simple examples. It would however be unlikely that such a simple form of microbial interaction would be responsible for the failure of experienced microbiologists to achieve axenic cultures of the component partners. It would be relatively straightforward to dissect the relationship and to provide the appropriate substrates to individual strains. The situation would be less tractable if the links between the organisms involved nutrients at low, non-substrate concentration levels or were comprised of signal-type interactions. The occurrence of the phenomenon of swarming among bacteria is distributed throughout the phylogenetic tree (Harshey 1994) and indicates that inter-cellular signalling is widespread, at least between the same species. A bacterial consortium held together by inter-species signals would be difficult to recognise and even more so to isolate and maintain. There is widespread distribution among the Gram-negative bacteria of communication by a series of N-acyl homoserine lactones which allow species to modulate population growth and to respond to environmental change (see Swift *et al.*. 1994). Perhaps the best known example of this being the way in which *Vibrio fischeri* in symbiotic association with certain fish or squid have small-molecule organised autoinduction of their bioluminescence at high cell densities. In natural populations signalling processes would be a means by which large numbers of organisms can co-ordinate their response to stimuli and amplify inputs from their environment.

In considering the non-culturable organisms that exist in oligotrophic environments such as the open ocean and having looked at some reasons why such bacteria have not been brought into culture, it becomes reasonable to ask if they have attributes which in any way basically differ from those associated with culturable organisms. It is a common experience that after the isolation of an axenic strain its maintenance in culture is accompanied by an increase in maximum growth rate. This adaptation to laboratory conditions is thought, of course, to be the selection of sub-strains more suited to the conditions applied, these usually involving higher nutrient availability and sometimes slightly elevated temperatures. The question is, does this adapting isolated strain differ from related organisms of the same physiological group that remained unisolated? Put in real terms, is the phycoerythrin-containing *Synechococcus* WH7803 isolated by Brand, described by Waterbury *et al.* (1986) and now growing to rather high cell densities at division times down to 18 hours in several laboratories, truly representative of the large number of phycoerythrin-containing oceanic *Synechococci* that occur in most ocean provinces? Could it be that most of those organisms in their natural environment are 'unculturable' and that what we are working with is an atypical representative which differs in having some property that allowed its isolation? Other than alterations in growth rate and cell density of batch cultures the apparent physiology - phototrophic growth, nutrient preferences *etc* - are the same as what we know of the natural population although the temperature optimum has been raised by several degrees. We know very little about what controls population size in oligotrophic environments other than the constraints imposed by nutrient limitation. There may be specific signals between bacteria that set limits to the density to which organisms can grow. Alternately the accumulation of excreted metabolic products in the immediate environment may be self-limiting to the culture that produces them. Mutational events that occur after the isolation of an axenic strain and during its maintenance in culture may alter the response to such self-generated controls over the population density. Such a scenario would lead to exponential growth proceeding over a greater period in the laboratory strain relative to its predecessor in the natural environment. I have used as an example an organism with which I have direct experience but similar questions could be posed about heterotrophic and chemotrophic bacteria isolated from acutely oligotophic environments which also show increases in growth rate after periods of laboratory culture.

It is not necessary to argue that a distinction between cultured organisms and the populations from which they were derived does not sometimes occur in bacteria from nutritionally rich environments. What may be important is the division time; bacteria with growth rates that may vary considerably are probably more amenable to isolation and culture if only because they can exploit the offered elective conditions. It is difficult to get robust information on the ranges over which bacterial growth rate can be varied. In principle infinitely slow division rates can be measured in chemostats and other continuous culture devices; however, at low growth rates problems such as wall-adhesion become more critical and make accurate measurement of growth impossible. There are even greater problems in measuring the growth rate of natural populations (see Kemp, 1995) and so far this has not been achieved for component bacteria, although in principle this should be possible with species-specific rRNA probes. There is considerable evidence to suggest that there are minimum values of growth rate that can be maintained in chemostat cultures. Using the versatile heterotroph *Aerobacter aerogenes*, Tempest *et al.*. (1967) showed that with various nutrient limitations the doubling time reached a minimum at about 100 hr (about 130 times less than the maximum rate) and at these dilution rates a large proportion of the bacteria were non-viable. The proportion of non-viable cells increased steadily as division time increased, reaching about 50% at minimum growth rate. The concept of the non-viable bacterial cell and the ways in which the viability index (θ, the probability that a newly formed cell will be viable) may be calculated in batch as well as chemostat cultures, is discussed by Pirt (1975). Even under optimum conditions a proportion of divisions produce a non-viable cell; the reasons for these are unknown. If a number (y_v) of bacteria divide and produce $2y_v$ daughter cells of which a fraction (θ) are viable then $2\theta y_v$ viable cells and $2(1-\theta)y_v$ dead cells are produced with each generation. Thus, when 1% of non-viable cells are produced at each division the viability index is 0.99. It follows, therefore, that when $\theta = 0.5$ the number of viable cells in a chemostat culture remains constant and values above and below that figure indicate whether the number of cells in a culture will increase or decrease. In a perfectly exponentially growing culture θ will be 1.0 although it is clear that this is unlikely to be achieved in most laboratory cultures, let alone natural populations, for anything other than the shortest periods of time. It would seem that these considerations, derived from chemostat cultures of rapidly growing heterotrophic bacteria, have relevance to how we think about the unculturable populations that inhabit oligotrophic environments. In particular, we need to know more about the relationship between growth rate and the value of θ for such organisms in their natural environment. However since a very high proportion of bacterial cells in ocean water appear to be unculturable an explanation based on θ values could not be the whole answer unless very low values of θ obtained due to very low growth rates compared with those obtained with *A. aerogenes* above.

There are several well documented examples of microbial species which under certain circumstances will give rise to cells that are not capable of further division. The spore mother cell in bacilli and the heterocyst in cyanobacteria are bacterial examples of cells that develop as a response to particular environmental changes and have a specialised function associated with a defined life-time followed by necrosis, a term more usually employed in discussing the cells of eukaryotic tissues. Let us consider one example in a little more detail. The heterocyst develops in many species of filamentous cyanobacteria when the supply of fixed-nitrogen becomes limiting. It is morphologically and physiologically specialised and serves as the sole site of nitrogen-fixation in the cyanobacterial filament, the products of which are exported to adjacent vegetative cells which then grow and divide in the normal manner (Wolk 1982; Adams 1992). In species of the cyanobacteria *Anabaena* there are mechanisms which determine the number of heterocysts that form in a filament and locate these with precision through the vegetative cell filament so as to facilitate the equal distribution of fixed-nitrogen to all the vegetative cells that comprise that filament. This is known because, at least in the species studied, the inter-heterocystous vegetative cells divide at the same rate. But the mature heterocyst does not divide: it is an example of a cell that has altered its genetic expression such as to preclude any further division. The consequence of this is that the remaining vegetative cells in the filament survive an otherwise absolute shortage of available nitrogen. The behaviour of the cell that develops to become a heterocyst may correctly be described as altruistic in that it sacrifices the possibility of its own propagation so that cells with an identical genetic endowment may survive the period of nitrogen starvation. The selective value of this strategy to the cyanobacterium in being able to survive intermittent interruptions in the supply of fixed nitrogen is evident and implies a selective control over whether cells will proceed or not to division. This selective control over heterocyst numbers must be highly effective since too many heterocysts, because they cannot divide, can be shown to result in population reduction in succeeding generations over a relatively short time span (Adams and Carr, 1981).

There is no reason to argue that the processes which lead to differentiation and consequent cessation of cell division in heterocyst formation are comparable to those which lead to a proportion of new cells in slow-growing cultures to be non-viable. It is, however, important to remember that the occurrence of non-viable bacteria in slow growing cultures described above was a statistically measurable event in the culture as a whole. There was no information regarding at what point a new cell became non-viable and it may well be that non-viability was acquired during cell life-time of a bacterial cell (as distinct from at daughter-cell formation) with a frequency that increased with increasing nutrient limitation/decreasing growth rate.

In addition to the widespread effects on the biochemistry of bacterial cells, nutrient deprivation causes changes in their morphology and size and examples of this may be readily seen in marine bacteria (Kjelleberg *et al.*., 1993; Morita, 1982, 1985) and this is consistent with measurements made on natural populations. The reduction in size of non-growing cells appears to be a general bacterial response, although the detailed inverse relationship between size and growth rate has been accurately measured in only a limited number of strains. On starvation some heterotrophic strains produce what have been termed 'ultramicrocells' which are considerably smaller than vegetative cells but still have measurable metabolic activity, at least in initial period following their formation. Ultramicrocells are formed by a process of reductive division, and have volumes down to only 1% that of rapidly growing organisms. The formation and subsequent reversion to vegetative cell state of ultramicrocells can form part of bacterial life cycles driven primarily by nutrient availability. The extent to which such forms contribute to observed populations of bacteria in ocean waters is not known but is probably considerable.

Cyanobacteria exhibit a variety examples of morphological deviation from normal growth pattern, greater than those found among other bacteria (Adams, 1992). The most dramatic in terms of alteration of cell size occur in the pleurocapsalean cyanobacteria; an account of many new isolates of these was given in detail by Waterbury and Stanier (1978). For example, these workers describe a marine strain of *Dermocarpa* in which spherical cells of up to 30μm diameter undergo synchronous multiple fission producing, typically, 500 to 800 baeocytes (daughter cells) of less than 2μm diameter. The baeocyte, a term devised by Waterbury and Stanier meaning small cell, in due course commences growth to become, over some hundreds of hours, a large cell ready for another round of synchronous multiple fission. The environmental controls over this process in the natural habitat of the organism, as well as in the laboratory, remain to be explored. Whether the baeocyte form of this and other pleurocapsalean cyanobacteria contribute significantly to the observed marine microbiota is not known.

Both heterotrophic prosthecate bacteria as well as some cyanobacteria can develop forms in which extended cell shapes occur. These have been termed the 'shut down' or 'growth precursor cell' and clearly are associated with an increase in surface to volume ratio (Dow *et al.*. 1983). There is evidence that these changes are accompanied by the development of polarity in cell growth and by the presence of DNA in a condensed format and by unequal cell division. The latter is a characteristic feature of other processes of differentiation in bacteria including those of heterocyst development in cyanobacteria and sporulation in *Bacillus* species.

It is evident from the brief description of the examples above that the process of cell division of many microorganisms in their natural environment, and of some in laboratory culture, differs from that which produces two equal daughter cells, the usually accepted form of growth, at least for unicellular bacteria. How do these considerations align with the idea of θ, the viability index arising from heterotrophic chemostat cultures discussed above? Values of θ less than 1 in oligotrophic natural populations may arise not from the production of non-viable cells *per se*. If bacterial cell division produce one 'normal' daughter cell and one daughter that was so programmed as to take a longer (perhaps much longer) period to reach its division, then the growth of the population would be non-exponential. Such a unequal division rule is known to operate in some species of *Anabaena* which always divide to produce a large and a small daughter cell (Mitchison and Wilcox 1972). If the recognition that cell division rates may be genetically programmed so as to be less than maximal is extended widely to organisms existing in oligotrophic environments, and present to a lesser or greater extent in all organisms (*ie* both daughter cells) some of the difficulties in isolation may be explained. It may be useful to try and discern an underlying pattern in these growth forms and cell cycles and an approach to this could be to consider the process of 'apoptosis'. This term was applied by Kerr *et al.*. (1972) to the physiological regulation of mammalian tissue size in response to an injury or stress, such responses being distinct from necrosis but still resulting in cell destruction but with a quite distinct and ordered mechanism. *Apoptosis*, from the Greek meaning 'falling off', is a process of programmed cell death that is being increasingly used to describe some cellular changes in human tissues and is supported by ultrastructural and biochemical studies (see Tomei and Cope 1991). There is evidence that in many tissues there exists an a potential for an active, genetically programmed process of cell destruction and that this is employed to achieve a homeostatic function in response to some form of assault on a tissue. Apoptosis is also thought to play a role in the developmental biology of tissue formation. There is as yet no clear mechanism by which apoptosis is induced but it does appear that, in some examples at least, synchronously with the initial morphological changes there is double stranded-cleavage of nuclear DNA at the linker regions between nucleosomes, producing regular sized DNA fragments recognisable on gels.

Clearly programmed cell death is an integral part of some microbial responses to environmental shifts, heterocyst differentiation for example. Is there however a more general role for the bacterial equivalent of apoptosis in explaining the behaviour of oligotrophic and very slow growing populations and in the relative lack of success in culturing? The decrease in viability index with declining growth rate of *Aerobacter aerogenes* described above is fully reversible, thus when optimum conditions are restored, the remaining viable cells become capable of exponential growth - that is, of achieving a viability index approaching 1. But let us suppose that some (most?) of the bacteria observed microscopically in sea water have a low viability index programmed into their genetic composition. If we consider the ocean to be chemostat (probably an erroneous supposition in real space and time) a viability index of 0.5 would yield steady state numbers. An ability to vary the viability index to a relatively small degree when nutrients become available and growth rate increases would produce intermittent variation in numbers over a limited range, compatible with what is observed. An effect of such a programmed and fixed viability range would be to preclude any exponential growth during attempted isolation even under the most favourable conditions. In batch culture, which is how most attempts at isolation are conducted, lack of exponential growth would not in itself prevent success but would certainly make it more difficult. But some organisms have been isolated from open sea water and these are capable of a logarithmic growth. It may be that what has been isolated in such cases are variants that have, presumable by endogenous mutation, escaped from the constraints of a genetically programmed, relatively low viability index. The implication of this would be that some (many?) natural bacterial populations are incapable of the exponential phase of growth that has been thought to be characteristic of bacteria at least since the time of Buchanan (1918). One could postulate that the bacteria that have been isolated and maintained in culture from such an environment are atypical in that they have become like the accepted paradigm with respect to bacterial growth: that is, capable of exponential growth when nutrients are freely available and they have become like this by mutational release from constraints over their division rates. If this is correct, it follows that a low viability index (or some form of programmed cell death such that exponential growth does not occur) must have selective value under the conditions from which such organisms were isolated. Since the low nutrient flux in oligotrophic environments is often constant the selective advantage of having highly variable and rapid growth rates may be minimal and may be disadvantageous in terms of commitment of resource to macromolecular synthesis needed for only episodic function. If what is argued above is true it is not without implications for those of us that try to understand natural populations by studying isolated strains of those populations. I think that it means that cultivated strains will be more variable because the opportunity for exponential growth would facilitate the propagation of variants more adapted to the culture conditions and hence their ultimate domination and exclusion of the originally isolated form.

References

Adams DG and Carr NG (1981) The developmental biology of heterocyst and akinete formation in cyanobacteria. *CRC Crit Rev Microbiol* **9:** 45-100

Adams DG (1992) Multicellularity in cyanobacteria. In *Prokaryotic structure and function: a new perspective.* Mohan S, Dow C., Cole JA (eds) Cambridge University Press, Cambridge *Soc Gen Micro Sym* **47:** 341-384

Bianchi M, Marty D, Teyssie J-L, Fowler SW (1992) Strictly aerobic and anaerobic bacteria associated with sinking particulate matter and zooplankton fecal pellets. *Mar Ecol Prog Ser* **88:** 55-60

Buchanan RE (1918) Life phases in a bacterial culture. *J Infect Dis* **23:** 109-125

Doolittle WF and Brown JR (1994) Tempo, mode, the progenote, and the universal tree. *Proc Natl Acad Sci USA* **91:** 6721-6728

Dow C, Whittenbury R, Carr NG (1983) The 'shut down' or 'growth precursor' cell - an adaptation for survival in a potentially hostile environment. In *Microbes in their natural environment.* Slater JH, Whittenbury R, Wimpenny JWT(eds) Cambridge University Press, Cambridge *Soc Gen Microbiol Sym* **34:** 187-247

Giovannoni SJ, Mullins TD, Field KG (1995) Microbial diversity in oceanic systems: rRNA approaches to the study of unculturable microbes *This volume* 217-248

Harshey TM (1994) Bees aren't the only ones: swarming in Gram-negative bacteria. *Mol Micro* **13:** 389-394

Karl DM and Tilbrook BD (1994) Production and transport of methane in oceanic particulate organic matter. *Nature* **368:** 732-734

Kemp PK (1995) Can we estimate bacterial growth rates from ribosomal RNA content? *This volume* 279-302

Kerr JFR, Wyllie AH, Currie AR (1972) Apoptosis: a basic biological phenomenon with wide-ranging implications in tissue kinetics. *Br J Cancer* **26:** 239-257

Kjelleberg S, Flardh KBG, Nystrom T, Moriarty DJW (1993) Growth limitation and starvation in bacteria. In *Aquatic microbiology: an ecological approach.* Ford TE (ed) Blackwell Scientific Publications ,Oxford 298-320

Logan NA (1994) *Bacterial systematics,* Blackwell Scientific Publications, Oxford pp263

Mitchison GJ and Wilcox M (1972) A rule governing cell division in *Anabaena. Nature* **239:** 110-111

Morita RY (1982) Starvation-survival of heterotrophs in the marine environment. *Adv Microb Ecol* **6:** 171-178

Morita RY (1985) Starvation and miniturisation of heterotrophs, with special emphasis on maintenance of the starved viable state. In *Bacteria in their natural environments.* Fletcher MM, Floodgate GD (eds) Academic Press, London 111-130

O'Donnell AG, Goodfellow M, Hawksworth DL (1994) Theoretical and practical aspects of the quantification of biodiversity among microorganisms. *Phil Trans R Soc Lond B* **345:** 65-73

Oremland RS (1979) Methanogenic activity in plankton samples and fish intestines: a mechanism for *in situ* methanogenesis in ocean surface waters. *Limnol Oceangr* **24:** 1136-1141

Slater JH and Bull AT (1982) Environmental microbiology: biodegradation. *Phil Trans R Soc Lond B* **297:** 575-597

Swift S, Bainton NJ, Winson MK (1994) Gram-negative bacterial communication by *N*-acyl homoserine lactones: a universal language? *Trends Microbiol* **2:** 193-198

Tempest DW, Herbert D, Phipps PJ (1967) Studies on the growth of *Aerobacter aerogenes* at low dilution rates in a chemostat. In *Microbial physiology and continuous culture*. Powell EO, Evans CGT, Strange RE, Tempest DW (eds) H.M.S.O. London 240-254

Tomei LD and Cope FC (1991) Apoptosis: the molecular basis of cell death. *Curr Comm Cell Mol Biol* **3**: 1-322

Waterbury JB, Watson SW, Valois FW, Franks DG (1986) Biological and ecological characterization of the marine unicellular cyanobacterium *Synechococcus.* In Photosynthetic picoplankton. Platt T and Li WKW (eds) *Can Bull Fish Aquat Sci* **214:** 71- 120

Waterbury JB and, Stanier RY (1978) Patterns of growth and development in pleurocapsalean cyanobacteria. *Microbiol Rev* **42:** 2-44

Weller R, Bateson MM, Heimbach BK, Kopcyzski ED, Ward DM (1992) Uncultivated cyanobacteria, *Chloroflexus*-like inhabitants, and spirochete-like inhabitants of a hot spring microbial mat. *Appl Environ Microbiol* **58:** 3964-3969

Woese CR (1987) Bacterial evolution. *Microbiol Rev* **51:** 221-271

Wolk C P (1982) Heterocysts. In *The biology of cyanobacteria* Carr NG and Whitton BA (eds) Blackwell Scientific Publications, Oxford 359-386

List of Participants

Dr S M Allison
Department of Molecular and Cell Biology, University of Aberdeen,
Aberdeen, AB9 1AS
UK

Dr J W Ammerman
Center of Marine Biotechnology,
University of Maryland
Baltimore, MD 21202,
USA

Dr Ó S Andrésson
Institute for Experimental Pathology,
University of Iceland,
Keldur, IS-112 Reykjavík
Iceland

Ms T Anning
Plymouth Marine Laboratory
Citadel Hill,
Plymouth, PL1 2PB
UK

Dr Farooq Azam
Scripps Institute of Oceanography
UCSD 0202,
La Jolla, CA 92093
USA

Dr M Balode
Latvian Academy of Sciences,
Laboratory of Marine Biology
3 Miera Street, Salaspils, LV 2169
Latvia

Dr E Berdalet
Institut de Ciències del Mar
P Joan de Borbó s/n, 08039 Barcelona
Spain

Dr J A Berges
Brookhaven National Laboratory
Building 318, Upton, NY 11973
USA

Prof. Henry Blackburn
Department of Ecology & Genetics,
University of Aarhus
Ny Munkegade, DK 8000 Aarhus C
Denmark

Dr Peter Bot
National Institute for Coastal and Marine Management
PO Box 20907, 2500 EX The Hague
The Netherlands

Dr Phil Boyd
Department of Oceaonography,
University of British Columbia
Vancouver C,V6T 1ZA
Canada

Dr Gunnar Bratbak
Department of Microbiology, University of Bergen
Jahnebakken 5, N 5007 Bergen
Norway

Dr C Brussaard
Netherlands Institute for Sea Research
P O Box 59, 1790 AB Den Burg, Texel
The Netherlands

Dr Berit Buchholz
MPI for Marine Microbiology
Fahrenheitstr 1, 28359 Bremen
Germany

Dr H B Büyükisik
Ege University, Fisheries Faculty, Dept of Hydrobiology
35100 Bornova, Izmir
Turkey

Dr J I Calderón-Paz
Institut de Ciències del Mar
Passeig Joan de Borbó s/n, 08039
Barcelona
Spain

Dr Lisa Campbell
Dept of Oceanography,
University of Hawaii
Honolulu, HI 96822
USA

Prof. Noel Carr
Department of Biology,
University of Warwick
Coventry, CV4 7AL
UK

Dr S C Cary
Department of Microbiology,
Oregon State University
Corvallis, OR 97331-3804
USA

Dr R Casotti
Stazione Zoologica, 'Anton Dohrn'
Villa Comunale, I 80121 Naples
Italy

Dr Urania Christaki
Station Marine d'Endoume
Rue de la Batterie des Lions,
13004 Marseille
France

Dr Keith E Cooksey
Department of Microbiology,
Montana State University
Bozeman, MT 59717-0352,
USA

Dr M Cooper
Department of Molecular and Cell Biology, University of Aberdeen
Aberdeen
UK

Ms Kirsten Donald
Department of Biological Sciences,
University of Warwick
Coventry, CV4 7AL
UK

Mr P Cummins
Plymouth Marine Laboratory
Prospect Place, The Hoe,
Plymouth, PL1 3DH
UK

Dr Paul Falkowski
Oceanographic and Atmospheric Sciences Division, Brookhaven National Laboratory
Upton, New York 11970
USA

Dr Richard J Geider
College of Marine Studies, University of Delaware
Lewes, DE 19958-1298
USA

Dr M Giordano
Via Placida n. 6
98121 Messina
Italy

Dr Stephen Giovannoni
Department of Microbiology, Oregon State University
Corvalis, OR 97331
USA

Dr L Guiliano
Universita' Degli Studi di Messina
C.da Papardo Salita Sperone, 31 S. Agata di Messina, 98166 Messina
Italy

Dr M E Hatipoglu
Middle East Technical University, Institute of Marine Sciences
P O Box 28, 33731 Erdemli ICEL
Turkey

Dr Mikal Heldal
Institutt for mikrobiologi, Universitetet i Bergen
Jahnebakken 5, N 5020 Bergen
Norway

Dr Robert P Hirt
Group of Microbiology, The Natural History Museum
Cromwell Road, London, SW7 5BD
UK

Dr Jean Houmard
Institut Pasteur, Département de B.G.M.,
28 rue du Docteur Roux, 75724 Paris Cedex 15
France

Dr Aaron Kaplan
Department of Botany, Hebrew University of Jerusalem
Givat Ram 91904, Jerusalem
Israel

Dr Ricardo Guerrero
Departament de Microbiologia, Universitat de Barcelona
Av. Diagonal 645, E-08028 Barcelona
Spain

Mr R. Hastings
University of Liverpool, Department of Genetics and Microbiology
P O Box 147, Liverpool, L69 3BX
UK

Dr Ian Head
Department of Fossil Fuels and Environmental Geochemistry, University of Newcastle
Newcastle upon Tyne, NE1 7RU
UK

Dr Will D Hiorns
University of Liverpool, Department of Genetics and Microbiology
P O Box 147, Liverpool, L69 3BX
UK

Dr Andrew Holmes
University of Warwick, Biological Sciences
Coventry, CV4 7AL
UK

Dr Ian Joint
Plymouth Marine Laboratory, Prospect Place, The Hoe, Plymouth, PL1 3DH,
UK

Dr Paul Kemp
Oceanographic and Atmospheric Sciences Division, Brookhaven National Laboratory
Upton, New York 11970
USA

Dr Lee J Kerkhof
Institute of Marine and Coastal Sciences,
Rutgers, the State University of New Jers
PO Box 2, New Bunswick, NJ 08903-023
USA

Dr V Kisand
Tartu University,
Institute Zoology and Hydrobiology
46 Vaneminse St, Tartu EE2400
Estonia
Dr Luigi Lazzara
Laboratorio di Ecologia,
Universita' Degli Studi di Firenze,
Via P.A. Micheli n. 1 -50121, Firenze
Italy

Mr Senjie Lin
Marine Sciences Research Center,
State University of New York at Stony
Brook,
Stony Brook, N Y 11794-5000
USA
Dr Nick Mann
Department of Biology, University of
Warwick
Coventry, CV4 7AL
UK
Dr E Marañon
Dept B.O.S. Ecologia,
Universidad de Oviedo
Spain

Prof Anton Markos
Charles University
Prague 1,
The Czech Republic

Dr R Massana
Instituto de Ciències del Mar
Passeig Joan de Borbó s/n, E-08039
Barcelona
Spain

Dr A E Kideys
Middle East Technical University,
Institute of Marine Sciences
P O Box 28, 33731, Erdemli ICEL
Turkey

Mr M LaMontagne
Boston University Marine Program,
Marine Biological Laboratory
Woods Hole, Massachusetts 02543
USA
Dr William Li
Biological Oceanography Division,
Bedford Institute of Oceanography
PO Box 1006, Dartmouth, Nova Scotia,
B2Y 4A2, Canada

Dr Wolfgang Löffelhardt
(Institute for biochemistry and
molecular cell biology
Vienna Biocenter, A-1030 Wien, Dr
Bühr-Gasse 9, Vienna,
Austria
Dr Serge Y Maestrini
CREMA Houmeau, CNRS IFREMER
BP5 , 17137 L'Houmeau
France

Dr P de Marco
Biological Sciences,
University of Warwick,
Coventry, CV4 7AL
UK
Dr V Martin Jezequel
Station Biologique
Place Teissier, 29680, Roscoff
France

Dr Linda Medlin
Alfred Wegener Institute, Postfach
120164
Bremerhaven 2850
Germany

Prof M J Merrett
School of Biological Sciences,
University College of Swansea
Singleton Park, Swansea, SA2 8PP,
Wales
UK

Mr C L Moyer
Dept of Oceanography,
University of Hawaii
Honolulu, HI 96822
USA

Dr Colin Murrell
Department of Biology,
University of Warwick
Coventry, CV4 7AL
UK

Ms J Newman
Dept of Biological Sciences,
University of Warwick
Coventry, CV4 7AL,
UK

Dr Y Nikolaev
Dept Molecular and Cell Biology,
University of Aberdeen
Aberdeen, AB9 1AS
UK

Dr M Olaizola
RSA
T P 272, I-21020
Ispra (VA)
Italy

Ms M P Olsen
Department of Microbial Ecology,
University of Aarhus
DK 8000 Aarhus C
Denmark

Dr G D'Onofrio
Statione Zoologica, 'Anton Dohrn'
Villa Comunale I,
80121 Napoli
Italy

Prof A Oren
The Hebrew University of Jerusalem,
Divison of Microbial and Molecular
Ecology, Jerusalem 91904
Israel

Ms Sílvia Pelegrí
Department of Microbial Ecology,
University of Aarhus
DK 8000 Aarhus C
Denmark

Dr Richard Powell
Recombinant DNA Group
University College Galway
Ireland

Dr O Prasil
Academy of Sciences of Czech Republic
379 81 Trebon,
Czech Republic

Dr F Rassoulzadegan
Station Zoologique URA - CNRS 716
B P 28,
06230 Villefranche-sur-Mer
France

Dr Colin Reynolds
Institute of Freshwater Ecology
The Ferry House, Far Sawry,
Ambleside, Cumbria LA22 0LP
UK

Dr Bo Riemann
Water Quality Institute
Agern Alle 11, DK-2970 Horsholm
Denmark

Ms Gabriella Rocap
Massachusetts Institute of Technology
Room 48-320, Cambridge, MA 02139
USA

Kristen M Romans
Marine Sciences Research Center, State University of New York at Stony Brook
Stony Brook, N Y 11794-5000
USA

Dr V Rousseau
University of Brussels
Campus de la Plaine, CP 221,
B 1050 Brussels
Belgium

Dr Dave Scanlan
Dept of Biological Sciences,
University of Warwick
Coventry, CV4 7AL,
UK

Dr A Shalapyonok
Institute for the Biology of Southern Seas, 2 Nakhimov Ave, Sevastopol
335011, Crimea
Ukraine

Dr Melvin Simon
Biology Division,
California Institute of Technology
Mail Code 147-75, Pasadena, CA 91125
USA

Dr Gary Smerdon
Plymouth Marine Laboratory
Prospect Place, The Hoe, Plymouth,
PL1 3DH
UK

Dr J Stein
The Agouron Institute
505 Coast Boulevard South, Suite 400,
La Jolla, California 92037
USA

Mr Grieg F Steward
Scripps Institution of Oceanography
La Jolla,
California 92093
USA

Dr W Stolte
Netherlands Institute for Sea Research
P O Box 59, 1790 AB Den Burg,
Texel
The Netherlands

Mr Marcelino Suzuki
College of Oceanography -
Oregon State University
Corvallis, OR 97331-5503
USA

Dr K R Timmermans
Netherlands Institute for Sea Research,
P O Box 59, 1790 AB Den Burg,
Texel
The Netherlands

Dr Ena Urbach
Massachusetts Institute of Technology
Room 48-320,
Cambridge, MA 02139
USA

Dr Daniel Vaulot
Observatoire Oceanologie de Roscoff
Place George Tessier, 29680 Roscoff
France

Dr M A Voytek
Marine Sciences,
University of California
Santa Cruz, CA 95064
USA

Dr Bess Ward
University of California, Santa Cruz,
Marine Sciences Program
Santa Cruz, CA 95064
USA

Mr William H Wilson
Dept of Biological Sciences,
University of Warwick
Coventry, CV4 7AL,
U K

Dr Barbara Wigglesworth-Cooksey
Department of Microbiology,
Montana State University
Bozeman, MT 59717-0352
USA

Dr Jonathan Zehr
Department of Biology,
Rensselar Polytechnic Institute
Troy, NY 12180
USA

Dr A Zingone
Stazione Zoologica, 'A. Dohrn'
Villa Comunale, 80121 Naples
Italy

Subject index

NATO ASI Series G

Vol. 1: **Numerical Taxonomy.** Edited by J. Felsenstein. 644 pages. 1983. (out of print)

Vol. 2: **Immunotoxicology.** Edited by P. W. Mullen. 161 pages. 1984.

Vol. 3: **In Vitro Effects of Mineral Dusts.** Edited by E. G. Beck and J. Bignon. 548 pages. 1985.

Vol. 4: **Environmental Impact Assessment, Technology Assessment, and Risk Analysis.** Edited by V. T. Covello, J. L. Mumpower, P. J. M. Stallen, and V. R. R. Uppuluri. 1068 pages.1985.

Vol. 5: **Genetic Differentiation and Dispersal in Plants.** Edited by P. Jacquard, G. Heim, and J. Antonovics. 452 pages. 1985.

Vol. 6: **Chemistry of Multiphase Atmospheric Systems.** Edited by W. Jaeschke. 773 pages. 1986.

Vol. 7: **The Role of Freshwater Outflow in Coastal Marine Ecosystems.** Edited by S. Skreslet. 453 pages. 1986.

Vol. 8: **Stratospheric Ozone Reduction, Solar Ultraviolet Radiation and Plant Life.** Edited by R. C. Worrest and M. M. Caldwell. 374 pages. 1986.

Vol. 9: **Strategies and Advanced Techniques for Marine Pollution Studies: Mediterranean Sea.** Edited by C. S. Giam and H. J.-M. Dou. 475 pages. 1986.

Vol. 10: **Urban Runoff Pollution.** Edited by H. C. Torno, J. Marsalek, and M. Desbordes. 893 pages. 1986.

Vol. 11: **Pest Control: Operations and Systems Analysis in Fruit Fly Management.** Edited by M. Mangel, J. R. Carey, and R. E. Plant. 465 pages. 1986.

Vol. 12: **Mediterranean Marine Avifauna: Population Studies and Conservation.** Edited by MEDMARAVIS and X. Monbailliu. 535 pages. 1986.

Vol. 13: **Taxonomy of Porifera from the N. E. Atlantic and Mediterranean Sea.** Edited by J. Vacelet and N. Boury-Esnault. 332 pages. 1987.

Vol. 14: **Developments in Numerical Ecology.** Edited by P. Legendre and L. Legendre. 585 pages. 1987.

Vol. 15: **Plant Response to Stress. Functional Analysis in Mediterranean Ecosystems.** Edited by J. D. Tenhunen, F. M. Catarino, O. L. Lange, and W. C. Oechel. 668 pages. 1987.

Vol. 16: **Effects of Atmospheric Pollutants on Forests, Wetlands and Agricultural Ecosystems.** Edited by T. C. Hutchinson and K. M. Meema. 652 pages. 1987.

Vol. 17: **Intelligence and Evolutionary Biology.** Edited by H. J. Jerison and I. Jerison. 481 pages. 1988.

Vol. 18: **Safety Assurance for Environmental Introductions of Genetically-Engineered Organisms.** Edited by J. Fiksel and V.T. Covello. 282 pages. 1988.

Vol. 19: **Environmental Stress in Plants. Biochemical and Physiological Mechanisms.** Edited by J. H. Cherry. 369 pages. 1989.

Vol. 20: **Behavioural Mechanisms of Food Selection.** Edited by R. N. Hughes. 886 pages. 1990.

NATO ASI Series G

Vol. 21: Health Related Effects of Phyllosilicates.
Edited by J. Bignon. 462 pages.1990.

Vol. 22: Evolutionary Biogeography of the Marine Algae of the North Atlantic.
Edited by D. J. Garbary and G. R. South. 439 pages. 1990.

Vol. 23: Metal Speciation in the Environment.
Edited by J. A. C . Broekaert, Ş. Güçer, and F. Adams. 655 pages. 1990.

Vol. 24: Population Biology of Passerine Birds. An Integrated Approach.
Edited by J. Blondel, A. Gosler, J.-D. Lebreton, and R . McCleery.
513 pages. 1990.

Vol. 25: Protozoa and Their Role in Marine Processes.
Edited by P. C. Reid, C. M . Turley, and P. H. Burkill. 516 pages. 1991.

Vol. 26: Decision Support Systems.
Edited by D. P Loucks and J. R. da Costa. 592 pages. 1991.

Vol. 27: Particle Analysis in Oceanography. Edited by S. Demers. 428 pages. 1991.

Vol. 28: Seasonal Snowpacks. Processes of Compositional Change.
Edited by T. D. Davies, M . Tranter, and H . G. Jones. 484 pages. 1991.

Vol. 29: Water Resources Engineering Risk Assessment.
Edited by. J. Ganoulis. 551 pages. 1991.

Vol. 30: Nitrate Contamination. Exposure, Consequence, and Control.
Edited by I. Bogárdi and R. D. Kuzelka. 532 pages. 1991.

Vol. 31: Industrial Air Pollution. Assessment and Control.
Edited by A. Müezzinoğlu and M. L. Williams. 245 pages. 1992.

Vol. 32: Migration and Fate of Pollutants in Soils and Subsoils.
Edited by D. Petruzzelli and F. G. Helfferich. 527 pages. 1993.

Vol. 33: Bivalve Filter Feeders in Estuarine and Coastal Ecosystem Processes.
Edited by R. F. Dame. 584 pages. 1993.

Vol. 34: Non-Thermal Plasma Techniques for Pollution Control.
Edited by B. M. Penetrante and S. E. Schultheis.
Part A: Overview, Fundamentals and Supporting Technologies. 429 pages. 1993.
Part B: Electron Beam and Electrical Discharge Processing. 433 pages. 1993.

Vol. 35: Microbial Mats. Structure, Development and Environmental Significance.
Edited by L. J. Stal and P. Caumette. 481 pages. 1994.

Vol. 36: Air Pollutants and the Leaf Cuticle.
Edited by K. E. Percy, J. N. Cape, R. Jagels and C. J. Simpson. 405 pages. 1994

Vol. 37: Azospirillum VI and Related Microorganisms.
Edited by I. Fendrik, M. del Gallo, J. Vanderleyden, and M. Zamaroczy.
588 pages. 1995.

Vol. 38: Molecular Ecology of Aquatic Microbes.
Edited by I. Joint. 423 pages. 1995.

Springer-Verlag and the Environment

We at Springer-Verlag firmly believe that an international science publisher has a special obligation to the environment, and our corporate policies consistently reflect this conviction.

We also expect our business partners – paper mills, printers, packaging manufacturers, etc. – to commit themselves to using environmentally friendly materials and production processes.

The paper in this book is made from low- or no-chlorine pulp and is acid free, in conformance with international standards for paper permanency.